SEMIMETALS
1. Graphite and its compounds

MODERN PROBLEMS IN CONDENSED MATTER SCIENCES

Volume 20.1

NORTH-HOLLAND
AMSTERDAM · OXFORD · NEW YORK · TOKYO

SEMIMETALS
1. Graphite and its compounds

N.B. BRANDT
S.M. CHUDINOV

and

Ya.G. PONOMAREV

Department of Low Temperature Physics
Physical Faculty
Moscow State University
Moscow 119899, USSR

Translated from the Russian
by
O.A. ZILBERT

NORTH-HOLLAND
AMSTERDAM · OXFORD · NEW YORK · TOKYO

1988

ISBN: 0 080 97846 0

Published by:

North-Holland Physics Publishing
a division of
Elsevier Science Publishers B.V.
P.O. Box 103
1000 AC Amsterdam
The Netherlands

Sole distributors for the USA and Canada:
Elsevier Science Publishing Company, Inc.
52 Vanderbilt Avenue
New York, NY 10017
USA

Library of Congress Cataloging-in-Publication Data

Brandt, N.B. (Nikolaĭ Borisovich), 1923–
Semimetals.

(Modern problems in condensed matter sciences; v. 20.1)
Bibliography: v. 1. p.
Vol. 1– includes indexes.
Contents: 1. Graphite and its compounds.
1. Semimetals. I. Chudinov, S.M. (Sergeĭ Mikhaĭlovich), 1935– . II. Ponomarev, Ya.G., 1938– . III. Title. IV. Series.
QC611.8.S45B73 1987 530.4'1 87–13388
ISBN 0-080-97846-0 (v. 1)

Printed in The Netherlands

MODERN PROBLEMS IN CONDENSED MATTER SCIENCES

Vol. 1. SURFACE POLARITONS
V.M. Agranovich and D.L. Mills, *editors*

Vol. 2. EXCITONS
E.I. Rashba and M.D. Sturge, *editors*

Vol. 3. ELECTRONIC EXCITATION ENERGY TRANSFER IN CONDENSED MATTER
V.M. Agranovich and M.D. Galanin

Vol. 4. SPECTROSCOPY AND EXCITATION DYNAMICS OF CONDENSED MOLECULAR SYSTEMS
V.M. Agranovich and R.M. Hochstrasser, *editors*

Vol. 5. LIGHT SCATTERING NEAR PHASE TRANSITIONS
H.Z. Cummins and A.P. Levanyuk, *editors*

Vol. 6. ELECTRON–HOLE DROPLETS IN SEMICONDUCTORS
C.D. Jeffries and L.V. Keldysh, *editors*

Vol. 7. THE DYNAMICAL JAHN–TELLER EFFECT IN LOCALIZED SYSTEMS
Yu.E. Perlin and M. Wagner, *editors*

Vol. 8. OPTICAL ORIENTATION
F. Meier and B.P. Zakharchenya, *editors*

Vol. 9. SURFACE EXCITATIONS
V.M. Agranovich and R. Loudon, *editors*

Vol. 10. ELECTRON–ELECTRON INTERACTIONS IN DISORDERED SYSTEMS
A.L. Efros and M. Pollak, *editors*

Vol. 11. MEDIUM-ENERGY ION REFLECTION FROM SOLIDS
E.S. Mashkova and V.A. Molchanov

Vol. 12. NONEQUILIBRIUM SUPERCONDUCTIVITY
D.N. Langenberg and A.I. Larkin, *editors*

MODERN PROBLEMS IN CONDENSED MATTER SCIENCES

Vol. 13. PHYSICS OF RADIATION EFFECTS IN CRYSTALS
R.A. Johnson and A.N. Orlov, *editors*

Vol. 14. INCOMMENSURATE PHASES IN DIELECTRICS
(Two volumes)
R. Blinc and A.P. Levanyuk, *editors*

Vol. 15. UNITARY TRANSFORMATIONS IN SOLID STATE PHYSICS
M. Wagner

Vol. 16. NONEQUILIBRIUM PHONONS IN NONMETALLIC CRYSTALS
W. Eisenmenger and A.A. Kaplyanskii, *editors*

Vol. 17. SOLITONS
S.E. Trullinger, V.E. Zakharov and V.L. Pokrovskii, *editors*

Vol. 18. TRANSPORT IN PHONON SYSTEMS
V.L. Gurevich

Vol. 19. CARRIER SCATTERING IN METALS AND SEMICONDUCTORS
V.F. Gantmakher and Y.B. Levinson

Vol. 20.1. SEMIMETALS – 1. GRAPHITE AND ITS COMPOUNDS
N.B. Brandt, S.M. Chudinov and Ya.G. Ponomarev

Vol. 21. SPECTROSCOPY OF SOLIDS CONTAINING RARE EARTH IONS
A.A. Kaplyanskii and R.M. Macfarlane, *editors*

In preparation

Vol. 20.2. SEMIMETALS 2

Vol. 22. SPIN WAVES AND MAGNETIC EXCITATIONS
(Two volumes)
A.S. Borovik-Romanov and S.K. Sinha, *editors*

Vol. 23. OPTICAL PROPERTIES OF MIXED CRYSTALS
R.J. Elliott and I.P. Ipatova, *editors*

Oh, how many of them there
are in the fields!
But each flowers in its
own way —
In this is the highest achievement
of a flower!

Matsuo Bashó
1644–1694

PREFACE TO THE SERIES

Our understanding of condensed matter is developing rapidly at the present time, and the numerous new insights gained in this field define to a significant degree the face of contemporary science. Furthermore, discoveries made in this area are shaping present and future technology. This being so, it is clear that the most important results and directions for future developments can only be covered by an international group of authors working in cooperation.

"Modern Problems in Condensed Matter Sciences" is a series of contributed volumes and monographs on condensed matter science that is published by North-Holland Physics Publishing, a division of Elsevier Science Publishers. With the support of a distinguished Advisory Editorial Board, areas of current interest that have reached a maturity to be reviewed, are selected for the series. Both Soviet and Western scholars are contributing to the series, and each contributed volume has, accordingly, two editors. Monographs, written by either Western or Soviet authors, are also included. The complete series will provide the most comprehensive coverage available of condensed matter science.

Another important outcome of the foundation of this series is the emergence of a rather interesting and fruitful form of collaboration among scholars from different countries. We are deeply convinced that such international collaboration in the spheres of science and art, as well as other socially useful spheres of human activity, will assist in the establishment of a climate of confidence and peace.

The publishing house "Nauka" publishes the volumes in the Russian language. This way the broadest possible readership is ensured.

The General Editors of the Series,

V.M. Agranovich   A.A. Maradudin

INTRODUCTION

In classifying crystalline solids according to their ability to conduct electrical current one may distinguish two major groups: metals and dielectrics. In metals the concentration of free carriers is so high that the corresponding Fermi energy greatly exceeds the energy of the thermal lattice vibrations in the entire range of temperatures up to the melting point. In contrast, in dielectrics there are practically no free carriers, because in these materials the energy interval between the completely filled valence band and the free conduction band – the range of forbidden energy values – is very large, as compared to the energy of thermal vibrations of the lattice. Owing to this, the conduction band of a dielectric is practically empty at any temperature up to its melting point.

In between these two broad classes of substances, representing the limiting conductivity cases, there are three large groups of crystalline materials with intermediate properties. Adjacent to dielectrics and to metals, there are, respectively, semiconductors and semimetals, while in between these two groups there is a group of new, extremely interesting substances, called gapless semiconductors.

In semiconductors, the magnitude of the energy gap between the conduction band and the valence band is comparable to the energy of the thermal vibrations of the lattice. Accordingly, in intrinsic semiconductors the carrier concentration in the conduction band is determined by the temperature: the concentration is zero at $T = 0\,\mathrm{K}$, and it increases with increasing temperature. As this increase is exponential, the electrical conductivity of the material also increases, in spite of the decreasing mean free path of the carriers. While in semiconductors it is the energy gap that is small, in semimetals it is the Fermi energy of carriers that is low. Naturally, gapless semiconductors have no energy gap at all, and the conduction band and valence band in these materials have a common degeneracy point, which gives rise to their highly specific and very unusual properties.

Thus, in accordance with the above-mentioned classification, semimetals represent metals with an abnormally small (10^{-2}–10^{-5}) number of carriers (electrons and holes) per atom of the material in question. Because of the low carrier concentration, typical conductivity values of even the most perfect single-crystal specimens of semimetals are much lower than those of normal metals, such as Au, Ag, Cu, Be, and Mg. Furthermore, the smallness of Fermi energies of the carriers precludes treating semimetals as substances with a constant, temperature-independent carrier concentration. At high temperatures the concentration of carriers in semimetals begins to increase, though much slower than in semiconductors.

Semimetals belong to the class of so-called compensated metals, having an even number of valence electrons per unit cell of the crystal. The very presence of free carriers in semimetals at $T = 0\,K$ is due to the overlap between the conduction band and valence band, resulting in equal concentrations of electrons and holes. In the intrinsic (undoped) semimetals, the Fermi surfaces consist of an electron and a hole part confining equal volumes, which are small in comparison with size of the Brillouin-zone.

Thus the distinguishing feature of semimetals is the smallness of the characteristic energies (band-overlap energy, electron and hole Fermi energies) in their electron energy spectrum. In this respect, semimetals are similar to semiconductors with a narrow forbidden band, and also to gapless semiconductors. The smallness of the characteristic energies in the spectrum of semimetals is the reason why a number of their physical parameters have abnormal values. It has already been mentioned above that semimetals have very small carrier concentrations (10^{-2}–10^{-5} per atom of a substance or 10^{17}–$10^{20}\,cm^{-3}$) and very small Fermi-surface cross sections (10^{-42}–$10^{-40}\,g^2\,cm^2\,s^{-2}$). Small effective masses of electrons and holes (10^{-2}–10^{-3} times the free electron mass m_0) and high mobilities of the carriers (10^6–$10^7\,cm^2\,V^{-1}\,s^{-1}$) at low temperatures are also typical of semimetals. Their g-factors (10^2–10^3), magnetoresistance coefficients $\Delta\rho/(\rho_0 H^2)$ (10^{-2}–$10^{-6}\,Oe^{-2}$), thermal emf (10^{-4} V/deg) are all abnormally high and their magnetic susceptibility χ exceeds that of all other substances, except magnetics and superconductors. In some semimetals, the dielectric constant is abnormally large (up to 10^2 and even larger). Semimetals are also characterized by the high anisotropy of their lattices, and of their electrophysical, mechanical, and elastic properties.

Elementary semimetals are represented by the elements of the fifth group of the Periodic Table (As, Sb, Bi) and by one crystalline modification of carbon – graphite. Semimetals are also represented by iso-electronic solid solutions of group-V elements (with the exception of $Bi_{1-x}Sb_x$ alloys, which in the concentration range $0.065 \lesssim x \lesssim 0.23$ are semiconductors with a narrow forbidden band). Also, some compounds, such as GeTe, are semimetals.

Although there have been many publications on semimetals, their coverage in the review literature is far from uniform. Among the group-V semimetals, bismuth and its alloys with antimony have received the most study; whereas pure antimony and arsenic, as well as their alloys, have been rather neglected. In the last decade no review devoted to group-V semimetals have been published. All reviews known in the literature dealt only with pure bismuth, and the last of these reviews was published as far back as 1977 (see, e.g., Falkovskii 1968, Edelman 1970, 1976, 1977).

Graphite and its compounds have been much better covered in the review literature, which can be explained by their extremely versatile applications. Among the old publications, one should not fail to mention the monograph by Ubbelohde and Lewis (1960). General reviews on the properties of graphite were published by Reynolds (1968) and Moore (1973). The electronic properties of graphite were reviewed by Spain (1973, 1980) and Bundy (1980). Graphite intercalation compounds (GIC) were discussed in the reviews by Fischer and Thompson (1978), Dresselhaus and Dresselhaus (1981), and Solin (1982). The properties of graphite and GIC under pressure were dealt with by Clarke and Uher (1984). The properties of GIC were also discussed by Levy (1979), Tanuma (1983) and in Synthetic Metals **5** (3/4) [Special Issue Devoted to Low Dimensional Superconductors].

In addition to the reviews listed above, one should also mention the proceedings of the numerous conferences on GIC, recently held in France (1977, 1983), the Netherlands (1979), the USA (1980, 1982), Japan (1980), and Italy (1981)*.

The unique properties of semimetals, the lability of their energy spectrum under the influence of external forces have aroused considerable interest, both scientific and practical, in these materials. One should especially note that in condensed-matter physics semimetals can serve as highly convenient models for observation and study of various single-electron and many-body effects, which can be induced under extreme conditions (e.g., high pressure, great axial strains, high magnetic and electrical fields, changes in the composition of solid solutions, strong doping, and also a combination of some of these factors). Thus, in semimetals there have been discovered and investigated the topological electronic phase transitions of the $2\frac{1}{2}$ kind predicted by Lifshitz, which are due to a change in the topology and connectivity of the Fermi surfaces; in particular, electronic phase transitions of the metal–dielectric type, induced by changes in the composition of solid solutions, by pressure, or by axial stress. In semimetals the following phenomena were discovered: the magnetic freezing-out of impurity carriers and the localization of two electrons onto singly charged impurity centres in a high magnetic field, transitions of the metal–dielectric type in a magnetic field, electron–hole pairing and the formation of the excitonic insulator phase, the emergence of charge density waves (CDW) and the reduced metal phase. Of great interest are the discovery and investigation, in semimetals, of the gapless state and the band inversion due to pressure, the magnetic field or a change in the composition of alloys. The investigation of these new phenomena in semimetals has served as a basis for

*For the list of the conferences see p. 447.

novel practical applications of these materials, which, in turn, have considerably stimulated physical research.

It appears that the time has come to generalize the results obtained from the study of semimetals, and to survey recent developments in this fascinating field of solid-state physics in a monograph.

It seems appropriate to consider separately two groups of semimetals, one of which includes highly anisotropic layered materials and the other, the remaining semimetals. Accordingly, it seemed appropriate to divide this monograph into two volumes.

The necessity of such a separation naturally arises from the fact that in describing the electron energy spectrum and properties of layered materials, it is expedient to proceed from the analysis of these characteristics for a single layer, which represents a two-dimensional system, and only then pass to the quasi-two-dimensional and (highly anisotropic) quasi-three-dimensional systems by taking into account the interaction between the layers. This way of presenting the material is especially convenient and methodologically fruitful when dealing with questions concerning graphite, pyrocarbons and graphite intercalation compounds.

An entirely different approach is required when group-V semimetals and their alloys are discussed. Here, it seems more natural to proceed directly from three-dimensional concepts based on considering small deformations of the initial cubic lattice which give rise to a semimetallic electron spectrum with a small band overlap. It also seems that the presentation in two, rather independent, volumes will increase the audience of the book by enlarging the circle of those who may profit by reading it.

Last but not least, the successive publication of two volumes will make it possible to keep the material of each volume from getting outdated. In each volume of the present monograph, the authors consistently discuss the structural, mechanical, electrical, magnetic, optical and other properties of semimetals, their carrier-energy spectrum and its reconstruction under the influence of external forces and, finally, the vibrational spectrum and thermal characteristics of their lattices.

Much attention is paid to unsolved and controversial problems, as well as to effects and phenomena of fundamental importance to solid state physics. Each volume is concluded by considering unconventional and highly promising scientific and technological applications of the respective groups of semimetals.

The authors owe a debt of gratitude to Professor V.M. Agranovich, who carefully read through the whole manuscript and made helpful suggestions.

N.B. Brandt
S.M. Chudinov
Ya.G. Ponomarev

Moscow
May 1987

CONTENTS

Preface to the series vii
Introduction ix
Contents xiii

Chapter 1. General properties of carbon materials 1

Chapter 2. Crystal lattice of graphite and synthetic carbons 3

2.1. Chemical bonding in the carbon lattice 3
2.2. General remarks on the crystal structure of carbon materials 6
2.3. Perfect graphite lattice 7
2.3.1. A single graphite layer 7
2.3.2. Three-dimensional graphite lattices 8
2.4. Defects in real graphite lattices 11
2.4.1. Stacking faults 11
2.4.2. Defects of carbon-atom bonding in a layer 12
2.4.3. Foreign atoms 12

Chapter 3. Mechanical properties of graphite and carbon materials 15

3.1. General remarks on the elastic parameters of crystals with hexagonal symmetry 15
3.2. Models of interaction between atoms in a graphite crystal 18
3.3. Elastic moduli and force constants for the various kinds of graphite 21

Chapter 4. The energy spectrum of electrons and holes in graphite 29

4.1. The energy spectrum of electrons in a single graphite layer 29
4.2. The electron energy spectrum of three-dimensional graphite. The Slonzewski–Weiss–McClure model 35

4.3. The four-parameter model of the electron energy spectrum of three-dimensional graphite 42
4.4. Trigonal warping of the Fermi surface. Minority carriers in graphite 46
4.5. The electron energy spectrum of graphite in a quantizing magnetic field 50
4.6. Location of the Fermi level in graphite 60
4.7. The parameters of the Slonzewski–Weiss–McClure model 67
4.8. The electron energy spectrum of pyrocarbon materials 69

Chapter 5. Galvanomagnetic effects in graphite and carbon materials .. 73

5.1. Electrical conductivity of graphite 73
5.2. Galvanomagnetic effects in graphite in weak and semiclassical magnetic fields 77
5.3. Mechanisms of carrier relaxation in graphite 88
5.4. Galvanomagnetic effects in graphite in high quantizing magnetic fields and in the extreme quantum limit of fields 91
5.4.1. Magnetic freezing-out of impurity carriers in graphite 93
5.4.2. Magnetically induced transitions of graphite to the charge density wave (CDW) state 101
5.5. Peculiarities of galvanomagnetic effects in carbon materials 104

Chapter 6. Quantum oscillation effects 109

6.1. General remarks on quantum oscillation effects 109
6.2. Oscillation effects in monocrystalline graphite 109
6.3. Parameters of the electron energy spectrum of graphite obtained from oscillation measurements 122

Chapter 7. Optical and magneto-optical effects 129

7.1. Optical properties of graphite 129
7.2. Some remarks on magneto-optical phenomena in graphite 135
7.3. Cyclotron resonance in graphite 137
7.4. Magnetoreflection in graphite 146

Chapter 8. The effects of hydrostatic pressure on the electronic properties and energy spectrum of graphite and pyrocarbon materials 151

8.1. Optical properties of graphite under pressure 152
8.2. Magneto-optical measurements in graphite under pressure 157
8.3. Galvanomagnetic measurements in graphite and pyrocarbon materials under pressure 158

8.4. Oscillation effects in graphite under pressure 161
8.5. Transformation of the energy spectrum of graphite under pressure 167

Chapter 9. Magnetic susceptibility of graphite and carbon materials ... 175

9.1. Magnetic properties of single-crystal and pyrolytic graphite in weak and quantizing magnetic fields 175
9.2. Magnetic properties of graphite in the extreme quantum region of magnetic fields 183
9.2.1. Magnetic susceptibility of graphite in EQL magnetic fields 185
9.2.2. The relation between the field dependence of the magnetic susceptibility of graphite and the phenomenon of magnetic freeze-out 188
9.3. Diamagnetism of quasi-two-dimensional graphite 190
9.4. Magnetic properties of carbon materials 193

Chapter 10. Graphite intercalation compounds (GIC) 197

10.1. Concepts and properties 197
10.2. Mechanisms of formation of a periodic structure in GIC 200
10.3. Synthesis of GIC 204
10.4. Structure of GIC 208
10.5. Phase transitions in GIC 216
10.5.1. Transitions accompanied by a change in stage index, i.e., transitions of the form $n \rightleftarrows n-1$ 216
10.5.2. Transitions accompanied by a change in the order of succession of the intercalate layer (intercalate layer stacking) 216
10.5.3. Structural transitions associated with a change in the ordering of atoms or molecules in the intercalate layer (from commensurate to incommensurate relative to the graphite-layer structure) 216
10.5.4. Transitions to the superconducting state 218
10.5.5. Transitions accompanied by magnetic ordering 218
10.6. The energy spectrum of electrons and holes in GIC 219
10.6.1. Peculiarities of the energy spectrum of carriers in GIC 219
10.6.2. Models of the carrier energy spectrum in GIC of low stage indices in the tight-binding approximation 226
10.6.2.1. A TBA model for the spectrum of GIC with stage index one .. 227
10.6.2.2. A TBA model for the spectrum of GIC with $n = 2$ 228
10.6.2.3. A TBA model for the spectrum of GIC with $n = 3$ 230
10.6.3. The calculation of the energy spectrum of GIC of low stage indices from first principles 232
10.6.4. The rigid-band model for GIC with high n 235
10.7. Electrical conductivity of GIC 241
10.7.1. Peculiarities of the electrical conductivity in GIC 241
10.7.2. Electrical conductivity of GIC in the basal plane 245

10.7.3. Electrical conductivity of GIC along the c-axis 250
10.7.4. Temperature dependence of the conductivities σ_a and σ_c in GIC 251
10.8. Galvanomagnetic effects in GIC 255
10.9. Oscillation effects in GIC 260
10.9.1. General remarks on oscillation phenomena in GIC 260
10.9.2. Fermi-surface structure in GIC 263
10.9.3. Angular dependences of quantum oscillations in GIC 270
10.9.4. Trigonal warping of the Fermi surface and magnetic breakdown oscillations in GIC 273
10.10. Magneto-optical effects in GIC 280
10.10.1. General remarks 280
10.10.2. Magneto-optical spectra in GIC with high n 281
10.10.3. Peculiarities of magneto-optical effects in GIC 284
10.11. The effect of pressure on the structure and energy spectrum of GIC 289
10.11.1. General remarks 289
10.11.2. The effect of pressure on the energy spectrum of GIC with ICl and $CuCl_2$ for $n = 2$ 290
10.11.3. Pressure-induced staging transitions 294
10.12. Magnetic properties of GIC 299
10.12.1. GIC without magnetic ions 299
10.12.2. GIC containing magnetic ions (magnetic GIC) 303
10.12.2.1. Paramagnetic spin-glass transitions 304
10.12.2.2. Transitions leading to magnetic ordering in the intercalate layer 305
10.12.2.3. Transitions to phases with imperfect magnetic ordering 307
10.13. Superconductivity in GIC 309
10.13.1. General remarks 309
10.13.2. Classification of superconducting GIC 310
10.13.3. Pecularities in the superconduction properties of GIC 311
10.13.4. The nature of superconductivity in GIC 314
10.13.5. The effect of pressure on the superconduction properties of GIC 317

Chapter 11. *Lattice vibrations and the phonon spectrum of graphite and graphite-based compounds* 322

11.1. Models of the vibrational spectrum of graphite 322
11.1.1. Vibrational spectrum of the graphite lattice in models with four force constants 324
11.1.2. The axial-symmetric (AS) model of the vibrational spectrum of graphite 339
11.1.3. Graphite-lattice vibrations of small wave vectors (near the Γ-point – the Brillouin-zone centre) 345
11.2. Experimental investigation of the phonon spectrum of graphite 350
11.2.1. Inelastic coherent scattering of slow neutrons in graphite 350
11.2.2. Raman scattering and infrared reflectance studies on the lattice dynamics of graphite 355

11.3. Phonon spectra of GIC 363
11.3.1. The intercalation-induced change in the vibrational modes of pure graphite 365
11.3.2. Vibrational modes in the phonon spectrum of GIC due to the intercalate layers 377
11.3.3. Strongly coupled phonon modes – mixed phonon modes 385
11.3.4. Zone-folding effect in GIC 390
11.3.5. Changes in intercalate-layer structure and degree of ordering with temperature 394
11.4. Heat capacity of graphite and its intercalation compounds 396
11.4.1. Lattice heat capacity of graphite 397
11.4.2. Heat capacity of GIC 402

Chapter 12. Unsolved problems and applications 410

12.1. Research prospects for graphite, pyrocarbons, and GIC 410
12.1.1. Mechanical properties, lattice 410
12.1.2. Electron and hole energy spectra in graphite 410
12.1.3. Transport phenomena in graphite 412
12.1.4. The energy spectrum of carriers and the magnetic properties of pyrocarbon materials 412
12.1.5. Transport properties of pyrocarbons 412
12.1.6. The energy spectrum of GIC 413
12.1.7. Supermetallic conductivity of GIC 414
12.1.8. Superconductivity of GIC 414
12.2. Technological applications of graphite, pyrocarbons and GIC 415
12.2.1. Graphite and pyrocarbon materials 415
12.2.2. Graphite intercalation compounds – novel applications 416
12.2.3. Technological and industrial uses of GIC 416
12.2.3.1. GIC as battery and electrode materials 416
12.2.3.2. GIC as high-conductivity materials 418
12.2.3.3. Graphite fibres used to produce composite materials 419
12.2.3.4. Applications in chemical catalysis 420
12.3. Graphite and carbon materials in atomic technology 420
12.3.1. Properties of carbons used in building nuclear reactors 421
12.3.1.1. Neutron moderating and absorbing power 422
12.3.1.2. Heat resistance and strength properties at high temperature ... 423
12.3.1.3. Low penetrability 424
12.3.1.4. Chemical reactivity 426
12.3.2. Methods of reactor graphite production 429
12.3.2.1. Heat treatment and graphitization 429
12.3.2.2. Elimination of impurities 430
12.3.3. Radiation damage 433
12.3.3.1. Change in the crystal-lattice parameters of graphite 433

12.3.3.2. Defect formation 435
12.3.3.3. Changes in size stability and porosity 438
12.3.3.4. Changes in thermal and electronic properties 441

Proceedings of Conferences on GIC 447
References 449
Author index 467
Subject index 483
Cumulative index 489

CHAPTER 1

General Properties of Carbon Materials

In addition to the elements of the fifth group of the Periodic Table, their alloys and individuals compounds, semimetals include graphite, a crystalline allotropic form of carbon, the lamellar compounds of graphite, and various artificially made carbon materials.

Among the most important properties of carbon materials, one should mention their low specific gravity, chemical inertness preserved up to 700 or even 900 K, high antifrictional power, large neutron-attenuation factor, sublimation temperature above 3900 K, their ability to retain their mechanical characteristics at temperatures approaching (and sometimes exceeding) 3500 K, and the capacity of varying widely their resistivity in response to heat treatment or intercalation with various substances. It is precisely these outstanding properties that have ensured the spectacularly versatile applications of carbon materials in modern technology we are now witnessing. Technological needs have stimulated extensive studies of carbons aimed at producing hitherto unknown materials with pre-assigned properties.

From the scientific viewpoint carbon materials are objects of great interest, which is primarily due to the lamellated character of their crystal structure, entailing an extremely high anisotropy in their electric, magnetic, mechanical, and other characteristics.

The study of carbon materials involves multitudinous problems in physics and chemistry. Thus, the organic chemist may take interest in them as substances representing a limiting case of the various aromatic series, whereas the physicist regards them as substances which, depending on how they are produced and heat treated, may possess either semimetallic or semiconducting properties.

In recent years, carbon materials have been much studied. Especially intensive has been the research on the lamellar graphite compounds formed by intercalation of graphite with alkali metals, inorganic acids, their salts, halogens, etc. All these compounds exhibit a metallic character of conductivity and an higher anisotropy in their properties than do the nonintercalated carbons. The intercalated carbons are unique as construction materials and are highly effective as substitutes for metallic conductors.

Of all carbons, natural monocrystalline graphite has received the most study. As a result, an adequate theoretical model to describe the band structure of graphite has been developed, the parameters of this model being determined and refined from the data of many experiments. Artificial graphites are rather close, in their crystal structure and electronic properties, to natural monocrystalline graphite. Yet, owing to defects in their crystal structure and the block structure of their lattice, the electric and magnetic properties of synthetic graphites show some pecularities. In particular, certain pyrocarbon materials, in which the distance between the neighbouring carbon layers is greater than that in natural graphite, are characterized by the highest diamagmetic susceptibility among all known substances, with the exception of superconductors.

CHAPTER 2

Crystal Lattice of Graphite and Synthetic Carbons

2.1. Chemical bonding in the carbon lattice

Small specimens of natural graphite represent nearly perfect single crystals, whose structure was first studied by X-ray techniques as far back as 1917 by Hall and later (in 1924) by Bernal. The accuracy of their original data was steadily improved upon in subsequent studies. As a result, the crystal structure of graphite and other forms of carbon has by now been studied quite comprehensively.

The structural features of carbon materials and, specifically, of graphite are largely due to peculiarities in the character of the bonding between carbon atoms, which we will discuss in more detail.

In the periodic table, carbon is placed in the fourth group, with atomic number 6 and atomic weight 12.011. The neutral carbon atom contains six electrons in a $1s^2 2s^2 2p^2$ configuration, which describes the ground state of the carbon atom when carbon is divalent. The ionization potential of the carbon atom is 11.217 eV. As is known, the distribution of the s-electrons in carbon is spherically symmetric, whereas that of the p-electrons resembles a dumb-bell with an infinitely thin handle, its centre being on the nucleus. The shape of these spatial distributions is altered when the wave functions for the respective states are hybridized. The shape of a spatial distribution depends on the type of the sp-hybridization, which, in turn, is determined by the number of p-electrons being hybridized with a single s-electron. As the number of p-electrons participating in the hybridization increases, the statistical weight of the s-wave function decreases, and the shape of the spatial distribution is altered.

In most chemical compounds, carbon is known to be tetravalent. This results from a transition of a single electron from the 2s- to the 2p-state. The excited state of the carbon atom is characterized by an electron configuration of the $1s^2 2s^1 2p_x^1 2p_y^1 2p_z^1$ type. In such a configuration, spatial eights formed by the p-electrons are elongated in the directions of the rectangular axes (x, y, z), with the origin on the nucleus.

A transition of the carbon atom to the excited $1s^2 2s^1 2p^3$ state requires that an energy of about 4 eV be expended. This is more than compensated for by the energy released in the formation of the bonds which tetravalent carbon atoms exhibit in various chemical compounds. These bonds arise from the overlap between the wave functions that couple the electrons of two neighbouring atoms to one another. In all innumerable organic compounds, tetravalent carbon forms bonds which arise at one or another degree of hybridization of the wave functions for the s- and p-electrons. These bonds are reasonably well described within either of the following three models: tetragonal, trigonal, and digonal.

The tetragonal model was proposed to account for the equivalence of the four carbon bonds in methane-type molecules. In this model, the carbon atom is placed at the centre of a tetrahedron, the four equivalent bonds making the same angle with each other and being directed towards the corners. Such bonds arise from the mixing of the states of one s-electron and three p-electrons. Such a situation is referred to as the sp^3-hybridization (Coulson 1961). In this case, the spatial electron distribution has the shape of four asymmetric eights, whose major axes make the same angle with each other (fig. 1a). The tetragonal model describes rather well carbon-atom bonds in diamond crystals and in so-called limiting hydrocarbons.

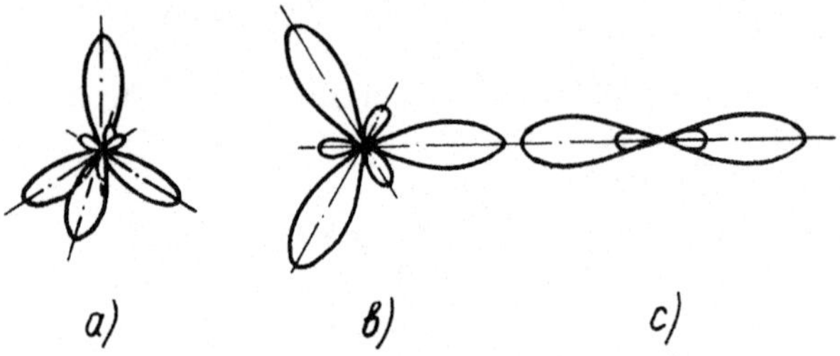

Fig. 1. Spatial distribution of the electrons in the carbon atom forming the σ-bonds for various types of the sp-hybridization: (a) sp^3, (b) sp^2, (c) sp.

In nonlimiting organic compounds with double bond, in aromatic compounds and in graphite, carbon atoms occur in a state described by the trigonal model. In such substances, the carbon atom has only three equivalent bonds, which all lie in the same plane and make an angle of 120° with each other. These bonds arise from the mixing of the states of one s-electron and those of two p-electrons (the sp^2-hybridization), and are called σ-bonds. The spatial distribution of electrons forming σ-bonds represents three asymmetric spatial eights, whose major axes lie in the same plane and make an angle of 120° with each other (fig. 1b). The fourth valence p-electron is then called a π-electron. The corresponding electron density distribution has the shape of a symmetric eight elongated in the direction perpendicular to the plane of σ-bonds. The overlap of the π-electrons belonging to the neighbouring carbon atoms gives rise to π-bonds. Accordingly, in the trigonal model, each carbon atom is linked to other atoms by one π-bond and three σ-bonds.

In molecules with triple bond, such as acetylene (HC≡CH), hydrocyanic acid (HC≡N) and some others, the carbon atom is in a third state, which is commonly described by the digonal model. The triple bond is then formed by two π-bonds and one σ-bond. Such a state of the carbon atom originates from sp-hybridization, in which the state of one s-electron is mixed with that of one p-electron, thereby giving rise to two hybrid electron clouds in the shape of two asymmetric dumb-bells. The angle between two σ-bonds is then equal to 180°, the bonds themselves being directed along one of the coordinate axes (fig. 1c). The electron clouds formed by the remaining two p-electrons are elongated in the direction of the other two rectangular axes, which are normal to the σ-bond. The overlap between these electrons belonging to the neighbouring atoms gives rise to two π-bonds.

Studies of infrared spectra, combination scattering spectra, and the energy of formation of carbon compounds, as well as X-ray analyses, have shown that in different substances the carbon–carbon bonds have different length, electron density distribution, and force constants. The most comprehensive classification of carbon–carbon bonds was proposed by Tatewskii (1953) and Brenstein (1962). It is based on the concepts of the bond multiplicity and the valence state of the carbon atom. Multiple bonds are due to the involvement of π-electrons. When double and triple carbon–carbon bonds are formed, they grow shorter, stronger, their interaction energy increasing. The change in bond length may be caused by the varying extent of hybridization of the s- and p-electrons. The larger the contribution of the s-electron to the hybrid bond, the shorter and stronger it becomes. Naturally, the statistical weight of the s-wave function in the hybridized state is lower in the sp^3-hybridization than in the sp-hybridization. By comparing the lengths of carbon–carbon bonds, one may get an idea about their strength and calculate the values for the linkage force constants. (By the force constant ϕ we here mean the factor of proportionality between the increment of bond length Δl, and the applied force, F, i.e., $F = \phi \Delta l$; hence, ϕ is measured in N/m or dyne/cm). Thus, for the benzene molecule and for a layer of atoms in graphite, the force constants of carbon–carbon bonds are equal to 725.2 and 568.4 N/m, respectively (Coulson 1961); their length being about 1.4 Å. The force constant of the interplanar bonds in the graphite lattice, whose length is about 3.35 Å, has the value of about 19.6 N/m, which is practically 30 × as low as that of the force constant for the bonds between graphite atoms in a layer. We indicate for comparison that in the structure of a linear carbon polymer, called carbine, the shortest distance between atomic strings is 2.95 Å. According to Kasatochkin et al. (1967), the force constant of the interstring bonding in carbine is close to 49 N/m.

At present, there is no unanimity regarding the nature of interlayer bonding in graphite. Some believe it to be of the Van Der Waals type, others regard these bonds as metallic, while a third group treats them as weakened chemical bonds of the covalent type. The complexity of the problem lies in the fact that, although the interlayer bonds in graphite do possess some of the features typical of the bond types mentioned above, their peculiarity precludes an unequivocal classification.

The bond strength is best described by a value of the binding energy. i.e., the energy that must be spent to break the bond. Binding-energy values are calculated from data on the heats of formation, combustion and decomposition. The binding energy of carbon atoms in graphite can be estimated from the fact that a monatomic gas is produced when graphite evaporates. Hence it follows that the formation heat of atomic carbon coincides with its sublimation heat. According to Vedeneev (1962), the sublimation heats of carbon at temperatures of 0 and 298 K are equal to 169.96 and 170.9 kcal/g-atom, respectively, which corresponds to binding energy values of 7.38 and 7.42 eV per carbon atom. A calculation by Bernstein (1962) yielded values of 4.35–4.8 eV for the binding energy of carbon atoms in a graphite layer, and a value of 0.43 eV for the interaction of the neighbouring layers in graphite, calculated per carbon atom. Evidently, the total energy of a multiple bond is the sum of the σ- and π-bond energies. The calculation of the π-bond energy in a graphite layer made by Stepanov and Tatewskii (1961) gave a value of 1.3 eV per carbon atom. Therefore, the σ-bond energy in a graphite layer is 5.05–5.5 eV.

In crystals of graphite and diamond carbon atoms are in different hybrid states. The stability of these crystalline forms of carbon depends on the ambient conditions. According to Berman and Simon (1955), the equilibrium pressure for a graphite–diamond system constitutes 13.5 kbar at $T = 0$ K, increasing to 39.5 kbar as the temperature increases to 1200 K. At atmospheric pressure and a high temperature, graphite sublimes without passing through the liquid state. Liquid graphite exists at pressures above 100 bar and temperatures $\gtrsim$ 4000 K. Bundy (1963) cites the melting heat of graphite at a pressure of 48 kbar as equal to 25 kcal/g-atom.

2.2. *General remarks on the crystal structure of carbon materials*

The great diversity in structure of solid condensed compounds of carbon (diamonds, graphites, coals, cokes, soots, etc.), discovered by X-ray and electron diffraction studies, is due to the fact that the carbon atom may occur in various valence states and may form bonds of different types. Of the structures mentioned above, those of the three crystalline forms of carbon, i.e. diamond, graphite and carbine are the best known. The valence state of the carbon atom in these crystalline substances corresponds to the sp^3-, sp^2-, and sp-hybridization, respectively. Depending on the conditions of their formation, real structures may be significantly different from perfect, displaying a variety of intermediate structures. Also, in real structures, the statistical weight factors of the s- and p-electron wave functions contained in the hybrid functions may considerably differ from their values for the sp^3-, sp^2-, and sp-hybrid states.

In the sp^2-hybrid state, carbon atoms form lamellar structures, as it follows from the very character of bonds then arising. Here, the principal structural unit is represented by a layer of carbon atoms consisting of hexagonal rings. If such layers are regularly oriented relative to one another, there will arise a graphite lattice, whose type depends on the order of alternation of the layers. When there is no regularity in the orientation of layers with respect to the hexagonal axis (the six-fold symmetry axis normal to the layer), the structure is called, as suggested by Warren (1941), turbostratum.

2.3. *Perfect graphite lattice*

X-ray studies suggest that, in the lattice of monocrystalline graphite, carbon atoms occur in almost parallel layers, the distance between two neighbouring layers at room temperature being, according to Franklin (1951), $d_0 = 3.3538\,\text{Å}$. In each layer, carbon atoms are at a distance $a = 1.415\,\text{Å}$ from one another, forming a network of regular hexagons. Since the interplanar distance in graphite is nearly 2.5 times as long as the distance between the nearest neighbour atoms in a layer, a single layer of carbon atoms may essentially be regarded as an independent structural unit of the graphite lattice. Each layer is in fact similar to a large molecule of the type represented by the flat benzene molecule. The weak bonding between layers is responsible for a high anisotropy in many properties of graphite.

Thus an advisable course to start a detailed consideration of the graphite lattice structure is to survey the structure of a single layer composed of carbon atoms.

2.3.1. *A single graphite layer*

A monolayer of carbon atoms in graphite forms a two-dimensional periodic structure. In order to isolate a unit cell in such a layer, we shall divide the structure into regular rhombs by two systems of parallel lines, as shown in fig. 2a. To such a unit cell, there correspond two fundamental translation vectors, $\boldsymbol{a}_1$ and $\boldsymbol{a}_2$, with the same absolute values equal to $|\boldsymbol{a}_1| = |\boldsymbol{a}_2| = a_0 = \sqrt{3}a \sim 2.46\,\text{Å}$, the angle between these vectors being 60°. The area of such an elementary rhomb is equal to that of a hexagon, designated as S_0. This area is given by the scalar product of these two vectors:

$$S_0 = (\boldsymbol{a}_1 \cdot \boldsymbol{a}_2) = \tfrac{1}{2}\sqrt{3}\,a_0^2. \qquad (1)$$

There are two carbon atoms per unit cell in a layer. Consequently, if the total area of the layer is denoted by S, the number of atoms in the layer, N, is given

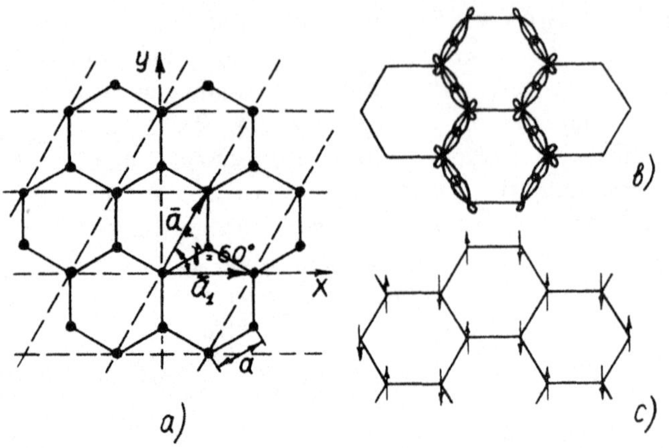

Fig. 2. Schematic representation of carbon atoms in a single graphite layer: (a) primitive translation vectors $\boldsymbol{a}_1$ and $\boldsymbol{a}_2$ and a unit cell of the single layer, (b) overlap of the wave functions in the formation of σ-bonds, (c) π-electron spin distribution in the graphite layer.

by

$$N = 2\frac{S}{S_0} = \frac{4S}{\sqrt{3}\,a_0^2}. \qquad (2)$$

The location of carbon atoms at the corners of a regular hexagon is a consequence of the equivalence of the three bonds of the carbon atom with its three nearest neighbours in the layer, which takes place in the sp^2-hybridization (fig. 2b). Atoms of a graphite layer are linked with each other not only by the directed σ-bonds, but also due to the overlapping of the wave functions for the unhybridized p-electrons. This overlap results in the formation of π-bonds isotropic in the plane of a graphite layer. We note that, in order that π-bonds be formed, it is necessary that the spins of the neighbouring atoms in the layer be antiparallel (fig. 2c). The semimetallic properties of three-dimensional graphite are associated with the collectivization of the π-electrons in a layer. The high mobility of these electrons causes the high electrical conductivity in the plane of a graphite layer. It is also responsible for the characteristic metallic lustre of graphite. Thus a single graphite layer in three-dimensional graphite is close in its properties to a quasi-two-dimensional metal. We note that the π-electron collectivization somewhat weakens the carbon–carbon bonds in a graphite layer, which results in an increase in length from a typical value of 1.39 Å to 1.417 Å.

2.3.2. *Three-dimensional graphite lattices*

X-ray studies have revealed that monocrystalline graphite may have either a hexagonal or a rhombohedral lattice. These two types of lattice differ from

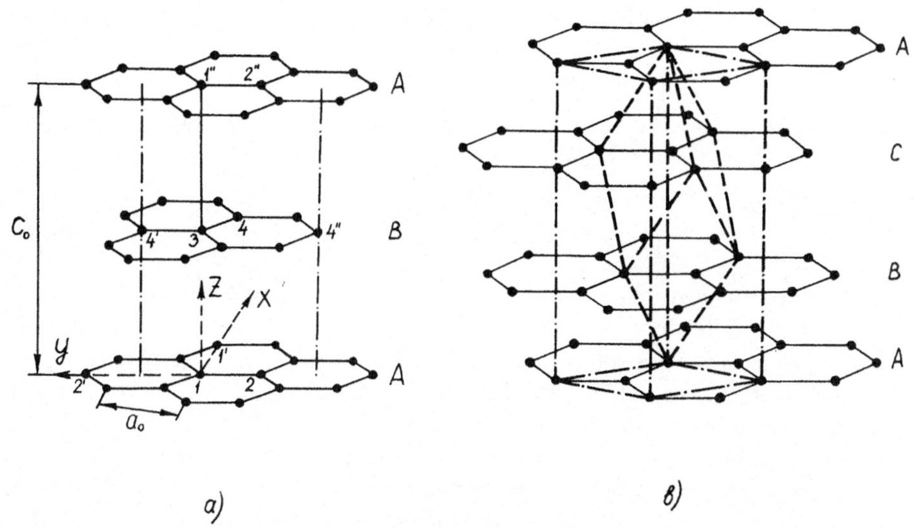

Fig. 3. Crystal structures of graphite (Wyckoff 1964) (a) hexagonal, (b) rhombohedral (the rhombohedron is formed by the dashed lines). The letters A, B, C denote non-equivalent graphite layers.

each other solely by the order of alternation of monolayers, the structure of each layer being the same. The hexagonal lattice, in which a row of atoms in one of the layers is exactly above (or under) the centres of the hexagons in the neighbouring layers, as shown in fig. 3a, is the most typical one for monocrystalline graphite. The order of stacking is shown by the alternation of two types of layers as follows: ABABA, the period of recurrence of one type being equal to the double distance, d_0, between two neighbouring layers.

In a hexagonal lattice, each successive layer is shifted relative to the preceding one in the direction parallel to the edge of a unit hexagon in its plane by a distance equal to the bond length, a, between carbon atoms in a layer. The hexagonal lattice for graphite was first proposed by Bernal (1924) and has been named after him.

A right-angle prism having a regular rhomb as its base serves as a unit cell for the hexagonal lattice of graphite. The fundamental translation vectors, $\boldsymbol{a}_1$, $\boldsymbol{a}_2$, $\boldsymbol{a}_3$ (the last of them being normal to the plane of a layer) of a unit cell have the following rectangular coordinates (x, y, z):

$$\boldsymbol{a}_1 = (a_0, 0, 0), \qquad \boldsymbol{a}_2 = (\tfrac{1}{2}a_0, \tfrac{1}{2}\sqrt{3}\,a_0, 0), \qquad \boldsymbol{a}_3 = (0, 0, c_0). \tag{3}$$

Their absolute values are: $|\boldsymbol{a}_1| = |\boldsymbol{a}_2| = a_0$, $|\boldsymbol{a}_3| = c_0 = 6.7076\,\text{Å}$. The translation vector $\boldsymbol{n}$ in the lattice is defined as $\boldsymbol{n} = n_1\boldsymbol{a}_1 + n_2\boldsymbol{a}_2 + n_3\boldsymbol{a}_3$, where n_1, n_2, n_3 are integers. The volume of a unit cell is thus given by

$$V_0 = ([\boldsymbol{a}_1 \times \boldsymbol{a}_2]\cdot\boldsymbol{a}_3) = \tfrac{1}{2}\sqrt{3}\,a_0^2 c_0. \tag{4}$$

A unit cell of the hexagonal lattice contains four carbon atoms. Atoms labelled 1, 2, 3, 4 in fig. 3a are all in the same unit cell, while those labelled 1, 1′, 1″, etc., are on the same sublattice. The elementary volume per carbon atom is equal to $\frac{1}{8}\sqrt{3}\,a_0^2 c_0$. If V stands for the volume of a crystal, then the number of carbon atoms in the crystal, N, is given by

$$N = \frac{8V}{\sqrt{3}\,a_0^2 c_0}. \tag{5}$$

Knowing the values of a_0 and c_0, one can readily calculate that the density of a perfect graphite crystal with a hexagonal lattice will be $\rho_0 = 2.265\,\mathrm{g/cm^3}$. This value is rather close to that for natural monocrystalline graphite. The hexagonal (Bernal-type) lattice should properly be included in the spatial symmetry group $P6_3/mmc$ (or, in the notation of Schoenflies, D_{6h}^4). It represents a perfect lattice of the most stable form of monocrystalline graphite.

The arrangement of atoms in the lattice of natural graphite differs somewhat from perfect. Its spatial symmetry is lower, being described by the group $P6_3\,mc$ (or C_{6v}^4).

As has already been mentioned, the rhombohedral lattice differs from the hexagonal one by the order of alternation of its layers. According to Boehm and Hofman (1955), the rhombohedral (or trigonal) lattice is essentially trilamellar and is characterized by the succession of layers of the form ABCABC ... (fig. 3b). This graphite lattice was first discovered by Finch and Wilman (1936) who X-rayed a number of specimens of both natural and synthetic graphite. These authors observed several diffraction lines which could not be explained on the basis of a hexagonal structure. And these additional lines could not be accounted for by either impurities or the diffusive X-ray scattering. Lipson and Stokes (1942) demonstrated that these diffraction lines were due to a previously unknown structure, with a different order of layers, as compared to the hexagonal lattice.

In rectangular coordinates, the fundamental translation vectors $\boldsymbol{a}_1, \boldsymbol{a}_2, \boldsymbol{a}_3$ of the unit cell of the trigonal lattice have the form

$$\boldsymbol{a}_1 = (a_0, 0, 0), \quad \boldsymbol{a}_2 = (\tfrac{1}{2}a_0, \tfrac{1}{2}\sqrt{3}a_0, 0), \quad \boldsymbol{a}_3 = (a_0, \tfrac{1}{3}\sqrt{3}a_0, \tfrac{1}{2}c_0). \tag{6}$$

The unit-cell volume is given by

$$V_0 = ([\boldsymbol{a}_1 \times \boldsymbol{a}_2]\cdot \boldsymbol{a}_3) = \tfrac{1}{4}\sqrt{3}a_0^2 c_0. \tag{7}$$

In the trigonal structure, there are two carbon atoms per unit cell. Therefore, the elementary volume per carbon atom is, as in the hexagonal lattice, $\frac{1}{8}\sqrt{3}\,a_0^2 c_0$.

The trigonal lattice of graphite finds its place is the spatial symmetry group $R\bar{3}m$ (D_{3d}^5). The fraction of the trigonal lattice in natural graphite may be as high as 20%. Matuyama (1956) indicated that the heat treatment of natural graphite at temperatures from 2000 to 3000°C, as well as treating it with hot, strong acids, results in a decrease of that fraction. This is suggestive of the higher stability of the hexagonal graphite lattice.

It should be noted that the formation of a hexagonal lattice is usually preceded by a variety of intermediate structures, showing very little, if any, regularity in layer alternation. In such intermediate structures, the period of layer repetition decreases with increasing perfection of the lattice from infinity (the turbostratum lattice) through a structure with three interlayer distances (trigonal) and eventually to two interlayer distances (hexagonal).

The great diversity of carbon structures is a consequence of the considerable mobility of layers of carbon atoms relative to one another which in turn is a result of the weak interaction between the layers. Such polymorphism is peculiar not only to graphite, but to all lamellar structures.

2.4. *Defects in real graphite lattices*

Real structures of graphites differ from the perfect structures described above by the presence of various defects associated with the breaking of the crystal structure periodicity. Even monocrystalline graphite has an imperfect, commonly mosaic, lattice.

Let us dwell very briefly on the main types of defects.

2.4.1. *Stacking faults*

The most common defect in the graphite lattice is the breaking of periodicity in the alternation of layers. Such defects are referred to as stacking faults. These defects are especially numerous when a crystal has been strained. Turbostratum materials, in which the periodicity of layer alternation, typical of the hexagonal or trigonal phases, is completely broken, are produced, for example, by decomposing the various aromatic hydrocarbons, followed by the precipitation of the reaction products onto a heated substrate. The average distance, $\bar{d}$, between the neighbouring graphite layers, which can be determined by X-ray techniques, provides a quantitative characteristic for the degree of disorder. Bacon (1951) established a relation between this quantity and the parameter p representing the fraction of graphite layers found in the turbostratum state. A perfect graphite lattice corresponds to $p = 0$ whereas a completely disordered lattice corresponds to $p = 1$. These quantities are empirically related by the following expression

$$\bar{d} = 3.440 - 0.086(1 - p) - 0.064(p - 1)p\,(\text{Å}). \quad (8)$$

By this formula, in a turbostratum material ($p = 1$) the average spacing $\bar{d} = 3.44$ Å (Bernal's lattice).

Artificial graphites represent intermediate structures, in which stacking faults are always present. Thus, according to Blackman (1960), for typical graphites used in nuclear reactors the average spacing $\bar{d} = 3.36$ Å, which corresponds to a degree of lattice disorder $p = 0.17$.

2.4.2. Defects of carbon-atom bonding in a layer

Another common type of defects in the structure of artificial carbon materials is represented by defects in atomic bonds in graphite layers. These include, in particular, vacant sites (or vacancies) in the hexagonal network of a carbon layer (Schottky defects) and also displacements of carbon atoms from lattice points to interstitial positions (Frenkel defects). An accumulation of Schottky defects may bring about a disruption of the carbon network leading to formation of the so-called "pliers". In the vicinity of such defects, the crystal lattice swells as a result of repulsion between the groups of atoms with broken bonds. Pliers often give rise to screw dislocations and other distortions of the carbon network. An accumulation of vacancies may also result in the formation of voids in graphite crystals. Bonfiglioli and Mojoni (1964) discovered lens-shaped defects with cross-sections ranging from 200 to 3000 Å and a thickness of several lattice parameters in natural graphite. Viewed through an electron microscope, such defects can be recognized from Newton's rings that fringe them. The closing of lens-shaped regions formed by the coagulation of vacancies gives rise to edge dislocations and dislocation loops. It has been found experimentally that the number of vacancies in a graphite crystal increases as it is heated to high temperatures. Henning (1965) reported the formation of vacancies and dislocation loops in the peripheral zone, as graphite crystals were heated in vacuum to between 1900 and 3400 K. Rasor and McClelland (1960) estimated the formation energy of a vacancy in graphite as (7.7 ± 0.5) eV, i.e., roughly the same as the sublimation heat (7.4 eV). One can readily estimate that, as a graphite crystal absorbs an energy of about 600 cal/g, there arises approximately one vacancy per every ten carbon atoms. Frenkel defects are formed when atoms are wedged into interstitial sites in the graphite lattice. Interstitials are placed between layers, causing their bending. A large number of interstitials may form associations, which leads to the parting of layers. Such defects are especially common in graphite irradiated with neutrons.

2.4.3. Foreign atoms

Defects in a graphite lattice arise when foreign atoms, which may not only enter into the space between layers, but also occupy vacant sites in carbon layers themselves, are incorporated into the lattice. The ability of atoms or even molecules to enter in large numbers into the interlayer regions of graphite is widely used to produce lamellar graphite compounds. Incorporation of foreign atoms into graphite will be discussed at length in ch.

10 in connection with the study of graphite intercalation compounds.

It will be noted that in the carbon network, interstitial impurity atoms play the part of electron donors or acceptors. Thus, the quintavalent atoms of phosphorus may serve as donors, while the trivalent atoms of boron may serve as acceptors. Point defects and dislocations may act as acceptors, too.

By their ability to perfect their structure under heat treatment, artificial carbon materials are grouped into graphitizable and nongraphitizable. The latter do not alter their structure when heated to temperatures above 2000°C, their interlayer spacing being $\tilde{d} = 3.70$ Å. Graphitizable materials begin to change their structure already at 1600–1800°C, their initial interlayer spacing being $\tilde{d} = 3.44$ Å. Although the methods of producing carbon materials may be different, the analysis of carbon particles from oil and pitch cokes, soot, refined graphites, as well as from pyrographite and kish-graphite*, indicates that all these particles have a very similar chemical composition, the fraction of impurities constituting merely several hundredths of a percent. The basic unit of all the materials listed above is a single graphite layer having one or another degree of imperfection. It should be noted that the variation of certain parameters even within a single technological process (e.g., a variation of the temperature of the substrate by about 20°C or some difference in the size of the chamber used for producing pyrocarbons) may significantly alter the structure and, as a consequence, bring about a considerable change in mechanical and electrophysical properties of carbon materials.

At present, in quantitative descriptions of various carbon structures the following parameters are used:

(a) the average distance between layers, $\tilde{d}$, whose value may vary from 3.400 to 3.354 Å;

(b) a dimensionless parameter γ, called the graphitization parameter, which gives the measure of three-dimensional order:

$$\gamma = \frac{3.400 - \tilde{d}}{0.086}, \tag{9}$$

where 3.440 Å is the interlayer distance of the least-ordered structure, 0.086 Å is the difference between the values for the interlayer distances in the least-ordered structure and in monocrystalline graphite;

(c) the average statistical size of the regions of coherent scattering in the basal plane (L_a) and in the plane perpendicular to it (L_c); these parameters are

*Pyrographite (or pyrolytic graphite) is an artificial carbon material produced by precipitation of carbon from the gaseous phase in the course of the thermal decomposition of methane in vacuum, and kish-graphite is produced by crystallization of carbon from an iron melt. One of the varieties of pyrolytic graphite is highly oriented pyrolytic graphite (HOPG) produced in the additional graphitization of pyrographite by heating to 2800–3000°C under pressure. Of all the artificial carbon materials produced so far, HOPG comes closest in its structure to perfect single-crystal graphite.

determined from the broadening of X-ray lines:

$$L_{a,c} = \frac{K\lambda}{\beta \operatorname{Cos} \theta}, \tag{10}$$

where λ and θ are the wavelength and angle of reflection of X-ray, respectively: K is the form factor of X-ray lines, β is the line's half-width. Morant (1970) established that the size L of the regions of coherent scattering is in good agreement with that of crystallites;

(d) the concentration of layer stacking faults, which can be determined by a procedure proposed by Kurdyumov and Piljankevich (1968) and Kurdyumov (1972);

(e) the degree of preferred orientation of crystallites, $n(\varphi)$, which is determined from the relative intensity of X-ray lines

$$n(\varphi) = I(\varphi)/I(0), \tag{11}$$

φ being the angle between the normal to a suitable plane and the preferred directions.

When graphitizable materials are heat treated, the perfection of their crystal lattice increases, which manifests itself in a decrease of the value of the parameter $\bar{d}$ and an increase in the values of γ, L_a, L_c, and $n(\varphi)$.

CHAPTER 3

Mechanical Properties of Graphite and Carbon Materials

3.1. General remarks on the elastic parameters of crystals with hexagonal symmetry

A description of the mechanical properties of crystals may involve two different approaches: macroscopic (or phenomenological) and microscopic.

The former consists in introducing the strain tensor, ε_{ij}, and the stress tensor, σ_{ij}, and in considering their relationship, which characterizes the elastic properties of a solid.

In the latter approach, one considers the potential energy of elastic deformation of a crystal, expressible in terms of the force constants, ϕ, of the bonds between the atoms in the lattice.

Let us examine at greater length each of these approaches, with reference to a description of the mechanical properties of crystals with hexagonal symmetry, including graphite and carbon materials. It is convenient to start the examination with a macroscopic description of the elastic properties, i.e., with a discussion of the strain and stress tensors.

It is known that the strain tensor, ε_{ij}, describes the change in the shape and volume of a crystal under the action of external forces, and represents a symmetric second-rank tensor: $\varepsilon_{ij} = \varepsilon_{ji}$. The diagonal components of this tensor, ε_{ii}, define the tensile or compressive strains in the direction of the three rectangular axes ($i = 1, 2, 3$), while the off-diagonal components, ε_{ij} ($i \neq j$), describe the shearing strain. In the general case, the number of independent components ε_{ij} is equal to six.

In a strained crystal, its separate parts are no longer in equilibrium with one another. As a result, there arise stresses, which tend to bring all parts of the crystal back into equilibrium. These stresses are described by a symmetric second-rank tensor: $\sigma_{ij} = \sigma_{ji}$. Its diagonal components, σ_{ii}, describe stresses along the axes $i = 1, 2, 3$, while the off-diagonal components, σ_{ij} ($i \neq j$) describe shearing or tangential stresses. In the general case, the number of independent components, σ_{ij}, is also equal to six.

For small deformations, the components of the strain tensor ε_{ij} are expressed through those of the stress tensor σ_{ij} using a linear relation between them. This is the famous Hookian law, which may be represented in either of the two following forms:

$$\varepsilon_{ij} = s_{ijkl}\sigma_{kl}, \tag{12}$$

$$\sigma_{ij} = c_{ijkl}\varepsilon_{kl}, \tag{13}$$

where the repeated indices imply summation. The coefficients, s_{ijkl} and c_{ijkl}, which characterize a crystal structure, are called elastic constants and elastic moduli, respectively. Because of the symmetry of tensors ε_{ij} and σ_{ij}, the coefficients just introduced are, in the most general case, symmetric in the first and second pairs of indices, as a result of which the number of the independent components of s_{ijkl} or c_{ijkl} will in general be equal to 36. Actually, this number is still smaller, being only 21. In order to show this, we shall write the relations (12) and (13) in a different form by replacing the first and second pairs of indices in s_{ijkl} and c_{ijkl} by a single index, in accordance with the following scheme proposed by Nye (1964),

$$11 \to 1; \quad 22 \to 2; \quad 33 \to 3; \quad 23 = 32 \to 4; \quad 13 = 31 \to 5; \quad 12 = 21 \to 6. \tag{14}$$

Then the above relations assume the form

$$\varepsilon_i = s_{ij}\sigma_j, \tag{15}$$

$$\sigma_i = c_{ij}\varepsilon_j, \tag{16}$$

where the repeated indices also imply summation. In the adopted notation, the tensors ε_{ij} and σ_{ij} will be replaced by the corresponding vectors in a six-dimensional space, while the tensors of fourth rank s_{ijkl} and c_{ijkl} will be replaced by second-rank tensors in a six-dimensional space. Let us now calculate the elastic strain energy of a crystal. The strain components σ_i acting upon the faces of a cube cut out of a crystal will evidently do the elementary work:

$$dE = \sigma_i d\varepsilon_i = c_{ij}\varepsilon_j d\varepsilon_i, \tag{17}$$

where the relation (16) has been taken into account. Integrating (17) yields the energy E of elastic deformation per unit volume:

$$E = \tfrac{1}{2}c_{ij}\varepsilon_i\varepsilon_j \tag{18}$$

From (17) it follows that

$$c_{ij} = \frac{\partial^2 E}{\partial\varepsilon_i \partial\varepsilon_j} = c_{ji}. \tag{19}$$

From the symmetry of the tensor c_{ij} there also follows the symmetry of the tensor s_{ij}. It is due to the symmetry of the elastic constants that their number reduces from 36 to 21. The further reduction in this number can be achieved when considering particular crystal symmetries. One can readily show that for crystals with hexagonal symmetry, to which graphite belongs, the number of independent elastic constants is equal to five. As such constants, one may choose, for example, the elastic moduli c_{11}, c_{12}, c_{13}, c_{33}, and c_{44}. Moreover, it is known (see, e.g., Nye 1964) that for crystals with hexagonal symmetry $c_{22} = c_{11}$, $c_{13} = c_{23}$, $c_{55} = c_{44}$, $c_{66} = \frac{1}{2}(c_{11} - c_{12})$, while the remaining elastic moduli are zero. There are similar relations also for the elastic constants s_{ij}. Thus the tensor of the elastic moduli for graphite may be written in the form of a matrix:

$$c_{ij} = \begin{pmatrix} c_{11} & c_{12} & c_{13} & 0 & 0 & 0 \\ c_{12} & c_{11} & c_{13} & 0 & 0 & 0 \\ c_{13} & c_{13} & c_{33} & 0 & 0 & 0 \\ 0 & 0 & 0 & c_{44} & 0 & 0 \\ 0 & 0 & 0 & 0 & c_{44} & 0 \\ 0 & 0 & 0 & 0 & 0 & c_{66} \end{pmatrix}. \tag{20}$$

For the stresses σ_1 (or σ_2) along the x- (or y-) axis in the basal plane, c_{11} (c_{22}) is the elastic modulus in the x- (y-) direction, c_{12} (c_{21}), in the direction of y (x); and c_{13} (c_{23}), in the direction of the z-axis. For the stress σ_3 along the z-axis, c_{33} is the elastic modulus in the z-direction; c_{31}, in the x-direction; c_{32}, in the y-direction.

The elastic moduli c_{44} and c_{35} are related to the tangential stresses σ_4 and σ_5, respectively. The elastic modulus c_{66} is a linear combination of c_{11} and c_{12}.

For crystals with hexagonal symmetry, the following relations between the components c_{ij} and s_{ij} are valid:

$$c_{11} + c_{12} = s_{33}/s, \quad c_{13} = -s_{13}/s, \quad c_{33} = (s_{11} + s_{12})/s,$$
$$c_{44} = 1/s_{44}, \quad c_{66} = 1/s_{66}, \tag{21}$$

where

$$s = s_{33}(s_{11} + s_{12}) - 2s_{13}^2.$$

In a microscopic treatment of the mechanical properties of a crystal, the potential energy Φ of elastic deformation is commonly introduced in the harmonic approximation as a quadratic form in the components $u_i(m, \boldsymbol{n})$ $(i = 1,2,3)$ of the displacement vectors of the mth atom occupying the unit cell defined by the translation vector $\boldsymbol{n}$. The coefficients of that quadratic form are the components of the tensor of the force constants $\phi_{ij}(m, \boldsymbol{n}; m', \boldsymbol{n}')$ characterizing the interaction of the mth atom in the $\boldsymbol{n}$th cell with the m'th atom in the $\boldsymbol{n}'$th cell. The force constants, $\phi_{ij}(m, \boldsymbol{n}; m', \boldsymbol{n}')$ are given by the second derivatives of the elastic energy Φ with respect to the displacement vector components calculated in the equilibrium position:

$$\phi_{ij}(m, \boldsymbol{n}; m', \boldsymbol{n}') = \left.\frac{\partial^2 \Phi}{\partial u_i(m, \boldsymbol{n})\, \partial u_j(m', \boldsymbol{n}')}\right|_{\substack{u_i(m, \boldsymbol{n}) = 0, \\ u_j(m', \boldsymbol{n}') = 0.}} \tag{22}$$

Evidently, the elastic moduli c_{ijkl} (or the elastic constants s_{ijkl}) of a crystal may be expressed in terms of the force constants $\phi_{ij}(m, \boldsymbol{n}; m', \boldsymbol{n}')$.

In the general case, the number of force constants for a macroscopic crystal may be infinitely large. However, it is obvious that the values of $\phi_{ij}(m, \boldsymbol{n}; m', \boldsymbol{n}')$ should decrease with increasing distances between the atoms whose interaction is described by these quantities. This means that $\phi_{ij}(m, \boldsymbol{n}; m', \boldsymbol{n}')$ decreases with increasing $|\boldsymbol{n} - \boldsymbol{n}'|$. For this reason, in describing the elastic energy of a crystal, one may restrict oneself to the interaction between a finite number of rather close neighbours, i.e., describe the interaction between atoms in the crystal by using some crude model. In any model, however, the set of force constants should satisfy certain general requirements. Thus, from definition (22) and the continuity requirement for the elastic energy, Φ, it follows that the tensor of the force constants is symmetric with respect to the permutations $i \rightarrow j$ and $m, \boldsymbol{n} \rightarrow m', \boldsymbol{n}'$:

$$\phi_{ij}(m, \boldsymbol{n}; m', \boldsymbol{n}') = \phi_{ji}(m', \boldsymbol{n}'; m, \boldsymbol{n}). \tag{23}$$

The invariance of ϕ under rigid-body translations entails that

$$\sum_{m', \boldsymbol{n}'} \phi_{ij}\ (m, \boldsymbol{n}; m', \boldsymbol{n}') = 0. \tag{24}$$

The number of force constants is further restricted by the requirements that Φ be invariant (a) under rotation of the crystal as a whole, (b) under the symmetry-group transformations, and the requirement (c) that the crystal structure correspond to an equilibrium atomic configuration. In any model that describes the interaction between a limited number of neighbouring atoms, the conditions (23), (24), as well as requirements (a), (b) and (c), allow the number of independent force constants to be reduced to a minimum.

3.2. *Models of interaction between atoms in a graphite crystal*

The simplest model to describe the interaction between atoms in a graphite lattice was proposed by Young and Koppel (1965). This model takes advantage of the circumstance that the force of interaction between atoms in a graphite layer is much greater than that between two layers. The model contains four force constants, μ, k, μ', and k', which correspond to the interaction between the nearest neighbours both in a graphite layer and in neighbouring layers. The Young–Koppel model has however the disadvantage that the graphite lattice turns out to be unstable when layers are displaced relative to one another (the elastic component c_{44} turns out to be zero).

More detailed models have been developed by Nicklow et al. (1972) and by Mani and Ramani (1974). These models take into account the interaction with the third-nearest neighbour in the plane of a graphite layer, as well as the interaction between the nearest-neighbour atoms situated in neighbouring layers. The models contain twelve independent force constants, which, however, is insufficient even for a rough description of interaction between atoms in a real crystal of graphite. Indeed, as Nicholson and Bacon (1977) have shown, the restriction to only twelve force constants leads to a nonvanishing stress in the lattice, which requires that an external force be applied to the lattice to maintain its equilibrium. Nicholson and Bacon (1977) proposed instead a model, which is free from the above disadvantages, yet contains a minimum number of force constants. The model takes into account the atomic

Table 1

Particular types of atomic interactions in a graphite lattice and the corresponding matrices.

Interacting atoms	Matrix of the force-constant tensor for atoms M and N shown in fig. 3a
Three 1st-nearest neighbours at distance $r_1 = a = a_0/\sqrt{3}$, *e.g.*, $M = 1$, $N = 2$ (in-plane)	$\begin{pmatrix} A_1 & 0 & 0 \\ 0 & B_1 & 0 \\ 0 & 0 & C_1 \end{pmatrix}$
Six 2nd-nearest neighbours at $r_2 = a_0$, e.g. $M = 1$, $N = 1'$ (in-plane)	$\begin{pmatrix} A_2 & D_2 & 0 \\ -D_2 & B_2 & 0 \\ 0 & 0 & C_2 \end{pmatrix}$
Three 3rd-nearest neighbours at $r_3 = \frac{2}{3}a_0$, e.g., $M = 1$, $N = 2'$ (in-plane)	$\begin{pmatrix} A_3 & 0 & 0 \\ 0 & B_3 & 0 \\ 0 & 0 & C_3 \end{pmatrix}$
Two 4th-nearest neighbours at $r_4 = \frac{1}{2}c_0$, e.g., $M = 1$, $N = 3$ (1st-nearest out-of-plane neighbours)	$\begin{pmatrix} A_4 & 0 & 0 \\ 0 & A_4 & 0 \\ 0 & 0 & C_4 \end{pmatrix}$
Twelve 5th-nearest neighbours of of atom 1 at $r_5 = (\frac{1}{3}a_0^2 + \frac{1}{4}c_0^2)^{1/2}$, e.g., $M = 1$, $N = 4'$	$\begin{pmatrix} A_5 & 0 & E_5 \\ 0 & B_5 & 0 \\ F_5 & 0 & C_5 \end{pmatrix}$
Six 5th-nearest neighbours of atom 2 of sublattice 4, e.g. $M = 2$, $N = 4''$	$\begin{pmatrix} A'_5 & 0 & E'_5 \\ 0 & B'_5 & 0 \\ E'_5 & 0 & C'_5 \end{pmatrix}$

interaction with nearest neighbours up to the third in the plane, as well as with second-nearest neighbours in two adjacent layers. Second-neighbours in plane interactions between atoms on sublattice 1 (see fig. 3a) and similar interactions between atoms on sublattice 2 have been assumed to be exactly equivalent. In order to describe such an interaction in the graphite lattice, one should use twenty-one independent force constants. Table 1 presents five types of third-rank matrices (in indices α, β), which fully determine the force-constant tensor ϕ_{ij} in the Nicholson–Bacon model. Each matrix describes the interaction between neighbouring atoms placed at a definite distance r_γ ($\gamma = 1,2,3,\ldots,6$) in the graphite lattice. Atoms are denoted by the symbols M and N assuming values 1, 2; 1′, 2′, 1″, 2″ (see the indexing of lattice atoms in fig. 3a). The nonzero matrix components are subscripts designated as A, B, C, D, E, F with indices ranging from 1 to 5 and also with or without the prime.

The matrices that describe the interaction between the neighbouring atoms at distances r_1, r_3, and r_4 from one another do not contain off-diagonal elements.

In order to go over from the Nicholson–Bacon model to the model of Nicklow et al. and to that of Mani and Ramani, one should set the elements of the two lower matrices in table 1 with index 5 equal to zero, because the corresponding atomic interactions are omitted from the latter models. The reduction of the Nicholson–Bacon model to the Young–Koppel one also implies that all force constants with indices 3 and 4, except C_4, should turn to zero. Then the force constants μ, k, μ', k' used in the latter model are expressed in terms of the nonzero matrix elements in table 1 as follows:

$$\begin{aligned} k &= A_1, \qquad \mu = \tfrac{1}{6}B_1 = -\tfrac{4}{3}A_2 = 4B_2 = \tfrac{4}{3}\sqrt{3}\,D_2, \\ k' &= C_4, \qquad \mu' = \tfrac{3}{2}C_1 = -9C_2. \end{aligned} \tag{25}$$

In the Nicholson–Bacon model, the elastic moduli c_{11} and c_{66} are expressed through the force-constant tensor components:

$$\begin{aligned} c_{11} &= F + \frac{1}{2\sqrt{3}\,c_0}[(3A_1 + B_1) + (6A_2 + 18B_2) + (12A_3 + 4B_3) \\ &\qquad + (3A'_5 + B'_5)], \\ c_{66} &= F + \frac{1}{2\sqrt{3}\,c_0}[(A_1 + 3B_1) + (18A_2 + 6B_2) + (4A_3 + 12B_3) \\ &\qquad + (2A_5 + 6B_5) + (A'_5 + 3B'_5)], \end{aligned} \tag{26}$$

where

$$F = \frac{\sqrt{3}\,[(Z + 2Q + \tfrac{8}{3}A_4)X^2 + (Z + 2Q + 4P)Y^2 - 2(Z - 2Q)XY]}{16c_0[3ZQ + 4A_4P + (Z + 2Q)(A_4 + \tfrac{3}{2}P)]},$$

$$X = A_1 - B_1 - 2(A_3 - B_3) + 2(A_5 - B_5) - 2(A'_5 - B'_5),$$
$$Y = A_1 - B_1 - 2(A_3 - B_3) - 2(A_5 - B_5),$$
$$Z = A_1 + B_1 + A_3 + B_3,$$
$$Q = A_5 + B_5,$$
$$P = A'_5 + B'_5.$$

The small elastic moduli are given by

$$c_{33} = \frac{c_0}{\sqrt{3}\, a_0^2}(C_4 + 6C_5 + 3C'_5),$$

$$c_{44} = \frac{c_0}{\sqrt{3}\, a_0^2}[A_4 + 3(A_5 + B_5) + \tfrac{3}{2}(A'_5 + B'_5)],$$

$$c_{13} = -\frac{2}{a_0}(D_5 + E_5 - D'_5) - c_{44}. \tag{27}$$

The above-mentioned conditions (a), (b), (c), and eq. (24), which should be satisfied by the force-constant tensor, make it possible to eliminate the constants C_1, C_2, C_3, E'_5 and F_5 from eqs. (26) and (27). Also, one can demonstrate that the elastic moduli are independent of the force constant D_2. Thus in the Nicholson–Bacon model, the elastic moduli c_{ij} of graphite are expressed through 15 force constants describing the interaction between the atoms in the lattice.

We note that the model is currently held to be the most complete and exact one within which the elastic properties of graphite can be described microscopically.

3.3. Elastic moduli and force constants for the various kinds of graphite

Because of a high anisotropy in properties of graphite, the diversity of its crystal lattices, and the presence of numerous and varied defects in graphite crystals, the elastic moduli c_{ij} quoted by different authors are widely different.

First data on the elastic moduli of graphite were obtained by Rilay (1945), who made use of the experimental results obtained by Basset (1941) for the compressibility and thermal expansion of natural graphite and utilized the equation of state due to Grüneisen. These data only made it possible to determine the sum of the moduli, $c_{11} + c_{12}$, and also the moduli c_{13} and c_{33} separately.

Bowman and Krumhansl (1958) were the first to determine separately the

Table 2
Elastic moduli c_{ij} for various graphite specimens (in units of 10^{11} dyne/cm^2).

Material	c_{11}	c_{12}	$c_{11}+c_{12}$	c_{13}	c_{33}
Canadian natural graphite	—	—	—	—	3.56
Reactor graphite	—	—	—	—	—
Pyrolytic graphite	—	—	—	—	3.9 ± 0.4
HOPG	144 ± 20	—	—	—	3.71 ± 0.05
Natural graphite	91	—	—	—	3.65
Natural graphite	106	—	—	—	3.65
HOPG	—	—	—	—	3.69
Natural graphite from Madagascar	—	—	—	—	3.56
Graphitized soot	—	—	—	—	2.88
Reactor graphite	—	—	—	—	3.56
Pyrolytic graphite	—	—	—	—	2.03
Natural graphite	116	29.0	145	10.9	4.7
Natural and synthetic graphite	113	28.25	141.25	—	1.8
Natural graphite	—	—	85.4	6.28	2.63
Natural graphite	—	—	150	11	4.48
HOPG	—	—	—	—	3.50 ± 0.07
	—	—	—	—	3.50 ± 0.07
HOPG	—	—	—	—	

References:
[1] Komatsu (1958).
[2] Dolling and Brockhouse (1962).
[3] Nicklow et al. (1972).
[4] Mani and Ramani (1974).
[5] Blackslee et al. (1970).
[6] Seldin and Nezbeda (1970).
[7] Leung et al. (1981a).
[8] Komatsu (1964).
[9] Van der Hoeven and Keesom (1963).

values for c_{11} and c_{12}, as well as for c_{33} and c_{44}, in natural graphite from the measurements of its heat capacity and compressibility. A detailed study of the elastic properties of perfect crystals of natural graphite was made by Spence (1961), who determined a practically complete set of elastic moduli. The elastic properties of perfect HOPG crystals were investigated by Nicklow et al. (1972). Apart from the other elastic moduli, these authors determined for the first time the c_{66}-component. The same component was found for natural graphite crystals by Blackslee et al. (1970), Seldin and Nezbeda (1970), Mani and Ramani (1974). In order to gain a deeper insight into the nature of the interlayer and intralayer bonding forces in graphite, it is of importance that one should know the exact values of the elastic moduli, especially those of the

c_{44}	c_{66}	Crystallite size (Å)		Notes	Ref.
		In the basal plane	Along the c-axis		
0.405	—	—	—	—	[1]
0.292	—	—	—	—	[1]
0.42 ± 0.2	—	—	—	—	[2]
0.46 ± 0.02	46	—	—	—	[3]
0.4	43	—	—	Sound-velocity measurement	[4]
0.4	44	—	—	—	[5, 6]
0.4572	—	—	—	—	[7]
0.452	—	10 000		Few stacking faults	[8, 9]
0.0702	—	90	125	Many stacking faults	[8]
0.312	—	240		Few stacking faults	[8]
0.0702	—	200	265	Many stacking faults	[8]
0.23	—	—	—	—	[10]
0.23	—	—	—	—	[11]
—	—	—	—	—	[12, 13]
—	—	—	—	—	[14, 15]
0.0353 ± 0.0014	—	—	—	—	[16]
0.274 ± 0.011	—	—	—	—	[5]
0.024 ± 0.43	—	—	—	—	[17, 18]

[10] Spence (1961).
[11] Bowman and Krumhansl (1958).
[12] Rilay (1945).
[13] Basset (1941).
[14] Bridgman (1947, 1948, 1952).
[15] Kabalkina and Vereschagin (1960).
[16] Hwang (1983).
[17] Gauster (1972).
[18] Green et al. (1973).

small components c_{33} and c_{44}. Accordingly, the elastic moduli of various crystals of graphite have repeatedly been determined using the most accurate techniques available (for example, ultrasonic measurements). Thus, Dolling and Brockhouse (1962) determined c_{33} and c_{44} for perfect single-crystal specimens of pyrographite from the neutronographic data on lattice vibrations. The small elastic moduli c_{33} and c_{44} for HOPG were found from the ultrasonic measurements by Blackslee et al. (1970), Gauster (1972), Nicklow et al. (1972), Green et al. (1973), and Hwang (1983).

The data on the values of c_{ij} for various kinds of graphite which are available in the literature are summarized in table 2.

Let us give a brief analysis of these data. The elastic moduli c_{11}, c_{12}, and c_{13}

appear to depend weakly on the quality of specimens studied, different experimental techniques yielding values that agree rather well with one another. If we disregard the data of the earliest study by Basset (1941), in which the compressibility of graphite was overestimated by a factor of about 1.5, as compared with the more accurate data obtained in subsequent studies (accordingly, the elastic moduli calculated by Rilay (1945) on the basis of Basset's data are ~1.5 times smaller) and if we also ignore $c_{11} = (144 \pm 20) \times 10^{11}$ dyne/cm^2 determined comparatively crudely for HOPG by Nicklow et al. (1972), then the discrepancy between the results obtained by different workers turns out to be relatively small.

For natural graphite, the most probable values for the elastic moduli (in units of 10^{11} dyne/cm^2) are as follows: $c_{11} = 100$–115, $c_{12} = 28$–29, $c_{13} = 10$–11.

Let us examine separately the data on the small moduli c_{33} and c_{44}. First of all, we note that, contrary to Komatsu's assumption (1955, 1958), there is no correlation between the size and quality of crystallites, on the one hand, and the magnitude of the modulus c_{33}, on the other hand. It can be seen from table 2 that, in spite of the great diversity of materials studied and methods of investigation used, all obtained values of c_{33} deviate but little from a definite (average) value. If one disregards the data obtained with less perfect specimens and less accurate techniques, then the value of c_{33} for rather good single crystals of graphite is $(3.5$–$3.7) \times 10^{11}$ dyne/cm^2. This is confirmed by the results of Hwang (1983) who studied a large number of perfect HOPG specimens with the purpose of determining the small moduli c_{33} and c_{44}. His study has shown that the magnitude of c_{33} is practically independent of the degree of perfection of single crystals. Much more sensitive to the state of the object under study is the magnitude of the c_{44} modulus.

Thus, in studying the elastic constants of polycrystalline carbons, Kelly (1964) showed that the significant decrease of the c_{44} value in certain specimens of graphite could be explained by the presence of dislocations. Baker and Kelly (1964) point out that dislocations may reduce the value of c_{44} by more than an order of magnitude. By contrast, the elimination of dislocations brought about as a result of graphite layer ordering due to high-temperature annealing leads to an increase in the value of the elastic modulus c_{44}.

Accordingly, in contrast to c_{33}, the value of c_{44} at room temperature varies greatly from specimen to specimen from 0.07×10^{11} to 0.46×10^{11} dyne/cm^2. However, for the most perfect specimens with a small number of defects this range considerably narrows. The average values of c_{44} for good single crystals of natural graphite and HOPG range from 0.27×10^{11} to 0.46×10^{11} dyne/cm^2, the upper limit being apparently characteristic of the most perfect specimens. The temperature dependences of the small elastic

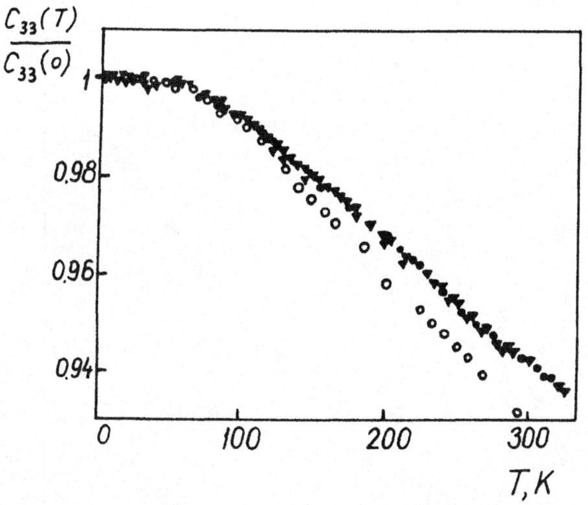

Fig. 4. The normalized elastic constant c_{33} [$= c_{33}(T)/\ c_{33}(0)$] for HOPG: ▼ ● determined by ultrasonic techniques; ○ the lattice energy contribution to c_{33} calculated from the thermal expansion data for a simple Lennard–Jones interlayer potential model. The difference is attributed to the anharmonic contribution (Hwang 1983).

moduli c_{33} and c_{44} have been studied by Gauster and Fritz (1974) and Hwang (1983). These authors used an ultrasonic technique to measure c_{33} and c_{44} in HOPG specimens over a wide temperature range from 4 to 300 K. As the temperature is lowered, there is a slight increase in c_{33} values, but in the range $T < 70$ K the value of c_{33} is practically temperature-independent. The total variation of c_{33} over the whole temperature range from 4 to 300 K does not exceed 6% (see fig. 4). We note that, in obtaining the T-dependence of c_{33}, Hwang (1983) introduced the correction for the thermal expansion of the lattice by taking into account the change in the lattice constants a_0 and c_0 in accordance with the data obtained by Bailey and Yates (1970) and by Hardcastle and Zabel (1981) (see fig. 5). The sensitiveness of c_{44} to the state of a specimen under investigation is apparent in the temperature dependence of this modulus. In the most perfect specimens, c_{44} increases by several percent (practically linearly) as the temperature is lowered from 295 to 4 K. In the less perfect specimens, c_{44} slightly decreases as the crystal is cooled down. The temperature cycling of specimens results in an irreversible change in c_{44}, which also suggests a strong dependence of this quantity upon the number of dislocations present. The c_{66}-component of the elastic-moduli tensor in natural single-crystal graphite was measured for the first time independently by Blackslee et al. (1970), who used ultrasonic techniques in their measurements. The value of c_{66} they determined is 46×10^{11} dyne/cm^2. In the subsequent

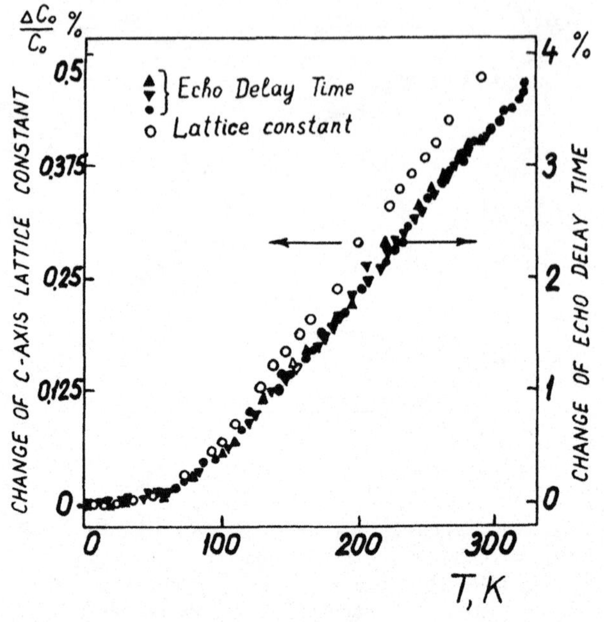

Fig. 5. The percentual change of the echo delay times normalized to the low-temperature value for the *c*-axis longitudinal sound in HOPG. Also indicated in this figure are the *c*-axis thermal expansion data determined by X-ray methods. (Hardcastle and Zabel 1981.)

studies by Nicklow et al. (1972) and by Mani and Ramani (1974), who measured c_{66} in specimens of HOPG and natural graphite, rather similar values of this component have been obtained.

It will be recalled that in crystals with the hexagonal symmetry the c_{66}-component is a linear combination of the c_{11}- and c_{12}-components: $c_{66} = \frac{1}{2}(c_{11} - c_{12})$. If one takes the values $c_{11} \simeq 110$ and $c_{12} \simeq 28$ (in units of 10^{11} dyne/cm^2), then the corresponding value for c_{66} must equal $\simeq 41$. This estimate of c_{66} agrees rather well with the data of ultrasonic measurements. This implies that the independent measurements of the large elastic moduli c_{11}, c_{12}, c_{66} in graphite are mutually consistent and supplement one another.

Let us briefly discuss calculations of the elastic moduli in graphite. First of all, we note that the order of magnitude of the elastic moduli c_{33}, c_{44} and the character of their dependence on temperature, observed experimentally for perfect single crystals, may reasonably well be described by approximating the interaction between graphite layers with the Lennard–Jones potential:

$$\varphi(r) = \text{const.}\left(\frac{1}{r^6} - \frac{r_0^6}{2r^{12}}\right), \tag{28}$$

where r is the distance between atoms, r_0 is a quantity having the dimension of length. Calculations of c_{33} and c_{44} using the force constants A_4, C_4, A_5, B_5, C_5, B'_5, C'_5 [see eqs. (25) and (26)] have been made, within this simple model, by Kelly and Duff (1970), Green and Spain (1973), and Hwang (1983). These calculations yield the correct order of magnitude for c_{33} and c_{44}, and qualitatively describe the relatively small increase in the values of the elastic moduli with lowering temperature. Such a simple description of elastic moduli in graphite proves to be adequate, evidently because of the fact that the direct electrostatic interaction between carbon ions in this material is highly screened by conduction electrons and, therefore, provides a relatively small contribution to the bonding forces in the lattice.

Pairwise and valence-type potentials were used (separately or in combination) to give a qualitative description of the bonding forces in the graphite lattice. In such a description, valence-type potentials are especially convenient because their use will automatically assure the fulfilment of equilibrium conditions in the lattice.

Table 3
Force constants (10^5 dyne/cm) for graphite used in various models.

Constant	Ref. [1]	Ref. [2]	Ref. [3]	Refs. [4, 5]
A_1	3.23662	3.62	0.04977	6.71
B_1	2.7995	1.99	6.25941	2.88
C_1	2.3195	1.99	0.46863	1.95
A_2	−0.33533	−0.52	0.48933	−0.36
B_2	0.9144	1.33	0.6026	0.12
C_2	−0.38533	−0.52	0.07423	−0.325
D_2	0.22517	0	0.31758	0.208
A_3	−0.1883	−0.037	−0.22779	0
B_3	0	0.288	−0.22779	0
C_3	0	0.288	0	0
A_4	0.00529	0.0077	0.00626	0
C_4	0.00997	0.058	0.05724	0.0557
A_5	0.00433	0	0	0
$B_5 = B'_5$	−0.0025	0	0	0
C_5	0.00362	0	0	0
E_5	−0.00259	0	0	0
F_5	−0.01612	0	0	0
A'_5	−0.00053	0	0	0
C'_5	0.00848	0	0	0
F'_5	0.00465	0	0	0

*References:
[1] Nicholson and Bacon (1977).
[2] Nicklow et al. (1972).
[3] Mani and Ramani (1974).
[4] Yoshimori and Kitano (1956).
[5] Young and Koppel (1965).

Nicholson and Bacon (1977) have successfully used pairwise potentials to describe interactions with second-nearest out-of-plane neighbours ($r = r_5$, see table 1). The force constants for the graphite lattice were evaluated by least-squares fittings to all the experimental data available at the time, including those on the phonon spectrum, Raman scattering and elastic moduli.

The force constants thus obtained for the various models are presented in table 3. Attention is called to the significant difference between the same force constants (with subscripts 1, 2, 3, 4) in the models by Nicklow et al. (1972), on the one hand, and Mani and Ramani (1974), on the other hand, the total number of the constants in these models coinciding. This difference indicates the ambiguity of determination of the force constants by fitting to experimental data. This ambiguity is mainly due to the relative scarcity and low accuracy of the data and also to a considerable spread in parameter values measured on graphite specimens of various nature.

CHAPTER 4

The Energy Spectrum of Electrons and Holes in Graphite

4.1. The energy spectrum of electrons in a single graphite layer

Since the bonding forces of carbon atoms in the plane of a graphite layer are two orders of magnitude as great as those between atoms in neighbouring layers, it will be quite reasonable to treat the behaviour of electrons in graphite as being two-dimensional, taking then account of the interaction between layers as a correction.

Such an approach towards the consideration of the electronic structure in graphite was adopted in all original studies in which the energy spectrum of electrons and holes in graphite was calculated. Following these works, we shall now consider the energy spectrum of charge carriers in an isolated graphite layer.

A unit cell of the two-dimensional graphite lattice (see fig. 2a) contains two nonequivalent carbon atoms.

The first Brillouin zone for two-dimensional graphite is a regular hexagon (fig. 6), whose most characteristic points are its centre Γ, the upper corners K and K', and the centres of the lateral edges M and M'. As has already been mentioned, in the sp^2-hybrid state of the carbon atom, occurring in graphite, three out of the four valence electrons participating in the formation of σ-bonds are described by the wave functions which represent linear com-

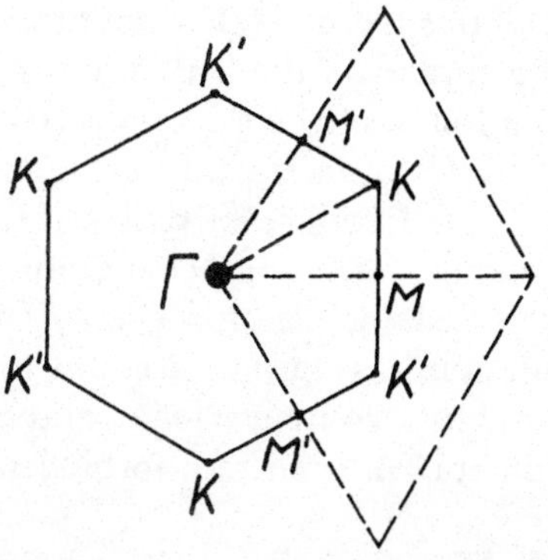

Fig. 6. The Brillouin zone for a two-dimensional graphite model.

binations of atomic wave functions of the s- and p-type. These three hybridized wave functions may, according to Haering and Mrozowski (1960), be written as

$$\begin{aligned}\Psi_1 &= \tfrac{1}{3}\sqrt{3}(\psi_{2s} + \sqrt{2}\psi_{2p_x}),\\ \Psi_2 &= \tfrac{1}{3}\sqrt{3}(\psi_{2s} - \tfrac{1}{2}\sqrt{2}\psi_{2p_x} + \tfrac{1}{2}\sqrt{3}\psi_{2p_y}),\\ \Psi_3 &= \tfrac{1}{3}\sqrt{3}(\psi_{2s} - \tfrac{1}{2}\sqrt{2}\psi_{2p_x} - \tfrac{1}{2}\sqrt{3}\psi_{2p_y}),\end{aligned} \tag{29}$$

where ψ_{2s}, ψ_{2p_x}, ψ_{2p_y} designate the respective atomic wave functions. (The Cartesian x- and y-axes lying in the plane of the graphite layer; see fig. 2a). The fourth electron of the carbon atom is described by the atomic wave function ψ_{2p_z} and takes part in the formation of π-bonds (the z-axis being perpendicular to the graphite layer). Using the tight-binding (TB) approximation, Wallace (1947) made a very simple calculation on the electron spectrum of a single graphite layer. He showed that the σ-orbitals, which lie in the plane of the layer and which are made up by the wave functions ψ_{2s}, ψ_{2p_x}, and ψ_{2p_y}, are completely filled and do not contribute to conductivity. The latter is wholly determined by π-electrons, described by the wave function of the p_z-symmetry.

The energy states of σ- and π-electrons in a two-dimensional graphite crystal form eight bands, six of which are bonding and antibonding bands of the σ-type and two belong to the π-type. Since the capacity of each band is $2N$, where N is the number of unit cells in the crystal, whereas the number of valence electrons is equal to $8N$, only four of the bands available turn out to be filled (these are referred to as valence bands).

The three filled σ-bands have the lowest energy followed, in order of increasing energy, by the two π-bands. It is precisely these bands that lie in the vicinity of the level of the chemical potential in two-dimensional graphite.

In his electron spectrum calculation, Wallace (1947) took into consideration only the overlap integrals of π-electrons for the nearest neighbours by treating them as unknown parameters, but disregarded the non-orthogonality of the wave functions for the various atoms. In this way, Wallace conceived some general ideas about the band structure of two-dimensional graphite and its main properties. These were further developed in a number of later works. The most important inference of this theory is that both the filled and the empty π-band are degenerate in the corners of the Brillouin zone (the points K and K' in fig. 6). Such a conclusion follows from the symmetry of a graphite layer and does not depend upon the kind of interaction between carbon atoms. At the absolute zero temperature, the Fermi level passes through a common point (degeneracy point) of π-bands. A similar conclusion was also reached by Corbato (1957), who calculated the electron energy spectrum for a graphite layer in the TB approximation, taking into account the overlap integrals for nine nearest neighbours. Thus two-dimensional graphite turns out to be a peculiar gapless semiconductor with the zero forbidden band at the K points of the Brillouin zone, rather than a semimetal. At $T = 0$ K, three σ-bands and one π-band are completely filled in its spectrum. The upper π-band, which is contiguous to the filled π-band at the degeneracy points, and the next three upper σ-bands are entirely empty.

As an illustration, fig. 7a shows the results of one of the most exact, recent calculations on the structure of the σ- and π-bands of the graphite layer along directions in the Brillouin zone between high-symmetry points, made by Nagayoshi et al. (1976a). For their calculations, these authors used a combination of the orthogonalized plane wave (OPW) method and the TB approximation. Two fitting parameters were determined by comparing with the data of optical measurements in three-dimensional graphite for transitions characteristic for the band structure of the graphite layer. The most important parameters characterizing the energy spectrum of the graphite layer are the following energy intervals:

(1) $\varepsilon(K_6) - \varepsilon(\Gamma_1^+)$, the width of the valence band;

(2) $\varepsilon(\Gamma_5^+) - \varepsilon(\Gamma_2^-)$, the overlapping of the σ- and π-valence bands;

(3) $\varepsilon(\Gamma_6^-) - \varepsilon(\Gamma_5^+)$, the energy gap between σ-bands at the point Γ;

(4)–(7) $\varepsilon(M_4^-) - \varepsilon(M_3^+)$, $\varepsilon(M_1^+) - \varepsilon(M_3^-)$, $\varepsilon(M_2^-) - \varepsilon(M_3^+)$, $\varepsilon(M_4^-) - \varepsilon(M_1^+)$ are the energies of the (π–π)-, (σ–σ)-, and (σ–π)-optical transitions at the points M [see table 4 and fig. 7a,b; the symbols for the terms have been taken from the papers by Nagayoshi et al. (1976a)].

Attention is drawn to the fact that a general model for a graphite layer will be useful also in the case when one does not look into the details of the band structure in the neighbourhood of characteristic points in the Brillouin zone, but rather takes interest in the gross structure of the spectrum for the whole Brillouin zone, which is necessary, for instance, for an analysis of optical data.

In particular, the band spectrum of a graphite layer calculated from first principles enables one to give a fairly good interpretation of the energy of the optical transitions in three-dimensional graphite at which the electric-field vector of an electromagnetic wave is directed parallel to the basal plane of the crystal. As an illustration of the foregoing, in table 4 the energies of characteristic optical transitions obtained by calculating the band spectrum of a graphite layer are compared to the experimental optical data.

As can be seen from table 4, there is a rather good agreement between the experimental and calculated energies, except for the energy of the $(\sigma - \pi)$-transition (6) at the point M. One should, however, bear in mind that the above transition is rather uncommon for the model of a single graphite layer, because it is observed in the spectrum of the imaginary part of the dielectric function when electromagnetic waves are polarized normally to the plane of the graphite layer.

The explicit form of the π-electron dispersion law in the single graphite layer model can be obtained by using the (kp)-method. In this manner, Slonzewski and Weiss (1958) have shown that the dependence of the energy ε of π-

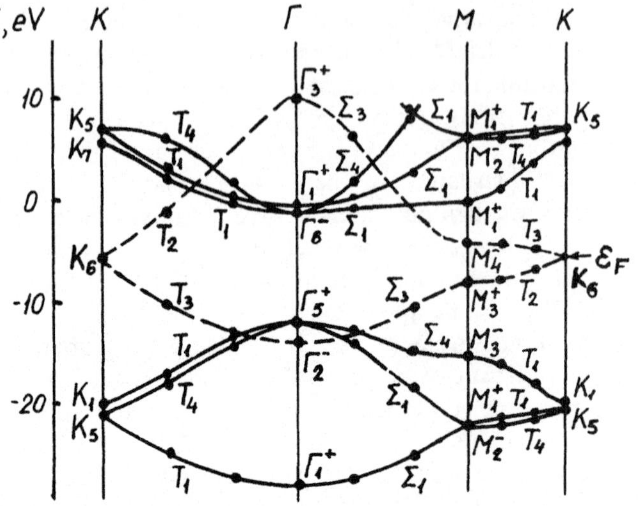

ε, eV
K
Γ
M
K
10
0
-10
-20
ε_F

a)

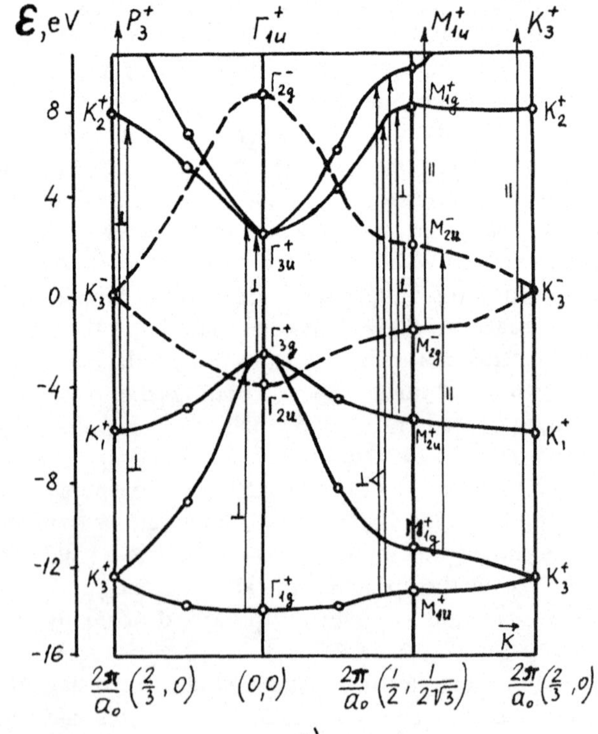

ε, eV
8
4
0
-4
-8
-12
-16
$\frac{2\pi}{a_0}\left(\frac{2}{3},0\right)$
(0,0)
$\frac{2\pi}{a_0}\left(\frac{1}{2},\frac{1}{2\sqrt{3}}\right)$
$\frac{2\pi}{a_0}\left(\frac{2}{3},0\right)$

в)

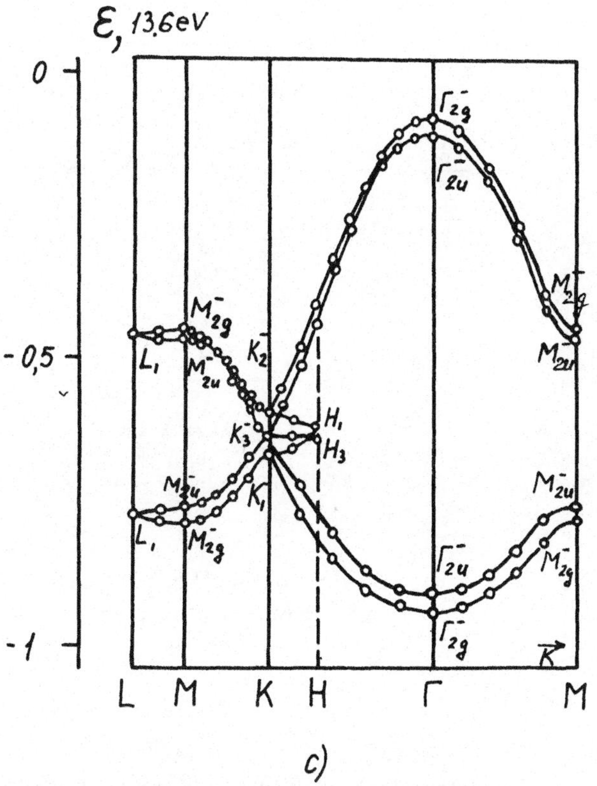

Fig. 7. (a) The band structure for a monolayer of graphite calculated by Nagayoshi et al. (1976a); solid and dashed lines represent σ- and π-bands, respectively. (b) The two-dimensional graphite bands calculated according to the interpolation scheme of Bassani and Parravicini (1967); dashed and solid lines represent σ- and π-bands, respectively; arrows indicate main optical transitions, the symbols $\|$ and $\perp$ denoting the polarization of the transitions $\boldsymbol{E}\|c$ and $\boldsymbol{E} \perp c$, (c) π-bands for three-dimensional graphite (as calculated by Doni and Parravicini (1969) and by Greenaway et al. (1969)).

electrons on the wave-vector $\boldsymbol{k}$ in the vicinity of the common points of the two π-bands (the degeneracy points K or K' in the Brillouin zone) is a linear function:

$$\varepsilon(k) = \varepsilon_K \pm \frac{\hbar}{m_0} p_0 k, \tag{30}$$

where the wave-vector $\boldsymbol{k}$ is measured from the degeneracy points, ε_K is the energy at the degeneracy point of the π-bands, m_0 is the mass of a free electron, p_0 is a constant with the dimension of momentum. The structure of π-bands of a graphite layer near the K-point is schematically shown in fig. 8. According to

Table 4
Comparison of calculated and experimental optical transition energies.

Energy interval (eV)	Calculation		Experimental optical data
	Ref. [1]	Ref. [2]	Refs. [4–6]
(1) $\varepsilon(K_6) - \varepsilon(\Gamma_1^+)$	19.5	23.3	15-20
(2) $\varepsilon(\Gamma_5^+) - \varepsilon(\Gamma_2^-)$	2.4	2.2	—
(3) $\varepsilon(\Gamma_6^-) - \varepsilon(\Gamma_5^+)$	12.2	11.3	10
(4) $\varepsilon(M_4^-) - \varepsilon(M_3^+)$	4.6	3.9	4.5
(5) $\varepsilon(M_1^+) - \varepsilon(M_3^-)$	16.3	15.7	14.5
(6) $\varepsilon(M_2^-) - \varepsilon(M_3^+)$	21.3	14.8	11.0
(7) $\varepsilon(M_4^-) - \varepsilon(M_1^+)$	16.5	18.4	15.5

References:
[1] Painter and Ellis (1970).
[2] Nagayoshi et al. (1976a).
[3] Taft and Philipp (1965).
[4] Sagawa (1966).
[5] Zeppenfeld (1968).
[6] Klucker et al. (1974).

the TB calculation of Slonzewski and Weiss (1958), the value of p_0 in eq. (30) taking account of interactions only between the nearest neighbours is given by

$$p_0 = \tfrac{1}{2}\sqrt{3}\,\frac{m_0}{\hbar}\,\gamma_0 a_0, \tag{31}$$

where γ_0 is the resonance integral for the π-orbitals of the nearest neighbours in a graphite layer, calculated using the wave functions $\psi_{2p_z}(r)$. The estimated value of this integral is between 2 and 4 eV. The spin–orbit interaction

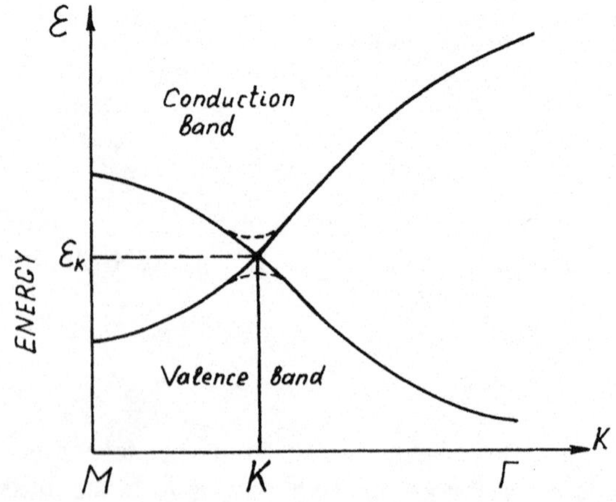

Fig. 8. Schematic representation of electron energy as a function of the wave vector in the π-bands of the graphite layer, with and without spin–orbit interaction (dashed and solid lines, respectively).

eliminates the degeneracy at the K-points. Yet, the resulting spacing between the levels turns out to be less than 10^{-4} eV and may be neglected even at very low temperatures (down to about 1 K).

4.2. The electron energy spectrum of three-dimensional graphite. The Slonzewski–Weiss–McClure model

The number of carbon atoms in a unit cell of three-dimensional graphite (see fig. 3a) is twice that in a unit cell of two-dimensional graphite. Therefore, each electron state of a single graphite layer corresponds, for any $\boldsymbol{k}$, to two states in the three-dimensional graphite lattice.

The first Brillouin zone for a hexagonal lattice of three-dimensional graphite is shown in fig. 9. It represents a regular hexahedron. The characteristic points of high symmetry are its centre, Γ; the corners H and H'; the centres of the top and bottom faces, A; the centres of the side faces, M; the centres of the side edges, K and K'; the centres of the edges of the top and bottom faces.

Using the methods of perturbation theory, Slonzewski and Weiss (1958) and McClure (1957) calculated the band structure of three-dimensional graphite. The influence of the various types of graphite layer stacking (AAA ...., ABAB ..., ABCABC ...) on the electronic structure has been theoretically investigated by McClure (1969) and Samuelson et al. (1980). They have shown that a change in stacking type results only in shifting the energy levels in the vicinity of the Fermi level. As is indicated by the calculations, the electron energy spectrum near the Fermi level may be described by a model including

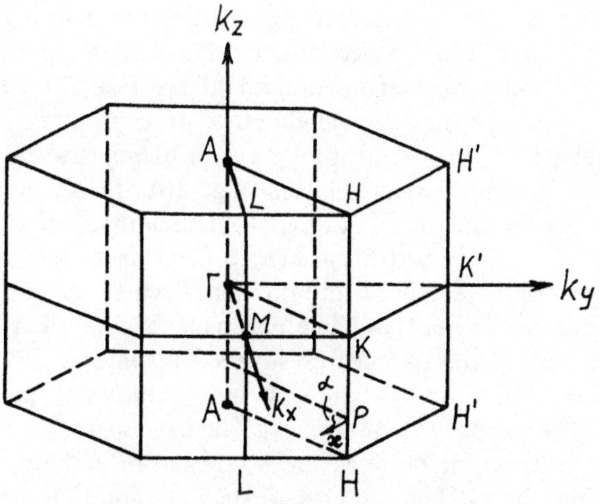

Fig. 9. The Brillouin zone of graphite.

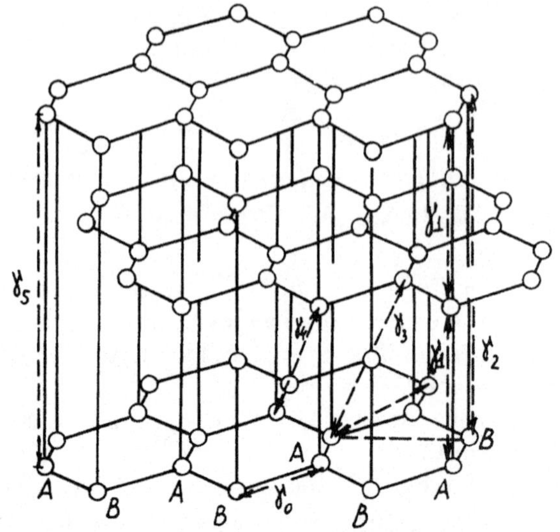

Fig. 10. Correspondence (dashed line) between the SWMC-model parameters γ_i and the interaction between the individual atoms in the graphite lattice.

seven parameters: γ_0, γ_1, γ_2, γ_3, γ_4, γ_5, and $\gamma_6 \equiv \Delta$, which define the interaction energy of the various carbon atoms in the graphite lattice. The correspondence between these parameters and the interaction of particular atoms in the graphite lattice is shown in fig. 10.

The seven-parameter model developed by Slonzewski, Weiss and McClure is now generally accepted and is commonly referred to as the SWMC-model.

The physical meaning of the SWMC-model parameters γ_i is as follows. All these parameters have the dimension of energy and fall into two groups: in-plane and out-of-plane. The first group contains only one parameter, namely, the parameter γ_0 that has been introduced above [see (31)] to describe the interaction between the nearest neighbours in a graphite layer. All the remaining parameters γ_i ($i = 1, \ldots, 6$) are out-of-plane, describing interactions between atoms found in various layers (fig. 10). Thus, the parameter γ_1 corresponds to the interaction of A-type atoms from two neighbouring layers placed one under the other; this parameter determines the width of the π-bands at the K-points of the Brillouin zone, which is equal to $4\gamma_1$. An estimation yields $\gamma_1 \sim (0.2\text{–}0.4)$ eV. The parameter γ_2 describes the interaction between B-type atoms from two nearest layers. This interaction determines the coupling between the π- and σ-bands. It also determines the magnitude of overlapping of π-bands, which equals $2|\gamma_2|$. The estimated γ_2 ranges from -0.02 to 0.02 eV. The interaction between an A-type atom and the nearest B-type atom is determined by γ_3. The presence of such coupling lowers the symmetry of the interaction Hamiltonian. The z-axis becomes a three-fold symmetry axis and there arises the trigonal warping of the surfaces of constant energy. An estimation gives $\gamma_3 \sim (0.1\text{–}0.4)\,\mathrm{eV}$. The parameter γ_4 depends on the interaction between atoms belonging to two different layers. This interaction gives rise to the difference in properties between the conduction and valence bands, i.e., to the "nonmirror" character of the bands. According to estimations, $\gamma_4 \sim (0\text{–}0.25)\,\mathrm{eV}$. The interaction between the nearest A-type atoms placed in the neighbouring equivalent layers is described by the parameter γ_5, whose value is close to that of γ_2. The presence of γ_5 in the model under discussion leads to a small relative shift of the band boundaries. Finally, the parameter $\gamma_6 \equiv \Delta$ is associated with the difference in the position of A-type and B-type atoms. Its value lies within the interval 0–0.1 eV.

The electron dispersion law for graphite can be obtained within the SWMC-model by the (kp)-method, which is based on the symmetry properties of the lattice. This yields the wave-vector dependence of the electron energy in some small region of the Brillouin zone. In the SWMC-model, such regions are represented by the neighbourhood of the six vertical edges HKH or $H'K'H'$ of the Brillouin zone (fig. 9). Thus the SWMC-model essentially extends the dispersion law (30) for graphite-layer electrons, which has been obtained by the (kp)-method, to the three-dimensional case.

The eigenvalues of electron energy can be found by solving the fourth-order secular equation

$$\hat{H}(\varepsilon_i - \varepsilon_i^0) = 0, \tag{32}$$

provided that the above interactions described by γ_i parameters are included by the (kp)-method. In this equation, the Hamiltonian matrix is written as;

$$\mathbf{H} = \begin{pmatrix} \varepsilon_1^0 & 0 & H_{13} & H_{13}^* \\ 0 & \varepsilon_2^0 & H_{23} & -H_{23}^* \\ H_{13}^* & H_{23}^* & \varepsilon_3^0 & H_{33} \\ H_{13} & -H_{23} & H_{33}^* & \varepsilon_3^0 \end{pmatrix}. \tag{33}$$

Here we have used the following notation:

$$\begin{aligned} &\varepsilon_1^0 = \Delta + \gamma_1\Gamma + \tfrac{1}{2}\gamma_5\Gamma^2, \quad \varepsilon_2^0 = \Delta - \gamma_1\Gamma + \tfrac{1}{2}\gamma_5\Gamma^2, \quad \varepsilon_3^0 = \tfrac{1}{2}\gamma_2\Gamma^2, \\ &H_{13} = \tfrac{1}{2}\sqrt{2}(-\gamma_0 + \gamma_4\Gamma)\mathrm{e}^{+i\alpha}\zeta, \quad H_{23} = \tfrac{1}{2}\sqrt{2}(\gamma_0 + \gamma_4\Gamma)\mathrm{e}^{+i\alpha}\zeta, \\ &H_{33} = \gamma_3\Gamma\mathrm{e}^{+i\alpha}\zeta, \end{aligned} \tag{34}$$

where α is the angle between the direction of the vector $\boldsymbol{k}$ and the ΓK direction in the Brillouin zone (see fig. 9), $\Gamma = 2\cos\xi$, ξ and ζ are dimensionless wave vectors in, respectively, the direction of the c-axis and in the basal plane: $\xi = \tfrac{1}{2}k_z c_0$, $\zeta = \tfrac{1}{2}\sqrt{3}\,a_0\sqrt{k_x^2 + k_y^2}$; k_z is measured from the point K; k_x and k_y, from the HKH edge of the Brillouin zone. Equations (33) and (34) are valid only in a rather close vicinity of the vertical edges HKH or $H'K'H'$ of the Brillouin zone. The eigenvalues of the Hamiltonian energy given by (33) are the roots of the fourth-order equation

$$\begin{aligned}
&(\varepsilon_1^0 - \varepsilon)(\varepsilon_2^0 - \varepsilon)(\varepsilon_3^0 - \varepsilon)^2 \\
&\quad - \frac{\hbar^2 k^2 p_0^2}{m_0^2}\Big[(\varepsilon_1^0 - \varepsilon)(\varepsilon_3^0 - \varepsilon)(1+\nu)^2 + (\varepsilon_2^0 - \varepsilon)(\varepsilon_3^0 - \varepsilon)(1-\nu)^2 \\
&\qquad\qquad + 4\frac{\gamma_3^2}{\gamma_0^2}\cos^2\xi\,(\varepsilon_1^0 - \varepsilon)(\varepsilon_2^0 - \varepsilon)\Big] \\
&\quad - 2\frac{\hbar^3 k^3 p_0^3}{m_0^3}\frac{\gamma_3}{\gamma_0}\cos\xi\,\cos 3\alpha\big[(\varepsilon_1^0 - \varepsilon)(1+\nu)^2 - (\varepsilon_2^0 - \varepsilon)(1-\nu)^2\big] \\
&\quad + \left(\frac{\hbar k p_0}{m_0}\right)^4 (1+\nu)^2(1-\nu)^2 \\
&\quad = 0,
\end{aligned} \tag{35}$$

where $\nu = 2(\gamma_4/\gamma_0)\cos\xi$, p_0 is given by (31). Equation (35) gives the form of the dependence of electron energy on wave-vector in the SWMC-model, the specific form of this dependence being determined by the sign and magnitude of the γ_i parameters contained in this equation.

The above estimates of these parameters are rather crude, giving merely the order of their magnitude. Particularly crude is the estimation of the smallest parameters of the SWMC-model, i.e., $\gamma_2, \gamma_5, \gamma_6$. It has so far been impossible to determine reliably even the sign of the parameter γ_2, let alone its value. And this in spite of the fact that γ_2 determines the structure of the electron spectrum of graphite in the immediate vicinity of the Fermi level, because the overlapping of the conduction and valence bands is equal to $2|\gamma_2|$. The sign of γ_2 determines the position of the electron and hole extrema on the edges HKH or $H'K'H'$ of the Brillouin zone (for $\gamma_2 < 0$, electron extrema are at the K-points, hole extrema are near the H-points; when the sign of γ_2 is reversed, their positions are interchanged).

There are two approaches towards the determination of an adequate set of the γ_i parameters within the SWMC-model. One is to use the experimental data on the structure of the spectrum and the characteristics of charge carriers in graphite. The other is to directly calculate the SWMC-model parameters on the basis of general theoretical concepts. The former approach was first adopted by McClure (1957) who calculated the values of γ_1, γ_2, and Δ at various γ_0 from the data of Shoenberg (1952) and of Berlincourt and Steele (1955), which had been obtained in studying the dHvA effect. McClure found the parameter γ_2 to be positive and concluded that the hole extrema were at the centres of the side edges of the Brillouin zone (at the points K). A positive

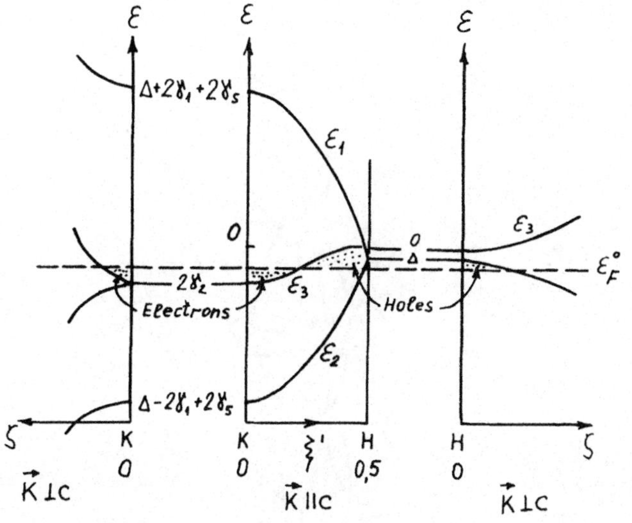

Fig. 11. Energy as a function of wave vector in the SWMC-model ($\xi' = \xi/\frac{1}{2}\pi = k_z c_0/\pi$).

value of γ_2 (0.02 eV) has been found also by Painter and Ellis (1970) who calculated γ_2 within a variational LCAO-method using Bloch states.

Subsequently, using the more accurate experimental data obtained by McClure (1970), Mrozowski (1971), Wills et al. (1971), and Toy et al. (1977), it has been found that the parameter γ_2 is negative, while γ_1, and Δ are positive. A negative value of γ_2 (−0.019 eV) has been found also by Nagayoshi et al. (1976a) who calculated the electron spectrum of graphite from first principles.

The character of the dispersion law in bands for the parameter values $\gamma_1 > 0$ and $\gamma_2 < 0$ is illustrated in fig. 11. Two out of the four solutions of eq. (35) are degenerate along the *HKH* edge of the Brillouin zone and correspond to the ε_3 root. The remaining, nondegenerate solutions are designated as ε_1 and ε_2. The degeneracy of the ε_3-levels is removed as one departs from the *HKH* edge ($k_x, k_y \neq 0$, $\zeta \neq 0$), which can be seen in the left-hand part of fig. 11 (the dependences of ε on ζ in the plane $k_z = \xi = 0$). In the BZ corners (points H, H', $\xi = \pm\frac{1}{2}\pi$) the levels ε_1 and ε_2 are degenerate and remain so when one departs from a corner in the plane of the upper or lower face of the Brillouin zone (see the right-hand part of fig. 11). At $\gamma_2 < 0$, the electron extrema are located at the *K*-points, which are centres of thin cigar-shaped electron pockets elongated along the *HKH* edges. Above and below these pockets, two thin hole pockets are placed symmetrically, whose centres are near the *H*-points. Such a position of the electron and hole surfaces of constant energy in

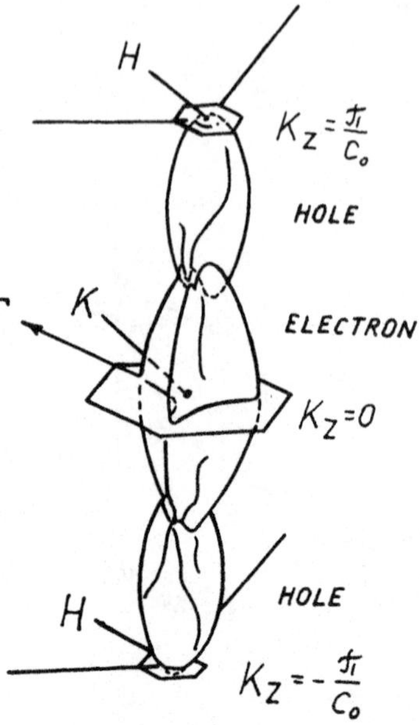

Fig. 12. The Brillouin zone and Fermi surfaces for graphite. The Fermi surfaces are drawn schematically, being magnified in the horizontal direction by a factor of about 5, in order to show the trigonal warping more clearly.

graphite is consistent with the experimental data obtained by Cooper et al. (1970) in studying the effect of neutron irradiation upon the Shubnikov–de Haas oscillations, as well as with the galvanomagnetic and thermomagnetic data for graphite, and is at present generally accepted.

The general form of the Fermi surface in graphite is shown in fig. 12. Attention is called to the fact that the position of electron extrema at the K-points agrees with the model of "nearly free electrons" and with the construction of the Fermi surfaces for graphite by the method proposed by Harrison (1966). An analysis of eq. (35) reveals that the transformation of a gapless semiconductor, such as two-dimensional graphite, into a semimetal is mostly due to the parameters γ_2 and Δ. Taking account of Δ eliminates the degeneracy that took place in the corners of the two-dimensional Brillouin zone. The overlapping of the conduction band (π^*-antibonding states) with the valence band (π-bonding states) is, as already mentioned, $2|\gamma_2|$. Thus three-dimensional graphite is a semimetal with an equal density of electrons and holes. The small number of electrons and holes is due to the smallness of $|\gamma_2|$. The negative value of γ_2 corresponds to the model in which the effective electron masses are greater than the effective hole masses. In broad outline, the structure of the σ- and π-bands of three-dimensional graphite coincides with the band structure of the graphite layer shown in fig. 7. Numerous calculations on the electron spectrum of three-dimensional graphite indicate that taking into account the interaction between carbon layers in graphite mainly affects the band structure in the immediate vicinity of the Fermi level in an energy interval of about 0.1 eV and results, as already mentioned, in the transformation of the spectrum of a gapless semiconductor with a linear dispersion law into the spectrum of a semimetal with a small overlapping between the π-bands. The transformation of the electron spectrum of two-dimensional graphite to that of three-dimensional graphite was considered from first principles in a number of theoretical papers. Wallace (1947), Bassani and Parravicini (1967), and Painter and Ellis (1970) calculated the electron spectrum of graphite by TB and LCAO methods. The pseudopotential method was used by Van Haeringen and Junginger (1969) to calculate the band structure of graphite. Tsukada et al. (1972), Nagayoshi et al. (1976a,b), and Zunger (1978) used the combined OPW–TB-method in their electron spectrum calculations. Attention should be paid to the fact that Nagayoshi et al. (1976a,b) have in fact developed a new method of calculating the energy spectrum of lamellar crystals. The gist of their method is that the electron wave function is constructed as a combination of two-dimensional orthogonalized plane waves (in the plane of a layer) and one-dimensional Bloch functions (in the direction normal to the layer). These Bloch functions satisfy the one-electron Schroedinger equation with a one-dimensional potential obtained by averaging in the plane of the layer over the three-dimensional potential $V(\boldsymbol{r})$. This method assures a good approximation in the direction of the z-axis and is free from the difficulties inherent in LCAO-like methods in treating the continuous states of atoms and the overlaps between far-separated atomic orbitals. In these numerical calculations, a model potential was employed that represented a superposition of the screened Coulomb atomic potentials $V_a(\boldsymbol{r})$. The potential $V_a(\boldsymbol{r})$ was, in turn, chosen as the self-consistent potential for the carbon atom that was proposed by Herrmann and Skillman (1963).

The method proposed by Nagayoshi et al. (1976a,b) to calculate the band structure of graphite has been discussed in more detail than the other methods primarily because it is, at present, one of the most accurate methods. The accuracy of their calculations, involving 32 Bloch functions and 48 two-dimensional plane waves, constitutes $\sim 10^{-1}$ eV for the position of remote levels, $\sim 10^{-2}$ eV for the position of extrema near the Fermi level, and $\sim 5 \times 10^{-3}$ eV for the difference between the energies of the levels near the Fermi level. Moreover, this method has allowed a rather exact calculation of the parameters γ_i ($i = 0, 1, \ldots, 6$) of the SWMC-model, four of them ($\gamma_1, \gamma_2, \gamma_5, \gamma_6$) being determined by a direct calculation of the energy levels ε_1^0, ε_2^0, and ε_3^0 [see (34)] at various values of k_z. The parameter values thus found have the following values (in eV): $\gamma_1 = 0.32$, $\gamma_2 = -0.019$, $\gamma_5 = 0.021$, $\gamma_6 \equiv \Delta = 0.017$. The remaining parameters ($\gamma_0, \gamma_3, \gamma_4$) were found by solving the dispersion equations in the directions of the k_x- and k_y-axes, the solutions being fitted by the results of calculations according to the SWMC-model. This procedure has given the following values (in eV) for those parameters: $\gamma_0 = 2.73$, $\gamma_3 = 0.29$, and $\gamma_4 = 0.15$. In table 5, the parameters calculated by Nagayoshi et al. (1976a,b) are compared with the γ_i values determined by other authors from experimental data.

Concluding this subsection, we note that, on the basis of calculations on the electron spectrum of graphite from first principles, there have recently been made several analyses of the transformation of the electronic structure of graphite under pressure (Nagayoshi 1977) and with intercalation (Inoshita et al. 1977, Holzwarth and Rabii 1977, Holzwarth et al. 1978a, Ohno et al. 1979).

4.3. *The four-parameter model of the electron energy spectrum of three-dimensional graphite*

The surface of constant energy described by eq. (35) of the SWMC-model does not possess the cylindrical symmetry around the vertical *HKH* edge of the Brillouin zone. The k_z-axis for such surfaces is a three-fold symmetry axis, which is suggested by the term with a factor of $\cos 3\alpha$ in eq. (35). When the cross-sections of surfaces of constant energy formed by the plane normal to the k_z-axis are not circular, this is called trigonal warping. The warping vanishes if the parameter γ_3 is set equal to zero. Equation (35) then reduces to two second-order equations describing four branches of the electron energy spectrum:

$$\varepsilon_1^{\pm} = \tfrac{1}{2}(\varepsilon_1^0 + \varepsilon_3^0) \pm \left\{\tfrac{1}{4}(\varepsilon_1^0 - \varepsilon_3^0)^2 + \left(\frac{\hbar p_0}{m_0}\right)^2 k_\rho^2\left[1 - \frac{\gamma_4}{\gamma_0}\Gamma\right]^2\right\}^{1/2}, \tag{36}$$

and

$$\varepsilon_2^{\pm} = \tfrac{1}{2}(\varepsilon_2^0 + \varepsilon_3^0) \pm \left\{\tfrac{1}{4}(\varepsilon_2^0 - \varepsilon_3^0)^2 + \left(\frac{\hbar p_0}{m_0}\right)^2 k_\rho^2\left[1 + \frac{\gamma_4}{\gamma_0}\Gamma\right]^2\right\}^{1/2}, \tag{37}$$

where $k_\rho = \sqrt{k_x^2 + k_y^2}$.

The surfaces of constant energy corresponding to (36) and (37) are the surfaces of rotation around the *HKH* edges. If $\gamma_3 = 0$ is assumed, it would also be reasonable to omit the small parameters γ_4 and γ_5, bearing in mind that γ_4, $\gamma_5 \ll \gamma_0$, γ_1. As a result, eqs. (36) and (37) reduce to one equation, which contains only the parameters γ_0, γ_1, γ_2, and γ_6. This equation is known as the dispersion law in the four-parameter model for the band spectrum of three-dimensional graphite:

$$\varepsilon_{1,2}^{\pm} = \tfrac{1}{2}(\varepsilon_{1,2}^0 + \varepsilon_3^0) \pm \left\{\tfrac{1}{4}(\varepsilon_{1,2}^0 - \varepsilon_3^0)^2 + \left(\frac{\hbar}{m_0} p_0\right)^2 k_\rho^2\right\}^{1/2}. \tag{38}$$

Equation (38) describes reasonably well the dispersion law of electrons and holes in graphite in the entire range of variation of the wave vector k_z, except for the close vicinities of the corners of the Brillouin zone and those regions in which electron and hole surfaces make contact and in which the terms previously ignored should be taken into account.

Consider the electron and hole surfaces of constant energy described by eq. (38). Their cross sections formed by planes perpendicular to the *HKH* edge represent circles of radius

$$k_{\rho 1,2} = \frac{m_0}{\hbar p_0}\left[\varepsilon^2 - \varepsilon(\varepsilon^0_{1,2} + \varepsilon^0_3) + \varepsilon^0_{1,2}\varepsilon^0_3\right]^{1/2}. \tag{39}$$

Therefore, the cross sections of the electron (S_e) and hole (S_h) surfaces as functions of energy ε and wave vector k_z are given by

$$S_e = \frac{\pi m_0^2}{p_0^2}\frac{(\frac{1}{2}\gamma_2\Gamma^2 - \varepsilon)(\Delta - \gamma_1\Gamma - \varepsilon)}{(1+\nu)^2}, \tag{40}$$

$$S_h = \frac{\pi m_0^2}{p_0^2}\frac{(\frac{1}{2}\gamma_2\Gamma^2 - \varepsilon)(\gamma_1\Gamma - \varepsilon + \Delta)}{(1-\nu)^2}. \tag{41}$$

Formulae (40) and (41) are more exact than eq. (38), because they include the terms containing the parameter $\gamma_4 = \frac{1}{2}\nu\gamma_0$ which have been omitted in (38).

From the above formulae it follows that, irrespective of the value of ε, the maximum of S_e occurs at $k_z = 0$, i.e., at the *K*-points (fig. 12), while that of S_h takes place at two symmetrical, about *K*, points on *HKH* edges for $\xi = \pm\xi^h_{max}$, where

$$\xi^h_{max} = \frac{1}{\pi}\arccos\left[\frac{-(\Delta - \varepsilon) + \sqrt{(\Delta - \varepsilon)^2 + 6(\gamma_1^2/\gamma_2)\varepsilon}}{6\gamma_1}\right] \simeq \arccos(\varepsilon/6\gamma_2)^{1/2}, \tag{42}$$

and is dependent on ε.

One can get a clear idea of the Fermi surfaces for electrons and holes within the four-parameter model by considering the contours of the cross section of these surfaces by a plane passing through the *HKH* edge of the Brillouin zone. These contours are determined by the dependence $k_\rho = f(k_z)$. Figure 13 shows a family of such contours for electrons and holes taken from the review of the electronic properties of graphite by Spain (1973). The curves are constructed for several values of energy ε. For clarity the scale of the k_ρ-axis is increased fivefold. The energies (positive for electrons and negative for holes) are measured from the edges of the conduction and valence bands. At small values of the wave vector, eq. (38) reduces to a quadratic dispersion law for electrons and holes:

$$\varepsilon_e = \varepsilon^0_3 + \left(\frac{\hbar p_0}{m_0}\right)^2 \frac{1}{|\varepsilon^0_2 - \varepsilon^0_3|}k_\rho^2, \tag{43}$$

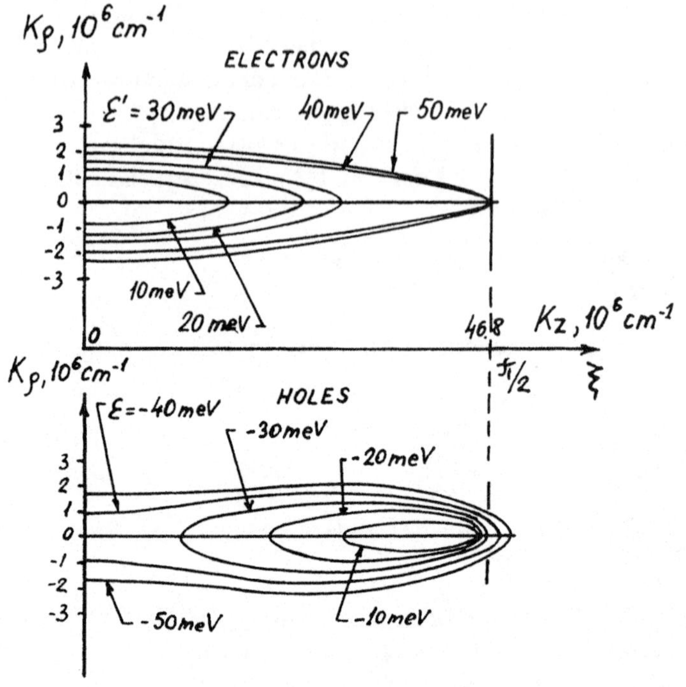

Fig. 13. Contourlines of constant energy for electrons and holes in the plane containing the HKH edge of the Brillouin zone. The contours have been obtained using the four-parameter model for the spectrum of graphite; $\xi = \frac{1}{2}\pi$ corresponds to the upper edge of the Brillouin zone, and $\xi = 0$ corresponds to the plane normal to HKH, and passing through the K-point. For clarity the axes of k_ρ and k_z are scaled differently from each other. The energy ε' for electrons differs from ε by the value of the overlap, $\varepsilon_\pi = 2|\gamma_2|$, between the conduction band and the valence land: $\varepsilon' = \varepsilon - 2|\gamma_2|$ (Spain 1973).

$$\varepsilon_h = \varepsilon_3^0 - \left(\frac{\hbar p_0}{m_0}\right)^2 \frac{1}{|\varepsilon_1^0 - \varepsilon_3^0|} k_\rho^2. \tag{44}$$

This entails that the effective masses of electrons m^e and holes m^h at the edges of the conduction and valence bands, which describe the motion of charge carriers in the plane of the graphite layer, are isotropic and depend only on the magnitude of the wave vector k_z along the edge of the Brillouin zone

$$m^{e,h} = m_0^2 \frac{\gamma_1 \Gamma}{2p_0^2} = \frac{m_0^2}{2p_0^2} \gamma_1 \cos(\tfrac{1}{2} k_z c_0). \tag{45}$$

Since electrons occur near the K-points ($k_z = 0$), their masses turn out to be greater, as it follows from (45), than the hole masses.

In the four-parameter model, the velocities of electrons and holes (v_e, v_h) in the basal plane are isotropic, too. Thus the electron velocity at the edge of the conduction band and the hole velocity at the edge of the valence band are given by

$$v_{e,h} = \frac{2\hbar p_0^2}{m_0^2}\frac{k_\rho}{|\varepsilon_{1,2}^0 - \varepsilon_3^0|} = \frac{2\hbar p_0^2}{m_0^2\gamma_1 \cos(\frac{1}{2}k_z c_0)}k_\rho. \tag{46}$$

At high values of the wave vector k_ρ, the effective masses $m^{e,h}$ begin to depend on the value of k_ρ, while the velocities $v_{e,h}$ tend to a maximum constant value

$$v_{e,h\,max} = \frac{p_0}{m_0}, \tag{47}$$

which corresponds to the asymptotic behaviour of the linear dispersion law in the limit of high k_ρ.

If the previously neglected small parameter γ_4 is taken into account, then the effective masses of holes become slightly greater, while those of electrons become slightly smaller than their masses given by eq. (45):

$$m^e = \frac{m_0^2}{2p_0^2}\gamma_1 \cos(\tfrac{1}{2}k_z c_0)\frac{1}{(1+v)^2}, \tag{48}$$

$$m^h = \frac{m_0^2}{2p_0^2}\gamma_1 \cos(\tfrac{1}{2}k_z c_0)\frac{1}{(1-v)^2}. \tag{49}$$

The small magnitude of the interaction between graphite layers enables one to neglect, in the first-order approximation, the motion of carriers in the direction normal to the layers. This corresponds to a situation in which the Fermi surfaces in graphite are approximated by cylinders whose axes coincide with the side edges of the Brillouin zone. This approximation, suggested by Klein (1964), is highly convenient for calculating the transport properties of graphite in the basal plane. The dispersion laws for electrons and holes may then be written in a very simple form:

$$\varepsilon_e = \frac{\hbar^2}{2m^e}k_\rho^2, \tag{50}$$

$$\varepsilon_h = \varepsilon_\pi - \frac{\hbar^2}{2m^h}k_\rho^2, \tag{51}$$

where ε_π is the band overlap energy equal to $2|\gamma_2|$ and the effective masses m^e and m^h are regarded as independent of k_z.

Another simple model of the electron spectrum of graphite is the parabolic ellipsoidal model, which is convenient for approximate calculations. In this model, the surfaces of constant energy are approximated by prolate revolution ellipsoids in the direction of the k_z-axis. The dispersion laws will then also have a simple form:

$$\varepsilon_e = \tfrac{1}{2}\hbar^2\left[\frac{k_\rho^2}{m_1^e} + \frac{k_z^2}{m_3^e}\right], \tag{52}$$

$$\varepsilon_h = \varepsilon_\pi - \tfrac{1}{2}\hbar^2\left[\frac{k_\rho^2}{m_1^h} + \frac{k_z^2}{m_3^h}\right]. \tag{53}$$

The value of the ratio of the major axis of such an ellipsoid to the minor one is close to 15. According to Spain (1973), the values of the effective electron and hole masses in the ellipsoidal parabolic model are

$$m_1^e = 0.055\,m_0, \qquad m_3^e = 3\,m_0, \qquad m_1^h = 0.04\,m_0, \qquad m_3^h = 6\,m_0, \tag{54}$$

while the band overlap energy is taken to be $\varepsilon_\pi = 36$ meV.

It should be noted that, although the ellipsoidal parabolic model is simple and covenient for rough estimates, the range of its validity and its predicting power are rather small. The model is especially ineffective in calculating the thermoelectrical and thermomagnetic properties of graphite or carbon materials, and also in interpreting effects due to the shift of the Fermi level.

4.4. *Trigonal warping of the Fermi surface. Minority carriers in graphite*

The breaking of the cylindrical symmetry of the Fermi surface in graphite, which occurs when the interaction characterized by the parameter γ_3 is taken into account, may be described by treating this parameter as a small perturbation, as was first suggested by McClure (1957). In this way, one can obtain from eq. (35) at small k_ρ the following expression for dispersion law:

$$\varepsilon = \varepsilon_3^0 + \left(\frac{\hbar p_0}{m_0}\right)^2 \frac{1}{|\varepsilon_2^0 - \varepsilon_3^0|} k_\rho^2$$

$$\pm \frac{\hbar p_0}{2m_0} k_\rho \cos \tfrac{1}{2}k_z c_0 \left\{1 + \frac{2k_\rho}{k_S}\sin 3\alpha + \frac{k_\rho^2}{k_S^2}\right\}^{1/2},$$

where

$$k_S = \frac{\gamma_1\gamma_3}{2a_0\gamma_0^2}\cos^2(\tfrac{1}{2}k_z c_0) = \tfrac{3}{2}\frac{m_0}{\hbar p_0}\frac{\gamma_1\gamma_3}{\gamma_0}\cos^2(\tfrac{1}{2}k_z c_0). \tag{56}$$

Just as in the four-parameter model, the small terms containing the parameters γ_4 and γ_5 have been omitted in eq. (55). The quantity k_S, which has the dimension of wave vector, determines the size of the region in the Brillouin zone beyond which the trigonal warping of the surface of constant energy

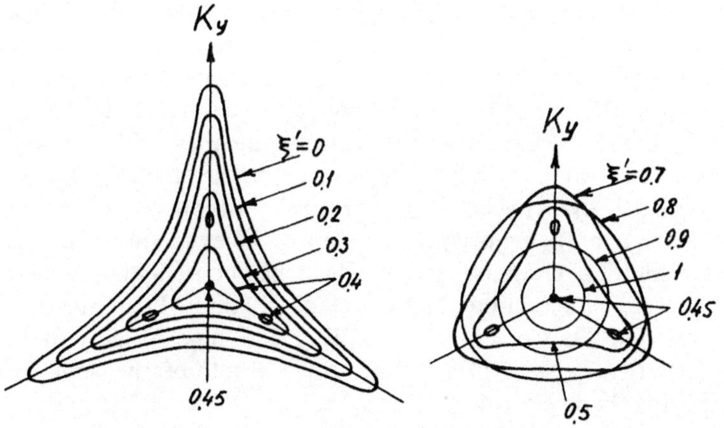

Fig. 14. Sketch of the constant-energy surfaces at the Fermi energy for various values of ξ' ($\xi' = k_z c_0/\pi$, where k_z is the z-component of wave vector and c_0 is the lattice constant, $\xi' = \xi/\frac{1}{2}\pi$). Band parameters are for set B of table 5 (Dillon and Spain 1978).

cannot be neglected. According to eq. (55), the warping is particularly distinct near the K-points, at which $\cos(\frac{1}{2}k_z c_0) \simeq 1$. At the same time, near the points H we have $|\cos(\frac{1}{2}k_z c_0)| \ll 1$ and the Fermi surface has practically circular cross sections. Figure 14, taken from a paper by Dillon and Spain (1978), illustrates the cross sections of the electron and hole Fermi surfaces formed by planes parallel to the basal one. The trigonally warped Fermi surface of graphite was shown in fig. 12.

The Fermi surface has a very complicated structure in those regions where the electron surfaces are in contact with the hole ones. According to the SWMC-model, in these regions there should arise three characteristic features (so-called "outrigger pieces" of the Fermi surface; see fig. 12), whose existence has not yet been confirmed experimentally.

We note that the parameter γ_3 is not, in fact, small (as compared with the parameters γ_0 and γ_1). Therefore, the expansion in γ_3, with the aid of which eq. (55) has been derived, does not permit a sufficiently correct description of the details of the Fermi surface in graphite. This is true primarily of the region of contact between the electron and hole parts of the Fermi surface and also of the neighbourhood of the corners of the Brillouin zone (i.e., points H and H'). In order to describe the Fermi surface in graphite more accurately, one should use the complete equation, eq. (35), of the SWMC-model, as was done by Dresselhaus and Mavroides (1966).

It should also be stressed that the structure of the electron spectrum of graphite near the points H and H' is highly sensitive to the values of the small parameters of the SWMC-model, in particular to the relationship between γ_2 and Δ. Proceeding from eq. (35), one can show that in the case when γ_2 and Δ are negative (the condition $\gamma_2 < 0$ corresponds to the π-band overlap in graphite) and the inequality $|\Delta| < \varepsilon_F^h$ holds (ε_F^h being the Fermi hole energy measured from the edge of the valence band ε_3 near the points H of the Brillouin zone), there appear additional hole extrema associated with the level ε_2, which give rise to the so-called minority holes. The holes associated with the extremum of the level ε_3 are then called majority holes. If the inequality $|\Delta| < \varepsilon_F^h$ holds, the density of the majority holes is higher than that of the minority holes. The relative position of the energy levels at which there are both majority and minority holes is shown in fig. 11. The spin–orbit splitting of energy levels, which has consistently been taken into account in the interaction Hamiltonian of the SWMC-model first by Dresselhaus and Dresselhaus (1965), plays a significant part in the formation of minority carriers in graphite. As a result of the spin–orbit interaction, the doubly degenerate level ε_3 splits into two sublevels spaced by $\sim 2 \times 10^{-4}$ eV near the points K. Owing to the same cause, the degeneracy of the levels ε_1 and ε_2 at the points H is removed, there arising a splitting of about 4×10^{-4} eV. As a result of the spin–orbit splitting near the points H, the Fermi surface for minority holes, which has the shape, of a prolate revolution ellipsoid around the k_z-axis, is separated from the narrow part ("neck") of the Fermi surface for majority holes (fig. 15).

It should also be mentioned that McClure (1956) and Noziers (1958) made the suggestion that near the points K in the Brillouin zone there might be additional electron extrema in the graphite spectrum. Subsequently, Dresselhaus (1974) calculated the structure of π-bands in graphite near the points K in the Brillouin zone on the basis of the SWMC-model taking account of trigonal warping. He has demonstrated that taking into account the parameter γ_3 gives rise to additional extrema near the points K. The existence of

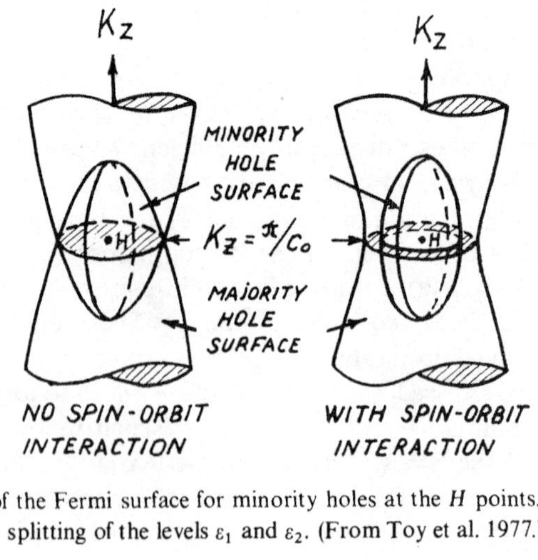

Fig. 15. Emergence of the Fermi surface for minority holes at the H points, due to the spin–orbit splitting of the levels ε_1 and ε_2. (From Toy et al. 1977.)

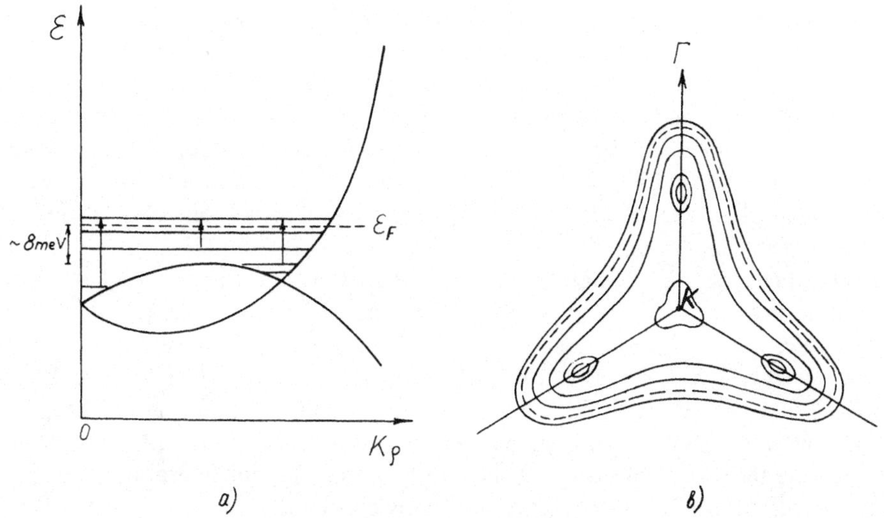

Fig. 16. The effect of trigonal warping on energy dispersion near the K-point. (a) Schematic Landau levels with transitions from central and leg levels, and from ordinary levels. The dispersion is shown along the axis from K to Γ. (b) Corresponding orbits in $\boldsymbol{k}$-space (Doezema et al. 1979).

these extrema should lead to the appearance in a magnetic field of a special system of levels, whose energy is proportional to $H^{1/2}$. These levels, which have been called "leg levels", are schematically shown in fig. 16a, illustrating the dependence of the energy on the wave vector in the $K\Gamma$-directions in the Brillouin zone. The electron orbits in $\boldsymbol{k}$-space, which correspond to these levels, are shown in fig. 16b.

The existence of the additional electron extrema in the graphite spectrum was experimentally found by Platts et al. (1977) and Doezema et al. (1979) in studying magneto-optical transitions associated with leg levels.

Thus on the basis of a general analysis of the SWMC-model one may conclude that near the Fermi level in graphite there are twelve, relatively deep, hole and six electron extrema, with which majority holes and electrons are associated, and also twelve shallow hole and eighteen shallow electron extrema, with which minority holes and electrons are associated. Therefore, in accordance with the SWMC-model, the Fermi surface in graphite should, in general, consist of six majority electron and twelve majority hole surfaces of constant energy, as well as of twelve minority hole and eighteen minority electron surfaces of small area. The model is insensitive to the existence of two groups of majority carriers (whether they be electrons or holes), but much more sensitive to the existence of groups of minority carriers, associated with the small parameters of the model. In particular, the minority-electron group considered above may exist only provided that the Fermi level passes below the saddle point in the spectrum (see fig. 16a). Otherwise, the presence of additional electron extrema near the points K would merely result in a more complicated relief of the conduction-band bottom for majority electrons.

However, the small parameters of the SWMC-model are not yet reliably known, on the one hand, and are the most structure-sensitive (i.e., capable of changing significantly with varying degree of perfection of graphite single crystals (see sect. 4.7)), on the other hand. Hence, from the viewpoint of experimental study of the Fermi surface in graphite, one can draw the following important conclusions:

(1) Since the above-mentioned general variant of the Fermi surface in graphite may be realized only at some definite values for the SWMC-model parameter, most of which (but not all) have been determined rather accurately and reliably, it is generally not clear whether such a model for the Fermi surface can be fully realized even in absolutely perfect single-crystal graphite.

(2) Since the SWMC-model parameters, especially the small ones, which determine the "fine" structure of the energy spectrum, considerably depend on the degree of perfection of a graphite single crystal and on its lattice parameters, various variants of minority-carrier surfaces of constant energy may in principle be realized in materials with various degrees of structural perfection. For example, there may be realized solely electron or solely hole surfaces for minority carriers. The same is true of those parts that connect the electron and hole surfaces for majority carriers: they may be present in one but absent in another material.

Accordingly, one cannot, it seems, speak of a single Fermi surface that describes the energy spectrum of charge carriers in different kinds of graphite. It would be more correct to speak of the Fermi surface for a particular material with definite lattice parameters, definite degree of structural perfection and known concentrations of donor- and acceptor-type impurities.

In connection with the foregoing, the problem of restoring some abstract Fermi surface on the basis of the results obtained in various investigations, mostly made with specimens having different (and uncontrollable) levels of doping and imperfection, as well as with unknown lattice parameters, can hardly be regarded as worthwhile. It is expedient to speak only of some common, characteristic, stable features of the energy spectrum of graphite and to study the character of its change as the lattice parameters and the degree of doping are purposefully varied.

4.5. The electron energy spectrum of graphite in a quantizing magnetic field

The electron energy levels of graphite in a quantizing magnetic field may be found most simply in the semiclassical approximation. In order to quantize the electron spectrum, one can then make use of the Lifshitz–Onsager formula that relates the area S of the cross section of the Fermi surface by a plane perpendicular to the magnetic field $\boldsymbol{H}$ to the field strength and the quantum number n:

$$S = \frac{2\pi e\hbar}{c} H(n + \delta). \tag{57}$$

Here n is an integer ($n = 0, 1, \ldots$) and δ is a constant of the order of unity. For high quantum numbers ($n \gg 1$), eq. (57) is rigorously valid.

Within the SWMC-model, the simplest way to obtain an analytical expression relating the electron energy ε to the quantum numbers n is to use eq. (57) disregarding the effect of trigonal warping. In this case, the cross sections of the Fermi surface at $\boldsymbol{H}\|Z$ are simply circles of radius k_ρ, whose value is given by (39). Assuming in (57) that $S = \pi\hbar^2 k_\rho^2$ and inserting the k_ρ from (39), we arrive at the semiclassical expression for the quantized spectrum ε_n of graphite at $\boldsymbol{H}\|Z$ and $\gamma_3 = \gamma_4 = 0$:

$$\frac{2e\hbar}{c} H(n + \delta) = \frac{m_0^2}{p_0^2}[\varepsilon_n^2 - \varepsilon_n(\varepsilon_{1,2}^0 - \varepsilon_3^0) + \varepsilon_{1,2}^0\varepsilon_3^0]. \tag{58}$$

Hence

$$\varepsilon_n^{\pm}(1, 2) = \tfrac{1}{2}(\varepsilon_{1,2}^0 + \varepsilon_3^0) \pm \left[\tfrac{1}{4}(\varepsilon_{1,2}^0 - \varepsilon_3^0)^2 + \frac{2p_0^2\hbar e}{m_0^2 c} H(n + \delta)\right]^{1/2}. \tag{59}$$

The semiclassical expressions given by (58) and (59) may be extended to the case when the magnetic field makes an angle θ with the c-axis.

On the basis of the SWMC-model, an expression for electron energy levels in the magnetic field valid at any values of n can be derived only when the field is directed along the c-axis and when trigonal warping is neglected ($\gamma_3 = 0$). Under these conditions, the problem has been solved by McClure (1960), Inoue (1962), Dresselhaus and Dresselhaus (1965), Dresselhaus (1974), and Nakao (1976). According to these authors, the matrix of the Hamiltonian defining the electron energy levels in the magnetic field in states with quantum number n has the form

$$\mathscr{H}(n, k_z) =$$

$$\begin{pmatrix} \varepsilon_n^{(1)} & 0 & -(\tfrac{1}{2}B)^{1/2}(1 - \nu)n^{1/2} & -(\tfrac{1}{2}B)^{1/2}(1 - \nu)(n + 1)^{1/2} \\ 0 & \varepsilon_n^{(2)} & (\tfrac{1}{2}B)^{1/2}(1 + \nu)n^{1/2} & -(\tfrac{1}{2}B)^{1/2}(1 + \nu)(n + 1)^{1/2} \\ -(\tfrac{1}{2}B)^{1/2}(1 - \nu)n^{1/2} & (\tfrac{1}{2}B)^{1/2}(1 + \nu)n^{1/2} & \varepsilon_n^{(3)} - \mu H & 0 \\ -(\tfrac{1}{2}B)^{1/2}(1 - \nu)(n + 1)^{1/2} & -(\tfrac{1}{2}B)^{1/2}(1 + \nu)(n + 1)^{1/2} & 0 & \varepsilon_n^{(3)} + \mu H \end{pmatrix} \tag{60}$$

where use is made of the following notation:

$$\varepsilon_n^{(1)} = \Delta + \gamma_1 \Gamma + \tfrac{1}{2}\gamma_5 \Gamma^2 + (n + \tfrac{1}{2})\mu H = \varepsilon_1^0 + (n + \tfrac{1}{2})\mu H,$$

$$\varepsilon_n^{(2)} = \Delta - \gamma_1 \Gamma + \tfrac{1}{2}\gamma_5 \Gamma^2 + (n + \tfrac{1}{2})\mu H = \varepsilon_2^0 + (n + \tfrac{1}{2})\mu H,$$

$$\varepsilon_n^{(3)} = \tfrac{1}{2}\gamma_2 \Gamma^2 + (n + \tfrac{1}{2})\mu H = \varepsilon_3^0 + (n + \tfrac{1}{2})\mu H,$$

$$\Gamma = 2\cos\xi = 2\cos(\tfrac{1}{2}k_z c_0) \quad (k_z \text{ being measured from the point } K)$$

$$B = \gamma_0^2 \tfrac{3}{2} a_0^2 (eH/c\hbar),$$

$$\mu = e\hbar/m_0 c = 2\mu_B \quad (\mu_B \text{ is the Bohr magneton}).$$

The energy levels ε_n in the magnetic field are described by the secular equation obtained from the diagonalization of matrix (60). If the paramagnetic terms $(n + \frac{1}{2})\mu H$ are neglected, the equation assumes the form

$$B^2 n(n+1) - (n + \tfrac{1}{2})B\left[\frac{\varepsilon_n - \varepsilon_1^0}{(1-\nu)^2} + \frac{\varepsilon_n - \varepsilon_2^0}{(1+\nu)^2}\right](\varepsilon_n - \varepsilon_3^0)$$

$$+ \frac{(\varepsilon_n - \varepsilon_3^0)^2(\varepsilon_n - \varepsilon_1^0)(\varepsilon_n - \varepsilon_2^0)}{(1-\nu^2)^2} = 0, \tag{61}$$

and represents a fourth-order equation in ε_n. This equation does not reduce to two second-order equations, as it happens at $H = 0$ for equation (35), which falls into two equations, eqs. (36) and (37), when the effect of trigonal warping is neglected.

Equation (61) may approximately be solved in the semiclassical limit at $n \gg 1$. The result of such a solution extends eq. (59) to the case of $\gamma_4 \neq 0$ and allows one to find that the constant δ of the semiclassical quantization of the electron spectrum within the SWMC-model is equal to $\frac{1}{2}$:

$$\varepsilon_n^{\pm(1,2)} = \tfrac{1}{2}(\varepsilon_{1,2}^0 + \varepsilon_3^0)$$

$$\pm\left[\tfrac{1}{4}(\varepsilon_{1,2}^0 - \varepsilon_3^0)^2 + \frac{2p_0^2\hbar e}{m_0^2 c}\{1 - (-1)^{1,2}\nu\}H(n + \tfrac{1}{2})\right]^{1/2}. \tag{62}$$

Near the points H of the Brillouin zone, $\cos\xi \simeq 0$ and eq. (61) may be solved exactly:

$$\varepsilon_n^{\pm} = \tfrac{1}{2}\Delta \pm \left[\tfrac{1}{4}\Delta^2 + 2\frac{p_0^2\hbar e}{m_0^2 c}H(n, n+1)\right]^{1/2}, \tag{63}$$

where $(n, n+1) = n$ or $(n+1)$.

Near the extreme quantum limit of fields at low quantum numbers n, eq.

(61) may be solved approximately. These approximate solutions, found by McClure (1960), are:

$$\varepsilon_n = \varepsilon_3^0 + \frac{p_0^2 \hbar e}{m_0^2 c} H[-(n+\tfrac{1}{2})(\omega_1 + \omega_2) \pm \{(n+\tfrac{1}{2})^2(\omega_1 - \omega_2)^2 + \omega_1\omega_2\}^{1/2}], \tag{64}$$

and

$$\varepsilon_n = \varepsilon_{1,2}^0 + 2\frac{p_0^2 \hbar e}{m_0^2 c} H(n+\tfrac{1}{2})\omega_{1,2}, \tag{65}$$

where

$$\omega_1 = \frac{(1-\nu)^2}{\varepsilon_1^0 - \varepsilon_3^0}, \qquad \omega_2 = \frac{(1+\nu)^2}{\varepsilon_2^0 - \varepsilon_3^0} \quad \text{for } \varepsilon_3^0 \neq \varepsilon_1^0,\ \varepsilon_2^0, \quad \nu = 2(\gamma_4/\gamma_0)\cos\xi.$$

When the quantum number $n = 0$, the positive root of (64) is represented by $\varepsilon_0 = \varepsilon_3^0$. This means that in the conduction band the zero Landau level is not shifted by a half-quantum of the cyclotron frequency $\frac{1}{2}\hbar\omega_c$ (i.e., the zero Landau level coincides with the edge of the conduction band at $H = 0$). This result was first obtained by McClure (1956), who considered a two-dimensional model of graphite. Finally, eq. (61) can be solved for the quantum number n:

$$(n+\tfrac{1}{2})B = \tfrac{1}{2}(\varepsilon_n - \varepsilon_3^0)\left[\frac{\varepsilon_n - \varepsilon_1^0}{(1-\nu)^2} + \frac{\varepsilon_n - \varepsilon_2^0}{(1+\nu)^2}\right] \pm \left\{\left[\tfrac{1}{2}(\varepsilon_n - \varepsilon_3^0)\left(\frac{\varepsilon_n - \varepsilon_1^0}{(1-\nu)^2} - \frac{\varepsilon_n - \varepsilon_2^0}{(1+\nu)^2}\right)\right]^2 + \tfrac{1}{4}B^2\right\}^{1/2}. \tag{66}$$

The k_z-dependence of the electron and hole energy levels in the magnetic field (Landau subbands) is implicitly given by eq. (61) or eq. (66). Using these equations, one can obtain the dependence of ε_n on k_z at various n by numerical calculations for known values of the parameters γ_i of the SWMC-model. Figure 17 illustrates the first twelve of Landau subbands ($n = 0, 1, \ldots, 11$) in the field $H = 50$ kOe, which were calculated by Dresselhaus and Mavroides (1964).

It should be noted that, near the point K, only the ε_3-band is close to the Fermi level. This band suffers in the magnetic field a perturbation of $\Delta\varepsilon$, so that

$$\varepsilon_n|_{k_z=0} = \varepsilon_3^0 + \Delta\varepsilon, \tag{67}$$

where $|\Delta\varepsilon| \ll |\varepsilon_3^0 - \varepsilon_{1,2}^0|$.

As has been shown by Mendez et al. (1980b), the Landau levels at the points K of the Brillouin zone may be described on the basis of eq. (61) to within the

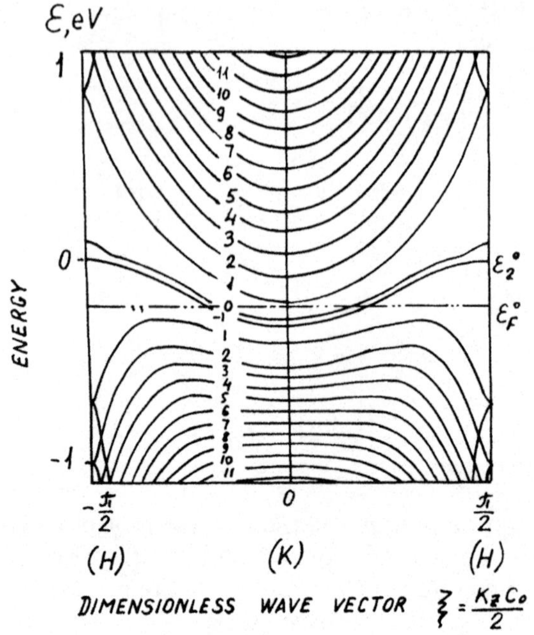

Fig. 17. First twelve electron and hole Landau levels in the SWMC-model without trigonal warping, which have been constructed using a numerical solution of eq. (61) for $H = 50\,kOe$ (Dresselhaus and Mavroides 1964).

second order in $\Delta\varepsilon$ by the following expressions

$$\varepsilon_v(n_v) = \varepsilon_3^0 - (n_v + \tfrac{1}{2})\hbar\omega_v^*, \qquad \varepsilon_c(n_c) = \varepsilon_3^0 + (n_c + \tfrac{1}{2})\hbar\omega_c^*, \tag{68}$$

where ε_3^0 is the edge of the band at the points K,

$$\omega_v^* = \frac{eH}{m_v^h c}, \qquad \omega_c^* = \frac{eH}{m_c^e c}, \tag{69}$$

and $n_{c,v} = 0, 1, 2, \ldots$ are the quantum numbers of the Landau levels in the conduction band and the valence band, respectively. In this approximation, the cyclotron masses m_c^e and m_v^h in the conduction band and the valence band are directly expressed in terms of the SWMC-model:

$$\frac{1}{m_v^h} = \frac{1}{m^*}(1 - 4\beta), \qquad \frac{1}{m_c^e} = \frac{1}{m^*}(1 + 4\beta), \tag{70}$$

where

$$1/m^* = \tfrac{3}{4}\frac{a_0^2\gamma_0^2}{\hbar\,\gamma_1} \quad \text{and} \quad \beta = \frac{\gamma_4}{\gamma_0} - \frac{2\gamma_2 - 2\gamma_5 - \Delta}{8\gamma_1}.$$

The parameter 4β defines the difference between the cyclotron masses of electrons and holes.

As has already been noted in the preceding subsection, taking account of trigonal warping leads to a shift of Landau levels, which is especially pronounced near the K-points in the Brillouin zone where trigonal warping gives rise to a qualitatively new kind of levels in the magnetic field. The problem of the influence of trigonal warping upon the Landau levels in graphite was first formulated by Inoue (1962) and Schroeder et al. (1970), who solved it by perturbative methods by expanding in the parameter γ_3. They have shown, in particular, that the correction, to the values of energy levels at the points K of the Brillouin zone with small quantum numbers ($n = 1, 2$) is of the order of $(\gamma_3/\gamma_0)^2\gamma_1 \sim 10^{-3}$ eV and may be regarded as small. At large quantum numbers n and arbitrary k_z, the effect of trigonal warping on the position of the Landau levels can no longer be taken into account by expanding in γ_3. For a correct consideration of the effect of the parameter γ_3 upon the energy levels in the magnetic field, it is necessary to use a 36×36 Hamiltonian matrix. Gupta and Wallace (1972) have modified the expansion in γ_3 used by Schroeder et al. (1970) to achieve a better convergence of the series in the vicinity of the points K in high magnetic fields. However, they have drawn the conclusion that at realistic values of the trigonal warping parameter γ_3 no form of expansion in γ_3 is capable of giving a correct description of the Landau levels in graphite. The failure of perturbative methods to describe Landau levels is associated with the difficulty of interpreting within this theory the additional energy extrema located beyond the HKH edges of the Brillouin zone, which arise from trigonal warping. At sufficiently large values of γ_3, a transformation of the energy spectrum near edges of the Brillouin zone takes place. As a result of this transformation, there appear saddle points along three symmetrical directions in the plane perpendicular to an edge of the zone (the $K\Gamma$ direction being one of them). This rearrangement of the spectrum due to the interaction described by γ_3 has been found independently by several authors who used the methods of calculation without expansion in γ_3.

Dresselhaus (1974) has directly obtained the energy levels in the magnetic field by the semiclassical Bohr–Sommerfeld quantization for electron-orbit areas (see (57)). Later, Nakao (1976) and Nagayoshi et al. (1976a) have established the presence of saddle points in the spectrum by calculating the band structure by a new method combining OPW and TB approaches (see sect. 4.2). The calculation indicates that away from the edge of the zone the double degeneracy of the ε_3-band is removed and ε_3 splits into two subbands. The dependence of the energy of these subbands on the wave vector k_ρ is qualitatively the same, however, only for $\gamma_3 = 0$; otherwise, there will be a significant difference between these dependences for the wave vector directed along $K\Gamma$ and KM. The effect of trigonal splitting on the qualitative character of the electron dispersion relation is illustrated in fig. 18. It can be seen that

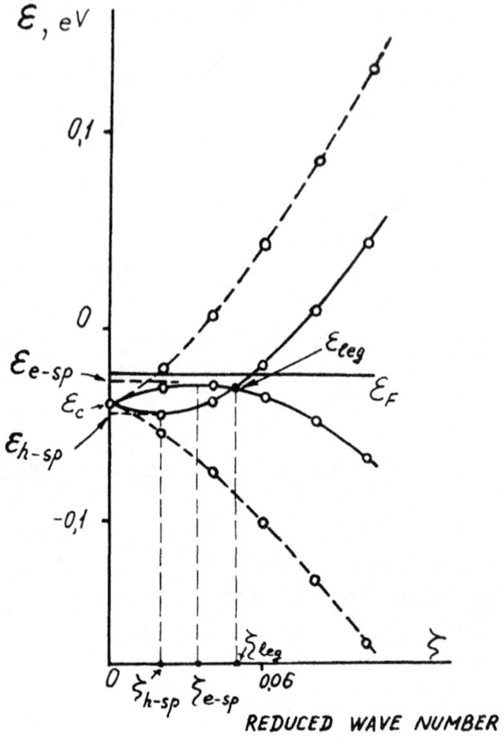

Fig. 18. Calculated energy band structure around the K-point. Solid and dashed lines correspond to $\alpha = 0$ (K-Γ direction) and $\alpha = \frac{1}{3}\pi$ (K-M direction), respectively. The value of the Fermi energy ε_F indicated here is calculated from the SWMC-parameters evaluated by Nagayoshi et al. (1976a) (see also table 5).

along $K\Gamma$-directions at reduced wave vector values $\zeta = \zeta_{e-sp}$ and $\zeta = \zeta_{h-sp}$ there appear off-axis extrema, to which there correspond saddle points on the surface of constant energy. As has been shown by Dresselhaus (1974), the electron and hole leg orbits in the magnetic field are associated with the intersection of two parabolas at $\zeta = \zeta_{leg}$. Admittedly, the spin–orbit splitting removes the ε_3-band degeneracy at $\zeta = 0$ and $\zeta = \zeta_{leg}$. But at realistic values of $\gamma_3 \sim 0.3$ eV, the difference in energy between the off-axis extrema and $\varepsilon = \varepsilon_c$ is two orders of magnitude greater than the value of the splitting ($\sim 2 \times 10^{-4}$ eV), so that in considering leg orbits the spin–orbit splitting may be ignored. The position of the saddle points in the spectrum, as well as the values of ε_c and ε_{leg}, depend on the wave vector k_z. These dependences, shown in fig. 19, can be divided into different energy regions, in which some or other types of electron and hole orbits manifest themselves. Figure 19 clearly shows that the effect of trigonal warping diminishes with increasing k_z and practically

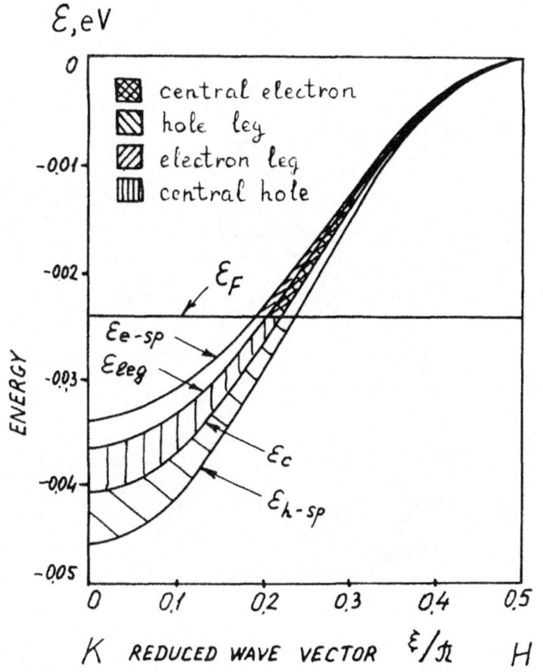

Fig. 19. Plot of the k_z-dependence of the critical points ε_c, ε_{leg}, ε_{e-sp}, ε_{h-sp}. The cross hatching gives the various regions of energy where the special orbits can occur. The majority-carrier Landau levels occupy the region above and below the cross-hatched region. (Dresselhaus 1974.)

vanishes as the H points are approached. The application of a magnetic field gives rise to three types of Landau levels. At $\varepsilon > \varepsilon_{e-sp}$, there occur the ordinary Landau levels for majority electrons, which are characteristic of graphite when trigonal warping is neglected (see, e.g., fig. 17). These levels are described by eq. (61) for $\gamma_3 = 0$ and also with the help of expansions in γ_3, as in Schroeder et al. (1970) and Gupta and Wallace (1972). The other two types of Landau levels occur in the following energy regions: $\varepsilon_{e-sp} > \varepsilon > \varepsilon_{leg}$, leg-electron Landau levels and $\varepsilon_{e-sp} > \varepsilon > \varepsilon_c$, central-electron Landau levels. These types of levels are unobtainable by perturbation theory expansions in γ_3. Three analogous types of levels arise for holes:

$\varepsilon_{h-sp} > \varepsilon$,	ordinary hole Landau levels;
$\varepsilon_{leg} > \varepsilon > \varepsilon_{h-sp}$,	leg-hole Landau levels; and
$\varepsilon_c > \varepsilon > \varepsilon_{h-sp}$,	central-hole Landau levels.

In high magnetic fields ($H \gtrsim 20\,\text{kOe}$), the Landau levels of the majority electrons and holes are dominant. If one disregards the ranges of the wave vector k_z at which the Fermi surfaces for electrons and holes are in contact

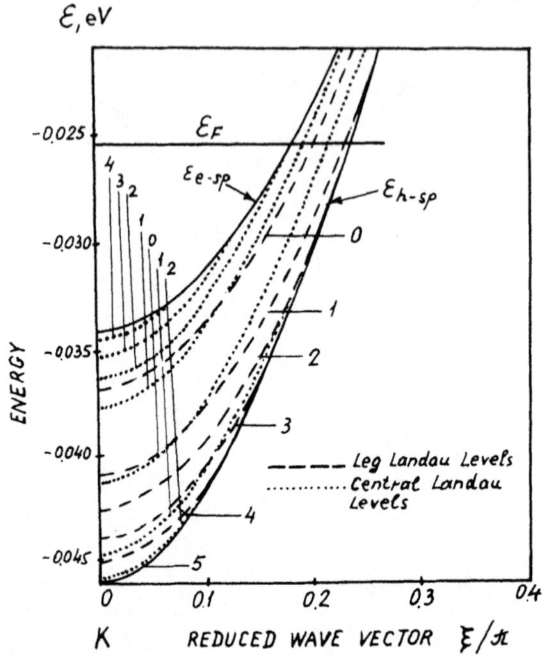

Fig. 20. The k_z-dependence of some of the leg and central Landau levels for $H = 3\,\mathrm{kOe}$, $\boldsymbol{H}\|c$-axis (Dresselhaus 1974).

with each other (see fig. 12), the electron and hole orbits in such fields are reasonably well described by the expressions ensuing from the SWMC-model for $\gamma_3 = 0$ or obtainable from expansions in γ_3. For $H < 20\,\mathrm{kOe}$, the leg- and central-type Landau levels play a significant part. This is illustrated in fig. 20, which shows the k_z-dependence of the various Landau levels for $H = 3\,\mathrm{kOe}$ (in the calculation, the SWMC-model parameters from column D of table 5 have been used, see below). This figure depicts six leg Landau levels and seven central Landau levels. One can see that the Landau levels discontinuously terminate at those values of ξ where they become equal in energy to $\varepsilon_{\mathrm{e-sp}}$ or $\varepsilon_{\mathrm{h-sp}}$. As the magnetic field increases, the distance between the levels becomes greater and an ever smaller number of levels occur between the cut-off energies $\varepsilon_{\mathrm{e-sp}}$ and $\varepsilon_{\mathrm{h-sp}}$. In a constant magnetic field, the spacing between the neighbouring leg (or central). Landau levels decreases with increasing quantum number. Using the Bohr–Sommerfeld quantization conditions, Dresselhaus (1974) has obtained an expression for the energy of the central electron $(+)$ and hole $(-)$ Landau levels with number n:

$$\varepsilon_n^{\pm(\mathrm{c})} \simeq 2\gamma_2 \cos^2 \xi \pm \frac{4\gamma_3}{\lambda} \cos \xi \left[\frac{eH}{ch}(n + \delta_\mathrm{c})\right]^{1/2}, \tag{71}$$

where $\lambda = \frac{2}{3}\sqrt{6}/a_0$ and δ_c is the quantization constant. The analogous expression for the leg electron (hole) Landau levels with the number n will be of the form

$$\varepsilon_n^{\pm(\mathrm{leg})} \simeq \varepsilon_{\mathrm{leg}}(\xi) \pm \alpha(\xi)\left[\frac{2eH}{c\hbar}(n + \delta_{\mathrm{leg}})\right]^{1/2}, \tag{72}$$

where δ_{leg} is the quantization constant, the dependence $\varepsilon_{\mathrm{leg}}(\xi)$ is shown in fig. 19, and $\alpha(\xi)$ is a slowly varying function of ξ. The most significant feature of the energy dependence for the central and leg Landau levels is that it has the form of $\varepsilon \sim A + BH^{1/2}$. Another feature, due to the existence of so-called special Landau levels, is the cut-off of the Landau levels for the majority electrons and holes at energies $\varepsilon = \varepsilon_{\mathrm{e-sp}}$ and $\varepsilon = \varepsilon_{\mathrm{h-sp}}$. This cut-off is abrupt and the levels reach $\varepsilon_{\mathrm{e-sp}}$ and $\varepsilon_{\mathrm{h-sp}}$ with a finite slope. The special leg and central levels can be supported between these cut-off energies. As the magnetic field decreases the number of special levels that may exist between the cut-off energies increases. However, the levels with the lowest quantum numbers ($n_c = 0$, $n_{\mathrm{leg}} = 0$) are independent of the magnetic field and remain unchanged up to the highest magnetic fields attainable. The cut-off of Landau levels in graphite is illustrated in fig. 21, which shows how several ordinary and special Landau

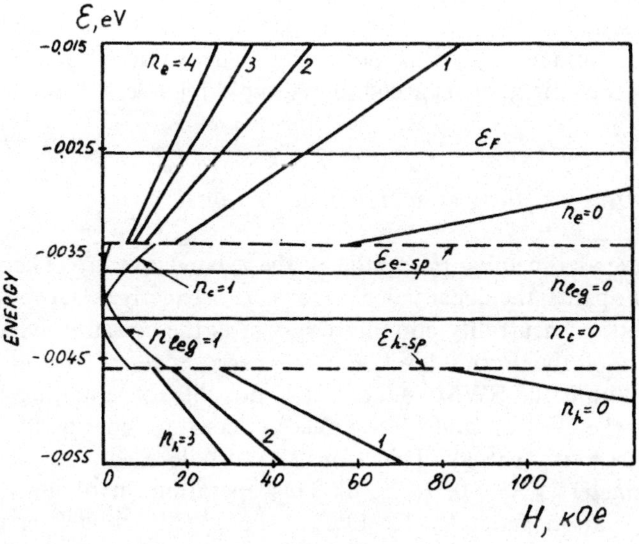

Fig. 21. Magnetic field dependence of the Landau levels at the K-point. The special levels exhibit a parabolic dependence on H and can be supported only for $\varepsilon_{\mathrm{h-sp}} < \varepsilon < \varepsilon_{\mathrm{e-sp}}$. The more usual linear dependence on H holds for the majority-carrier Landau levels in the energy region where they can be supported: $\varepsilon > \varepsilon_{\mathrm{e-sp}}$ and $\varepsilon < \varepsilon_{\mathrm{h-sp}}$. All levels are terminated discontinuously when they cross $\varepsilon_{\mathrm{e-sp}}$ and $\varepsilon_{\mathrm{h-sp}}$ (Dresselhaus 1974).

levels depend on magnetic field. The magnetic fields at which the Landau levels with numbers $n_c = 1$, $n_{leg} = 1$ are cut-off at the K-points constitute 2.2 kOe for the electron legs, 11.2 kOe for the hole legs, 13.9 kOe, for the central electrons, and 7.5 kOe for the central holes. The discontinuities of the Landau levels at ε_{e-sp} and ε_{h-sp} become somewhat modified by magnetic breakdown phenomena. Due to the magnetic breakdown between orbits including the central cross section, the majority electron (hole) levels extend into the energy region $\varepsilon_{e-sp} \gtrsim \varepsilon \gtrsim \varepsilon_{h-sp}$. Away from the saddle-point energies, the breakdown orbits broaden and become less intense. Similarly, the leg and central Landau levels become broader and less intense as their energy approaches the saddle-point limits. The special Landau levels could be directly observed in graphite doped with acceptor-type impurities, in which the Fermi level ε_F is displaced downwards into the energy interval $\varepsilon_{e-sp} < \varepsilon_F < \varepsilon_{h-sp}$. In this case, the special Landau levels may cross the Fermi level. So far as we know, however, oscillatory effects in doped graphite have not yet been considered from such a viewpoint. The question of what is the character of the features of the density of states when the leg and central Landau levels vanish in passing through the cut-off energies ε_{e-sp} and ε_{h-sp} in the magnetic field also remains open. The experimental evidence for the leg levels has been obtained by Platts et al. (1977) and Doezema et al. (1979) who studied the magneto-optical transitions at the points K at energies between 3 and 45 meV (see sect. 7.4). Since the specific character of Landau levels is determined by the trigonal warping parameter, the data of Doezema et al. (1979) have allowed a very accurate determination of this parameter. The value of γ_3 thus obtained, (0.315 ± 0.015) eV, is now regarded as the most reliable.

4.6. Location of the Fermi level in graphite

Graphite owes its metallic properties to the π-band overlap. Therefore, in the absence of dopants, the density of electrons, n_e, is exactly equal to that of holes, n_h. The electrical neutrality condition $n_e = n_h$ is the equation which describes the position of the Fermi level on the energy scale of the graphite band spectrum. Within the SWMC-model, it is possible, by equating the volumes confined by the electron and hole surfaces of constant energy, to calculate the values for the Fermi energy of electrons ε_F^e and holes ε_F^h at some known values of the parameters γ_i $(i = 0, 1, \ldots, 6)$. This operation, involving the complete dispersion relation given by eq. (35) of the SWMC-model, is rather complicated and can only be performed using a numerical method. The analytical dependence of the Fermi energy ε_F on the parameters of the dispersion relation can be obtained only for the simpler models.

In particular, using the parabolic approximation within the four-parameter model, one can readily show that (for $\Delta = 0$) the electron and hole densities are given by

$$n_e = n_h = \frac{64\gamma_1|\gamma_2|}{9\sqrt{3}\,\pi^2\gamma_0^2 a_0^2 c_0}. \tag{73}$$

In this case, the position of the Fermi level relative to the extremum of the ε_3-band at the point H is determined by the value of the Fermi energy

$$\varepsilon_F = -\tfrac{4}{3}|\gamma_2| = -\tfrac{2}{3}\varepsilon_\pi, \tag{74}$$

where $\varepsilon_\pi = 2|\gamma_2|$ is the band-overlap energy in graphite. Arkhipov et al. (1963) have shown that within the model being considered the position of the Fermi level is independent of temperature. A weak dependence of ε_F on T is observed only in the case when the parameter $\gamma_4 \neq 0$. This weak dependence is associated with the temperature dependence of the electron and hole densities of states and is described, according to Kechin (1969), by the formula

$$\varepsilon_F = -\tfrac{4}{3}|\gamma_2| + \frac{8\gamma_4 kT}{\gamma_0}\pi\int_0^{\pi/2}\log\left|2\cosh\frac{\varepsilon_3^0 - \varepsilon_F}{kT}\right|\cos^2\xi\,d\xi. \tag{75}$$

The variation of ε_F with temperature found from eq. (75) is shown by the dashed curve in fig. 22. This dependence is compared with a similar temperature dependence obtained by Ono and Sugihara (1966) within the four-

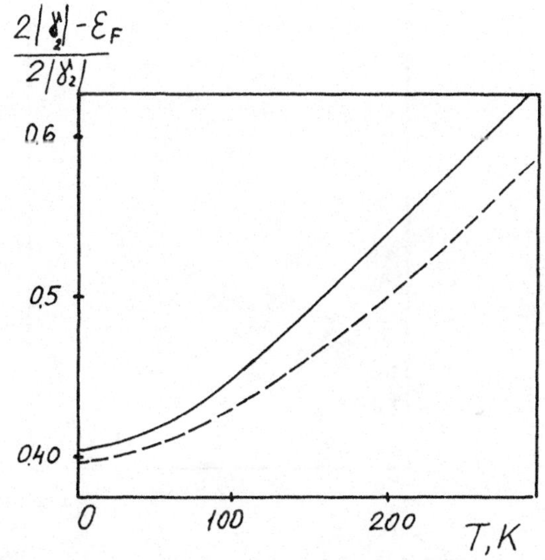

Fig. 22. Temperature dependence of the reduced Fermi-level value in graphite. The dashed curve represents a calculation by Kechin (1969), the solid curve, represent a calculation by Ono and Sugihara (1966). Both calculations having been done using the four-parameter model for the spectrum.

parameter model for $\gamma_4 \neq 0$ without the parabolic approximation (the solid line in fig. 22). For clarity the quantity $(2|\gamma_2| - \varepsilon_F)/2|\gamma_2|$ is plotted as ordinate in fig. 22. This quantity represents the relative change in the Fermi energy measured from the edge of the conduction band (in calculating the volumes confined by the surfaces of constant energy, the minority carriers were ignored).

One may expect that in a quantizing magnetic field the Fermi energy will also vary because the electron density of states and the hole density of states differently depend on the field. The associated change in the quantity $(2|\gamma_2| - \varepsilon_F)/2|\gamma_2|$ has been calculated by Sugihara and Ono (1966). The results of their calculations on pure graphite and also on graphite doped with donor and acceptor impurities are presented in fig. 23. The kink points on the curves correspond to the magnetic fields at which the Landau levels of electrons and holes cross the Fermi level. We note that the calculated values of these fields disagree with the position of peaks on the experimental oscillation curves for graphite. This discrepancy is the most marked for small quantum numbers; for instance, the electron level with $n = 1$ crosses the Fermi level in the field $H \simeq 70$ kOe, which is much higher than the calculated value $\simeq 38$ kOe. The

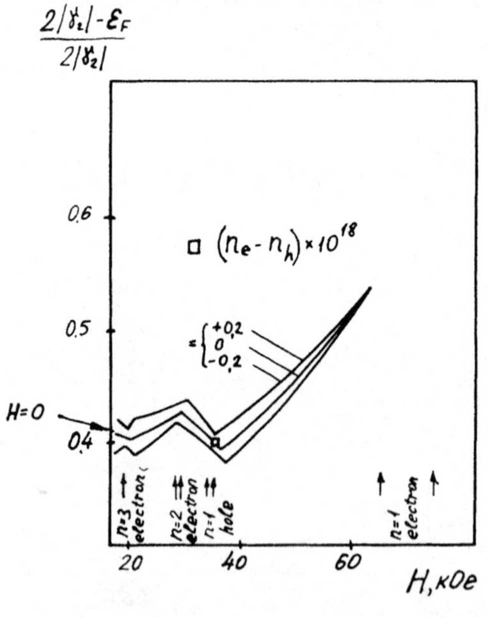

Fig. 23. Magnetic field dependence of the Fermi level in graphite at liquid-helium temperatures. The results of a calculation by Sugihara and Ono (1966) for natural graphite and for graphite doped with donor or acceptor impurities with a concentration of 0.2×10^{18} cm^{-3}. Vertical arrows indicate the experimental magnetic-field values corresponding to the crossing of the Landau levels by the Fermi level. (According to Woollam (1970a).)

discrepancy between the calculated dependence of ε_F on H and the experimental data is due to the neglect of the spin splitting of the Landau levels in the calculation. Sugihara and Ono (1966) have predicted a decrease of the band overlap in graphite resulting from the shift of their boundaries in the extreme quantum limit of magnetic fields. Such a transformation of the energy spectrum of three-dimensional graphite in high fields should give rise to the electron transition semimetal–semiconductor induced by magnetic field. This transition has been observed in bismuth–antimony alloys by Brandt and Svistova (1970).

A more rigorous calculation of the behaviour of the Fermi level in graphite in high quantizing fields and in the extreme quantum limit of fields at $H \| C$ has been made by Kapustin and Meilikhov (1975) who proceeded from eq. (61) and the electrical neutrality condition. The Fermi energy ε_F was determined by solving eq. (61) numerically in fields up to 2.3×10^6 Oe, using the parameters γ_i of the SWMC-model for the energy spectrum given by Gupta and Wallace (1972): $\gamma_0 = 2.85$; $\gamma_1 = 0.31$; $\gamma_2 = \gamma_5 = -0.0185$; $\gamma_4 = 0.19$ (all in eV). The value of the parameter $\gamma_6 \equiv \Delta$ was varied in the calculation from -0.025 to 0.025 eV. These authors have shown that the results of calculations of ε_F and n are highly sensitive to the value of the Δ chosen. It has been found that in the extreme quantum limit of the magnetic field (i.e., $H \gtrsim 70$ kOe for graphite if $H \| c$) the position of the Landau level with $n = 0$ for majority holes does not depend, for $\Delta > 0$, upon the magnetic field and coincides with the highest point of the valence band ($\varepsilon = \varepsilon_3$), while the one for majority electrons is shifted upwards on the energy scale for all values of k_z, save for $k_z = \pi/c_0$ (the point H of the Brillouin zone) where the energy remains constant and equals Δ.

With such a field dependence of the zero Landau levels in a rather high magnetic field $H = H_{M \to S}$, the π-band overlap in graphite should be removed and an energy gap should appear between the conduction band and the valence band (the metal–semiconductor transition in a magnetic field). At $\Delta = 0.025$ eV, the field necessary for the transition is $H_{M \to S} \simeq 800$ kOe.

For $\Delta < 0$, the behaviour of the zero Landau levels is entirely different from that for $\Delta > 0$. It can be shown that in the former case there is a singularity $k_z = k_z^*$ on the k_z-axis. At this singularity, eq. (61) has a double root $\varepsilon = \varepsilon_3$ and the zero Landau levels of majority electrons and holes are contiguous at all values of the magnetic field. The position of this singularity is given by: $\Gamma^* = \Gamma(k_z^*) = 2 \cos \frac{1}{2} k_z^* c_0 \simeq -\frac{1}{2} \Delta \gamma_0 / \gamma_1 \gamma_4$. Since the parameters γ_0, γ_1, γ_4 are positive and since $0 \leqslant 2 \cos \frac{1}{2} k_z c_0 \leqslant 2$, the degeneracy point $k_z = k_z^*$ mentioned above exists only for the values of Δ that satisfy the inequality -0.165 eV $\leqslant \Delta \leqslant 0$. For $k_z < k_z^*$ and $\Delta < 0$, the zero Landau level for holes is immobile, just as it was for $\Delta > 0$, whereas the electron Landau level moves. By contrast, for $k_z > k_z^*$ it is the zero hole level that moves, while the zero electron level is immobile. If one neglects the spin splitting of the levels, this should lead to the appearance of a gapless state induced by the magnetic field at $H = H_{GS}$. For $\Delta = -0.025$ eV, the required H_{GS} field should be $\simeq 10^6$ Oe. For $\Delta < -0.165$ eV the levels stop being in contact with each other, but now it is the zero electron level that does not move, while the zero hole one shifts in the magnetic field, which as for $\Delta > 0$, may result in the metal–semiconductor transition.

In calculating the dependence of the Fermi energy ε_F on the magnetic field, Kapustin and Meilikhov (1975) put the g-factor to be equal to 2, because the exact value for spin splitting in graphite is unknown, while g-factor values found from the theory (McClure and Yafet 1961) and experiments (Wagoner 1960) concerned with the spin resonance of conduction electrons are only slightly different from two. And the high spin splitting of the majority electron and hole levels with $n = 1$, observed by Woollam (1970a), may be explained by a rapid change in the Fermi energy in this range of magnetic fields. As a result, the periodicity in the succession of oscillation peaks changes so radically that the above peaks may erroneously be ascribed to the high spin splitting.

The dependences of ε_F on H for $\Delta = 0.025$ eV in the range of quantizing magnetic fields $H \lesssim 70$–80 kOe, obtained by Kapustin and Meilikhov (1975), are illustrated in fig. 24a. Similar dependences in fields up to 2.3×10^6 Oe at various values of Δ are presented in fig. 24b.

The field dependences of the Fermi energy make it possible to calculate the carrier-density in the magnetic field. The results of such a calculation for various Δ are shown in fig. 25. The calculation indicates that the density of majority carriers increases almost linearly with magnetic field at $70\ \text{kOe} \leqslant H \leqslant 400$ kOe and in this field range it depends but weakly on the temperature and the choice of the value of Δ. In a field $H \simeq 400$ kOe, this

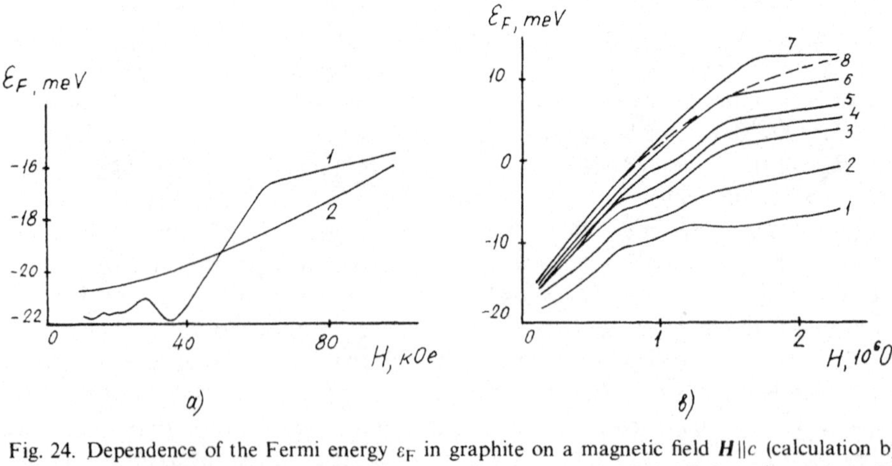

Fig. 24. Dependence of the Fermi energy ε_F in graphite on a magnetic field $\boldsymbol{H} \| c$ (calculation by Kapustin and Meilikhov 1975): (a) $\Delta = 25$ meV, T (in K): curve, 1, 4.2; 2, 77. (b) Δ (in meV); 1 − 25; 2 − 15; 3 − 5; 4 0; 5 + 5; 6 + 15, 7, 8 + 25. For curves 1 − 7, $T = 4.2$ K while for curve 8, $T = 77$ K.

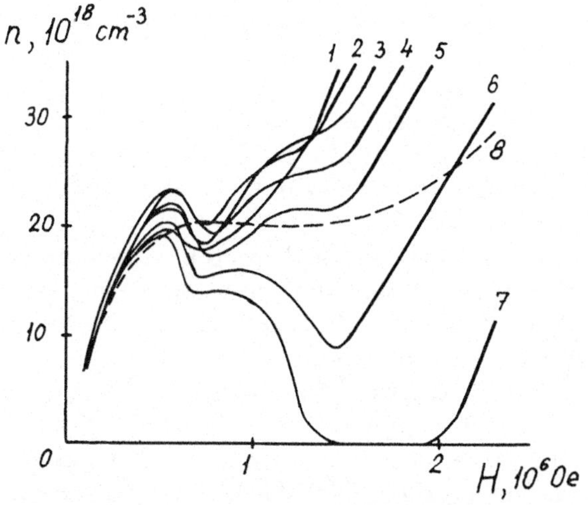

Fig. 25. Magnetic field dependence of carrier concentration n in graphite at $T = 4.2$ K (solid curves 1–7) and $T = 77$ K (dashed curve 8) calculated by Kapustin and Meilikhov (1975) for the following values of $\gamma_6 \equiv \Delta$(meV): curve 1 −25, 2 −15, 3 −5, 4 0, 5 +5, 6 +15, 7,8 +25.

density is 8 times higher than for $H = 0$. Such an increase can be attributed to the increased density of states at the Landau levels in the magnetic field and the higher degree of state degeneracy by $\sim eH/c\hbar$.

In fields above 500 kOe the situation is entirely different. Taking account of the spin splitting shifts the point of the metal–semiconductor transition towards higher fields (in this case, for $\Delta = 0.025$ eV, $H_{M\to S} = 1.25 \times 10^6$ Oe). But in fields $H > H_{M\to S}$, the reverse transition (semiconductor metal) turns out to be possible at a certain value of $H = H_{S\to M}$ associated with the intersection of the hole level 0^+ with the electron level 0^- at the points H of the Brillouin zone (at $k_z = \pm\pi/c_0$), where the position of the zero Landau levels is independent of the magnetic field if the spin splitting is disregarded. The field $H_{S\to M}$ of such a transition is given by the relation $\frac{1}{2}(g_e + g_h)\mu_B H_{S\to M} = \Delta$, where g_e, g_h are the g-factors for the majority electrons and holes, μ_B is the Bohr magneton. At $\Delta = 0.025$ eV, $H_{S\to M} = 2.16 \times 10^6$ Oe. The magnetically induced semiconducting state exists in graphite only for $\Delta > 0.017$ eV in the field range $H_{M\to S} \leqslant H \leqslant H_{S\to M}$, the maximum value for the forbidden band at $\Delta = 0.025$ eV not exceeding $\sim 6 \times 10^{-3}$ eV. Accordingly, in the semiconducting range, the carrier density n strongly depends on the temperature and for $T = 77$ K (and $\Delta = 0.025$ eV), the dependence of n upon the field exhibits a relatively weak dip, as shown in fig. 25. Taking account of spin splitting for $\Delta < 0$ results in the appearance of the overlap between the conduction and valence bands at the point $k_z = k_z^*$, which increases with magnetic field, and the gapless state does not arise.

In such a case, in magnetic fields up to $\sim 1.5 \times 10^6$ Oe, the Fermi energy follows the displacement of the zero Landau levels for majority electrons at the points K of the Brillouin zone (0^- and 0^+ for $H \lesssim 800$ kOe and 0^- for 800 kOe $\lesssim H \lesssim$ 1500 kOe), because their motion is determined not only by the spin-splitting magnitude but also by the displacement of the unsplit zero level, while the zero level of majority holes is immobile if the spin splitting is ignored.

After the electron levels 0^- and hole levels 0^+ cross the Fermi level at the point K, the displacement of the latter will be determined by the behaviour of the former levels near the point H. Since near this point one of the unsplit levels is immobile, while the other moves much slower than at the point K, the Fermi energy varies slower with the magnetic field than it does at $\Delta > 0$ (compare curves 1 and 2 with 6 and 7 in fig. 25). The dips of the field dependence of the carrier density and the inflexion point on the dependence of ε_F which takes place for $H \sim$ 700–800 kOe are due to the crossing of the Fermi level by the electron level 0^- and the hole level 0^+.

As has been mentioned above in sect. 4.5, the trigonal warping affects least the Landau levels with small quantum numbers ($n = 0, 1$). Taking account of the effect of trigonal warping on the above field dependences of ε_F and n [e.g., by the method developed by Gupta and Wallace (1972)] does not alter the shape of the curves shown in figs. 24 and 25, but leads to an increased field for the metal–semiconductor transition, $H_{M\to S}$. The semiconducting state will then exist at the higher values of Δ and in a narrower range of magnetic fields. At the same time, taking account of the parameter γ_3 does not affect the magnitude of the reverse transition field $H_{S\to M}$, because there is no trigonal warping at the H points.

We note that the variation of density n and Fermi energy ε_F in a high magnetic field, calculated by Kapustin and Meilikhov (1975), suggests that these parameters should be highly sensitive to the values of Δ and g-factors at the points K and H of the Brillouin zone. However, the values of Δ and g-factors at these points have not yet been determined conclusively. Therefore, the study of the magnetically induced transitions from metal to semiconductor and back in graphite may provide a tool for determining the parameters Δ, g_e, and g_h.

Recently, there have been published, however, the experimental data that throw some doubt upon the possibility of reducing continuously the band overlap in graphite to zero under the influence of a magnetic field, thereby inducing a metal–dielectric transition. The thing is that in studying the galvanomagnetic effects in graphite in fields $\sim$300 kOe Tanuma et al. (1981a) have observed an abnormal increase in magnetoresistance, interpreted by Yoshioka and Fukuyama (1981) as a transition to the charge density wave (CDW) state. Such a transition is equivalent to the formation of the dielectric phase in a quasi-one-dimensional gas of charge carriers in graphite. The formation of this phase precludes the observation of a metal–dielectric transition in the single-particle spectrum of graphite associated with the displacement of the band boundaries in the magnetic field, which leads to a steady removal of the π-band overlap. The abnormally high magnetoresistance of graphite in strong magnetic fields will be discussed in more detail later on (see sect. 5.4).

4.7. *The parameters of the Slonzewski–Weiss–McClure model*

The values of the parameters γ_i $(i = 0, 1, \ldots, 6)$, which describe the carrier dispersion relation in the SWMC-model, were determined by various workers from the condition of the best fit of the relevant experimental data to the calculated values for the extremal cross sections of the Fermi surface, cyclotron masses, position of the peaks of magneto-optical oscillations, etc. Table 5 presents the values for the parameters γ_i and the Fermi energy ε_F measured from the ε_3^0-extremum at the points H of the Brillouin zone, as found by various authors.

Table 5

Values of γ_i and ε_F (in eV) according to various authors and for various sets, corresponding to the different models used

Parameter	Set					
	A	B	C	D	J	N	M
γ_0	2.85	3.2	3.11	3.124	3.18	2.73	3.16 ±0.05 [a]
γ_1	0.30	0.397	0.392	0.377	0.40	0.32	0.39 ±0.01 [b]
γ_2	−0.02	−0.0202	−0.0201	−0.021	−0.0207	−0.019	−0.20 ±0.002 [c]
γ_3	0	0.29	0.29	0.276	0.30	0.29	0.315 ±0.015 [d]
γ_4	0	0.132	0.124	0.120	0.18	0.15	0.044 ±0.024 [e]
γ_5	0	0.0098	0.0234	−0.003	−0.006	0.021	0.038 ±0.005 [b]
$\gamma_6 \equiv \Delta$	0.006	0.0221	−0.0049	−0.002	0.005	0.017	0.008 ±0.002 [a]
ε_F	−0.026	−0.0223	−0.0235	—	−0.0240	−0.0213	−0.024 ±0.002 [d]

[a] Toy et al. (1977).
[b] Misu et al. (1979).
[c] Soule et al. (1964).
[d] Doezema et al. (1979).
[e] Mendez et al. (1980a,b).

Table 5 gives seven sets of the parameters γ_i, for which we have used the following notation:

Set A: a simplified model developed by Anderson et al. (1967), where the parameters γ_3, γ_4, γ_5 are set equal to zero and the values of the remaining are taken from McClure (1960).

Sets B, and C: the model of Dillon et al. (1977), where γ_0 and γ_3 are fixed and the remaining parameters are found by their varying to obtain the best agreement with the electroneutrality equation $n_e = n_h$. The distinction between the sets B and C is as follows. In calculating the SWMC-model parameters it has been assumed that there is only one group of minority carriers, i.e., minority holes at the points H. However, since the value for the extremal cross section of the minority hole surface of constant energy formed by a plane perpendicular to the c-axis has not yet been determined unambiguously, it was taken to be $S_{m1} = 0.79 \times 10^{-42}$ $g^2\,cm^2/s^2$ when calculating the carrier density in the electroneutrality equation for the set B, and $S_{m2} = 0.36 \times 10^{-42}$ $g^2\,cm^2/s^2$, for the set C.

Set D is the parameter set obtained by Dresselhaus (1974) by fitting to the data of low-frequency magneto-optical measurements.

Set J stands for the model of Johnson and Dresselhaus (1973), in which the parameters γ_i are found from optical data.

Set N is the set of parameters computed by Nagayoshi et al. (1973, 1976a,b) who developed a combined OPW–TB technique to be able to determine the electron spectrum of graphite.

Set M represents the set derived from the most recent data on the SWMC-model parameters obtained by Toy et al. (1977), Misu et al. (1979), Doezema et al. (1979), Mendez et al. (1980a,b).

As can be noticed from this table, the small parameters (γ_4, γ_5, γ_6) diverge most in different determinations, while the remaining parameters have not so different values.

Table 6 shows the values of the cross sections for the Fermi surfaces for the majority electrons (S_e), majority holes (S_h), minority holes (S_m), associated cyclotron masses (m^e, m^h, m^m) for $\boldsymbol{H}\|z$, and carrier densitites $n_e = n_h$. All these values have been computed on the basis of the SWMC-model with the γ_i values given in table 5 (symbols A, B, C, J, N corresponding to the sets from table 5). The Fermi-surface cross sections and carrier densitites are given in units of 10^{-42} $g^2\,cm^2/s^2$ and $10^{18}\,cm^{-3}$, respectively. Similar experimental data from various sources are presented in tables 11 and 12 (see sect. 6.3).

Table 6

Fermi-surface cross sections, cyclotron masses, and carrier densities calculated using the γ_i values from table 5.

Parameter	Set				
	A	B	C	J	N
S_e	7.1	7.09	7.09	7.24	6.58
S_h	4.99	5.02	5.04	5.14	4.81
S_m	0.80	0.76	0.35	0.53	0.10
m_c^e/m_0	0.060	0.057	0.059	0.059	0.061
m_c^h/m_0	0.033	0.043	0.043	0.042	0.047
m_c^m/m_0	0.006	0.006	0.004	0.004	0.003
$n_e = n_h$	2.4	2.2	2.2	2.3	2.0

We should like to draw the reader's attention to the fact that the values of S_e, S_h, m^e, m^h and n_e, n_h in table 6 agree much better with each other than do the values of the original parameters of the SWMC-model from the various sets. This can be ascribed to the relatively low sensitivity of the cross sections and effective masses of the majority carriers, as well as of the total carrier density, to the choice of the small parameters [see, e.g., eqs. (40), (41), (48), (49), and (73)]. However, both the cross sections and the effective masses of the minority carriers vary widely. This is indicative of a considerable uncertainty in the currently accepted interpretation of the nature of minority carriers (the location of the associated extrema in the Brillouin zone, the number of their groups, the origin of their spectrum, etc.).

4.8. The electron energy spectrum of pyrocarbon materials

As has already been mentioned (see sect. 2.4), the temperature at which carbon materials have been produced or subsequently heat-treated is one of the main characteristics determining their structure.

At relatively low temperatures of heat treatment (up to 1300°C), the carbon materials thus produced represent organic polymer compounds. A carbon polymer consists of macroscopic molecules, but no regular structure of graphite is formed at this level of values of heat treatment temperature (HTT). As the HTT increases from 1300 to 2800°C, a graphite-like structure evolves. Klein (1962) has shown that the size of crystallites of the carbon materials produced in this temperature range varies from 100 to 1000 Å, the distance between carbon layers being 3.425 to 3.35 Å. According to Morant (1970), the crystal structure of carbon materials produced at temperatures above 3000°C is identical with that of monocrystalline graphite.

No unified model for the electron spectrum of carbon materials with the graphitization parameter $\gamma < 1$ has been developed as yet, while the existing models fail to account for all abnormalities in their electrophysical properties; specifically, negative magnetoresistance, negative temperature coefficient of electrical resistance, abnormal magnetization, and other peculiarities.

A model for the electron spectrum of extremely disordered (glassy) carbon ($\gamma = 0$) has been developed by Harker (1967). In his model, glassy carbon is treated as a molecular crystal, in which charge carriers move, under the influence of the external field, by the so-called hopping conductivity. According to the Harker model, the resistivity of glassy carbon should exponentially increase as the temperature is lowered and at low temperatures the variable range hopping becomes possible. For this kind of conductivity, the temperature dependence of resistivity expressed by the Mott relation,

$$\rho = \rho_0 \exp\left(\frac{\text{const.}}{T}\right)^{1/4}, \tag{76}$$

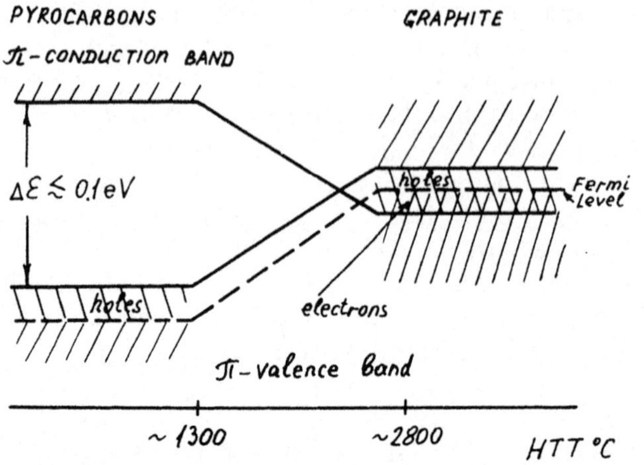

Fig. 26. Reconstruction of the energy spectrum of carbons, from semiconductor to semimetal, with rising HTT (Mrozowski 1950, 1952).

is typical. The results of the experimental studies of pyrolytic carbons performed by Adkins et al. (1970), Carmona et al. (1974), Hishiyama et al. (1978) indicate that the Harker model describes reasonably well the electron spectrum of these materials at an HTT < 1300°C.

In order to explain the electronic properties of carbons heat-treated at temperatures ranging from 1300 to 2800°C (turbostratum and pyrolytic carbons) several models have been advanced. Mrozowski (1950, 1952) made the assumption that in these materials there exists an energy gap between the conduction and valence bands. In conformity with this assumption, the change in pyrocarbon structure that is brought about by raising the HTT gives rise to band overlapping and the semimetallic phase (fig. 26). This is consistent with the data of Carmona et al. (1974) who observed the nonmetal-metal transition in pyrocarbons as the HTT was raised above 1300°C. The authors ascribe the above transition to the Anderson-type localization of charge carriers which arises as a certain degree of disorder in the system is reached. The results of an EPR study by Arnold and Mrozowski (1968) and the results obtained by Tojoda and Mrozowski (1969) and Mrozowski (1971) in their investigation of the Hall effect in disordered carbon materials suggest that the band overlap already arises in an early stage of graphitization at the HTT ~ 1500°C. Lutkov et al. (1976) have conjectured that carbon materials with HTT > 1300°C apparently represent a combination of ordered and glassy carbons, the latter being found at the marginal zones of the ordered crystal lattice. In the ordered regions, resistivity increases linearly with temperature, whereas that of glassy carbon exponentially decreases with rising temperature. The amount of change of the resistivity of synthetic graphite with temperature depends on the content of glassy carbon in it. From this viewpoint, it is precisely a high content of glassy carbon in pyrocarbons that is responsible for the negative temperature coefficient of resistivity. If the content of glassy carbon is insignificant or zero, the temperature coefficient of the resistivity of such carbon materials will be zero or positive. It should be noted, however, that this hypothesis has not yet been confirmed experimentally on the basis of a quantitative estimation of glassy carbon content in carbon materials.

A simple two-band model has been proposed by Klein (1964) to describe the electron spectrum of ordered carbon materials on the assumption that the electronic structure of highly graphitized carbons is described by two overlapping parabolic π-bands. The magnitude of the band overlap, ε_π, and the effective masses of electrons (m^e) and holes (m^h) are the fitting parameters of the model. A downward shift of the Fermi level due to the reduced number of electrons, partially precipitated in traps, is additionally assumed in the model to describe the spectrum of the disordered carbons. The Klein model has gained a wide acceptance and, in particular, has successfully been used to describe the properties of pyrolytic graphite, heat-treated at high temperatures, whose resistivity has a positive temperature coefficient. However, in attempting to describe the electron spectrum of relatively disordered carbon materials obtained at comparatively lower HTT and characterized by a weakly negative temperature coefficient of resistivity, the Klein model no longer agrees with experiment. Saha et al. (1970) have succeeded in describing the electron spectrum of those materials by taking into account the effect of the finite size of crystallites and making the a priori assumption that the distance between crystallites is greater than the crystallite size in the basal plane.

Finally, it should be mentioned that Bobrovnikov and Kotosonov (1978) have shown, in an EPR study, that in materials with an interplanar distance of 3.42 Å one may neglect the effect of the weak interaction between the layers on the electron density of states distribution in a layer. Therefore, the model of the electron spectrum of graphite developed by Wallace (1947) can be applied to such materials. This model has been used with some success to account for the electrical and magnetic properties of carbon materials. Yet, it is incapable of explaining the negative magnetoresistance (NM), which is a property common to many carbon materials. Thus there emerges the following picture of the transformation of the energy spectrum in pyrocarbon materials with the increasing HTT.

In the early stages, for HTT lower than 1300°C, no metallic state is formed. In the amorphous phase, the electrons are localized as in a molecular crystal, the conductivity having the character of thermal activation. As the HTT increases, the metallic phase begins to form; however, the electrons remain localized because of a too high degree of disorder (the Anderson-type localization). The metallic conductivity arises for HTT about 1500°C when, as a result of the increased degree of order, the threshold of mobility is lowered and a transition from the nonmetallic to metallic state takes place. In this stage and with a further increase in HTT, it becomes essential to take into account the inhomogeneity of the system under study (a mixture of the crystalline and the amorphous phase) and the intercrystallite scattering of charge carriers. For a high degree of graphitization, but at HTT below 3000°C, when the temperature dependence of the conductivity acquires a distinctly metallic character, the Klein two-band model (Klein 1964) works satisfactorily. Finally, at HTT above 3000°C, the energy spectrum of single-crystal graphite is formed.

We note that, when explaining the NM of carbon materials, one should remember that the mechanism of this phenomenon may be associated with the collective and quantum interference effects manifesting themselves near the electron localization threshold in disordered systems, rather than with the features of the single-particle spectrum of these substances. Models of NM in carbon materials are discussed at greater length in sect. 5.5).

CHAPTER 5

Galvanomagnetic Effects in Graphite and Carbon Materials

5.1. Electrical conductivity of graphite

As is known, the electrical conductivity, σ, of metals with isotropic surfaces of constant energy is given by the Lifshitz formula

$$\sigma = \frac{2}{3}\frac{e^2}{(2\pi\hbar)^3}\sum_i S_{Fi} l_i \tag{77}$$

where S_{Fi} and l_i are the area of the surface of constant energy and the mean free path of charge carriers of the ith group, respectively, the summation being over all the groups of carriers. In a metal with only one group of carriers and a spherical Fermi surface, eq. (77) changes into the familiar formula of Drude–Lorentz

$$\sigma = \frac{e^2 n\tau}{m^*}, \tag{78}$$

where n is the total carrier density, m^* is the effective mass of carriers and τ is the momentum relaxation time expressed in terms of the mean-free path l and the Fermi velocity v_F: $\tau = l/v_F$. For a phenomenological description of the contribution σ_i to the conductivity due to one group of carriers, it is convenient to introduce the notion of carrier mobility μ_i, which represents the coefficient of proportionality between σ_i and the charge density en_i:

$$\sigma_i = en_i\mu_i. \tag{79}$$

In the simplest case of an isotropic metal with one group of carriers, the mobility is expressed through m^* and τ:

$$\mu = e\tau/m^*. \tag{80}$$

In an anisotropic metal, both mobility $\mu_{i\alpha\beta}$ and conductivity $\sigma_{i\alpha\beta}$ are symmetric tensors of second rank and are expressed through the relaxation time tensor $\tau_{i\alpha\beta}$ and the inverse effective mass tensor $(m_i^*)^{-1}_{\alpha\beta}$:

$$\sigma_{i\alpha\beta} = en_i\mu_{i\alpha\beta} = e^2 n_i \tau_{i\alpha\gamma}(m_i^*)^{-1}_{\gamma\beta} \quad (\alpha, \beta, \gamma = 1, 2, 3). \tag{81}$$

In a normal metal the Fermi energy constitutes several eV and the temperature of electron gas degeneracy is tens of thousand degrees Kelvin. For this reason the charge density in normal metals is temperature independent up to the melting point. The temperature dependence of metallic conductivity is then completely determined by the change with temperature in the mobility μ (or mean free path) of carriers. The Fermi energy in semimetals is much (two or three orders of magnitude) lower than that in metals. Because of this, the temperature dependence of semimetallic conductivity is determined by the change of both μ and n with temperature.

Apart from this, common to all semimetals, feature, the conductivity of graphite is characterized by a high anisotropy resulting from the lamellar crystal lattice of graphite. As far back as 1939, Krishnan and Ganguli established that the conductivity of graphite is practically two-dimensional; the conductivity σ_a along carbon layers in monocrystalline graphite exceeding the conductivity σ_c in the perpendicular direction (along the c-axis) by 3 or 4 orders of magnitude. Subsequent investigations have shown that the anisotropy of conductivity $\eta = \sigma_a/\sigma_c$ increases as single crystals become more perfect and decreases with increasing defect number and lattice disorder.

The anisotropy of conductivity in perfect single crystals of graphite, which can be as high as $\sim 10^5$, is much greater than the anisotropy of the surface of constant energy for electrons and holes or the anisotropy of their effective masses. Accordingly, the validity of eqs. (77)–(81), derived from the Boltzmann kinetic equation, becomes questionable when they are used to describe the conductivity σ_c. Indeed, the use of (77) in calculating σ_c for a known size of the Fermi surface of graphite leads to the electron and hole mean free paths becoming of the same order as the interplanar distance c_0 or less. The nature of this conductivity is considered later on. Attention is called to the fact that the high anisotropy of conductivity of monocrystalline graphite seriously complicates the conventional four-probe measurements of conductivity because of the sharply nonuniform distribution of the current in the crystal between the electrodes. In such a situation, it becomes necessary to study the potential distribution over the surface of the crystal and compare it to the solution of the Laplace equation with appropriate boundary conditions in order to determine σ_a or σ_c.

The most accurate measurements of σ_a and σ_c in graphite specimens of various perfection have been made by Primak and Fuchs (1954, 1956b). The values of the conductivities σ_a and σ_c at room temperature and those of their anisotropy $\eta = \sigma_a/\sigma_c$ for natural and artificial graphite are presented in table 7.

The temperature dependence of the conductivity of single crystals of natural graphite has been studied by Primak and Fuchs (1954) and also by Soule (1958a). Similar dependences for pyrolytic graphite and kish-graphite have been reported by Kawamura et al. (1977).

It has been found that in all types of graphite the temperature coefficient of resistivity in the direction of carbon layers, $\partial\rho_a/\partial T$ ($\rho_a = \sigma_a^{-1}$), is positive, as that of a normal metal. In graphites with sufficiently high values of the graphitization parameter $\gamma > 0.5$ [see definition (9)] the resistivity ρ_a at room temperature is of the order of $4 \times 10^{-4} - 4 \times 10^{-5}\,\Omega$cm. As the temperature drops to that of liquid helium, ρ_a becomes 10 to 50 times smaller. The dependence of ρ_a on T in graphites with $\gamma > 0.5$ displays two characteristic segments: $T \sim (20\text{–}80)$K and $T > 100$ K, in which the values of $\partial\rho_a/\partial T$ are significantly different. In each of these temperature ranges ρ_a exhibits a power-type dependence on T, the exponent decreasing rather rapidly for $T \sim 100$ K,

Table 7
Conductivities σ_a and σ_c and their anisotropy in graphite.

Kind of graphite	σ_a (Ω^{-1}cm^{-1})	σ_c (Ω^{-1}cm^{-1})	$\eta = \sigma_a/\sigma_c$	Ref.
Natural, highly perfect, single crystal	10^4	1	10^4	[1]
	1.01×10^4	1	1.01×10^4	[2]
	$(2.5\text{–}7) \times 10^3$	0.25–0.06	$10^4\text{–}10^5$ (anisotropy increased due to high-temperature annealing)	[3]
Perfect pyrographite	1.67×10^4	2.0	0.8×10^4	[4]
Kish-graphite	1.7×10	–	–	[5]
Perfect synthetic graphite	2.26×10^4	5.9	3.8×10^3	[6]
Imperfect crystal of natural graphite	2.63×10^4	100	260	[7]
	2.6×10^4	3.3×10^2	80	[8]
Pyrographite deposited at 2200°C	6.9	1.25	5.5	[9]
Pyrographite deposited at 2500°C	4.15	0.83	5.0	[10]
Pyrographite, heat-treated to 3000°C	20	3.85	5.2	[10]
Hot pressed pyrolytic graphite annealed at 3500°C	22.4	5.9	3.8	[11]

References:
[1] Krishan and Ganguli (1939).
[2] Dutta (1953).
[3] Bhattacharya (1959).
[4] Brown and Watt (1958).
[5] Kawamura et al. (1977).
[6] Spain et al. (1967).
[7] Primak and Fuchs (1954, 1956a).
[8] Wagoner (1960).
[9] Blackman et al. (1961).
[10] Klein (1962).
[11] Spain et al. (1967).

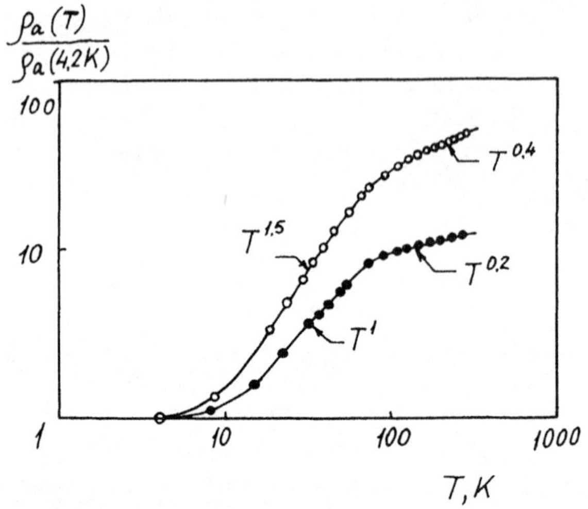

Fig. 27. Ratio of electrical resistivity to that at 4.2 K for: ○ kish-graphite (KG), ● compression-annealed pyrolytic graphite (PG). At 298 K the resistivity is about 6×10^{-5} Ohm cm for both specimens (Kawamura et al. 1977). For KG $\rho_a(298\,K)/\rho_a(4.2\,K) = 45$, and for PG, $\rho_a(298\,K)/\rho_a(4.2\,K) = 12$.

which is close to the electron-gas degeneracy temperature. Typical dependences of ρ_a on T for pyrolytic graphite and kish-graphite are illustrated in fig. 27.

The character of the temperature dependence of conductivity in the direction of the *c*-axis depends to a considerable extent on the perfection of the graphite lattice. According to Krishnan and Ganguli (1939) and Dutta (1953), in perfect single crystals of graphite with $\sigma_c \simeq 1\,\Omega^{-1}\,cm^{-1}$ at 300 K, σ_c decreases as the temperature falls to $T = (35\text{–}40)$ K.

In the temperature range $T < 35$ K, σ_c increases with decreasing temperature. Thus the dependence of σ_c on T is nonmonotonic, having a dip at $T \sim 35\text{–}40$ K.

In imperfect single crystals of graphite, for which typical values of σ_c at room temperature range from 10^2 to $10^3\,\Omega^{-1}\,cm^{-1}$, one can observe a steady increase in σ_c with decreasing temperature. This increase continues until the range of the liquid-hydrogen temperatures is reached, then the curve levels off (Primak and Fuchs 1954, 1956a).

As has already been mentioned the σ_a conductivity of graphite has, in contrast to its σ_c conductivity, a typically metallic character. Its interpretation is based on the concept of the motion of delocalized electrons along the layer. In contrast, the conductivity σ_c along the *c*-axis cannot be attributed to the motion of completely delocalized carriers. A marked increase of σ_c with decreasing size of crystallites in graphite suggests that the charge transport along the c-axis occurs predominantly along the boundaries of grains where the perfect lattice of graphite breaks down. On the other hand, the calculations and the experimental data on the conductivity σ_c in highly perfect single crystals of graphite indicate that in a perfect single crystal of graphite electrons may be quasi-localized in respect to motion along the c-axis.

The localization may be due to the fact that π-electrons, which are responsible for the conductivity σ_c, are separated from one another by rather low potential barriers $\Delta\varepsilon$. At the same time, electron localization in the direction of the c-axis may not be complete, because the assumption about a complete localization would be at variance with the shape of the temperature dependence of σ_c, which does not have a well-pronounced activation character.

Young (1968) has suggested that the localization of electrons and holes in a layer is incomplete, the degree of localization being determined by defects and by the breaking of the long-range order along the c-axis. Other workers (e.g., Spain 1971) ascribe the quasi-localization of π-electrons in layers to the interaction between π-electrons and electrons of atomic cores or to the correlation effects.

A fresh view of the mechanism of conductivity along the c-axis in graphite has been taken by Ono (1976) who considered the so-called Tamm states leading to the weakly coupled wave functions in the plane of a defect. In this context, one may find the probability of electron delocalization with rising temperature and substantiate the weak dependence of σ_c on T.

In conclusion, we should note that no quantitative theory of conductivity along the c-axis in graphite that agrees with experiment has been developed as yet.

5.2. *Galvanomagnetic effects in graphite in weak and semiclassical magnetic fields*

Galvanomagnetic measurements in semiclassical magnetic fields have been widely employed to investigate the electronic properties and energy spectrum of carriers in semimetals. In this way, the investigators have obtained, in particular, the data on densities and mobilities of various groups of carriers, as well as the data on their temperature dependence, which serve as a basis for the analysis of mechanisms of electron and hole relaxation in graphite.

Galvanomagnetic effects have been studied by many workers on various types of graphite. Kawamura et al. (1977) have employed alternating current for the first time to measure galvanomagnetic effects in graphite, which made it possible to eliminate the anomalous thermomagnetic effects due to the phonon drag mechanism discovered by Tsuzuku et al. (1972) and Sugihara et al. (1972).

In studying galvanomagnetic effects it has been found that at low temperatures $T < 50\,\mathrm{K}$ in magnetic fields to several kOe the transverse magnetoresistance measured in the basal plane $\rho_{a\perp}$ ($\boldsymbol{H}\|c$) in specimens of perfect single

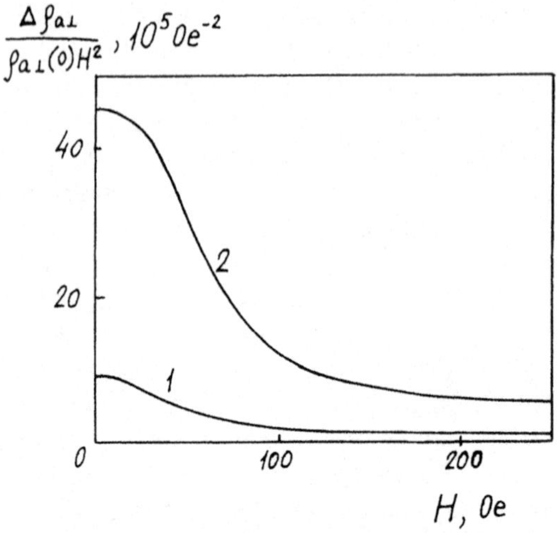

Fig. 28. Magnetic field dependence of the transverse magnetoresistance coefficient in kish-graphite at $T = 4.2$ K. Curves 1 and 2 refer to specimens with $\rho_a(300\,K)/\rho_a(4.2\,K)$ values equal to 15.1 and 18.2, respectively (Brandt et al. 1974).

crystals of natural graphite, pyrolytic graphite and kish-graphite exhibits a quadratic dependence on magnetic field, there being two such parts of the curve: one for $0 \leqslant H \lesssim 30$–50 Oe and the other for $H > 0.1$–0.2 kOe. To each of these segments of the curve for the dependence of $\Delta\rho_{a\perp}/\rho_{a\perp}(0) = [\rho_{a\perp}(H) - \rho_{a\perp}(0)]/\rho_{a\perp}(0)$ upon H there corresponds a horizontal part of the field dependence of the transverse magnetoresistance coefficient $\Delta\rho_{a\perp}/\rho_{a\perp}(0)H^2$. This dependence is illustrated in fig. 28 for two perfect specimens of kish-graphite in the field range $H \simeq 250$ Oe (see Brandt et al. 1974).

The low-field region in the quadratic part of the field dependence of $\rho_{a\perp}$ is usually associated with the contribution from minority carriers, which have small effective masses and high mobilities. A second quadratic part of the dependence is commonly regarded as being due to majority carriers.

Hall-effect measurements in graphite for $\boldsymbol{H} \| c$ indicate that in fields up to several kOe the Hall coefficient R_H strongly depends on magnetic field. Thus, it was Soule (1958a) who already established that at $T = 4.2$ K the Hall coefficient in single crystals of natural graphite is positive in fields $H < 1.2$ kOe, having a distinct peak at $H \simeq 0.3$ kOe. In fields $H \simeq 1.2$ kOe, the sign of R_H reverses. At $T = 77$ K, the Hall coefficient for the same crystal is positive in weak fields, but becomes negative beginning with $H \simeq 4.2$ kOe. Finally, at $T = 298$ K the Hall coefficient of natural graphite remains negative over the whole range of weak fields $H \lesssim 6$ kOe (fig. 29a). According to Spain et al. (1967), in HOPG specimens, R_H is negative at $T = 4.2$ K and positive at

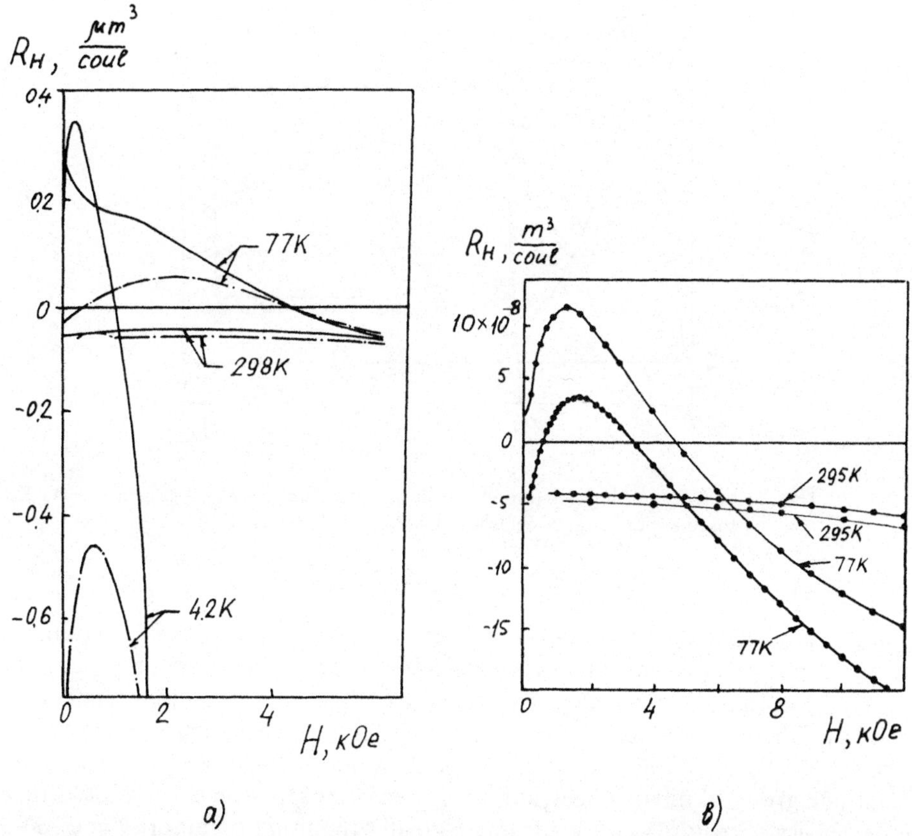

Fig. 29. Magnetic field dependence of the Hall coefficient R_H: (a) purified natural crystal (solid lines) (Soule 1958a) and HOPG (dot-dashed line) (Spain et al. 1967); (b) HOPG (Dillon et al. 1978a). Note the trends in R_H for HOPG at very low field, particularly the reversal of the trend for the natural crystal between 77 and 4.2 K. (Dillon and Spain 1978.)

$T = 77$ K in the magnetic-field range $0.3\,\text{kOe} \lesssim H \lesssim 4.2\,\text{kOe}$, reversing its sign at the end points of the above range. The Hall coefficient of HOPG coincides with that of natural graphite at $T = 298$ K (fig. 29a). Similar field dependences of R_H have been found in HOPG specimens at room and at liquid-nitrogen temperatures also by Dillon et al. (1978a) (fig. 29b). The data obtained by Brandt et al. (1974) and Kuvshinnikov (1980) indicate that the Hall conductivity σ_{xy} in perfect specimens of kish-graphite is positive at 4.2 K in very low fields, reversing its sign for $H \simeq 0.5$ kOe (fig. 30).

In the range of semiclassical magnetic fields the galvanomagnetic effects in graphite have been examined by Johnson (1960) and by McClure and Smith (1961). Their analyses were based on solving the Boltzmann equation in a

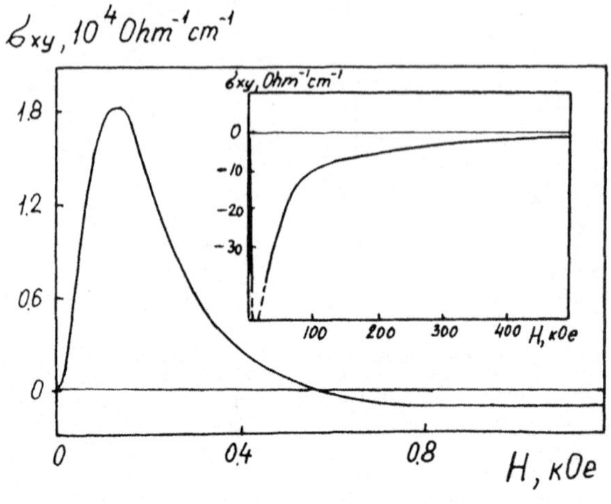

Fig. 30. The conductivity σ_{xy} in kish-graphite (KG) as a function of magnetic field ($T = 4.2$ K). Inset: a wide range of fields up to 500 kOe. (Brandt et al. 1974.)

magnetic field. These authors considered the case of a semimetal containing several groups of carriers and assumed that for each of groups the product of the cyclotron frequency and the relaxation time, $\omega_c\tau$, was independent of energy. Three groups of carriers are commonly assumed when the galvanomagnetic effects in graphite are considered: majority electrons, majority holes, and one group of minority carriers (electrons or holes). It is also assumed that the density of majority carriers is much greater than that of minority ones. The latter in turn have much higher mobilities at low temperatures.

It should be noted that the complicated character of field dependences of the galvanomagnetic effects in graphite may be described correctly only provided that minority carriers are assumed.

Semiclassical expressions for the magnetic conductivity tensor components σ_{xx} and σ_{xy} ($\boldsymbol{H} \| z$) for a metal containing N groups of carriers have the form

$$\sigma_{xx} = \sum_{i=1}^{N} \frac{\sigma_i^{(0)}}{1 + (H/H_i)^2}, \tag{82}$$

$$\sigma_{xy} = \sum_{i=1}^{N} \frac{\alpha_i e_i n_i c\, H/H_i^2}{1 + (H/H_i)^2}, \tag{83}$$

where $\sigma_i = e n_i \mu_i = e n_i c/H_i$ is the zero-field conductivity due to the ith species of carriers, $H/H_i = \omega_{ci}\tau_i = e\tau_i H/m_{ci}c$, n_i, m_{ci}, e_i, and μ_i are the density, cyclotron mass, charge and mobility of carriers of the ith group, respectively; the coefficient $\alpha_i \sim 1$ takes account of the difference between the mobilities contained in the conductivity and in the Hall mobilities (α_i being the Hall factor for the ith type of carriers). One can pass from the experimentally determined quantities ρ_{xx} and R_H to the magnetic conductivity tensor components by using the expressions

$$\sigma_{xx} = \frac{\rho_{xx}^{-1}}{1 + (R_H H/\rho_{xx})^2}, \tag{84}$$

$$\sigma_{xy} = \frac{\rho_{xx}^{-2} R_H H}{1 + (R_H H/\rho_{xx})^2}. \tag{85}$$

Owing to the fact that graphite is a compensated semimetal, its Hall coefficient is relatively small and the dimensionless quantity (Hall-angle tangent) $R_H H/\rho_{xx}$ is smaller than 0.2 in the whole range of semiclassical fields. Equations (84) and (85) will then reduce to

$$\sigma_{xx} \simeq \rho_{xx}^{-1}, \tag{86}$$

$$\sigma_{xy} \simeq \rho_{xx}^{-2} R_H H. \tag{87}$$

In order to give an averaged characteristic of the relaxation time of majority carriers in graphite, one can rather conveniently use the concept of average mobility $\bar{\mu}$, which may be found from the change in magnetoresistance $\Delta\rho_{a\perp}/\rho_{a\perp}(0)$ in the magnetic field. In fields $H > 1$–$2\,\mathrm{kOe}$, the minority carriers, whose effective masses are very small, do not contribute to the conductivity of graphite, which may then be treated in the two-carrier group approximation (majority electrons and majority holes). As has been shown by Noto and Tsuzuku (1975), the transverse magnetoresistance for $\boldsymbol{H} \| c$ will then be expressed in terms of the densities n_e and n_h and the average mobility as follows:

$$\frac{H^2}{\Delta\rho_{a\perp}/\rho_{a\perp}(0)} = \left(\frac{\bar{\mu}}{c}\right)^2 + \left(\frac{n_e - n_h}{n_e + n_h}\right)^2 H^2. \tag{88}$$

Therefore, $\bar{\mu}$ can be found by extrapolating the dependence of $H^2/[\Delta\rho_{a\perp}/\rho_{a\perp}(0)]$ upon H^2 to zero, which is illustrated in fig. 31, where the relevant dependences for kish-graphite obtained by Noto and Tsuzuku (1975) are used. Figure 32 shows the temperature dependences of average mobility $\bar{\mu}$ for pyrolytic graphite and kish-graphite obtained by Kawamura et al. (1977) in the manner described above. In the range from liquid-nitrogen temperatures to room temperatures the average mobility varies according to the relation $\bar{\mu} \sim T^{-1.2}$. A similar dependence for μ in natural graphite in the same temperature range was established earlier by Soule (1958a). In the range below the liquid-nitrogen temperatures, the T-dependence of $\bar{\mu}$ in pyrolytic graphite and kish-graphite becomes better pronounced, obeying the relation $\mu \sim T^{-1.6}$. At $T \leqslant 40\,\mathrm{K}$, this dependence tends to saturation (fig. 32). The semiclassical formulae (82) and (83) expressing the magnetic conductivity tensor components, enable one to find the carrier density n_i and carrier mobility μ_i by fitting to the experimental curves obtained in galvanomagnetic measurements.

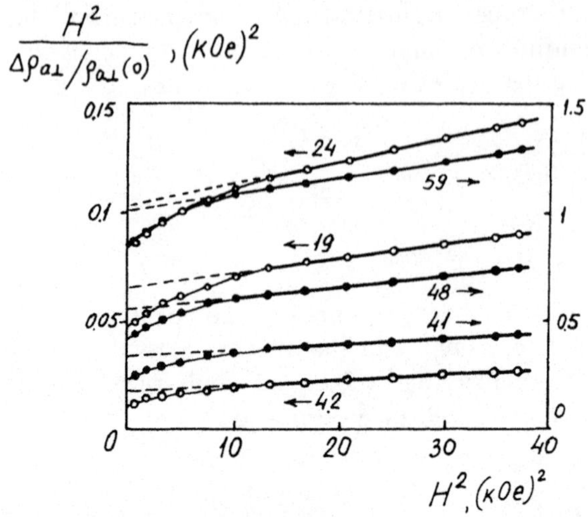

Fig. 31. Inverse magnetoresistance coefficient $H^2/(\Delta\rho_{a\perp}/\rho_{a\perp}(0))$ as a function of squared magnetic field for a perfect specimen of kish-graphite. The numbers at the curves indicate temperature in K (Noto and Tsuzuku 1975).

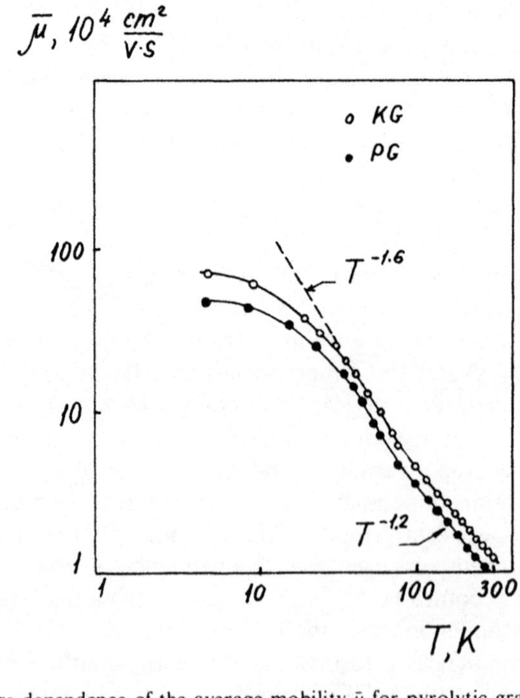

Fig. 32. Temperature dependence of the average mobility $\bar{\mu}$ for pyrolytic graphite (PG) and kish-graphite (KG). (Kawamura et al. 1977.)

Table 8
Carrier densities (in 10^{18} cm^{-3}) and mobilities (in cm^2 V^{-1} s^{-1}).

Type of graphite	T(K)	n_e	μ_e	n_h	μ_h	n'_e	μ'_e	n_h	μ_h	n	$\bar{\mu}$	Ref.*
Natural	300	7.04	0.113	7.04	0.101	5×10^{-4}	3.9	–	–	9.9	0.14	[1,2]
monocrystal-	77	2.24	0.638	2.19	0.733	–	–	3.3×10^{-3}	5.7	3.3	0.68	
line	4.2	2.92	8.39	2.88	10.4	2×10^{-1}	0.07	–	–	5.2	7.0	
Natural	300	7.36	0.109	7.22	0.099	–	–	0.57×10^{-2}	1.5	–	–	[1,2]
monocrystal-	77	2.46	0.486	2.39	0.653	–	–	8.7×10^{-2}	2.0	–	–	
line	4.2	2.14	5.91	2.08	6.45	0.6×10^{-1}	20	–	–	–	–	
Hot pressed	300	–	–	–	–	–	–	–	–	11.3	0.125	[3]
and annealed	77	1.98	0.582	1.98	0.62	2.44×10^{-2}	2.2	–	–	4.2	0.59	
	53	–	–	–	–	–	–	–	–	3.5	1.0	
	4.2	–	–	–	–	–	–	–	–	3.0	4.8	
Perfect crys-	4.2	2.42	4.61	2.32	5.66	2.25×10^{-2}	28.8	–	–	–	–	[4]
tal of kish-		1.72	8.24	1.68	8.64	–	–	–	–	–	–	
graphite[a]		1.48	9.1	1.46	10.2	1.73×10^{-2}	30.0	–	–	–	–	
HOPG (highly	300	–	–	–	–	–	–	–	–	13.5	0.118	[5]
oriented pyrolytic	77	–	–	–	–	–	–	–	–	4.25	0.61	
graphite)	4.2	2.46	0.49	2.39	0.65	0.87×10^{-2}	20.0	–	–	–	–	
HOPG	77	2.46	0.49	2.39	0.65	0.87×10^{-2}	20.0	–	–	–	–	[6]
	4.2	2.14	5.91	2.08	6.45	–	–	–	–	–	–	

[a] Several specimen

*References:

[1] Soule (1958a).
[2] McClure (1958).
[3] Spain et al. (1967).
[4] Brandt et al. (1974)
[5] Dillon et al. (1978a).
[6] Hishiyama et al. (1972).

Determination of n_i and μ_i at various temperatures has been done by many investigators. In processing the results, the presence of only one group of minority carriers is commonly assumed, but since it is unknown which carriers constitute this group, electrons and holes are considered separately. Table 8 gives the values of densities and mobilities of majority electrons (n_e, μ_e), majority holes (n_h, μ_h), minority electrons (n'_e, μ'_e), minority holes (n'_h, μ'_h), and also the total carrier density n and the average mobility $\bar{\mu}$ found from galvanomagnetic measurements in semiclassical fields in graphites of various types. All densities and mobilities in table 8 are expressed in units of $10^{18}\,cm^{-3}$ and $10^5\,cm^2\,V^{-1}\,s^{-1}$, respectively.

Up to now, we have examined the field dependences of galvanomagnetic coefficients for graphite in the semiclassical field range (see figs. 28, 29a,b, 30), assuming only three groups of carriers (majority electrons and holes and minority electrons or holes), but no trigonal warping of the Fermi surface of majority carriers. However, Dillon and Spain (1978) have shown that trigonal warping does affect the character of the field dependences of the diagonal, and especially off-diagonal, components of the magnetic conductivity tensor, $\sigma_{\alpha\beta}(H)$. It has been established that in the field range $H \sim (0.1\text{–}1)c/\mu$, where μ has a value close to that for the mobility of majority carriers, taking account of the parts of the electron and hole orbits with negative curvature with respect to the inner part of the orbit (concave parts of the Fermi surface in fig. 16b) leads to a change in the components of the tensor $\sigma_{\alpha\beta}(H)$ by more than 100%. The galvanomagnetic coefficients were calculated on the basis of the Boltzmann kinetic equation in the constant-relaxation-time approximation, using the formalism developed by McClure (1956).

The results of the calculation indicate that taking account of trigonal warping yields a basically nonmonotonic dependence of $\sigma_{\alpha\beta}$ on H in the range $H \sim (0.1\text{–}1)c/\mu$. The contribution of the negative-curvature parts of the orbit may be separated explicitly by dividing the component of $\sigma_{\alpha\beta}(H)$ calculated with trigonal warping taken into account (superscript t) by that calculated without trigonal warping (superscript c). Figure 33 illustrates the dependences of the Hall conductivity ratio $r = \sigma^{t}_{xy}/\sigma^{c}_{xy}$ for majority electrons and holes on the quantity $\omega_c\tau \equiv \mu H/c$ which have been obtained by Dillon and Spain (1978). The calculations indicate that for $\omega_c\tau \ll 1$ the contribution of the positive-curvature parts is dominant and $r \to 1$. In the range $\omega_c\tau \gg 1$, the component σ_{xy} tends to its classical limit and the ratio r tends to unity. The contribution of the negative-curvature parts is dominant in the intermediate range $0.1 \lesssim \omega_c\tau \lesssim 1$, manifesting itself in a drastic decrease of r. Clearly, the effect of trigonal warping is maximal for electrons (see figs. 12 and 14). It may lead to the reversal of the sign of σ_{xy} (fig. 33) and therefore simulate a transition from the electron-type conductivity to hole-type, and vice versa.

The strong influence of trigonal warping on the off-diagonal components of the tensor $\sigma_{\alpha\beta}$ can be explained by the fact that in a compensated semimetal

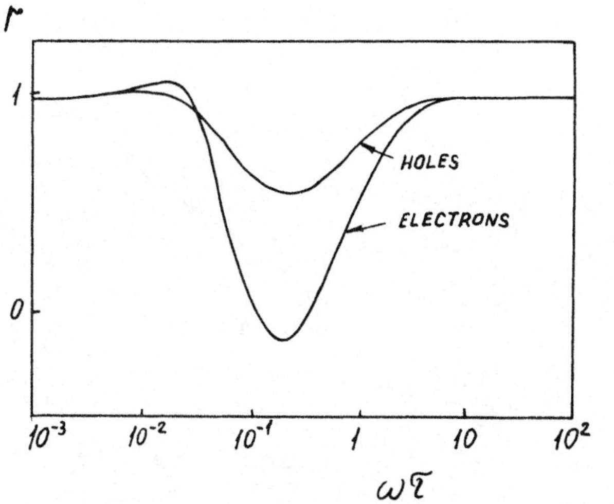

Fig. 33. Possible variation of the parameter $r = \sigma_{xy}^{\mathrm{t}}/\sigma_{xy}^{\mathrm{c}}$ as a function of $\omega\tau$. The case drawn shows a more pronounced dip for electrons than for holes, but this behaviour may be reversed for certain types of scattering. (Dillon and Spain 1978.)

the contributions from electrons and holes to these components enter with different signs and therefore even a small variation of these contributions may significantly alter the result. The variation of r for electrons and holes in a magnetic field (fig. 33) qualitatively correlates with the field dependence of the Hall coefficient R_{H} for HOPG and natural graphite (fig. 29a,b).

Thus from the calculation made by Dillon and Spain (1978) it follows that the neglect of the trigonal warping of the Fermi surface in an analysis of the field and temperature dependences of the galvanomagnetic coefficients of graphite may lead to substantial errors in the determination of densities and mobilities of individual groups of carriers.

Unfortunately, to our knowledge, no quantitative calculation of the parameters n_i and μ_i taking account of trigonal warping has been made to date.

Nevertheless, one may expect that the effects associated with trigonal warping will most affect a determination of characteristics of minority carriers (the small groups), so that the relevant part of the data presented in table 8 should be treated with caution.

We shall now discuss the temperature dependence of the carrier density in graphite. This dependence is caused by the removal of the electron gas degeneracy and by the thermal expansion of the lattice. The same causes are responsible for the variation of the Fermi electron (hole) energy in graphite (see fig. 22).

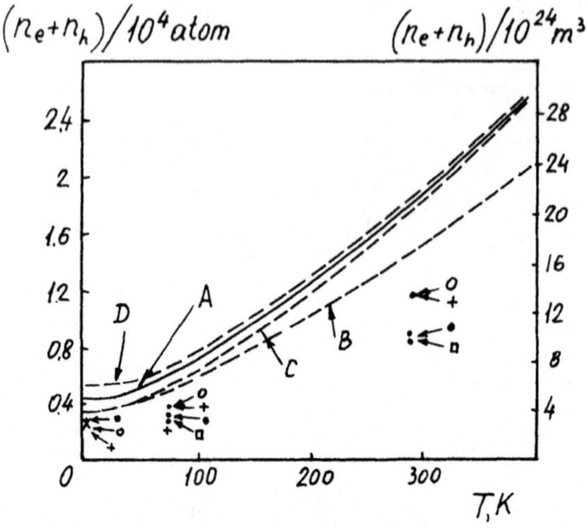

Fig. 34. The density of electrons and holes as a function of temperature including the parameter γ_3. Curves A,B,C,D are theoretical, based on the band-parameter sets given in table 5. Curve A corresponds to the set of parameters which describes best other electronic properties. Experimental points: ●, □ Soule (1958a); × Spain et al. (1967); +, ○ Dillon et al. (1978a). Theoretical curves by McClure. (See Dillon et al. 1978b.)

Galvanomagnetic measurements made over a wide temperature range indicate that in graphite one observes an increase in carrier density, which is the most marked above $T \simeq 100\,\text{K}$, i.e., close to the degeneracy temperature $\sim 120\,\text{K}$. Over the whole temperature range (from liquid-helium to room temperatures) the total carrier density increases by about a factor of two, from $\sim 5 \times 10^{18}\,\text{cm}^{-3}$ to $\sim 10 \times 10^{18}\,\text{cm}^{-3}$ (see table 8 and fig. 34).

The variation of n with temperature was calculated within the four-parameter model ($\gamma_3 = 0$) by McClure and Smith (1961), Ono and Sugihara (1966), and Kechin (1969). Within that model, the density of electrons and holes at low temperatures ($T \to 0$) is described by eq. (73). In the range above the degeneracy temperature, $kT > |\gamma_2| \sim \varepsilon_F$, the carrier density is determined solely by the parameters γ_0 and γ_1 of the SWMC-model and, according to Arkhipov et al. (1963), is proportional to T:

$$n = \frac{12.8\gamma_1 kT}{\sqrt{3\pi^2 a_0^2 \gamma_0^2 c_0}}. \tag{89}$$

The results of calculations using eq. (73) are in good agreement with the galvanomagnetic data obtained at low temperatures. Yet, the values of n calculated for high temperatures are overrated, being practically twice as high as the experimental values (see Spain 1973). Taking into account the thermal expansion of the lattice fails to eliminate this discrepancy.

Recently McClure has undertaken a calculation of the temperature dependence of n in graphite beyond the four-parameter model (see Dillon and Spain 1978). In this calculation, the Fermi-surface trigonal warping has been taken into consideration and four sets of the SWMC-model parameters have been utilized. In these sets, only the values of γ_1, γ_2 and γ_3 varied. The parameters found by McClure are given in table 9.

Table 9
SWMC-model parameters calculated by McClure. (After Dillon et al. 1978b.)

Set	SWMC parameters (in eV)						
	γ_0	γ_1	γ_2	γ_3	γ_4	γ_5	$\gamma_6 \equiv \Delta$
A	3.11	0.376	−0.0208	0.29	0.197	−0.01	−0.012
B	3.11	0.300	−0.0208	0.29	0.197	−0.01	−0.012
C	3.11	0.376	−0.017	0.29	0.197	−0.01	−0.012
D	3.11	0.376	−0.0208	0.232	0.197	−0.01	−0.012

Basically, these sets are close to set C from table 5. The calculated curves are shown in fig. 34, where the experimental values of n at various temperatures are also plotted. It can be seen that taking account of trigonal warping has failed to improve to any considerable extent the agreement between theory and experiment at high temperatures. At least some of the discrepancy between the experimental and calculated values of n can be attributed to the spread of carrier properties along the zone edge. However, it appears that the model for the dispersion of the relaxation time $\tau(k_z)$ required to eliminate the above discrepancy would be unrealistic. One may further assume that the SWMC-model parameters themselves significantly depend on temperature above 100 K. Finally, attention should be paid to the difference between the temperature dependences of the carrier density for monocrystalline kish-graphite (KG) and pyrolytic graphite (PG), whose lattices have different degrees of perfection. In any case, the difference between the experimental and theoretical values of n still remains a challenging problem. The temperature

Table 10
Temperature dependences of galvanomagnetic parameters and average carrier mobility in graphite. $H = 6.1$ kOe.

Parameter	Type of graphite		Temperature range
	Monocrystalline KG	PG	
$\bar{\mu}$	$T^{-1.6}$	$T^{-1.6}$	40–80 K
$\sigma_{xx}(0)$	$T^{-1.5}$	$T^{-1.0}$	20–100 K
$\sigma_{xx}(H)$	$T^{1.6}$	$T^{2.0}$	20–100 K

dependences of galvanomagnetic coefficients and average mobility in monocrystalline and pyrolytic graphite, which have been discussed above (see figs. 27, 32), are summarized in table 10.

According to McClure (1958), whose formulae were successfully used to analyze the experimental data obtained by Soule (1958b), the parameters of table 10 are related as follows:

$$\sigma_{xx}(0) = 1/\rho_a(0) \sim n\bar{\mu}, \qquad \sigma_{xx}(H) \sim n/\bar{\mu}. \tag{90}$$

From (90) and the data of table 10 one can conclude that:

(a) in monocrystalline KG, n is practically independent of T for T below 100 K, i.e., below the degeneracy temperature $\sim$120 K;

(b) in the less perfect PG, the carrier density slightly varies with temperature between 20 and 100 K according to the relation $n \sim T^{0.5\pm0.1}$, while the average mobility diminishes, as in KG, proportionally to $T^{-1.6}$. This structural feature of PG may be appropriately characterized, in contrast to KG, by a number of intercrystalline twist boundaries, where the stacking fault occurs in the form of crossed grids of extended screw dislocation ribbons. The carrier density of PG changing as $T^{0.5\pm0.1}$ is probably due to the stacking fault structure. Since the behaviour of conductivity in the range above the degeneracy temperature remains practically unchanged, the electronic structure of PG is essentially the same as that of KG, which has been confirmed by several previous studies. Therefore, the stacking faults seem to produce a slight perturbative effect giving rise to additional carries as the temperature is raised from 20 to 100 K.

5.3. Mechanisms of carrier relaxation in graphite

The main reason for the reduction of the average mobility $\bar{\mu}$ with increasing temperature is the electron and hole scattering by phonons. The high anisotropy of elastic properties associated with the lamellar structure of the graphite lattice enables one to consider two, practically independent, phonon systems, whose wave vectors $\boldsymbol{q}$ are parallel or perpendicular to the graphite layers. One system is represented by the so-called in-plane phonons having a high characteristics temperature ($\theta_{D1} \sim 2000$ K), the other, by so-called out-of-plane phonons whose characteristic temperatures are low ($\theta_{D2}^{TA} \sim 50$ K for the transverse acoustic vibrations and $\theta_{D2}^{LA} \sim 100$ K for the longitudinal ones). To these systems correspond different phonon propagation velocities in the long-wave limit, i.e. $v_{s1} \sim 2 \times 10^6$ cm/s and $v_{s2}^{TA} \sim v_{s2}^{LA} \sim 4 \times 10^5$ cm/s (Sugihara and Sato 1963). Since the size of the Fermi surface for electrons and holes in graphite is small, as compared to the reciprocal lattice vector, low-frequency phonons with energies of the order of $k\theta_{D2}^{LA,TA}$ will be sufficient for momentum relaxation if carriers move in the basal plane. Accordingly, the phonons with the lower characteristic temperature should play a significant part in the range below room temperature.

In connection with the aforesaid, the change in the exponent of the T-dependences of ρ_a (fig. 27) and of $\bar{\mu}$ (fig. 32) which is observed with increasing temperature, may be due not only to the removal of the electron gas degeneracy in graphite, but also to the change of the temperature dependence of the low-energy phonon number, as well as the changed effectiveness of electron and hole scattering by phonons of this type. The change in the character of scattering brought about by the removal of the electron-gas degeneracy is due to the fact that above the degeneracy temperature phonons may give rise to the intervalley scattering of electrons and holes, whose probability at the lower temperatures is exponentially small. As a result of the closeness of the degeneracy temperature to the characteristic temperatures θ_{D2}^{TA} and θ_{D2}^{LA}, the different mechanisms mentioned above, which influence the character of carrier scattering in graphite, come into play in practically the same temperature range between 80 and 100 K. The most rigorous theoretical calculations of the relaxation time for electron scattering by phonons in graphite have been carried out by Sugihara and Sato (1963) and by Ono and Sugihara (1966, 1968) on the basis of the SWMC-model spectrum.

The consideration of carrier scattering by longitudinal acoustical phonons which took no account of the trigonal warping of the Fermi surface has shown that the relaxation times associated with the in-plane and out-of-plane modes (τ_1, and τ_2, respectively) depend on temperature as $\tau_1 \sim T^{-1}$, and $\tau_2 \sim T^{-\delta}$, where $1 < \delta < 2$.

In this case, however, one cannot choose the electron–phonon coupling constants D_1 and D_2 for the two phonon modes so that there would be agreement between the temperature dependence of the total relaxation time $\tau_{e\text{-}ph} = \tau_1\tau_2/(\tau_1 + \tau_2)$ and the resistivity data over the whole temperature range under investigation. In order that the relation $\bar{\mu} \sim \tau \sim T^{-1.2}$, which takes place at $T > 80$ K (see fig. 32), be valid, the contants D_1 and D_2 should be equal to 16.2 and 3.73 eV, respectively. But this choice of D_i is unsuitable for the relation $\bar{\mu} \sim \tau \sim T^{-1.6}$, observed for $40 < T < 80$ K, to hold true. Since it is physically groundless to assume that the constants D_1 and D_2 suffer a sudden change at $T \sim 80$ K, the model adopted in the theory appears to be too simplified. The most important factors disregarded in the calculations are the inelastic phonon scattering, which becomes the most significant at low temperatures, and the trigonal warping of the Fermi surface, whose role also increases with decreasing temperature.

Actually, the importance of taking into account trigonal warping was emphasized already by Ono and Sugihara (1968) who demonstrated the heavy dependence of the electron–phonon coupling constants on the magnitude of γ_3. Thus, their values $D_1 = 16.2$ and $D_2 = 3.73$ eV, which correspond to $\gamma_3 = 0$, change to $D_1 \simeq 27.9$ eV and $D_2 \simeq 0$ for $\gamma_3 = 0.2$ eV. Moreover, since D_i

represents the change in energy at the band edge per unit strain, such a large value of D_1 suggests that the electronic structure is appreciably perturbed also by the in-plane defects, i.e., the widely-cored dislocations typical of graphite, which could be responsible for the difference between carrier constitutions in KG and PG.

Along with the electron and hole scattering from phonons, a noticeable contribution to relaxation in graphite may come from the scattering of carriers by one another. This kind of scattering is important for carrier densities $n \gtrsim n_c$, which, according to Fröhlich and Paranjable (1956), is given by

$$n_c = \frac{1}{4\pi} \frac{(m^*\eta)^{3/2} v_s^2 \kappa^2}{e^4 \tau_{e\text{-ph}} kT}, \tag{91}$$

where η is the electrochemical potential, κ the dielectrical constant, and $\tau_{e\text{-ph}}$ the relaxation time in electron scattering by phonons. An estimation made using eq. (91) at $T = 300\,K$ and parameter values corresponding to graphite yields a typical limiting density $n_c \sim 10^{14}\,cm^{-3}$.

Mutual carrier scattering is characterized by the relaxation times $\tau_{e\text{-}e}$, $\tau_{e\text{-}h}$, τ_{h-h}, proportional to T^{-2}. This kind of scattering may also appreciably contribute to the temperature dependence of the total relaxation time in graphite.

For a compensated semimetal, such as graphite, the effect of the mutual drag of electrons and holes, which move in opposite directions in the electric field, is important, too. McLean and Paige (1961) have demonstrated that the drag leads to momentum relaxation and significantly affects the electrical conductivity. The data of Sugihara (1966) indicate that the effect of drag changes most of all the mobility of minority carriers, reducing it by more than a factor of three.

In the low-temperature range $T < 40\,K$, where the temperature dependence of the resistivity ρ_a is saturated (fig. 27), scattering by ionized impurities, defects, grain boundaries, and interstitials becomes the dominant mechanism of momentum relaxation in the motion of carriers along the layer. The most effectual of them are two competing processes: scattering by ionized impurities and by crystallite boundaries. The relaxation time for electron and hole scattering by ionized impurities in graphite has been calculated by Boardman and Graham (1969). At an ionized-impurity density of $\sim 10^{18}\,cm^{-3}$, the estimated value of residual resistivity is $\rho_{a\,res} \sim 10^{-5}\,\Omega cm$, which has the same order of magnitude as the low-temperature resistivity of perfect natural graphite specimens.

Scattering by crystallite boundaries is dominant in synthetic pyrocarbon materials, which are composed of rather small crystallites. The relaxation time τ_B for this type of scattering is determined by the Fermi electron velocity v_F and by the average size L of a crystallite: $\tau_B \sim L/v_F$. The study of the residual conductivity in artificial graphites enables one to estimate, by using eq. (78), the average size of crystallites, which is $L = 1\,\mu m$ for the best specimens. This estimate is well consistent with the results of the electron microscopic measurements made by Morant (1970).

Thus the principal mechanisms of formation of the residual resistivity in graphites may be regarded as clarified in broad outline. However, until now there has been no systematic study of the dependence of the residual resistivity upon the concentration and kind of impurities and defects in natural graphite and upon the size of crystallites in pyrocarbon materials.

As for the temperature dependence of mobility, the situation here is more complicated. First of all, it is still unclear whether one may apply to graphite the so-called Koler rule: the residual resistivity and the temperature-dependent resistivity component are additive $[\rho = \rho_0 + \rho(T)]$. That is, the question of how the various imperfections of the crystal lattice may affect the T-dependence of ρ (or μ) remains unsettled.

The solution of this problem might assist in elucidating the mechanisms that shape the temperature dependence of the carrier mobility in graphite, thereby making for the development of a more general and rigorous scattering theory than the one available at present.

5.4. Galvanomagnetic effects in graphite in high quantizing magnetic fields and in the extreme quantum limit of fields

Galvanomagnetic coefficients for monocrystalline natural graphite in the extreme quantum limit of magnetic fields were measured first by McClure and Spry (1968). They have reported that in fields $\boldsymbol{H} \| c$ ranging from 80 to 140 kOe the transverse magnetoresistance $\rho_{a\perp}(H)$ increases linearly with field, while the Hall coefficient R_H is saturated for $H \gtrsim 90\,kOe$. The more accurate measurements made by Woollam (1970b) in permanent fields up to 150 kOe have shown that the dependence of $\rho_{a\perp}$ on H deviates from the linear behaviour and has a tendency to saturation.

Brandt et al. (1974) have measured $\rho_{a\perp}$ and R_H in perfect single crystals of kish-graphite in pulsed fields at the liquid-helium and liquid-nitrogen temperatures. They have found that the linear dependence of $\rho_{a\perp}$ on H changes to sublinear for $H \sim 100$–$120\,kOe$ and tends to saturation (fig. 35). The similar behaviour of the H-dependence of $\rho_{a\perp}$ at the liquid-helium temperature has been observed by Lowrey and Spain (1977) in HOPG specimens placed in permanent magnetic fields up to 240 KOe.

The results of the measurements made by Brandt et al. (1974) indicate that in single crystals of kish-graphite $\sigma_{xy}H$ is positive in fields $H < 100\,Oe$, but it rapidly increases with the field, passes through a peak for $H = 100\,Oe$, then decreases and reverses its sign for $H \simeq 500\,Oe$. In the range of negative values, $\sigma_{xy}H$ tends to saturation in fields $H \gtrsim 50\,kOe$ and remains practically con-

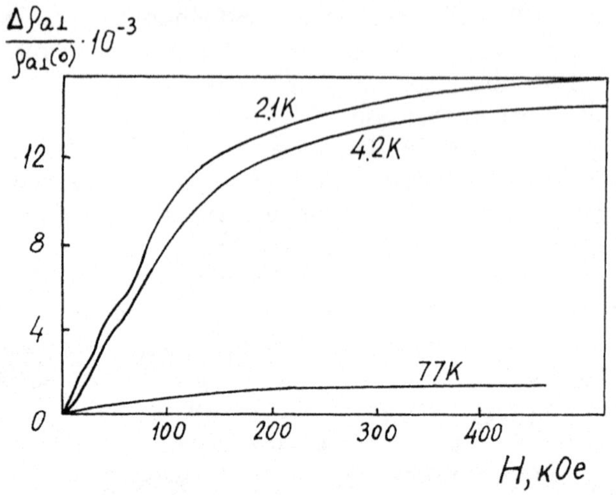

Fig. 35. Transverse magnetoresistance as a function of the magnetic field $\boldsymbol{H}\|c$ in perfect single-crystal kish-graphite at $T = 2.1$, 4.2 and 77 K. (Brandt et al. 1974.)

stant up to $H \sim 200$ kOe. Above that strength, there begins a sudden decrease of the absolute value of $\sigma_{xy}H$, continuing over the entire range of fields studied up to 520 kOe (fig. 36).

The linear dependence of $\rho_{a\perp}$ on H in graphite has been explained by McClure and Spry (1968) who proceeded from the concepts of carrier scatter-

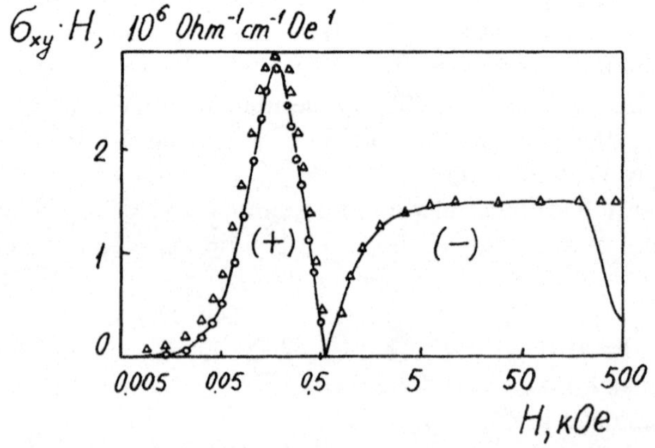

Fig. 36. The parameter $\sigma_{xy}H$ as a function of magnetic field $\boldsymbol{H}\|c$ in perfect single-crystal kish-graphite at $T = 4.2$ K (solid curve). (+) and (−) mark the regions of positive and negative Hall-coefficient R_H values, respectively. △ computer-calculated $\sigma_{xy}H$ values according to the two-band model; ○ values for the three-band model, disregarding the magnetic freezing-out of impurity carriers. (Brandt et al. 1974.)

ing governed by the screened Thomas–Fermi potential for ionized impurities taking account of the dependence of the screening radius upon the magnetic fields. They have shown that under these conditions the transverse magnetic conductivity in graphite has the form

$$\sigma_{xx} = \frac{\pi e(\Delta Z)^2 n_s c}{16H}, \tag{92}$$

where n_s is the number of charged scattering centres and $\Delta Z = Z - 4$ is the difference between the valence of the impurity atom and that of carbon. The linear behaviour of the dependence on H and the saturation of the Hall coefficient in fields up to 150 kOe enabled McClure and Spry (1968) to conclude that in the above range of fields there are no bound electron states due to impurities, whose degree of ionization would depend on magnetic field strength.

As is known, in the extreme quantum limit of magnetic fields, the non-diagonal component of the magnetic conductivity tensor is described by the Kubo relation (see, e.g., Abrikosov 1969):

$$\sigma_{xy} = \frac{ec(n_h - n_e)}{H}. \tag{93}$$

Therefore, the value of $\sigma_{xy}H$ is determined solely by the difference between the densities of electrons and holes, i.e., by the magnitude of the imbalance $\Delta n = n_h - n_e$. Thus in the region where $\rho_{a\perp}$ depends linearly on H, and where the parameter $\sigma_{xy}H$ is saturated, it is possible, using eqs. (92) and (93), to find the sum of the densities of ionized donor (n_s^+) and acceptor (n_s^-) scattering centres: $n_s = n_s^+ + n_s^-$ and also their difference $\Delta n = n_h - n_e = n_s^- - n_s^+$, which determines the imbalance, can be found. Hence, one can readily find the values of n_s^+ and n_s^- themselves. Such a calculation, carried out by Brandt et al. (1974) on the basis of measurements of $\rho_{a\perp}$ and $\sigma_{xy}H$, has shown that in perfect single crystals of kish-graphite for a majority carrier density $n_e \sim n_h \sim 2 \times 10^{18}\,\text{cm}^{-3}$, the densities of impurity centres of the donor and acceptor types constitute $n_s^+ \sim n_s^- \sim (0.4\text{–}0.9) \times 10^{18}\,\text{cm}^{-3}$.

5.4.1. Magnetic freezing-out of impurity carriers in graphite

The experimentally observed abrupt decrease of the magnitude of imbalance $\Delta n = n_h - n_e \sim \sigma_{xy}H$ in graphite (see fig. 36) may be attributed to the magnetically induced localization of impurity-type carriers on charged centres, which is called the magnetic freeze-out effect. This conclusion has been reached by Brandt et al. (1974) who were the first to observe the decrease of $\sigma_{xy}H$ in fields above 200 kOe.

The magnetic freeze-out effect discovered by Keyes and Sladek (1956) in n-

InSb alloys is well known in semiconductor physics [see, e.g., the review by Von Ortenberg (1973)]. The possibility of the magnetic freeze-out of impurity-type carriers in semimetals has been theoretically substantiated by Beneslavskii and Entralgo (1976) for substances with a sharply anisotropic electron spectrum. Earlier, Rashba and Edelstein (1969, 1970) called attention to the fact that in high magnetic fields near all critical points of the density of states in the spectrum of a semimetal (specifically, near saddle points) there should arise metastable Coulomb states. Although these are found against the background of the quasi-continuum states in other Landau subbands, the relative width of the metastable levels turns out, as Beneslavskii (1968) pointed out, to be much less than the energy of the levels themselves. Such levels are called quasi-localized or resonant.

It is known that the existence of bound states is greatly affected by the screening of charged centres by free carriers.

In the presence of this screening, electrons may be localized on impurity centres only provided that $a_B^* < r_D$, where a_B^* is the effective Bohr radius of the electron on a centre and r_D is the screening length. For a degenerate gas of charge carriers, the relation

$$r_D \simeq \tfrac{1}{2}(a_B^* n^{-1/3})^{1/2} \qquad (94)$$

is valid. For graphite, a high anisotropy of the carrier spectrum is typical. In the parabolic approximation, the spectra of electrons and holes in graphite are determined by two parameters: the longitudinal $m_3^{e,h}$ (along the c-axis) and transverse $m_1^{e,h}$ (along the layer) effective masses, whose values have been given above (in sect. 4.3). In accordance with this, charged impurity centres in graphite are characterized by two localization radii (effective Bohr radii):

$$a_{B\parallel}^{e,h} = \frac{\kappa_0 \hbar^2}{m_3^{e,h} e^2} \quad \text{(longitudinal radius), and}$$

$$a_{B\perp}^{e,h} = \frac{\kappa_0 \hbar^2}{m_1^{e,h} e^2} \quad \text{(transverse radius),} \qquad (95)$$

where κ_0 is the low-frequency dielectric constant (indices e and h refer to electrons and holes). Using the effective mass values given by (54) and $\kappa_0 = 9.1$ (see Taft and Philipp 1965), one finds

$$a_{B\parallel}^{e} \simeq 6\,\text{Å}; \qquad a_{B\parallel}^{h} \simeq 8\,\text{Å}; \qquad a_{B\perp}^{e} \simeq 90\,\text{Å}; \qquad a_{B\perp}^{h} \simeq 125\,\text{Å}. \qquad (96)$$

The volume of the region of carrier localization on an impurity centre constitutes $\sim(a_{B\perp}^2 a_{B\parallel})$, the characteristic size of the region being $\bar{a} \sim (a_{B\perp}^2 a_{B\parallel})^{1/3}$. Using this quantity, one can estimate the screening length by eq. (94) and show that for $H = 0$ in graphite there holds the inequality $\bar{a} > r_D$ and there arise no bound states on impurity-type centres.

Beneslavskii and Entralgo (1976) have shown that in a semimetal with a sharply anisotropic spectrum characterized by two greatly different components of the effective mass tensor, $m_3^{e,h}$ and $m_1^{e,h}$ ($m_3^{e,h} \gg m_1^{e,h}$), the binding energy ε_B of a carrier on a single impurity-type donor centre in those magnetic fields for which the inequality $a_{B\parallel}^{e} \ll \lambda \ll a_{B\perp}^{e}$ is valid coincides, as in the case of an unscreened potential, with the effective potential-well depth;

$$\varepsilon_B \simeq \frac{e^2}{\kappa_0 \lambda}\sqrt{\tfrac{1}{2}\pi}, \tag{97}$$

where $\lambda = \sqrt{c\hbar/eH}$ is the magnetic length. The above-mentioned inequality $a_{B\parallel} \ll \lambda \ll a_{B\perp}$ corresponds in graphite to the following magnetic-field ranges:

$$\begin{aligned} &10^7\,\text{Oe} \gg H \gg 4\times 10^4\,\text{Oe}, &&\text{for donor-type centres; and}\\ &1.75\times 10^7\,\text{Oe} \gg H \gg 8\times 10^4\,\text{Oe}, &&\text{for acceptor-type centres.} \end{aligned} \tag{98}$$

In these ranges, the volume of a region of electron (hole) localization on a centre, which is of the order of $a_{B\parallel}\lambda^2$, decreases inversely with the magnetic field. This decrease leads to the elimination of the overlap of the wave functions for electrons on neighbouring impurity centres. As a result, at low temperatures $kT < \varepsilon_F$ in the impurity subsystem of a semimetal there should occur a peculiar magnetic transition (a Mott-type one) similar to the transition which occurs in the impurity subsystem of a semiconductor in a magnetic field (see Fenton and Haering 1967). Taking into account the screening of impurity centres by free carriers in a semimetal does not alter the transition pattern qualitatively, but shifts the value of the magnetic field H_c at which electron localization on impurity centres begins, towards the higher fields.

The trapping of electrons and holes by impurity centres at $H > H_c$ should result in a change in the galvanomagnetic coefficients. The reason for the change in magnetoresistance is that, as a result of electron localization, the number of charged scattering centres n_s in eq. (92) decreases.

The change in $\sigma_{xy}H$ (or the Hall coefficient R_H) that accompanies the localization of electrons and holes in the magnetic field is due to the change in the magnitude of imbalance $\Delta n = n_h - n_e$ which should turn to zero in very high fields as a result of a complete localization of all impurity carriers. The magnetic freeze-out in graphite should manifest itself primarily by a change in the value of $\sigma_{xy}H$, which depends on the difference $n_h - n_e$. It is highly probable therefore that the steep decrease of the value of $\sigma_{xy}H$ in kish-graphite in fields $H > 200$ kOe (see fig. 36) is due to the magnetic freeze-out of impurity carriers. This supposition has been confirmed by Kuvshinnikov (1980) who measured the temperature dependences for $\sigma_{xy}H$ as proportional to the imbalance Δn. His curves for $\log(\Delta n)$ versus T^{-1} in the temperature range from 3.7 to 100 K for single crystals of kish-graphite in fields of 5–25 T are

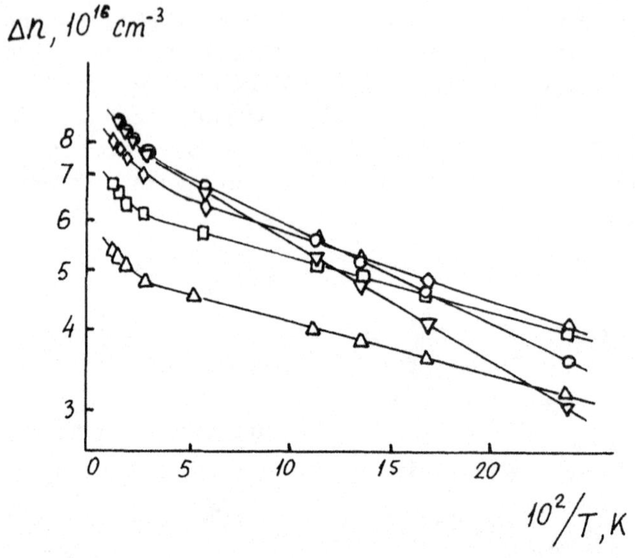

Fig. 37. Loss of carrier compensation as a function of inverse temperature for a specimen of kish-graphite at various magnetic field strengths: △ $H = 5$T, □ $H = 10$T, ◇ $H = 15$T, ○ $H = 20$T ▽ $H = 25$T. (Kuvshinnikov 1980.)

illustrated in fig. 37. From the shape of the curves of $\log(\Delta n) = f(1/T)$ one may assume that $\Delta n \sim \exp(-\varepsilon_A/kT)$, where the activation energy ε_A determines the binding energy ε_B of electrons on impurity centres. The increase in the slope of the curves with increasing field, which can be noticed in fig. 37, indicates an increase in binding energy in the magnetic field [see (97)]. The value of the binding energy ε_B determined from the slope of the curves is about 20 meV for $H = 10$T. This estimate agrees fairly well with the one ($\varepsilon_B \sim 25$ meV) obtained for the same field by eq. (97), which is valid for an isolated impurity centre without screening. Evidently, taking account of the screening should somewhat reduce the value of the binding energy.

Now it is of interest to discuss the possible character of change in the quantity $\sigma_{xy}H$ (see fig. 36) in higher magnetic fields than those considered so far; the more because there have already been developed methods for producing mega-Oersted magnetic fields.

The behaviour of $\sigma_{xy}H$ as a function of the field H in ultrahigh fields is determined by two factors: the number of the electrons trapped by ionized impurity centres and the field range that separates the beginning of the freeze-out of electrons onto the donor centres and holes onto the acceptor centres. Let us denote the concentration of the donor centres by n_s^+ and that of the acceptor ones by n_s^-. We shall first consider the case when each of the ionized impurity centres may capture only one carrier (electron or hole). We also assume that holes begin to freeze out onto acceptor centres only after the electron freeze-out is completed and that $n_s^+ > n_s^-$.

In the extreme quantum limit of magnetic fields, prior to the beginning of the magnetic freeze-out of carriers, the quantity $\sigma_{xy}H$ is proportional to the difference $n_h - n_e = n_s^- - n_s^+$, which is negative under the above assumption that $n_s^+ > n_n^-$.

As electrons are freezing out, their density decreases and, therefore, for some value of the magnetic field the difference $n_h - n_e$, as well as the quantity $\sigma_{xy}H$, which is proportional to that difference, will go to zero. This means that the electron number has become equal to the hole number. At the higher field values (but before the freeze-out of holes) $\sigma_{xy}H$ becomes positive and tends to a constant value corresponding to the situation in which all electrons are frozen out and $n_s^+ = 0$; free carriers being represented solely by holes with concentration n_s^-. In still higher fields, in which the freeze-out of holes has already started, $\sigma_{xy}H$ decreases to zero with increasing field. The general form of the field dependence of $\sigma_{xy}H$ for the case under consideration is shown by the solid curve in fig. 38a. If, however, the interval of fields separating the beginning of the electron freeze-out from that of the hole freeze-out is small, then both these processes occur either with a small shift in the field or practically in one and the same field range. In this case, the interval of fields in which $\sigma_{xy}H$ becomes positive may be very small or may even be absent, as is shown in fig. 38b by the solid curves 1 and 2.

Let us now consider the situation in which the impurity centres each may capture two carriers when a magnetic field is applied. The possibility of such a localization was first suggested by Beneslavskii et al. (1982). Their calculation has shown that in a substance with a sharply anisotropic spectrum in a magnetic field for which the inequality $a_{B\parallel} \ll \lambda \ll a_{B\perp}$ is satisfied, the binding energy of two electrons on a singly charged impurity centre is equal to $\simeq 1.3\varepsilon_B$, where ε_B is the binding energy of one electron on the centre, given by eq. (97). Hence it follows that in graphite, in the field range delineated by (98), the localization of two electrons on each donor centre is energetically more advantageous than that of one electron. In such a case, the character of the dependence of $\sigma_{xy}H$ on the magnetic field will be qualitatively different from the plausible dependence of $\sigma_{xy}H$ on H described above (the capture of a single carrier by an impurity centre of a corresponding sign). When each donor centre captures two electrons, the electron density decreases by $2n_s^+$ as compared with its value for $H = 0$. Accordingly, in hole localization on acceptor centres the hole concentration decreases by $2n_s^-$. As a result, after electrons and holes have become localized, the initial (for $H = 0$) value of the imbalance $n_n - n_h$ will reverse its sign, but its absolute value is unchanged (see inset of fig. 38), which shows the change in the density of electrons, Δn_e, and holes, Δn_h, due to doping in the zero magnetic field, as well as the correspond-

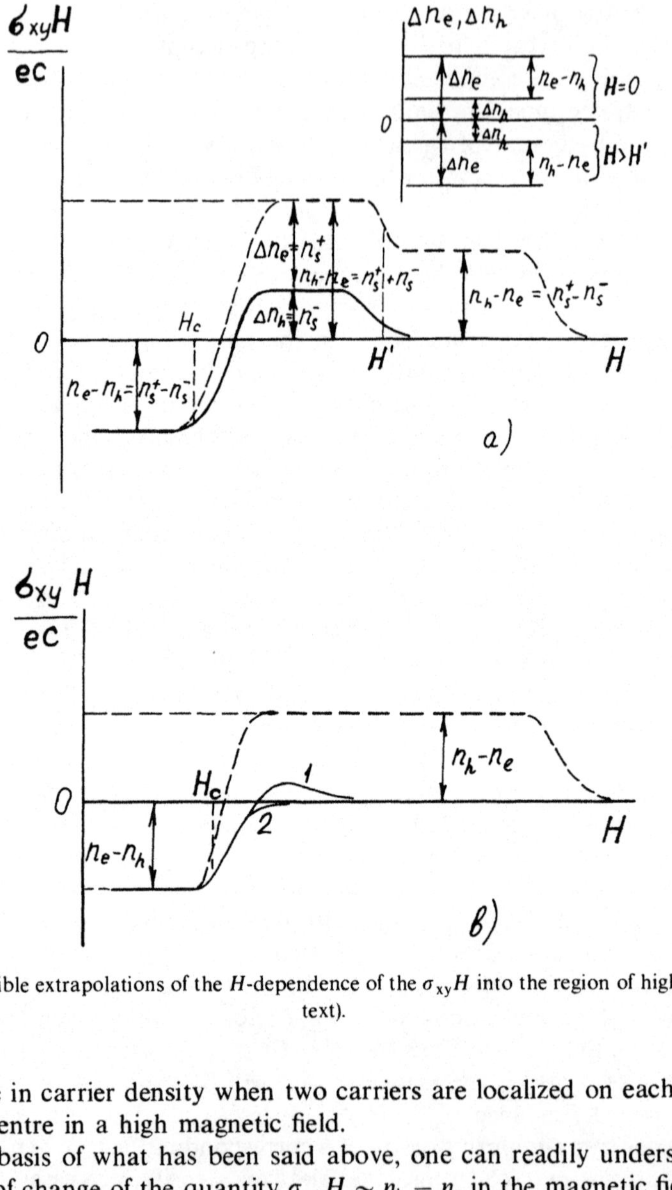

Fig. 38. Possible extrapolations of the H-dependence of the $\sigma_{xy}H$ into the region of high fields (see text).

ing change in carrier density when two carriers are localized on each charged impurity centre in a high magnetic field.

On the basis of what has been said above, one can readily understand the character of change of the quantity $\sigma_{xy}H \sim n_h - n_e$ in the magnetic field when the freezing-out of electrons and holes begins at different values of H. Let, for example, electrons begin to freeze out in lower fields than holes do. Upon electron localization (two electrons on each donor centre) the electron density decreases more than in the case of single-electron capture by $\Delta n_e = n_s^+$, while the hole density will remain unchanged. As a result, there will arise an excess of holes, equal to $n_s^+ + n_s^-$. The quantity $\sigma_{xy}H \sim n_h - n_e$ will then become proportional to $n_s^+ + n_s^-$. In the course of hole freeze-out, the hole density will decrease by $2n_s^-$ and the imbalance $n_h - n_e$ will become the same in absolute value as it is for $H = 0$, but with opposite sign: $n_h - n_e = n_s^+ - n_s^-$ (the dashed curve in fig. 38a). The imbalance assumes this value immediately if the processes of electron and hole localization (two carriers on a centre) occur in practically the same magnetic fields (the dashed curve in fig. 38a). What will occur when the magnetic field is further enhanced? One may expect that in the yet stronger magnetic fields, when the right-hand side of inequality (98) is invalid for both donor and acceptor centres and when the specific character of an anisotropic spectrum disappears, the impurity centres will be incapable of holding effectively a second electron at finite temperatures (see Beneslavskii et al. 1982). In such fields, the value of the imbalance (and of $\sigma_{xy}H$) will again go to zero, as is shown by the dashed curve in fig. 38a,b. It is also of interest to discuss one more possibility. For quasi-two-dimensional lamellar substances (such as graphite) with a sharply anisotropic spectrum in a magnetic field perpendicular to the layers, the creation of an electron–hole (electron–exciton) impurity complex consisting of an impurity centre and several electrons and holes would under certain conditions be energetically more advantageous than just the localization of two electrons on a single charged impurity centre. Such complexes may appear if the distance between two layers d_0 is small, as compared to the magnetic length $d_0 \ll \lambda$. In this case, it becomes advantageous, as Buzdin (1983) has shown, to add a hole to the two electrons already localized on a single charged centre [with a binding energy $1.3\varepsilon_B$, where ε_B is given by (97)]. The binding energy will then increase by $2 \times 0.7\varepsilon_B$ (the energy of attraction by the two electrons of the complex) and will decrease by $1.0\varepsilon_B$ (the energy of repulsion from the centre). As a result, the binding energy will increase from $1.3\varepsilon_B$ to $1.7\varepsilon_B$. One can readily see that a further attachment of an electron, a hole, etc. is possible. The binding energy of such electron–exciton impurity complexes with N holes and $N + 2$ electrons [$\varepsilon_B^-(N)$] and with $N + 1$ holes and $N + 2$ electrons [$\varepsilon_B^0(N)$] is given by:

$$\begin{aligned} \varepsilon_B^-(N) &= (1.3 + 0.7N)\varepsilon_B, \\ \varepsilon_B^0(N) &= (1.7 + 0.7N)\varepsilon_B. \end{aligned} \tag{99}$$

Physically, the formation of such complexes is associated with the possibility of disregarding the Pauli exclusion principle for those electrons and holes which are in different layers. The maximum number of particles in a complex, $N_{max} \sim \lambda/d_0$, is considerably greater than unity in the adopted approximation.

The origination of the complexes considered by Buzdin (1983) in graphite should lead, in contrast to the two-carrier localization, to a greater reduction of the carrier density in a magnetic field. As regards the behaviour of the function $\sigma_{xy}H$, it can easily be visualized on the basis of the above consideration of the cases of one- and two-carrier localization on each impurity centre, bearing in mind that the binding energy per particle is higher for the negatively charged complexes, so that, for $n_s^+ > n_s^-$, their formation also should lead to the reversal of the sign of the function $\sigma_{xy}H$ from negative to positive.

Since it is still obscure which of the cases considered above will occur in reality, it appears to be both interesting and tempting to measure the galvanomagnetic properties of graphite in mega-Oersted magnetic fields. To our knowledge, no such measurements have been made yet. Nevertheless, even now there are some experimental data that permit a discrimination between complexes with an odd or an even number of localized particles on the impurity centres. Complexes with an odd number consist of a donor centre plus an electron, two electrons plus one hole, three electrons plus two holes, etc. Even-number complexes are made up by a centre and two electrons, three electrons and one hole, four electrons and two holes, etc.

The magnetic properties of these two types of complexes will differ because of the cancellation of the spins of particles which may occur in the even-number complexes. As a result, the odd-number complexes should contribute paramagnetically (positively) to the magnetization of graphite. This contribution, provided by non-zero localized magnetic moments, should increase with increasing number of complexes in a magnetic field. When the formation of localized complexes comes to an end in a sufficiently high magnetic field, the magnitude of the paramagnetic addition to graphite magnetization will be determined only by the rate of the increase in the binding energy of the localized complexes in the field.

The paramagnetic contribution to magnetization in fields $H > 200\,\text{kOe}$, at which, according to the galvanomagnetic data, the value of $\sigma_{xy}H$ in graphite begins to decrease (i.e., the magnetic freeze-out of impurity carriers starts) has been detected experimentally by Kapustin et al. (1981) who measured the magnetic moment in specimens of kish-graphite in pulsed magnetic fields up to 380 kOe. The paramagnetic addition reached a peak value for $H \sim 300\,\text{kOe}$ and then decreased for fields $H \sim 380\,\text{kOe}$. The results of magnetization measurements in kish-graphite in high magnetic fields are discussed in more detail in sect. 9.2.

The presence of the paramagnetic contribution allows one to assume that in the course of magnetic freeze-out in graphite electron–exciton impurity complexes arise, consisting of an odd number of particles, because it is most likely that complexes consisting of an even number of particles have no localized moment. The decrease of the paramagnetic addition in fields $H \sim 300\,\text{kOe}$ appears to suggest that the formation of complexes in that range of fields is at an end.

The simplest complex whose formation may contribute paramagnetically to magnetization is that of a single, localized electron. Such a complex may originate in the course of a simple freeze-out of impurity carriers. Accordingly, magnetization data can mainly indicate the presence of magnetic freeze-out, but by no means are sufficient evidence for the emergence of complicated electron–exciton impurity systems.

5.4.2. Magnetically induced transitions of graphite to the charge density wave (CDW) state

Up to now, we have discussed the results obtained in studying single crystals of synthetic graphite (kish-graphite) of relatively low purity (with a charged-impurity density of $\sim 10^{18}\,\mathrm{cm}^{-3}$). When measuring the transverse magnetoresistance, $\rho_{a\perp}$, in natural single-crystal graphite of a high purity at liquid-helium temperatures Tanuma et al. (1981a) have found that, after a small decrease in the range 120–300 kOe, $\rho_{a\perp}$ suddenly (almost discontinuously) increases for field values $\sim$280–270 kOe by a finite value $\Delta\rho$ and then levels off or begins slightly to decrease. The field dependence of $\rho_{a\perp}$ just described is illustrated in fig. 39.

The "jump" of magnetoresistance in fields $\sim$300 kOe has been confirmed by Iye et al. (1982b) and Nakamura et al. (1983) who found that the critical value of the magnetic field, H_{cr}, at which the magnetoresistance begins to increase

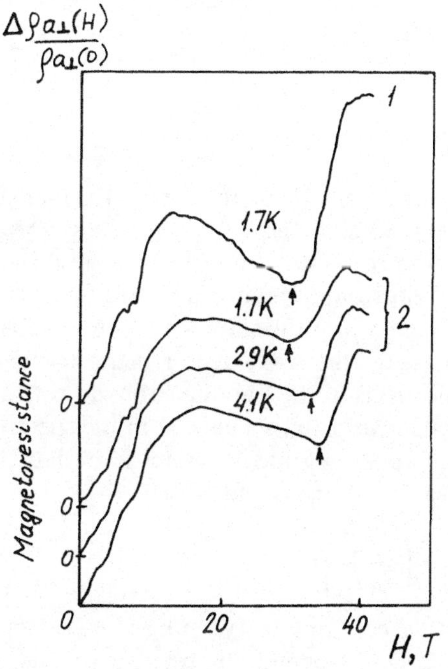

Fig. 39. Transverse magnetoresistance $\rho_{a\perp}(H)/\rho_{a\perp}(0)$ in two different samples of single-crystal graphite: 1, 507; 2, 514. The anomaly (critical field H_{cr}) is indicated by arrows. Sample 1 exhibits the anomaly more clearly. For sample 2, the T-dependence of magnetoresistance is given at three different temperatures: 1.7, 2.9 and 4.1.K. (Nakamura et al. 1983.)

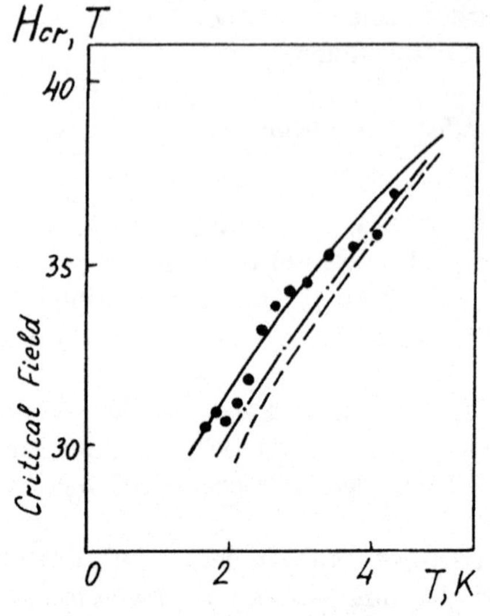

Fig. 40. Temperature dependence of the critical fields where the anomalies shown in fig. 39 (sample 2) appear. The solid curve is the least-squares fit to the data of Nakamura et al. (1983). The dot–dashed line shows the calculation of Yoshioka and Fukuyama (1981) and the dashed line is the extrapolation of the experimental curve determined by Iye et al. (1982a) (Nakamura et al. 1983).

abruptly, as does the magnitude of the jump $\Delta\rho$, increase with temperature from 1.7 to 4.5 K (figs. 40 and 41).

Yoshioka and Fukuyama (1981) ascribe this abnormal magnetoresistance to the heightening of electron correlations in strong magnetic fields, as a result of which the system passes into the state with a charge density wave (CDW) in the direction of the c-axis, the wave vector being equal to the double Fermi wave vector k_F^e for electrons in the Landau subband with number $n = 0$. Going over to the CDW-state gives rise to a gap 2Δ in the spectrum of the elementary excitations. This gap, appearing at the Fermi level, has the same temperature dependence as the gap in a superconductor

$$\Delta(T) = \Delta(0)\left[1 - \left(\frac{T}{T_c}\right)^2\right]^{1/2}, \tag{100}$$

where $2\Delta(0) = 3.5kT_c$. According to Iye et al. (1982b), the critical temperature of the transition T_c is exponentially related to the magnetic field. The transition to the CDW-state is essentially equivalent to the formation of the dielectrical phase in a quasi-one-dimensional gas of carriers in graphite in a high magnetic field.

The experimentally observed finite increase $\Delta\rho$ in magnetoresistance in a

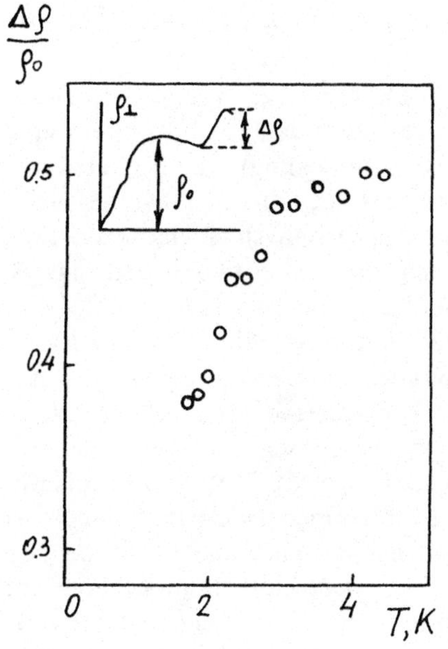

Fig. 41. Temperature dependence of the amount of jump by the anomaly. This amount is normalized to ρ_0, which is the increment of magnetoresistance from zero field to the field range where magnetoresistance is saturated around 12 T (sample 2). Inset specifies ρ_0 and $\Delta\rho$. (Nakamura et al. 1983.)

narrow range of magnetic fields, followed by saturation, enables one to suggest that no completely dielectrical state arises in graphite. Rather, there will be a transition to the partially dielectrical state with the fine structure which has already been predicted by Abrikosov (1973) for bismuth and which has been called "reduced metal". In the transition to the state of a reduced metal only some of the free carriers leave the spectrum of the initial system, whereas the rest of them remain in the degenerate state and take part in charge transfer at $T \to 0$K. Experimental evidence that reduced-metal states arise in $Bi_{1-x}Sb_x$ alloys in a magnetic field was obtained for the first time by Brandt and Chudinov (1972). As for graphite, there is only indirect evidence that a CDW-state is formed. The correctness of the proposed interpretation has to be corroborated by further studies to evince an additional structure with period $2\pi/2k_F^e$ along the c-axis. One method to study such a state is to observe the resonance absorption of ultrasonic waves by a periodic structure. It is also of interest to look at the temperature dependence of the ultrasonic absorptivity, which may have an anomaly due to fluctuations of the dielectric-phase order parameter Δ near the transition point to a CDW-state. A similar phenomenon was observed in bismuth in high magnetic fields (see, e.g., Mase and Sakai 1971, Mase et al. 1980).

5.5. Peculiarities of galvanomagnetic effects in carbon materials

Carbon materials represent a large family of graphite-like substances with a disturbed crystal lattice whose degree of disorder depends on the temperature of subsequent treatment (annealing) – or heat-treatment temperature (HTT).

The already known results, obtained by measuring the galvanomagnetic coefficients and other electrophysical characteristics of carbons, permit a division of carbon materials into two broad classes of homogeneous and inhomogeneous materials.

In all carbons whose inhomogeneity manifests itself at the semimacroscopic level when the size of a region of inhomogeneity is tens or hundreds times the lattice constant, one can observe even at room temperature a decrease of electrical resistance in the presence of the magnetic field or the effect of negative magnetoresistance (NM). Materials of this kind have been studied in considerable detail by Herring (1958, 1960) and Beer (1963). However, a discussion of inhomogeneous carbons would be beyond the scope of the present volume. In macroscopically homogeneous materials, which include polycrystalline graphites with fine grains (crystallites), highly defective artificial graphites, mosaic graphite structures, etc., one can observe both a decrease of resistance in magnetic field (NM) and its increase, or positive magnetoresistance (PM). In such substances, the NM-effect is typical of carbon materials heat-treated between 1500–2500°C and can be seen mostly at low temperatures. The PM-effect is common for synthetic graphites heat treated at 2800°C.

The first systematic study on the NM-effect in carbons has been made by Mrozowski and Chaberski (1956). For such studies, one usually takes homogeneous carbons consisting of preferably oriented graphite layers. Such materials are called pyrocarbons. The effect of negative magnetoresistance has received much study. The general appearance of the dependence of the transverse magnetoresistance of pyrocarbons with various HTT upon magnetic field at liquid-helium temperatures, obtained by Delhaes et al. (1974), is illustrated in fig. 42. The field dependences of the magnetoresistance of pyrocarbon with HTT = 2300°C, for which the NM effect is normally observed for $T < 100$ K, are shown in fig. 43. The dependences shown in figs. 42 and 43 are typical for all types of pyrocarbons, which differ only in their HTT values. The rise of HTT always leads to the change from NM to PM. And it is just as typical that in pyrocarbons with relatively low HTT (lower than 1300–1500°C) one can observe only the PM-effect whose magnitude depends quadratically on the magnetic field. The quadratic dependence of the resistance upon the field is characteristic also of NM at temperatures below the liquid-nitrogen temperature in the range of relatively low fields ($H \lesssim 1$ kOe). As the magnetic field increases, the dependence gradually changes to linear. The character of change in magnetoresistance $\Delta\rho_\perp/\rho_0$ of pyrocarbons at 77 and 300 K as a function of HTT is schematically portrayed in fig. 44. It

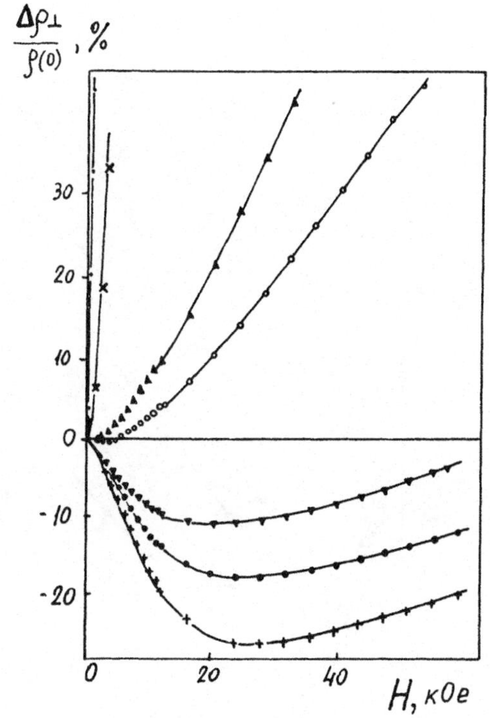

Fig. 42. Transverse magnetoresistance in pyrocarbon at liquid-helium temperature ($T = 4.2$ K) × HTT 2800°C, ▲ HTT 2600°C, ○ HTT 2560°C, ▼ HTT 2490°C, ● HTT 2300°C, + temperature of deposition – 2100°C (Delhaes et al. 1974).

should be noted that generally both the value of PM and the absolute value of NM decrease with increasing temperature. At the same time, the NM of pyrocarbons with HTT between 2250–2350°C changes to PM as the temperature is raised. Bright (1979) has observed the NM-effect in pyrocarbons at low temperatures and also oscillations of magnetoresistance at the same temperatures, which is indicative of the rather long mean free path of carriers in these materials.

Proceeding from the phenomenological expression for the conductivity $\sigma = en_e\mu_e + en_h\mu_h$, one may attribute the NM to both an increased carrier mobility and increased carrier density in the magnetic field. NM-models based on the latter factor have been developed by Hedgecock and Randorf (1970) and by Giovanini and Hedgecock (1972). These models assume an impurity zone with a higher density of states near the edge of the valence band. The increase in carrier density in a magnetic field is a consequence of the shift of the Fermi level towards the region with the higher density of states. In this model, the cause of a marked shift of the Fermi level in relatively low mag-

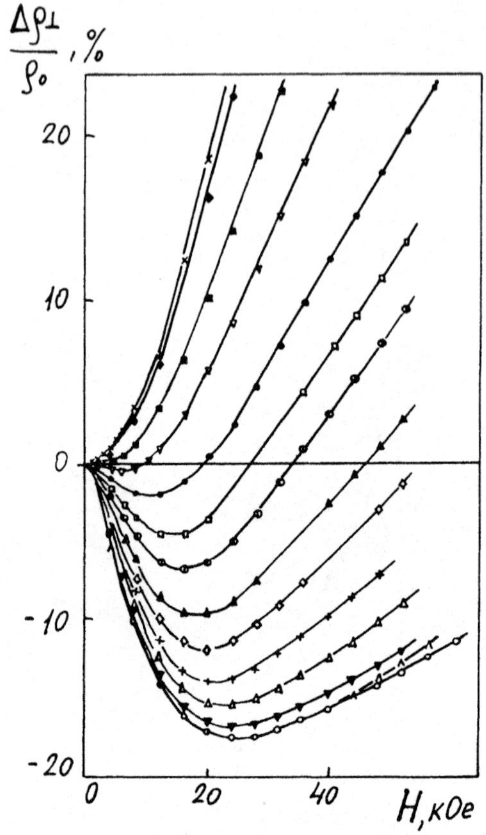

Fig. 43. Transverse magnetoresistance at various temperatures for pyrocarbon heat-treated at 2300°C; T (in K): × 280, ◆ 180, ■ 125; ▽ 100; ● 78; □ 61; ⌀ 56; ▲ 46; ◇ 35; + 26; △ 18; ▼ 10; ○ 4.3; ∧ 1.3. (Delhaes et al. 1974).

netic fields remains unexplained. Furthermore, so far as we know, the existence of the narrow impurity band in the required part of the spectrum of carbon materials has not yet been independently confirmed in experiment.

There exist NM-models based on the mechanism of an increased carrier mobility in the magnetic field due to an increase in the mean free path of the carriers (see, e.g., Fujita 1968, Dalton and Fujita 1971). This increase in the carrier mobility μ takes place as a result of the internal size effect, when the mean free path, limited for $H = 0$ by the crystallite size, increases in the magnetic field due to carrier-path curving. These models predict the quadratic dependence of $|\Delta\rho_\perp|/\rho_0$ on H in the NM regime, which agrees with experiment in the range of very weak magnetic fields.

Finally, we should mention the NM-model proposed by Toyozawa (1962),

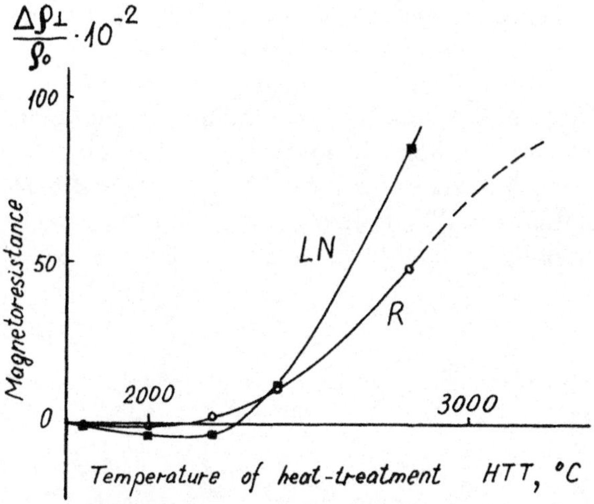

Fig. 44. Schematic drawing illustrating the influence of change in HTT on the magnetoresistance of pyrocarbon in the magnetic field $H = 12.7\,\text{kOe}$; LN, liquid-nitrogen temperature: R, room temperature. (Mrozowski and Chaberski 1956.)

in which an increased carrier mobility in the magnetic field is ascribed to the ordering of localized magnetic moments, leading to the weakening of carrier scattering. The quantitative analysis made by Delhaes (1971) has revealed, however, that quantitative agreement between the Toyozawa model and the relevant experimental data requires too high a concentration of magnetic moments, which is inconsistent with the data of EPR measurements. Moreover, Toyozawa's model has to overcome a great difficulty to explain the high anisotropy of NM observed in carbon materials.

Taking into account the quantum corrections for the conductivity of disordered materials in the magnetic field offers a fresh possibility for explaining the NM-effect. It is known that in such materials one can observe NM and a number of other anomalies in the behaviour of kinetic and thermodynamic parameters (heat conduction, heat capacity, magnetic susceptibility, etc.). According to Altshuler (1978), Altshuler and Aronov 1979) and Altshuler et al. (1980, 1981, 1982) in order to explain the temperature and field dependences of kinetic coefficients in disordered substances, one should take into account the inelastic interaction of electrons with one another, with phonons, or with other dynamic degrees of freedom.

It is known that the quantum corrections to conductivity are the largest in a metallic system near the threshold of electron localization due to the increased disorder (i.e., near the threshold of the Anderson-type localization). Under these conditions quantum corrections lead to the NM-effect even if the electron–electron interaction is neglected. If, however, this interaction is taken into account, it provides an additional contribution to the magnetoresistance, whose sign is determined by the sign of the electron–electron coupling constant. The analysis of the experimental data on NM in carbon materials has revealed their good agreement, both qualitative and quantitative, with the theory of quantum corrections. It quite may be that these corrections provide one of the principal mechanisms for NM in disordered carbon materials, in which the variation of HTT allows the degree of disorder to be varied near the threshold of mobility.

CHAPTER 6

Quantum Oscillation Effects

6.1. *General remarks on quantum oscillation effects*

As is known, quantum oscillation effects are caused by the periodic (in the inverse magnetic field) change in the electron or hole density of states brought about by the successive crossing of the Fermi level by the Landau levels. In semiclassical magnetic fields, the period of quantum oscillations $\Delta(H^{-1})$ in the inverse magnetic field is given, according to the Lifshitz–Onsager relation, by the area of the extremal cross section S_{extr} of the Fermi surface with a plane perpendicular to the magnetic field

$$\Delta(H^{-1}) = \frac{2\pi e\hbar}{cS_{\mathrm{extr}}}. \tag{101}$$

Therefore, quantum oscillation effects may yield direct and accurate information about the values of the extremal Fermi-surface cross sections corresponding to various orientations of the magnetic field, i.e., information about the shape and structure of the Fermi surface. Less accurately, quantum oscillation data enable one to determine the cyclotron masses m_c, the relaxation times τ and the effective g-factors of carriers.

In order to study the electron energy spectra of metals, semimetals and highly doped semiconductors, one commonly uses the results of measurements of the magnetoresistance oscillations, the so-called Shubnikov–de Haas (ShdH) effect, and magnetic susceptibility oscillations, the so-called de Haas–van Alphen (dHvA) effect. Apart from these effects, one can observe oscillations of other kinetic and thermodynamic parameters. All quantum oscillations of this type have a single nature.

6.2. *Oscillation effects in monocyrstalline graphite*

In our analysis of oscillation phenomena, we shall proceed from the most general model of the Fermi surface in graphite, considered in sect. 4.4.

Let us examine the oscillation frequencies that should be observed, within such a model, if the magnetic field is directed along the c-axis (fig. 12). They are as follows:

(1) The highest frequency f_e, which determines the cross section of the

majority-electron surface of constant energy by the plane $k_z = 0$ (the extremal majority-electron cross section S_e).

(2) The somewhat lower frequency f_h, which determines two equal extremal cross section, S_h, of the majority-hole surface of constant energy.

(3) The low frequency $f_{k_z = \pi/c_0}$, corresponding to the cross section of the "neck" of the majority-hole surface of constant energy at the point $\boldsymbol{H}$ of the Brillouin zone.

(4) The slightly lower frequency f^{h}_{m}, which determines the cross section at the point $\boldsymbol{H}$ of the minority-hole surface.

(5) The frequency f^{e}_{m}, which determines the extremal cross section of the minority-electron surface. This frequency is also low, but it is unknown whether it is higher or lower than the frequencies $f_{k_z = \pi c_0}$ and f^{h}_{m}.

(6) Finally, the frequency f_{out}, which is the lowest of all and which determines the cross section of the outrigger pieces connecting the surfaces of constant energy for majority electrons and majority holes.

Each of these frequencies has its own characteristic dependence on the angle θ between the direction of the magnetic field $\boldsymbol{H}$ and the c-axis, which makes it possible not only to classify the frequencies, but also to obtain direct information on the shape and structural details of corresponding surfaces of constant energy. Let us examine the angular dependences of the above frequencies in the order they have just been listed.

For the majority-electron surface of constant energy the frequency f_e should increase with increasing θ in the same way as, in a first approximation, it does for an ellipsoid strongly elongated in the direction of the c-axis. If the electron surface had no outrigger pieces by which it is connected with the hole surface, then nothing would preclude the observation of its maximum cross sections for the magnetic field $\boldsymbol{H}$ directed normally to the c-axis ($\theta = \frac{1}{2}\pi$). The presence of trigonal warping and outrigger pieces complicates the situation. Thus, because of trigonal warping the frequency and amplitude of oscillations must depend on the direction of the projection of the vector $\boldsymbol{H}$ upon the basal plane; and the above-mentioned dependences on angle φ, which determines the orientation of this projection in the basal plane, should be periodic with a period of 120°.

Because of the presence of outrigger pieces the oscillation amplitude should practically go to zero every time when the extremal cross section of the surface corresponding to a given direction of $\boldsymbol{H}$ will cross one of the outrigger pieces. A decrease (or disappearance) of oscillations will then be associated with the fact that in the vicinity of an outrigger piece the electron orbit passes close to a saddle point, at which the electron velocity goes to zero.

If, however, outrigger pieces are disrupted, one may expect an additional modulation of the amplitude and frequency of the oscillations when the projection of the magnetic field rotates in the basal plane with the period of 120°.

Since the structure of the majority-electron surface of constant energy is more complicated (especially near its pointed ends) than that of a simple ellipsoid, this should lead to a decrease of the oscillation amplitude near the surface's maximum cross sections, which greatly impedes their observation.

In the case of majority holes, the dependence of f_h on angle θ should be as follows. With increasing θ the frequency increases, in a first approximation, just as it does for an ellipsoid. But as soon as the cross section enters the region of the "neck" at $k_z = \pi/c_0$, the frequency should double, because in this region, the extremal cross sections will be approximately twice as high as before. Naturally, the oscillation amplitude should then drastically decrease not only because of the increased cross section, but also because of the closeness of the corresponding hole orbit to the saddle point on the surface of constant energy.

Thus, knowing the critical value of θ at which the oscillations due to majority holes drastically decrease or very probably disappear, one may determine the cross section of the "neck" by a plane perpendicular to the *c*-axis.

The following two low frequencies, $f_{k_z=\pi/c_0}$ and $f^{\rm h}_{\rm m}$, should be very close to each other, because in the SWMC-model their difference is given by the small value of the spin–orbit splitting energy at the points H (it is of the order of 2×10^{-4} eV). The difference between these frequencies may, in principle, be found from the beat frequency of low-frequency oscillations occurring for $\boldsymbol{H} \| c$-axis if a sufficiently large number of oscillation peaks is observed. Thus, for instance, in the case when these frequencies will differ from each other by 5%, one beat envelope will contain about 20 oscillation periods. However, one should bear in mind that if the extremal cross sections corresponding to these frequencies are very close to each other, then their difference will be undetectable in oscillations, because the Landau levels will be smeared in energy by $\delta\varepsilon \sim \hbar/\tau$ (where τ is the relaxation time for carriers). Therefore, for $\boldsymbol{H} \| c$-axis one may, in general, observe only one oscillation frequency (instead of the two, almost equal, frequencies $f_{k_z=\pi/c_0}$ and $f^{\rm h}_{\rm m}$). Even in the case, however, the difference between these frequencies should inevitably manifest itself when the magnetic-field vector $\boldsymbol{H}$ departs from the *c*-axis at some angle θ, because then their angular dependences will differ from each other in character. Indeed, for the minority holes the frequency $f^{\rm h}_{\rm m}$ should increase with increasing θ as it would do in the case of an ellipsoid, i.e., slower than for a cylinder approximating the ellipsoid in the extremal cross section for $\boldsymbol{H} \| c$-axis.

On the other hand, for the cross sections of the "neck" of the majority-hole surface, the frequency $f_{k_z=\pi/c_0}$ should increase much more rapidly (see fig. 15). Therefore, one may expect that the difference between these frequencies will rapidly increase as the angle θ increases. Simultaneously, the amplitude of the oscillations corresponding to the majority holes (the frequency $f_{k_z=\pi/c_0}$) should decrease faster than for the minority holes ($f^{\rm h}_{\rm m}$). As a result, with increasing θ the oscillations with the frequency $f_{k_z=\pi/c_0}$ may become unobservable. At the same time, it would be absolutely groundless to expect that the oscillations from the minority-hole ellipsoid should vanish up to $\theta = 90°$.

According to Dresselhaus (1974), the shape of the minority-electron surfaces of constant energy appears to be close to an ellipsoid. However, the orientation of such ellipsoids in the Brillouin zone and their anisotropy have not yet been determined. One may expect only that the angular dependences of the corresponding oscillation frequencies f_m^e should in general be determined by the cross sections of three ellipsoids turned through 120° relative to one another. For the lowest frequency f_{out}, associated with the narrow outrigger pieces connecting the surfaces for the majority electrons and holes, the disappearance of oscillations for the value of θ at which the cross sections limited by the outrigger-piece surface will sharply increase and will occupy some regions of the surfaces of majority electrons and holes should be expected. For lower values of θ, the oscillation frequency should increase somewhat more rapidly than for the cylinder approximation of the outrigger pieces.

Let us now turn to the results of experimental investigations of the oscillation effects in graphite and see whether and to what extent these results agree with the a priori picture outlined above on the basis of the general SWMC-model for the electron spectrum of graphite.

As the mean free path of carriers should be large in order to allow observation of quantum oscillations, these effects have been studied mostly in perfect single crystals of natural or synthetic graphite (kish-graphite and pyrolytic graphite). The ShdH- and dHvA-effects in graphite have received most study and are known in greater detail than the other oscillations observed in this substance (e.g., those of the Hall coefficient R_H, thermal electromotive force, acoustic absorptivity, as well as thermomagnetic and other types of quantum oscillations).

Some of the oscillatory effects observed in graphite are illustrated in figs. 45–49, which show typical curves for the ShdH effect (fig. 45), the dHvA effect (fig. 46), the Hall coefficient R_H (fig. 47), thermal electromotive force S_{xx} (fig. 48), and thermomagnetic oscillations (temperature oscillations of a specimen in the magnetic field under adiabatic conditions) (fig. 49).

The quantum oscillations of the various parameters of graphite are of the same nature and yield on the whole identical information about the electron energy spectrum. For this reason we shall, in what follows, not discriminate between the data on the various oscillation effects, putting emphasis on the results pertaining to the Fermi-surface structure and the values of certain characteristics of carriers in graphite.

In the earliest works devoted to the study of oscillation effects in graphite for $\boldsymbol{H}\|c$-axis, two oscillation periods were discovered: (1) $\Delta(H^{-1}) = 1.5 \times 10^{-5}$ and (2) $\Delta = 2.1 \times 10^{-5}\ \text{Oe}^{-1}$. The shorter period, corresponding to the maximum Fermi-surface cross section, was associated with the maximum cross

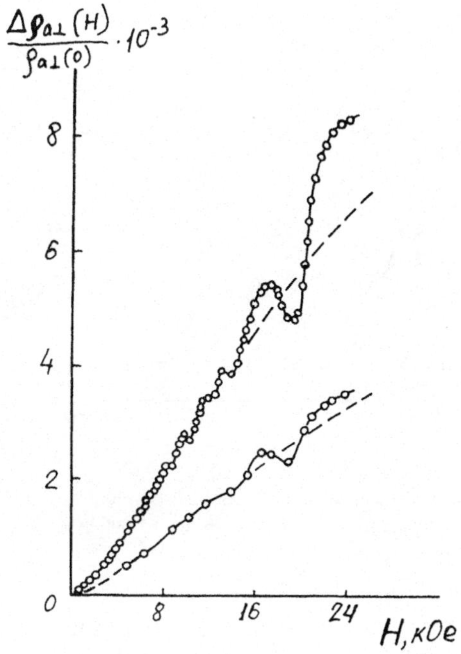

Fig. 45. Oscillations of the transverse magnetoresistance $\Delta\rho_{a_\perp}(H)/\rho_{a_\perp}(0)$ in two specimens of natural graphite of different degrees of perfection. $T = 4.2\ K$, $\boldsymbol{H}\|c$. (Soule 1958b.)

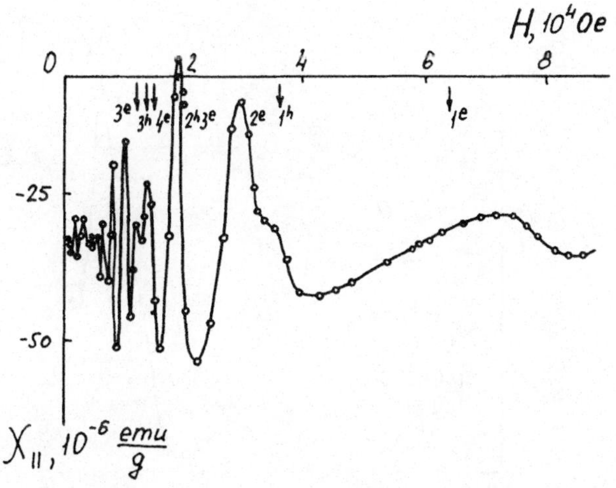

Fig. 46. Oscillations of the differential magnetic susceptibility in pyrolytic graphite for $\boldsymbol{H}\|c$ and $T = 4.2$ K. The fields at which the Landau levels for electrons (e) and holes (h) are crossed by the Fermi level are indicated by arrows. (Brandt et al. 1980b.)

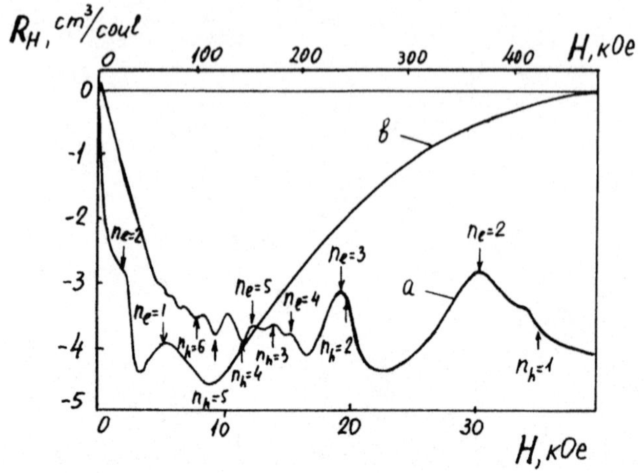

Fig. 47. Oscillations of the Hall coefficient R_H in kish-graphite as a function of magnetic field at $T = 4.2$ K: curve a, up to 40 kOe; curve b, up to 500 kOe. (Brandt et al. 1974.)

section for the majority-electron surface at the points K of the Brillouin zone, whereas the longer period was associated with the maximum cross section for the majority-hole surface near the points H. Oscillations with a much longer period, $\Delta(H^{-1}) = 13.5 \times 10^{-5}$ Oe^{-1} (practically, an order of magnitude longer), which correspond to minority carriers, were found first by Soule et al.

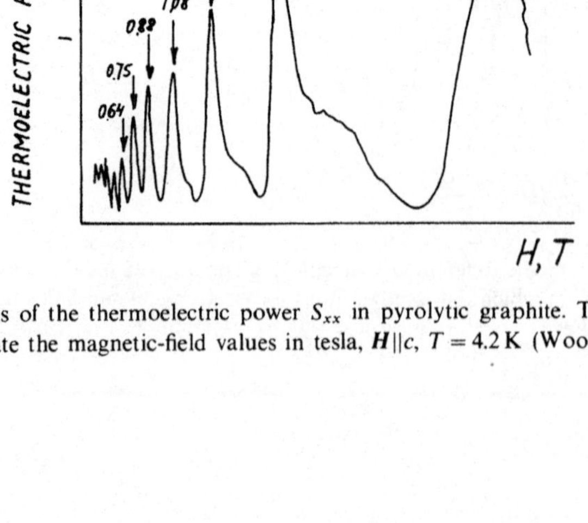

Fig. 48. Oscillations of the thermoelectric power S_{xx} in pyrolytic graphite. The numbers at the peaks indicate the magnetic-field values in tesla, $\boldsymbol{H} \| c$, $T = 4.2$ K (Woollam 1971c.)

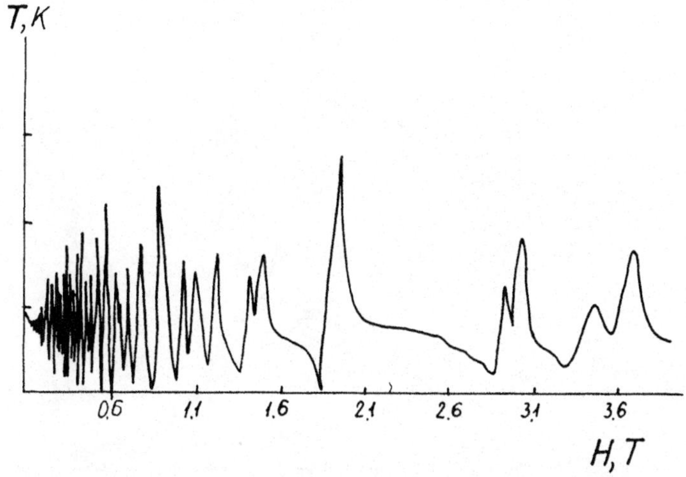

Fig. 49. Thermomagnetic oscillations in pyrolytic graphite. $\boldsymbol{H}\|c, T = 0.4\,\mathrm{K}$. (Chong Hon Yew 1979.)

(1964). Subsequently, Williamson et al. (1965), Anderson et al. (1967), and Woollam (1971a) have established that in weak magnetic fields ($H \lesssim 1$ or $2\,\mathrm{kOe}$) in highly perfect single-crystal graphite oscillations occur with a period $\Delta(H^{-1})$ equal to about $13.5 \times 10^{-5}\,\mathrm{Oe}^{-1}$ or $\simeq 30.3 \times 10^{-5}\,\mathrm{Oe}^{-1}$, or with both these periods simultaneously. To our knowledge, for $\boldsymbol{H}\|c$ no oscillations with other periods have been observed so far.

In order to answer the question of with which parts of the Fermi surface these oscillations are associated, one should examine their angular dependences (figs. 50–52). The character of these dependences leaves no doubt that high-frequency oscillations refer to the majority-electron and majority-hole surfaces of constant energy. The fact that the high-frequency oscillations have not been observed for angles θ exceeding 80° (fig. 52) may be due, as has already been noted above, to the complicated structure of the surfaces for this angular range. Therefore, from the angular dependences shown in figs. 50–52 nothing can be said about the fine structure of the surfaces in the region separating their electron and hole parts. The data on the anisotropy of the oscillation period due to the rotation in the basal plane of the magnetic-field component perpendicular to the c-axis are not available at present, either.

It will be noted that the low-frequency oscillations observed for $\boldsymbol{H}\|c$ cannot be associated with the cross sections of the "outrigger pieces" connecting the surfaces for the majority electrons and holes, because the cross-sectional areas corresponding to these frequencies are only about ten times as small as the areas of the extremal cross sections for the majority electrons and holes, whereas the average radii of these cross sections are only three times as short

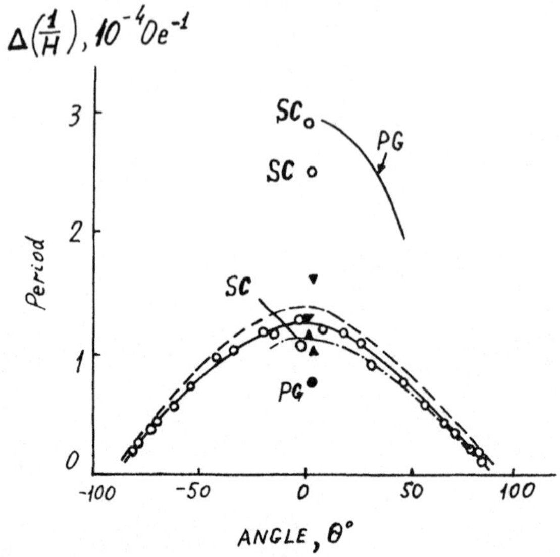

Fig. 50. Oscillation period $\Delta(H^{-1})$ for the minority carriers in single-crystal (SC) and highly ordered pyrolytic graphite (PG). Upper curve: ○ Williamson et al. (1965) and Soule (1964); — Woollam (1971a). Lower curves ○, ▼, ▲ Williamson et al. (1965); ● Flood (1969); -- Soule (1969); ——, -·- Woollam (1971a). The variation in period with θ, the angle between the c-axis direction and the magnetic field direction, is also shown. (Woollam 1971a.)

as those for the electron and hole ones. At the same time, according to the SWMC-model, the outrigger-piece cross sections should be at least one (or two) order of magnitude smaller than the cross sections corresponding to the low-frequency oscillations observed. This means that the oscillations due to the outrigger-piece cross sections should have periods one or two orders of magnitude longer. But since the low-frequency oscillations are observed in fields H less than 1–2 kOe, one would expect that the oscillations associated with the outrigger pieces exist for $H < 100$ Oe. For their observation, it is necessary to have specimens characterized by a very large mean free path for charge carriers, i.e., specimens of an exceptional, not yet attained, purity. Therefore, the nonobservation of oscillations due to outrigger pieces does not mean their nonexistence. Such oscillations are still unobserved because the outrigger pieces are either very narrow or are disrupted. If one disregards the fine structure of the majority-carrier surfaces, then one may conclude that they have the shape of ellipsoids strongly elongated in the direction of the c-axis, their anisotropy being 12.1:1 for holes and 17.3:1 for electrons.

In the angular range $\theta \lesssim 55°$, the angular dependences of the extremal cross sections and of the cyclotron masses of electrons and holes are well described by the relations S, $m_c \sim (\cos\theta)^{-1}$, which are typical of a surface of constant energy represented by a cylinder directed along the c-axis.

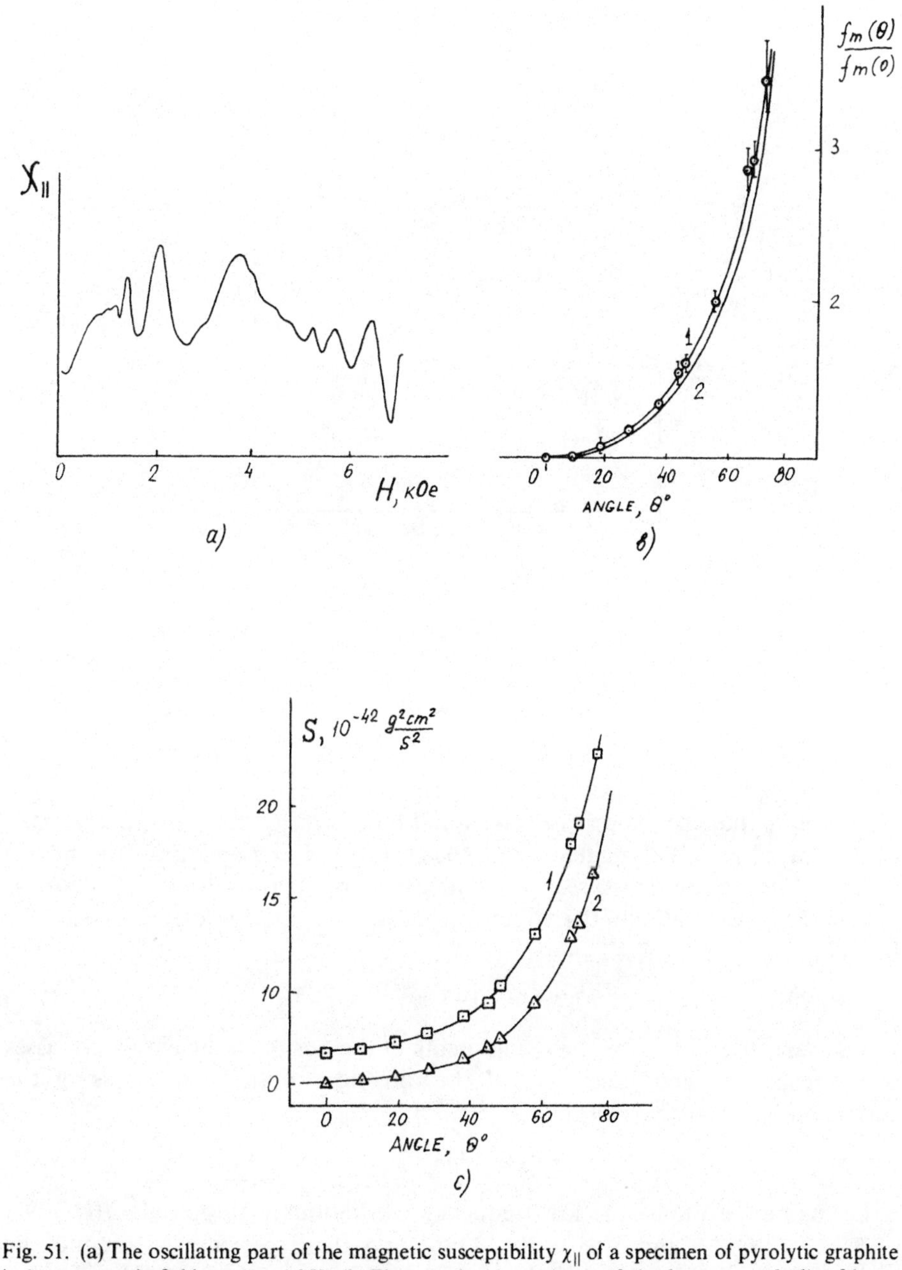

Fig. 51. (a) The oscillating part of the magnetic susceptibility $\chi_{\parallel}$ of a specimen of pyrolytic graphite in low magnetic fields at $T = 4.2$ K; (b) The angular dependence of the frequency, $f_m(\theta)$, of long-period oscillations corresponding to minority carriers in pyrolytic graphite in low magnetic fields (curve 1; ⊙ experimental points); the dependence $f_m(\theta) \sim \cos^{-1}\theta$ (curve 2); (c) Angular dependences of the extremal Fermi surface cross-sections in pyrolytic graphite: curve 1 ⊡ majority electrons, curve 2 △ majority holes. (Kuvshinnikov 1980.)

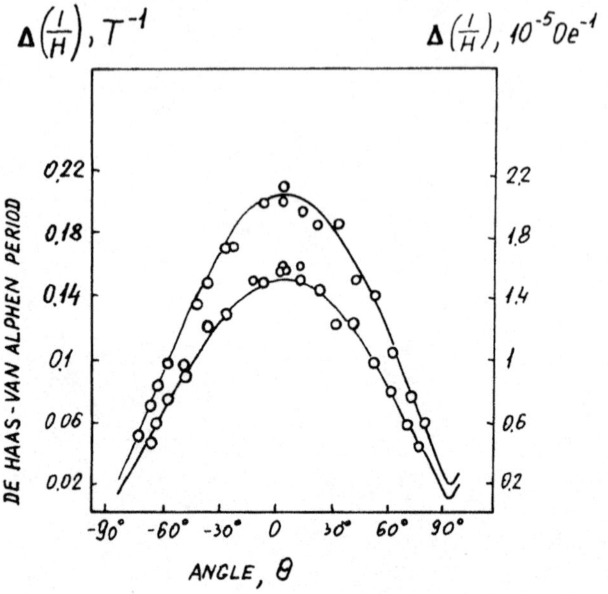

Fig. 52. Dependence of the majority carrier oscillation periods $\Delta(1/H)$ on θ, the angle between the magnetic field and the c-axis. (From Woollam 1971c.) A comparison of the data obtained by Soule et al. (1964) for natural single crystals (solid line) and those obtained by Woollam (1971c) for synthetic, highly ordered pyrolytic (PG3) crystals (open circles).

Assuming that the majority-electron and hole surfaces may be approximated by prolate ellipsoids elongated along the c-axis, and extrapolating the angular dependences of the cross sections and cyclotron masses to their values at $\theta = 90°$, one can obtain the following values for the cyclotron masses for $\boldsymbol{H} \perp c$:

$$m_c^e(90°) = 0.9\, m_0, \qquad m_c^h(90°) = 0.47\, m_0. \tag{102}$$

In this case, the values for the components of the tensor of the effective masses of electrons, m^e, and holes, m^h, in the direction of the major axis of the corresponding ellipsoid will be:

$$m^e(90°) \simeq 5.7\, m_0, \qquad m^h(90°) \simeq 14\, m_0. \tag{103}$$

Let us now examine the low-frequency oscillations with periods $\Delta(H^{-1}) \simeq 13.5 \times 10^{-5}\,\mathrm{Oe}^{-1}$ and $30.3 \times 10^{-5}\,\mathrm{Oe}^{-1}$. To these periods correspond the cross sections $S_{m1} \simeq 0.79 \times 10^{-42}\,\mathrm{g^2\,cm^2/s^2}$ and $S_{m2} \simeq 0.36 \times 10^{-42}\,\mathrm{g^2\,cm^2/s^2}$. The angular dependences of the low-frequency oscillations are illustrated in figs. 50 and 51b. As the angle θ between the direction of $\boldsymbol{H}$ and the c-axis increases, the cross section S_{m1} increases slower than according to the law

$S_{m1} \sim (\cos\theta)^{-1}$, as the cross section of a cylinder whose axis coincides with the c-axis would do. This means that, in a first approximation, the surface with the extremal cross section S_{m1} is that of an ellipsoid.

As for the cross section S_{m2}, the data on the character of its change with varying angle θ are more scarce. According to Kuvshinnikov (1980), who investigated long-period oscillations in monocrystalline pyrolytic graphite, with increasing θ the area of the cross section S_{m2} first increases somewhat more rapidly than the cross-sectional area of a cylindrical surface (fig. 51b). However, for $\theta \sim 50$–$60°$ the rate of increase in S_{m2} with angle appears to decrease, so that the difference between the dependence of S_{m2} on θ and the corresponding dependence for a cylinder or a greatly prolate ellipsoid becomes practically indistinguishable. Such a form of the θ-dependence of S_{m2} may suggest that the corresponding surface of constant energy is an indistinct dumb-bell (i.e., the one in which the cross section of the "neck" differs but insignificantly from the noncentral extremal cross sections) and with a small error may be approximated by the surface of a prolate ellipsoid.

Since the measurement of only one oscillational effect does not allow one to distinguish between the surfaces of constant energy of the electron- and hole-types, it appears to be impossible to draw any conclusion about the belonging of the surface with the extremal cross section S_{m2} to the former or latter type, on the basis of the oscillation data available. One can definitely state merely that the θ-dependence of S_{m2} presented in fig. 51b does not correspond to the angular dependence that could be expected for the cross section of the neck of the majority-hole surface at the point H in accordance with the SWMC-model for the graphite spectrum. Indeed, in this model, with increasing θ the angular dependence of the neck of the majority-hole surface should more and more rapidly depart from the angular dependence of an approximating cylinder; and in the angular range $\theta \sim 70°$ the difference between these curves should exceed 50%.

Hence one may conclude that the cross sections S_{m1} and S_{m2} may refer only to the minority carriers, electrons or holes, provided, of course, that the corresponding oscillations are indeed associated (as is commonly assumed) with the energy quantization in a magnetic field for the minority groups of carriers. Such an assumption made for both S_{m1} and S_{m2} may, however, give rise to some doubt. Indeed, Eberhard et al. (1975), who studied the angular dependences of oscillation frequencies due to majority electrons (f_e), majority holes (f_h) and the frequency f_{m1} that determines the cross section S_{m1}, have found that for any θ in the range from 0 to 80° the relation $f_e - f_h = 2f_{m1}$ is valid. Hence, they have concluded that the frequency f_{m1} arises from the simple effects of beats between the frequencies f_e and f_h, and, therefore, is not associated with any new group of carriers.

A similar explanation of the nature of oscillations with the frequency f_{m1} has been advanced also by Nakao (1976), who made the conjecture that the frequency f_{m1} represented the frequency of the oscillations arising due to the magnetic breakdown between the orbits of the majority and minority holes at the points H of the Brillouin zone. These magnetic breakdown phenomena could arise through a γ_3-induced mixing of the magnetic energy levels of the ε_3-bands when they cross the magnetic energy levels of the ε_1- and ε_2-bands near the point H.

In connection with the aforesaid, we would like to call the reader's attention to the fact that the peculiar character of oscillation phenomena in graphite, associated primarily with its manifestly complex crystal structure, has not yet been studied in sufficient detail. For the relevant data, the reader is referred to a paper by Kuvshinnikov (1980), who employed three different techniques [i.e., magnetoresistance $\rho_{a\perp}(H)$, surface impedance $Z(H)$, and magnetic susceptibility $\chi_{\parallel}(H)$ measurements] to study oscillation effects in pyrolytic and natural graphite at pressures up to 20 kbar. He has found that in pyrolytic graphite the oscillations of $\rho_{a\perp}(H)$, $Z(H)$, and their derivatives are monochromatic in fields $H \lesssim 50$ kOe over the entire pressure range investigated. At the same time, the oscillations of $\chi_{\parallel}(H)$ and its field derivatives measured in the same specimens represent a superposition of two almost identical frequencies, the amplitudes being also almost identical. The difference between the oscillation curves for $\partial Z(H)/\partial H$, $\rho_{a\perp}(H)$, and $\chi_{\parallel}(H)$ is illustrated in fig. 53. No such difference can be observed in single-crystal natural graphite. The observed features of oscillations of the kinetic and thermodynamical characteristics of pyrolytic graphite cannot be explained on the basis of the currently used concepts of the energy spectrum of this material.

One may thus conclude that many questions regarding the parameters of the minority-carrier group still remain open. But the very existence of at least one additional minority group, whose representatives are characterized by a lower density and a higher mobility than the majority carriers, appears to be beyond doubt. Indeed, the assumption that there exist only majority electrons and holes is at variance with the behaviour of the galvanomagnetic coefficients of graphite in the range of weak magnetic fields (see sect. 5.2) and also with the magneto-optical data (see ch. 7) and the data on the graphite-spectrum transformation under pressure (see ch. 8).

On the basis of the data available at present, the following structure of the Fermi surface for graphite seems the most probable (see fig. 12). In the neighbourhood of one of the edges of the Brillouin zone the Fermi surface consists of:

(1) the majority-electron surface of constant energy, which is a warped, prolate (along the c-axis) quasi-ellipsoid with its centre at the point K of the Brillouin zone and the ratio of its axes 17.3:1; no experimental data on the fine structure of this surface in the region of its greatest curvature are available;

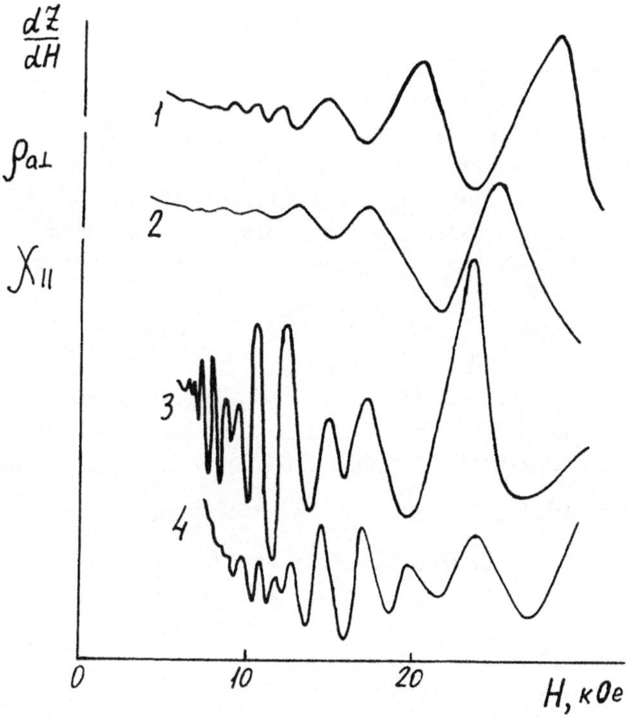

Fig. 53. Magnetic field dependences of: (1) $dZ/\partial H$, the surface impedance with respect to magnetic field; (2) $\rho_{a\perp}$, the oscillating part of magnetoresistance at a pressure of 1 bar, and (3,4) $\chi_{\|}$, the oscillating part of the magnetic susceptibility, at pressures of 4.9 and 17.3 kbar, respectively. $T = 4.2$ K and $\boldsymbol{H} \| C$ in pyrolytic graphite.

(2) two majority-hole surfaces of constant energy, which have the shape of warped dumb-bells; each of them may be approximated by two warped, greatly prolate (with their major axis along *c*-axis) quasi-ellipsoids with an axis ratio 12.1 : 1; these are connected by the "neck" with a cross section about ten times as small as their extremal cross sections for $\boldsymbol{H} \| c$. No experimental data on the structure of these surfaces in the region of their contact with the electron surface are available;

(3) two minority-hole surfaces of constant energy, apparently of ellipsoidal (or possibly, dumb-bell-like) shape, whose centres are at the points *H* of the Brillouin zone; no reliable experimental data on the anisotropy of these surfaces are available at present.

The question of the existence in graphite of the minority-electron group and the possible shape of the corresponding surfaces of constant energy remains open.

6.3. Parameters of the electron energy spectrum of graphite obtained from oscillation measurements

Table 11 summarizes the data on the extremal cross sections for the Fermi surface of natural and pyrolytic graphite for $\boldsymbol{H}\|c$, obtained from oscillation measurements. The oscillation data make it possible to calculate separately the densities of electrons and holes. Approximating the Fermi surface by an ellipsoid, one can calculate these densities by using the following expression:

$$n_{\mathrm{e,h}} = \tfrac{1}{3}\pi\left(\frac{4e}{\pi hc}\right)^{3/2} \eta N[\Delta_0(H^{-1})]^{3/2}, \tag{104}$$

where η is the surface anisotropy ($\eta \sim$ 12–17), N is the number of ellipsoids in the Brillouin zone, $\Delta_0(H^{-1})$ is the period of quantum oscillations for $\boldsymbol{H}\|c$.

For natural and pyrolytic graphite the calculation yields similar results:

$$n_{\mathrm{e}} \simeq 2.6 \times 10^{18}\,\mathrm{cm}^{-3} \quad \text{and} \quad n_{\mathrm{h}} \simeq 2.2 \times 10^{18}\,\mathrm{cm}^{-3}.$$

Table 12 gives the cyclotron masses m_{c} and the Dingle temperatures T_{D}* for carriers in graphite, determined from oscillation data (for comparison, the cyclotron masses found by the cyclotron resonance technique are also presented).

The results presented in tables 11 and 12 have been obtained by different authors who used different measuring techniques; yet, on the whole, they show a good agreement. The deviations of some results which go beyond the range of accuracy are most likely the consequence of the different parameters of the graphite specimens (lattice parameters, dopant concentrations, etc.) used in the measurements by the different workers. This point has already been discussed in sect. 4.4.

The sign of the carriers may be determined by comparing the oscillation curves for the magnetoresistance, the Hall electromotive force, and the thermal electromotive force. When a successive Landau level appears, one observes a peak in the density of states at the Fermi level and, therefore, the diagonal magnetoconductivity tensor components, σ_{xx}, σ_{yy}, and σ_{zz}, are maximal. In a compensated semimetal, such as graphite, the off-diagonal (Hall) magnetoconductivity tensor components are much smaller than the diagonal ones: $\sigma_{xy} \ll \sigma_{xx}$ or σ_{yy}.

As a result, the peaks of quantum oscillations of magnetoresistance correspond to the dips of the transverse magnetoresistance tensor components:

$$\rho_{xx} = \frac{\sigma_{xx}}{\sigma_{xx}^2 + \sigma_{xy}^2} \simeq \frac{1}{\sigma_{xx}}. \tag{105}$$

*The Dingle temperature T_{D} is related to the relaxation time for carriers at a Landau level by the expression $T_{\mathrm{D}} = \pi h/k\tau$.

Woollam (1970b, 1971c) has shown that when a sequential Landau level crosses the Fermi level, the Hall magnetoresistance tensor components ρ_{xy} and ρ_{xy} have been compared in fig. 54. One may also compare these curves with Similarly, as a Landau level crosses the Fermi level one can observe a hole dip and an electron peak in oscillations of thermal emf S_{xx}. Oscillations of ρ_{xx} and ρ_{xy} have been compared in fig. 54. One may also compare these curves with the curve for the oscillation dependence of S_{xx} in fig. 48.

Knowing the sign of carriers, one may in turn determine the sign of the SWMC-model parameter γ_2. To do so, one should know the position of oscillation extrema in high magnetic fields when the Landau electron and hole levels with quantum numbers $n = 2, 1$ cross the Fermi level.

Sugihara and Ono (1966) calculated the position of oscillation peaks for graphite in a magnetic field taking account of Fermi-energy variations. The dependence of ε_F on $\boldsymbol{H}$ insignificantly changes the period of quantum oscillations from its semiclassical value, but significantly affects the position of oscillation extrema in a magnetic field. At negative values, electrons are located at the points K of the Brillouin zone, and the Landau electron level with $n = 1$ should cross the Fermi level in fields $H \simeq 70\,\text{kOe}$, while the corresponding hole level should be observed for $H \simeq 35\,\text{kOe}$. The good agreement of the measured positions of oscillation peaks with the calculated ones (fig. 54) furnishes an additional, independent support for the correctness of the above classification of the electron and hole surfaces of constant energy for majority carriers in graphite (see fig. 12).

Oscillations in high magnetic fields enable one to determine yet another important characteristic of the spectrum, i.e., the magnitude of the spin splitting of Landau levels or the value of the effective g-factor for carriers. The magnitude of Landau-levels splitting $\Delta\varepsilon_s$ for opposite spin orientations may be expressed either through the parameter called the spin mass m_s or through the cyclotron mass m_c and the effective g-factor ($\tilde{g}$):

$$\Delta\varepsilon_s = \frac{e\hbar}{m_s c} H = \tilde{g}\frac{e\hbar}{2m_c c} H. \tag{106}$$

Let H_n^- and H_n^+ denote the values of the magnetic field corresponding to the emergence of the Landau upper and lower sublevels with quantum number n. It can readily be shown that, for a parabolic law governing carrier dispersion, the spin mass m_s and the effective g-factor may be expressed in terms of $H_n^{\pm}$ and m_c as follows:

$$m_s = \frac{m_c}{2n+1}\left(\frac{H_n^- + H_n^+}{H_n^- - H_n^+}\right), \tag{107}$$

$$\tilde{g} = 2\frac{m_c}{m_s} = 2(2n+1)\frac{H_n^- - H_n^+}{H_n^- + H_n^+}. \tag{108}$$

Table 11
Extremal Fermi-surface cross sections, S, in natural and pyrolytic graphite for $\boldsymbol{H} \| c$ obtained from oscillation measurement data.

Investigation method	Extremal cross section S for $\boldsymbol{H} \| c$ in units of 10^{-42} $g^2 cm^2/s^2$	Ref.*
Majority electrons		
ShdH effect under pressure	7.04	[1]
ShdH effect and Hall emf oscillations	6.60	[2]
	6.68 ± 0.24	[3]
	6.77 ± 0.17	[3]
	6.64	[4]
	7.23 ± 0.3	[5]
	6.9 ± 0.3 (pyrolytic)	[5]
	7.08	[6]
	6.52 ± 0.06 (pyrolytic)	[6]
ShdH effect and thermal emf oscillations	6.64	[7]
ShdH effect	7.04 ± 0.13	[8]
	6.53 (natural)	[9]
	6.59 (natural)	[9]
dHvA effect	6.45	[2]
	6.44	[10]
	7.04 ± 0.14	[11]
	6.64	[8]
	6.94	[12]
	7.09 ± 0.01	[13]
Majority holes		
ShdH effect	5.13 ± 0.1	[8]
	4.72	[9]
	4.86	[9]
ShdH effect and Hall emf oscillations	4.95	[2]
	5.00 ± 0.17	[3]
	5.10 ± 0.11	[3]
	4.85	[4]
	4.74 ± 0.20	[5]
	4.76 ± 0.20	[5]
	4.62 ± 0.21	[5]
	4.96	[5]
	4.8 ± 0.05	[6]
ShdH effect and thermal emf oscillations	4.85	[7]
dHvA effect	4.94	[2]
	4.63	[10]
	5.11 ± 0.05	[11]
	4.85	[8]
	5.16	[12]
	5.09 ± 0.005	[13]

Table 11 (cont'd)

Investigation method	Extremal cross section S for $\boldsymbol{H}\parallel c$ in units of 10^{-42} g^2 cm^2/s^2	Ref.*
Minority carriers		
dHvA effect	0.79 ±0.17	[8]
	0.42	[8]
	0.79 ± 0.005	[11]
	0.47 ± 0.015	[11]
	0.79 ± 0.05	[13]
ShdH effect	0.78	[14]
	0.36	[14]
ShdH effect and Hall emf oscillations	0.474	[4]
	0.35 ±0.030	[5]
	0.355±0.030	[5]
Hall emf and thermal emf oscillations	0.36 ± 0.01	[6]
	0.89 ± 0.07	[7]

*References:
[1] Itskevich and Fischer (1967).
[2] Berlincourt and Steele (1955).
[3] Soule (1958a,b).
[4] Hishiyama et al. (1972).
[5] Brandt et al. (1974).
[6] Kuvshinnikov (1980).
[7] Woollam (1971c).
[8] Soule et al. (1964).
[9] Cooper et al. (1971).
[10] Shoenberg (1952).
[11] Williamson et al. (1965, 1966).
[12] Spry and Scherer (1960).
[13] Anderson et al. (1967).
[14] Woollam (1971a).
[15] Galt et al. (1959).
[16] Inoue (1962).
[17] Noziers (1959).
[18] Suematsu and Tanuma (1972).
[19] Doezema et al. (1979).

Equations (108) and (107) are approximate; they are valid for rather small values of the Landau-levels spin splitting at which the change in Fermi energy occurring as the magnetic field increases from H_n^+ to H_n^-, may be neglected. The spin splitting in graphite is vividly illustrated in figs. 54 and 55. In order to estimate the effective g-factor for electrons and holes by formula (108), one may utilize the field values $H_1^- = 74$ kOe and $H_1^+ = 66$ kOe for electrons and $H_1^- = 36.43$ kOe and $H_1^+ = 34.12$ kOe for holes. This estimation yields:

$$\tilde{g}_\mathrm{e} = 0.343 \quad \text{and} \quad \tilde{g}_\mathrm{h} = 0.196, \tag{109}$$

hence it follows that in fields $\simeq 70$ kOe for electrons the spin splitting of Landau levels is about 5.8 times as small as the cyclotron splitting, whereas for holes in fields $\simeq 35$ kOe the cyclotron splitting is about 5.1 times as high as the spin splitting. The corresponding spin masses for electrons and holes will be

$$m_\mathrm{s}^\mathrm{e} = (0.33 \pm 0.01)m_0, \qquad m_\mathrm{s}^\mathrm{h} = (0.40 \pm 0.01)m_0. \tag{110}$$

Table 12
The cyclotron masses and the Dingle temperatures for carriers in graphite from oscillation measurements data.

Investigation method	m_c/m_0	T_D (K)	Ref.*
Majority electrons			
ShdH effect	0.057 ± 0.001	0.56	[8]
		0.62	[9]
		2.58	[9]
ShdH effect and Hall emf oscillations	0.057		[4]
	0.060	3.40	[3, 8]
	0.051 ± 0.005		[5]
	0.058 ± 0.003		[6]
dHvA effect	0.062		[2, 8]
	0.057	0.64	[2, 8]
	0.070	0.9 ± 0.6	[8, 10]
		0.6	[8, 10]
cyclotron resonance	0.066		[15]
	0.053		[16]
	0.066		[17]
	0.061 ± 0.002		[11]
	0.058 ± 0.001		[18]
	0.059 ± 0.001		[18]
Majority holes			
ShdH effect	0.039 ± 0.001	0.80	[8]
		1.70	[8]
		4.21	[9]
ShdH effect and Hall emf	0.030	3.60	[3, 8]
oscillations	0.39		[4]
	0.029 ± 0.005		[5]
	0.038 ± 0.002		[6]
dHvA effect	0.039	0.71	[2, 8]
	0.038		[2, 8]
	0.036	1.5	[8, 10]
cyclotron resonance	0.031		[15, 17]
	0.039 ($\hbar\omega$ = 3.68 meV)		[19]
	0.040 ($\hbar\omega$ = 10.45 meV)		[19]
	0.045 ($\hbar\omega$ = 14.7 meV)		[19]
	0.044 ($\hbar\omega$ = 15.8 meV)		[19]
Minority carriers			
ShdH effect and Hall emf oscillations	0.0038 ± 0.0004		[6]
dHvA effect		1.50 ± 0.10	[11]
	0.0038 ± 0.0004		[6]
	0.0023		[8]
	0.004 ± 0.0004		[11]

*References: see table 11.

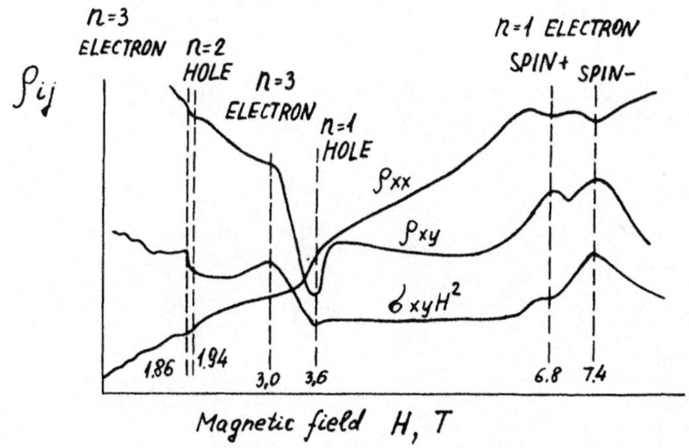

Fig. 54. Oscillation of the magnetoresistance ρ_{xx} and Hall-tensor component ρ_{xy} in the high-field region for pyrolytic graphite. The off-diagonal conductivity tensor component $\sigma_{xy}H^2$ is also shown. Minima in ρ_{xy} and $\sigma_{xy}H^2$ correspond to hole carriers, and maxima correspond to electron carriers. (Woollam 1971c.)

The fact that the cyclotron splitting for electrons and holes in graphite is greater than the spin splitting implies that in a magnetic field the conduction-band boundary (the 0^- level for electrons) and the valence-band boundary move towards each other on the energy scale, thereby reducing the band overlap. As has already been mentioned before (in sect. 4.6), this

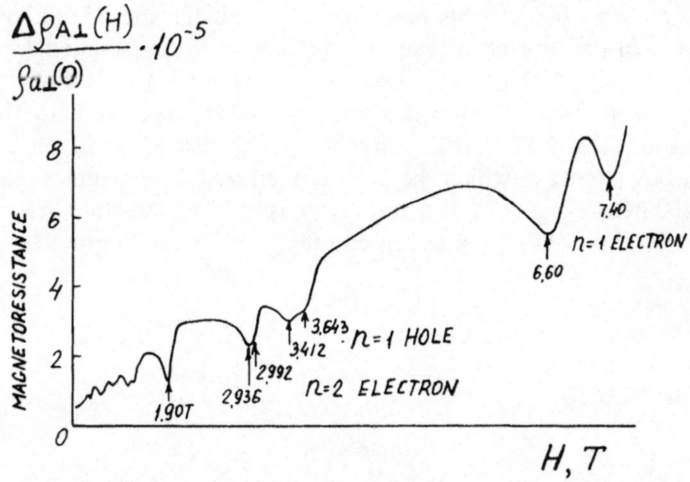

Fig. 55. Oscillation of magnetoresistance with magnetic field, showing the splitting of the Landau levels due to spin–orbit interaction for pyrolytic graphite (PG5) at $T = 1.1$ K. The numbers give the field value in tesla. (Woollam 1970a.)

behaviour of band boundaries should lead to the metal–semiconductor transition in a magnetic field.

According to Brandt and Svistova (1970), in the parabolic approximation, the expression for the transition field $H_{M\to S}$ is:

$$H_{M\to S} = \frac{\varepsilon_\pi^0}{\mu_B}\frac{1}{B}, \tag{111}$$

where

$$B = m_0\left(\frac{1}{m_c^e} - \frac{1}{m_s^e} + \frac{1}{m_c^h} - \frac{1}{m_s^h}\right)$$

$$= m_0\left[\frac{1}{m_c^e}(1 - \tfrac{1}{2}\tilde{g}_e) + \frac{1}{m_c^h}(1 - \tfrac{1}{2}\tilde{g}_h)\right], \tag{112}$$

and ε_π^0 is the band overlap for $H = 0$. Setting $\varepsilon_\pi^0 = 2|\gamma_2| \simeq 40$ meV and using the values of $m_c^{e,h}$ from table 12 and those of the g-factors as given by (109), we find that $H_{M\to S} \simeq 150$ kOe. The difference between this value and the value obtained in the more exact calculation of Kapustin and Meilikhov (1975) ($H_{M\to S} \simeq 1250$ kOe) clearly suggests the role of the nonparabolicity of the spectrum of graphite when its characteristics in high magnetic fields are calculated.

In conclusion, we should stress that the metal–semiconductor transition in graphite [as well as the inverse transition from semiconductor to metal taking place in higher magnetic fields predicted by Kapustin and Meilikhov (1975)], associated with the motion of band boundaries in its single-particle spectrum, appears to be observable if one uses specimens of relatively low purity and perfection, where there is no transformation of the spectrum to the CDW-state. It seems that the specimens investigated by Brandt et al. (1974) and by Kuvshinnikov (1980), in which the peak of the transverse magnetoresistance in fields ~300 kOe typical of the transition to the CDW-state has not been observed (see sect. 5.4.2), may be representative for such materials.

CHAPTER 7

Optical and Magneto-Optical Effects

7.1. Optical properties of graphite

The study of optical properties of solids enables one to obtain information about the general character of the electron energy spectrum over a wide energy region (see fig. 7).

It is convenient to describe the optical properties of a substance by using the complex dielectric constant $\kappa(\omega, \boldsymbol{q}) = \kappa_1(\omega, \boldsymbol{q}) + i\kappa_2(\omega, \boldsymbol{q})$, which is a function of the frequency ω and the wave vector $\boldsymbol{q}$, and which is related to the complex refractive index $(n + ik)$ by

$$\kappa_1 + i\kappa_2 = (n + ik)^2. \tag{113}$$

In an anisotropic medium, the dielectric constant is a second-rank tensor. In the most general case, the components of this tensor are defined as the sum of the contributions from the intraband and interband electronic transitions κ_{int}, from free carriers κ_{car}, and from phonon excitations κ_{lat}. The remaining contributions, manifesting themselves mainly at high frequencies, are of no consequence for semimetals and may be left unspecified when the optical properties of graphite are examined.

For a solid with hexagonal lattice symmetry, such as graphite, the dielectric-constant tensor has only two independent components $\kappa_\perp$, and $\kappa_\parallel$, which correspond to two polarization directions of the electric-field vector $\boldsymbol{E}$ in an electromagnetic wave, i.e., $\kappa_\perp$ for $\boldsymbol{E} \perp c$ and $\kappa_\parallel$ for $\boldsymbol{E} \parallel c$. The marked difference between the values of these components determines the high anisotropy in optical properties of graphite. The component κ_{int} in graphite is rather large in the IR, visible, and near UV parts of the spectrum, while the components κ_{car} and κ_{lat} are important only in the IR region. And everywhere, except in the region of extremely low frequencies $\hbar\omega \lesssim 0.05\,\text{eV}$, in which the optical properties of graphite are determined by the contribution of free carriers, the contribution of the intraband and interband transitions, κ_{int}, is dominant. The influence of these transitions in graphite upon its optical properties to the lowest frequencies is associated with the double degeneracy of the energy level ε_3 along the *HKH* edge of the Brillouin zone, which is removed only as a result of the spin–orbit interaction leading to the splitting of the ε_3-band by $\sim 2.2 \times 10^{-4}\,\text{eV}$.

In metals, the dielectric constant as a function of $\boldsymbol{q}$ and ω may be found by measuring the reflection coefficient R using the familiar Kramers–Kronig relationship between the real and imaginary parts of κ. Such a procedure for finding the dielectric constant of graphite has been developed by Philipp and Taft (1959). For large dielectric-constant values, the relation between R and κ is given by

$$1 - R = \tfrac{1}{4}\mathrm{Re}\left(\frac{1}{\sqrt{\kappa}}\right), \tag{114}$$

where $(1 - R)$ gives the fraction of the incident-wave energy absorbed in matter. Other methods of determining the optical characteristics of metals involve the study of the energy and momentum distributions of primary or secondary electrons due to the inelastic scattering of a beam of monochromatic electrons incident upon a thin plate of the substance under study. The angular distribution of the scattered primary electrons will then permit one to construct the energy-loss function (ELF) F_{EL}

$$F_{\mathrm{EL}} = -\mathrm{Im}\left(\frac{1}{\kappa}\right) = \frac{\kappa_2}{\kappa_1^2 + \kappa_2^2}. \tag{115}$$

At the same time, the energy distribution of the secondary electrons outgoing in the direction opposite to that of the primary beam offers a possibility of finding the secondary-electron emission (SEE) spectrum, which is also associated with the features of the dielectrical function $\kappa(\omega, \boldsymbol{q})$.

The reflection spectrum of graphite is schematically illustrated in fig. 56. The same figure also shows, for comparison, the reflection spectrum of glassy

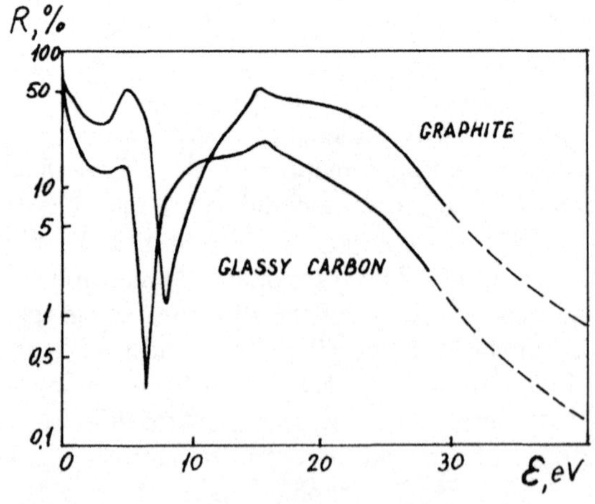

Fig. 56. The spectral dependence of the reflectance of graphite and glassy carbon. The extrapolations of the reflectance curves above 26 eV, used in the analysis of these data, are indicated by the dashed lines. (Taft and Philipp 1965.)

carbon. The distinguishing feature of the reflection spectrum due to graphite is an abrupt dip near an energy of $\hbar\omega \simeq 8\,\text{eV}$. In this part of the spectrum, a crystal of graphite is practically transparent (the "transparency window" of graphite). This interval of transparency divides the optical spectrum of graphite into two, qualitatively different regions of low ($0 \leqslant \hbar\omega \leqslant 8\,\text{eV}$) and high ($\hbar\omega > 8\,\text{eV}$) energies.

If the infrared region, where contributions provided by free carriers and phonon excitations are of importance, is left out of consideration, then in the low-energy region the optical spectrum of graphite is mainly determined by the contribution due to intraband and interband transitions for the π-band electrons. The wave functions for these electrons represent the $2p_z$ orbitals elongated along the c-axis, there being one π-electron per atom. It should be recalled that it is exactly the π-electrons that play the dominant role in the electrical conductivity of graphite. In the high-energy region $8\,\text{eV} \leqslant \hbar\omega \leqslant 25\,\text{eV}$, the optical spectrum of graphite is mainly determined by the interband transitions of electrons forming σ-bonds between the atoms in a carbon layer. The number of the σ-electrons per carbon atom is equal to three. A broad absorption peak near the energy $\hbar\omega \sim 15\,\text{eV}$ is due just to the interband transitions of σ-electrons. For graphite, the plasma frequency ω_p, which is determined by transitions of all electrons of the π- and σ-bands, is equivalent to an energy of $\hbar\omega_p \simeq 25\,\text{eV}$. Figure 57 shows the real (κ_1) and

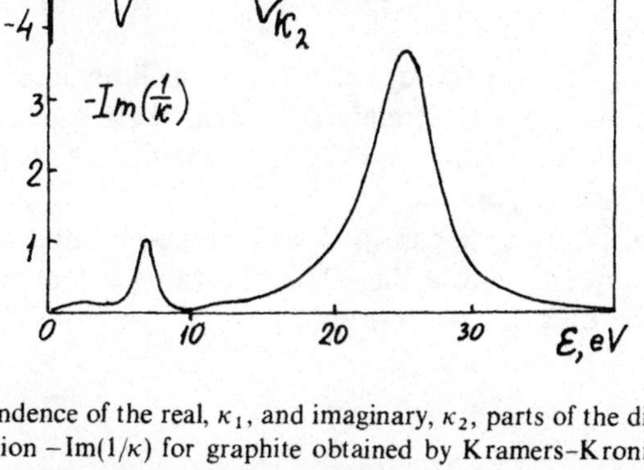

Fig. 57. The spectal dependence of the real, κ_1, and imaginary, κ_2, parts of the dielectric constant, and the energy-loss function $-\text{Im}(1/\kappa)$ for graphite obtained by Kramers–Kronig analysis of the curve of fig. 56. (Taft and Philipp 1965.)

imaginary (κ_2) parts of the dielectrical function of graphite and the EL function for this material as given by (115). Notice that in the region $\hbar\omega \leqslant 40$ eV the F_{EL} is characterized by two peaks, one at $\hbar\omega = 7$ and the other at $\hbar\omega = 25$ eV. The relatively low peak of F_{EL} at $\hbar\omega = 7$ eV corresponds, as Ichikawa (1958) first demonstrated, to the collective excitation of the π-electrons. In reality, this peak indicates the phenomenon of a hybrid resonance due to the cooperative behaviour of the π- and σ-electrons, because the latter take part in the screening effect and contribute positively to κ. Due to screening, the observed peak is at a considerably lower energy than it follows from a calculation for free π-electrons. The high resonance peak at $\hbar\omega \simeq 25$ eV is associated with plasma oscillations in which all four valence electrons of a carbon atom take part.

Summarizing, the features of the curves shown in fig. 57 represent three different effects:

(1) the free-electron contribution at very low energies, which determines the large negative value of κ_1;

(2) the contribution of interband transitions, which determines the non-monotonic behaviour of $\kappa_1(\hbar\omega)$ and $\kappa_2(\hbar\omega)$; and

(3) the plasma-oscillation contribution, to which the peaks of F_{EL} are due. The contribution from free carriers to the dielectric function is rather well described by the Drude relation (Drude 1902):

$$\kappa_{\mathrm{car}}(\omega) = -\frac{\omega_{\mathrm{p}}^2}{\omega(\omega + \mathrm{i}\tau^{-1})}, \tag{116}$$

where τ is the carrier relaxation time and ω_{p} is the plasma frequency.

Philipp and Ehrenreich (1963) have shown that in a crystal with atomic density N, the effective number n_{eff} of electrons per atom, taking part in transitions in the frequency range $0 \leqslant \omega \leqslant \omega_0$, can be estimated proceeding from the sum rule for the dielectric function

$$\left(\frac{2\pi^2 N e^2}{m^*}\right) n_{\mathrm{eff}} \simeq \int_0^{\omega_0} \omega\kappa_2(\omega)\,\mathrm{d}\omega \simeq -\int_0^{\omega_0} \omega\,\mathrm{Im}\left(\frac{1}{\kappa(\omega)}\right)\mathrm{d}\omega. \tag{117}$$

The dependence of n_{eff} on the upper integration limit in (117) is shown in fig. 58. One can distinctly see the frequency range where π-electrons are dominant in transitions. This is the range $\hbar\omega \lesssim 10$ eV, where $n_{\mathrm{eff}} \lesssim 1$. One can also notice that for $\hbar\omega \gtrsim 25$ eV, $n_{\mathrm{eff}} \simeq 4$.

Thus a crystal of graphite can (qualitatively) be treated as the superposition of two different media, whose dielectric constants are determined either only by π-electrons or only by σ-electrons:

$$\kappa = \kappa^{\pi} + \kappa^{\sigma}, \tag{118}$$

where

$$\kappa^{\pi} = \kappa_1^{\pi} + \mathrm{i}\kappa_2^{\pi}, \qquad \kappa^{\sigma} = (1 + \delta\kappa_1^{\sigma}) + \mathrm{i}\delta\kappa_2^{\sigma}.$$

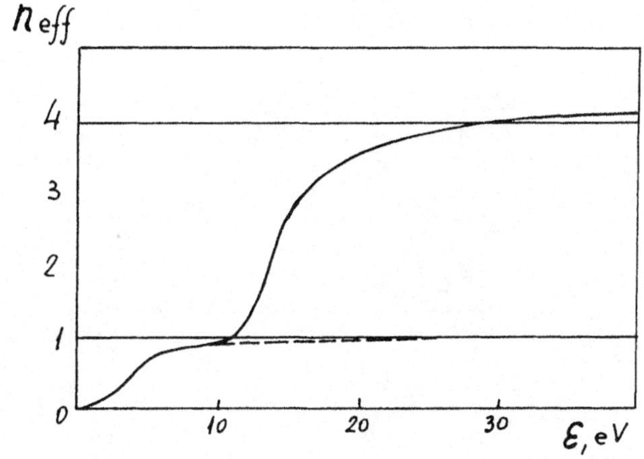

Fig. 58. The effective number, n_{eff}, of electrons per atom versus ε obtained from numerical integration of experimental κ_2 and $-\text{Im}(1/\kappa)$. The n_{eff} is defined by eq. (117). (Taft and Philipp 1965.)

In fact, however, the σ-electrons affect, via the screening effect, the position of the plasma resonance of π-electrons by shifting it downwards in energy by a factor of $[1 + \delta\kappa_1^{\sigma}(\omega)]^{1/2}$. As has been shown by Taft and Philipp (1965), taking this effect into account enables one to estimate the π-electron resonance energy as $\hbar\omega_p = 7.4\,\text{eV}$, in good agreement with experiment.

Much more information on the energy structure of graphite can be extracted from the optical spectra obtained with the polarized light.

Figures 59a,b,c,d show the most typical reflection spectra obtained for $\boldsymbol{E} \perp c$ and $\boldsymbol{E} \| c$ and also show the restored imaginary part of the dielectric function in the energy range $0 \leqslant \hbar\omega \leqslant 40\,\text{eV}$. As in the case of unpolarized light, one can see in the reflection spectrum for $\boldsymbol{E} \perp \text{c}$ two intense peaks at 5 and 15 eV, the "transparency window" at 8 eV, and several less distinct features. In the optical spectra for $\boldsymbol{E} \| c$, there are, in addition to the major peaks observed for $\boldsymbol{E} \perp c$, a doublet at 6 eV and peaks at 10, 18, and 32 eV. It is of interest to note that, according to optical data, the width of the forbidden band in the graphite energy spectrum at $\boldsymbol{E} \| \text{c}$ constitutes about 6 eV, i.e., in passing from $\boldsymbol{E} \perp c$ to $\boldsymbol{E} \| c$ the width of the forbidden band increases by more than five orders of magnitude (for $\boldsymbol{E} \perp c$, the forbidden band width is essentially determined solely by the magnitude of the spin splitting of the ε_3-level, equal to $\sim 10^{-4}\,\text{eV}$). However, graphite does not become a dielectric under these conditions, because for $\boldsymbol{E} \| \text{c}$ there is a weak absorption at $\hbar\omega < 6\,\text{eV}$, suggestive of the metallic character of the conductivity in graphite along the c-axis.

Let us now examine how much the structure of the optical spectra of graphite corresponds to details of its calculated band spectrum. The main optical transitions in the energy spectrum of graphite have been shown in fig.

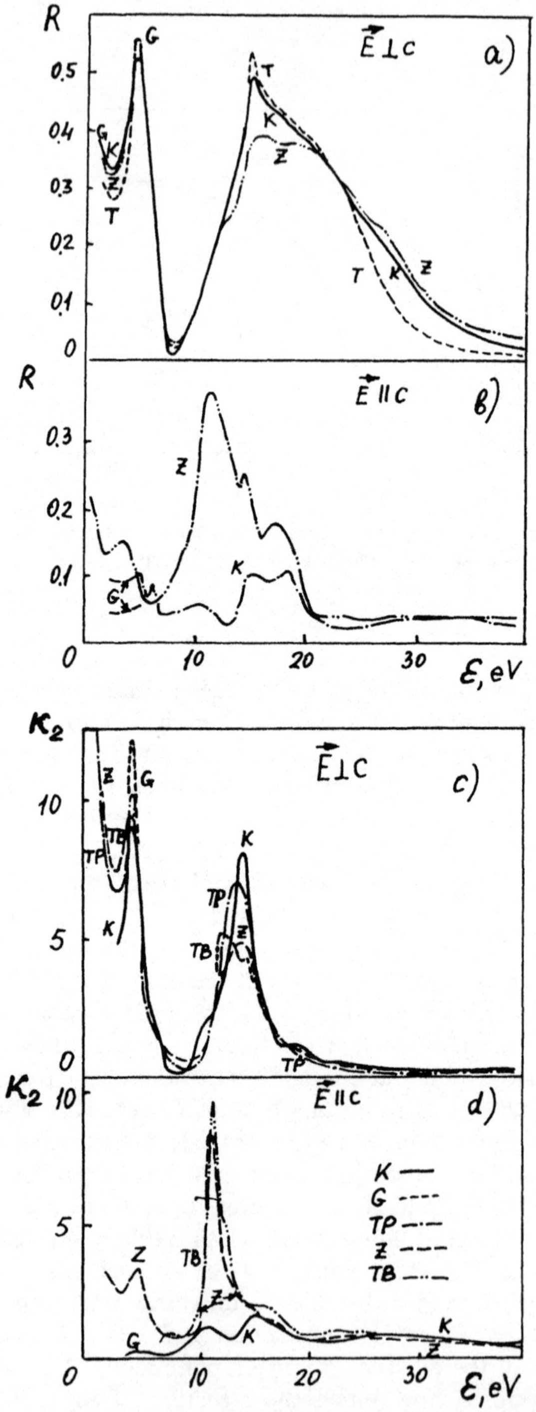

Fig. 59. Reflectance spectra for $E \perp c$ (a), $E \parallel c$ (b) and the energy dependence of κ_2 for $E \perp c$ (c) and $E \parallel c$ (d) for graphite according to the data of: Klucker et al. (1974), (K); Zeppenfeld (1968), (Z); Taft and Philipp (1965), (T, TP); Greenaway et al. (1969), (G); Tossatti and Bassani (1970), (TB).

7b and in table 4 (see sect. 4.1). We have already mentioned that the largest contribution to the structure of the optical spectra comes from the interband transitions corresponding to peaks of electron density of state in bands, i.e., transitions between those points of the Brillouin zone at which the derivatives $\partial\varepsilon/\partial \boldsymbol{k}$ are close to zero.

Let us mention the most characteristic features of the optical spectra of graphite for which it is possible to identify the corresponding optical transitions.

In the infrared region, the peaks in the reflection spectrum observed for $\hbar\omega = 0.108$ and $0.197\,\mathrm{eV}$ are associated, as suggested by Brillson et al. (1971) and Nemanich et al. (1977a), with the optical modes of lattice-atom vibrations perpendicular to the plane of the layer and in the plane of the layer, respectively. According to Bellodi et al. (1975) and Nakao (1979), the peak at $\hbar\omega \simeq 0.8\,\mathrm{eV}$ may be due to electron transitions at the point K of the Brillouin zone from the valence ε_2-band into the unfilled states that lie above the Fermi level in the conduction ε_3-band ($\hbar\omega = 0.706\,\mathrm{eV}$) and also to the transition from the occupied states in the valence ε_3-band to the free states in the ε_1-band ($\hbar\omega = 0.889\,\mathrm{eV}$). Johnson and Dresselhaus (1973) as well as Taft and Philipp (1965) attributed the major peak in the reflection spectrum at $\hbar\omega \simeq 5\,\mathrm{eV}$ to transitions between the π-band at the points M [see transition (4) in table 4]. Transitions between σ-bands give rise to peaks in the reflection spectrum near an energy of 13 eV [see transition (3) in table 4].

We shall now briefly list the most important optical transitions in graphite for $\boldsymbol{E}\|c$ and $\boldsymbol{E}\perp c$:

$\simeq 4.6\,\mathrm{eV}$; $M_{2g}^{-} - M_{2u}^{-}$, transition allowed for $\boldsymbol{E}\perp c$;

$\simeq 6\,\mathrm{eV}$; $\Gamma_{3g}^{+} - \Gamma_{3u}^{+}$, transition allowed for $\boldsymbol{E}\perp c$;

$\simeq 6\,\mathrm{eV}$; $M_{2u}^{-}, K_1^{+} - M_{2u}^{-}, K_3$, transitions forbidden at the points M and K and allowed along the line MK (with changing momentum) for $\boldsymbol{E}\|c$;

$\simeq 11\,\mathrm{eV}$; $M_{1g}^{+} - M_{2u}^{-}$ or $M_{2g}^{+} - M_{1u}^{+}$, transitions allowed for $\boldsymbol{E}\|c$;

$\simeq 14\,\mathrm{eV}$; $M_{2u}^{+} - M_{2g}^{+}$, transitions allowed for $\boldsymbol{E}\perp c$.

The most complete scheme of bands and optical transitions in graphite is presented in the article of Wills et al. (1974).

7.2. *Some remarks on magneto-optical phenomena in graphite*

Magneto-optical phenomena include effects due to electron transitions between the Landau levels in graphite in a magnetic field, which are induced by an electromagnetic excitation. The observation of such transitions allows one to determine directly the energy intervals between the Landau levels, and thus extract extensive information on the electron energy spectra of solids.

Resonance transitions between the Landau levels within a single band may occur when the electromagnetic wave frequency coincides with the cyclotron frequency ω_c for the respective group of carriers, or with harmonics of this frequency. The resonance effects in graphite in a magnetic field which result from intraband transitions are called the cyclotron resonance. Apart from these resonance effects, resonance transitions between Landau levels belonging to various bands are possible. At a fixed magnetic field, they may also take place at discrete values of the electromagnetic wave frequency, which is then determined, aside from the cyclotron frequencies of the carriers, by the magnitude of the energy gap between bands and by the position of the Fermi level. The interband resonance transitions in a magnetic field are responsible for the observed oscillating dependence of optical parameters of a metal [reflection or absorption coefficients, position of EL-function peaks (see (115)), etc.] upon the magnetic field strength or the wavelength of the incident light. In a narrower sense, the term "magneto-optical effects" describes precisely those phenomena which are due to interband transitions in a magnetic field. Experimentally, one may observe resonance transitions between the Landau levels associated with those points of the Brillouin zone at which the electron density of states is maximal. Such points coincide either with band extrema or with saddle points in the electron spectrum.

In graphite, magneto-optical transitions can be observed best at the points K and H on the vertical edges of the Brillouin zone (fig. 9), at which electron and hole extrema are located (see fig. 11). The Landau levels at these points for $\boldsymbol{H} \| c$ without trigonal warping are shown in fig. 17. In order to determine under what conditions resonance transitions in a magnetic field are observable, one may use, as a first approximation, the approximate expressions for Landau levels given by (68), which are, however, sufficiently exact in the neighbourhood of the points K (see sect. 4.5).

Evidently, a resonance transition will then occur if the energy of an electromagnetic-wave quantum $\hbar\omega$ satisfies the equation

$$\begin{aligned} \hbar\omega &= \varepsilon_c(n_c) - \varepsilon_v(n_v) \\ &= (n_c + \tfrac{1}{2})\hbar\omega_c^* - (n_v + \tfrac{1}{2})\hbar\omega_v^*. \end{aligned} \tag{119}$$

Condition (119) obeys the selection rules

$$\delta n = n_c - n_v = \pm 1, \tag{120}$$

which are valid only in the limit $\gamma_3 \to 0$. The plus and minus sign in (120) refer to the directions of the circular-wave polarization, which coincide with those of electron or hole precession in a magnetic field. Substituting into (119) the values of $\omega_{c,v}^*$ from (69) and those of $(m_c^{e,h})^{-1}$ from (70), we find that the condition for a resonance transition has the form

$$\hbar\omega = (n + 1 \mp 4\beta)\frac{e\hbar}{m^* c} H_n^{\pm}, \tag{121}$$

where m^* and β have been defined in (70) (see sect. 4.5), $n = n_c + n_v$ and $H_n^\pm$ are the resonance transition fields [the plus and minus sign have the same meaning as in (120)]. From (121) it follows that the inverse values of the resonance magnetic fields linearly depend on n:

$$\frac{1}{H_n^\pm} = \frac{1}{\hbar\omega}\frac{e\hbar}{m^*c}(n + 1 \mp 4\beta), \tag{122}$$

and, at a known value of $\hbar\omega$, the tangent of the slope angle determines the value of $1/m^*$.

The difference between the resonance fields for one and the same value of n but for different polarizations is given by:

$$\frac{H_n^+ - H_n^-}{H_n^+ + H_n^-} = \frac{4\beta}{n+1}, \tag{123}$$

which enables one to determine the parameter β from the dependence of its left-hand side upon $1/(n + 1)$. As shown by Schroeder et al. (1970), Dresselhaus (1974) and Nakao (1976), eqs. (119)–(123), which have been obtained in the parabolic approximation and with $\gamma_3 = 0$, are rather good approximations if they are used in describing the interband transitions at the points K of the Brillouin zone only in the limit of large quantum numbers. At small quantum numbers, equation (61) should be solved numerically.

The resonance transitions at the points H of the Brillouin zone may be denoted by the number of the corresponding Landau level in the conduction or valence band, because, according to Toy et al. (1977), transitions at this point are the superposition of transitions $n \to n - 1$ and $n + 1 \to n$ (i.e., transitions $n_v = n$; $n_c = n + 1$ and $n_v = n + 1$, $n_c = n$). Proceeding from the exact solution given by (63) of eq. (61), one can demonstrate that the transition energy at the points H and the resonance transition field H_n are related by an expression which depends only on the parameters γ_0 and Δ of the SWMC-model:

$$\hbar\omega = \tfrac{1}{2}\{[\Delta^2 + 4nbH_n]^{1/2} + [\Delta^2 + 4(n+1)bH_n]^{1/2}\}, \tag{124}$$

where

$$b = \tfrac{3}{2}e\frac{a_0^2\gamma_0^2}{\hbar c} \simeq 1.6 \times 10^{-3}(\text{eV})^2/\text{T}.$$

7.3. *Cyclotron resonance in graphite*

The first experiments to observe the cyclotron resonance in monocrystalline graphite were performed by Galt et al. (1956). These workers used a circularly polarized electromagnetic wave and observed the resonance using the Faraday configuration, in which the magnetic field $\boldsymbol{H}$ is parallel to the vector $\boldsymbol{q}$ of wave propagation and perpendicular to the crystal surface coinciding with the plane of the layer: $\boldsymbol{H}\|\boldsymbol{q}\|c$. The experimental study of the cyclotron resonance in graphite has stimulated a number of theoretical works, whose authors have calculated the magnetic field values for the fundamental resonance and its harmonics, transition matrix elements, line shapes and also obtained the selection rules for resonance transitions which have (with trigonal warping taken into account) the form

$$n = 1 + 3q, \tag{125}$$

where $q = 0, \pm 1, \pm 2, \ldots$; the plus and minus signs indicate, as above, the resonances due to electrons and holes, respectively.

The selection rules given by eq. (125) predict that, in a magnetic field H_l, resonances occur at frequencies ω which are given by

$$\omega = l\omega_c = (3q + 1)\omega_c, \tag{126}$$

whence

$$\pm l = \frac{m_c^{e,h} c}{eH_l}\omega,$$

where $l = 3q + 1$; $q = 0, 1, 2, 3, \ldots$, for electrons and $q = 0, -1, -2, \ldots$, for holes.

From (126) it follows that the electron cyclotron resonance should be observed for counterclock-wise circular wave polarization (viewed along the magnetic-field direction) for $H = H_c, \frac{1}{4}H_c, \frac{1}{7}H_c, \frac{1}{10}H_c$, etc., whereas the hole cyclotron resonance, for clockwise polarization and for $H = H_c, \frac{1}{2}H_c, \frac{1}{5}H_c, \frac{1}{8}H_c$, where H_c is the fundamental resonance field. Obviously, at an unchanged direction of circular wave polarization the reversal of the sign of the magnetic field makes it possible to go over from the hole resonances to the electron ones, and vice versa.

In order to determine the cyclotron mass, one usually constructs the dependence of the number of resonance harmonics, $l = 3q + 1$, on the reciprocal value of the corresponding magnetic fields, $1/H_l$ (see (126)), which represents a straight line with the dip-angle tangent equal to $(m_c^{e,h}c/e)\omega$.

The cyclotron resonance in highly perfect specimens of single-crystal graphite and pyrolytic graphite has received the most detailed study in the work done by Suematsu and Tanuma (1972). The resonance curves were recorded in a Faraday configuration ($\boldsymbol{H}\|\boldsymbol{q}\|c$) using circularly polarized electromagnetic waves with frequencies $\omega \simeq 23$ and 69 GHz. These authors obtained the field dependences of the first and second derivatives of the reflection coefficient R (or, what is the same, the absorption coefficient, equal to $1 - R$) of microwave energy in magnetic field at $T = 1.0$ and 1.6 K. The curves of $\mathrm{d}R/\mathrm{d}H = f(H)$ and $\mathrm{d}^2R/\mathrm{d}H^2 = f'(H)$ for the above specimens are shown in figs. 60a,b and

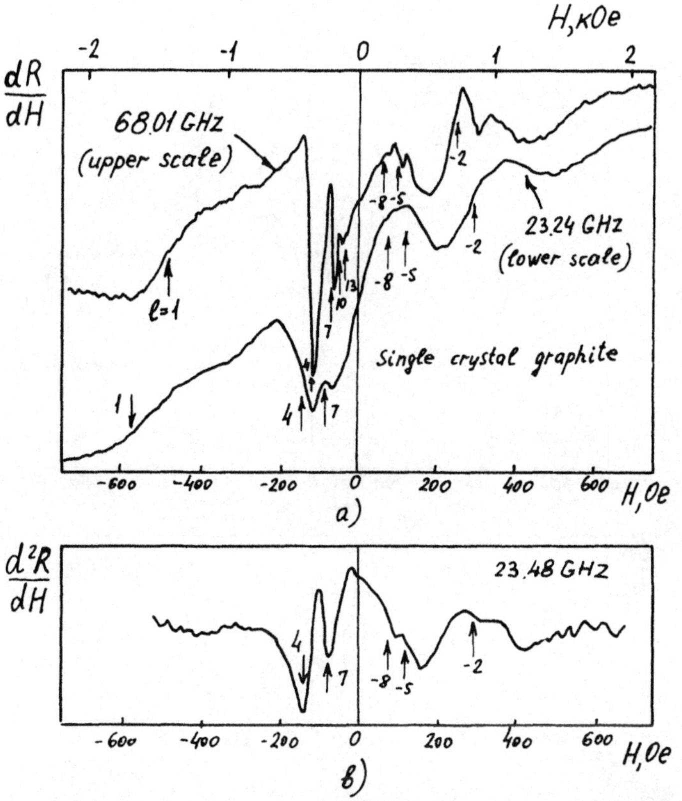

Fig. 60. The first (a) and second (b) field derivatives of microwave absorption, dR/dH and d^2R/dH^2, versus magnetic field $\boldsymbol{H}$ in single-crystal graphite. The measurements were done at 68.01 and 23.48 GHz with a magnetic field normal to the sample surface and along the c-axis. The negative-H and positive-H parts of the abscissa correspond to circularly polarized radiation sensitive to electrons and holes, respectively. The field scale used at 68.01 GHz is reduced by a factor of about 3 with respect to the scale at 23.24 GHz, so that the resonance field for all the absorption lines is on the same vertical, the other things being equal. The arrows indicate the fundamental ($l = 1$) and harmonic ($l = 3q + 1$) resonance fields. The oscillations observed at fields higher than 400 Oe on the d^2R/dH^2 curve are associated with the Shubnikov–de Haas effect for minority carriers. (Suematsu and Tanuma 1972.)

61a,b, respectively. In these figures the right-hand side on the abscissa (positive H) corresponds to the hole resonance and the left-hand side (negative H), to the electron one. For the convenience of comparison of corresponding resonance peaks, the scale of the magnetic fields for $\omega \simeq 69$ GHz is reduced by a factor of three in comparison with the scale for $\omega \simeq 23$ GHz. One can see that the resonances are much better resolved at the higher frequency, but the resonance curves are similar at both frequencies.

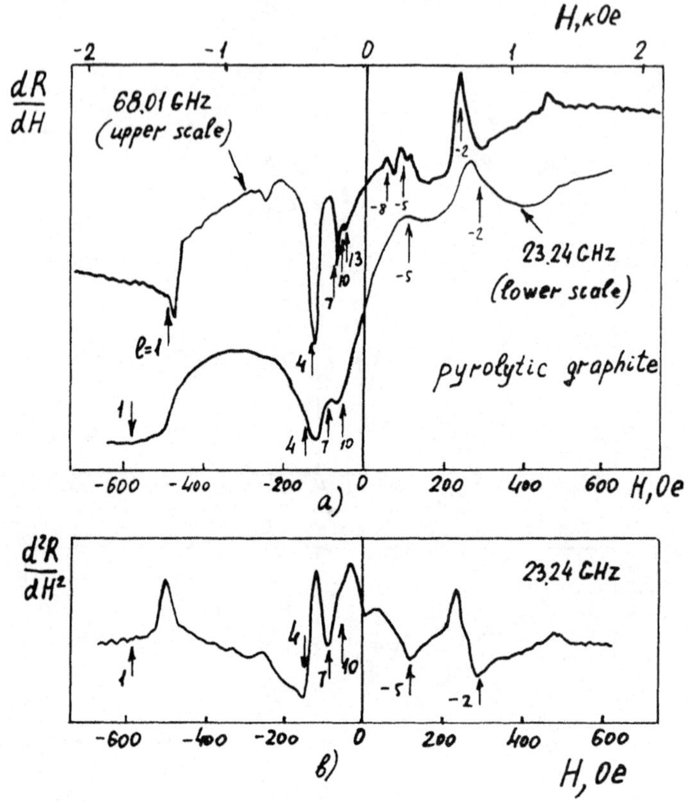

Fig. 61. The first (a) and second (b) field derivatives of microwave absorption, dR/dH and d^2R/dH^2, versus the magnetic field H in pyrolytic graphite. The measurements were done at 68.01 and 23.24 GHz by using circularly polarized radiation with the magnetic field applied normal to the sample surface and along the c-axis (see the caption of figure 60a, b). (Suematsu and Tanuma 1972.)

Figure 62a shows the resonance curves for single-crystal graphite (SCG) at various angle θ as obtained by Suematsu and Tanuma (1972). The corresponding angular dependences of peak positions for the various harmonics $l = 3q + 1$ are illustrated in fig. 62b. In the same figure are also shown solid curves that correspond to the relation

$$H_n(\theta) = \frac{H_n(0)}{\cos\theta}, \tag{127}$$

which corresponds to the Fermi surface represented by an open cylinder whose geometrical axis is directed along the c-axis. As can be seen from fig. 62b, for θ in the angular range from 0 to 70° the experimental angular dependences of resonance fields are described rather well by eq. (127).

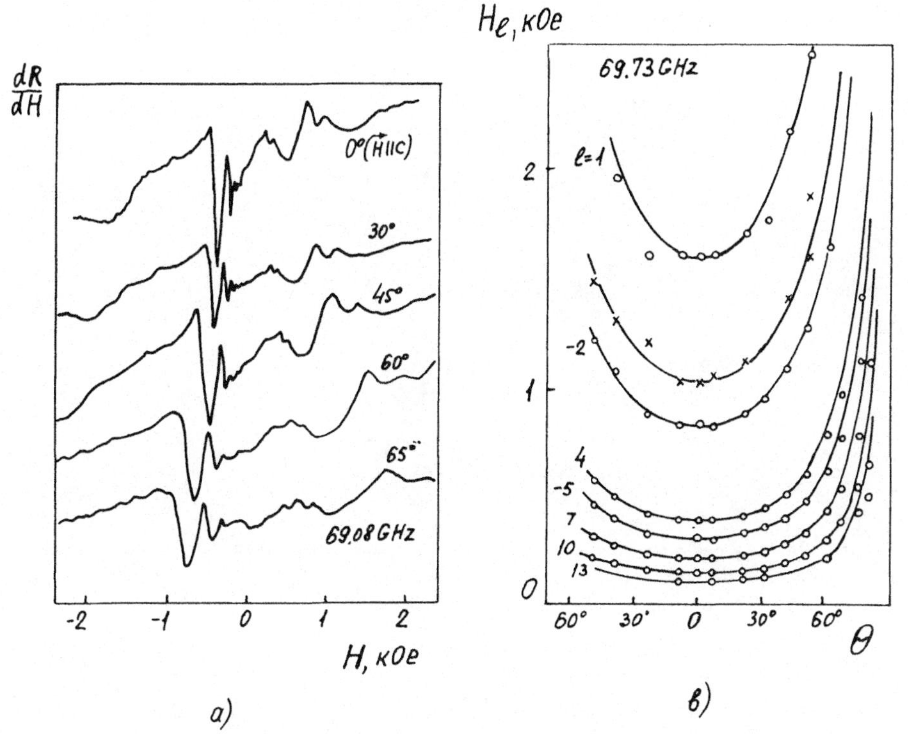

Fig. 62. (a) The variation of the first derivative of microwave absorption, dR/dH, with the magnetic field $\boldsymbol{H}$ for various angles θ between the magnetic field and the c-axis of the crystal. (b) Orientation dependence of the fundamental and harmonic resonances in SCG. Each solid line shows the $(\cos\theta)^{-1}$ dependence, where θ is the angle between the crystal's c-axis and the applied magnetic field. (Suematsu and Tanuma 1972).

The values for the cyclotron masses of carriers in graphite, obtained by different authors for a magnetic field oriented along the c-axis, have been given in table 12. The results of the calculation of the cyclotron masses of electrons m_c^e and holes m_c^h, made by Ushio et al. (1972) for various values of the wave vector k_z in the range from 0 to π/c_0, are shown in fig. 63. In the calculation, they used the following fixed values of the SWMC-model parameters γ_i: $\gamma_0 = 3.0$, $\gamma_1 = 0.392$, $\gamma_2 = -0.019$, $\gamma_4 = 0.193$, and $\Delta = -0.0002$ (eV). The parameter γ_3 was varied from 0.18 to 0.29 eV. The Fermi energy ε_F was taken equal to -0.023 eV. The electron cyclotron masses have been shown to change with varying parameter γ_3, the change being especially significant in the range of k_z values for which the electron Fermi surface makes contact with the hole Fermi surface (the region of "the outrigger pieces" of the Fermi surface). For the lower values of k_z the electron cyclotron mass weakly depends on k_z, which accounts for the sharpness of the experimentally observed electron cyclotron

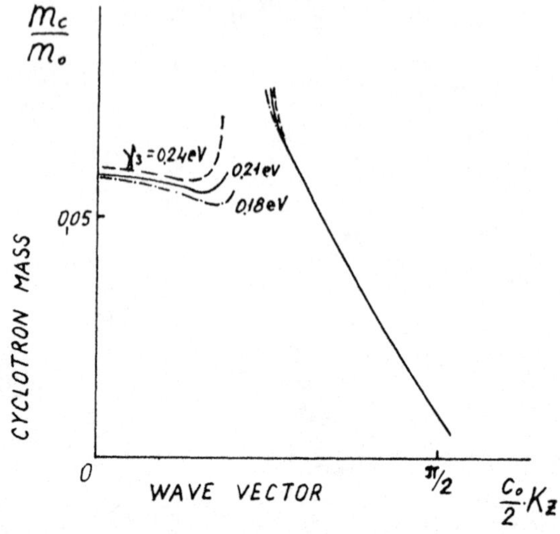

Fig. 63. Plots of cyclotron mass versus k_z for the case of $\gamma_3 = 0.18$ eV, 0.21 eV and 0.24 eV, respectively. The singularity in the mass distribution near $c_0 k_z/2 = \frac{1}{4}\pi$ corresponds to carriers at "outrigger" pieces of the Fermi surface. (Ushio et al. 1972.)

resonance peaks (see fig. 60a,b and 61a,b). In contrast to electrons, the hole cyclotron masses ($k_z > \pi/2c_0$) are practically independent of γ_3 and, in a first approximation, decrease linearly with increasing k_z. Because of this, hole resonance peaks are rather broad, which severely impedes a determination of the value of m_c^h corresponding to the extremal cross section for the Fermi surface of holes.

The parameter γ_3 has been determined by fitting the calculated resonance curve for $dR/dH = f(H)$ to the relevant experimental curve. The value for the warping parameter $\gamma_3 = (0.21 \pm 0.02)$ eV found from the best fit is considerably smaller than that found from the data of other experiments, in particular, the magneto-optical data.

Doezema et al. (1979), who observed the cyclotron resonance in graphite for frequencies in the far infrared range of electromagnetic oscillation ($\hbar\omega = 3.7$–15.8 meV) with circularly polarized waves, have detected some features of interband transitions suggesting the existence of saddle points in the spectrum near the points K of the Brillouin zone and the associated special Landau levels (see fig. 16a,b) which have been called leg levels. These experimenters observed a qualitative change of cyclotron resonance curves as the photon energy $\hbar\omega$ varied from 3.68 meV ($\lambda = 336.6$ μm) to 10.45, 14.7, and 15.8 meV ($\lambda = 118.6$, 84.3, and 78.5 μm). The relevant spectra for two directions of the circular wave polarization are illustrated in fig. 64a,b,c,d. The spectrum of fig. 64a is typical of the cyclotron resonance in graphite with the trigonally warped

Fermi surface. However, in spite of the fact that the photon energy $\hbar\omega = 3.68$ meV is insufficient for inducing transitions from leg levels, the high nonparabolicity of the spectrum affects the position of resonance fields. A comparison of figs. 64a,b,c,d indicates that as the photon energy increases the spectra become suddenly enriched in peaks due to additional transitions.

The position of resonance fields was calculated by Nakao (1976), who took into account the existence of saddle points and the associated leg levels. The structure of the Landau levels as calculated by Nakao, and the resonance transitions corresponding to $\hbar\omega = 10.45$ meV ($\lambda = 119\,\mu$m) are shown in fig. 65. Doezema et al. (1979) have fitted the resonance field values calculated by Nakao (1976) to the experimental values by varying the SWMC-model parameters. As a result, they have obtained a set of parameters γ_i that assures the agreement between the calculated and experimental field values to within 1% at three photon energies:

$$\gamma_0 = 3.12\,\text{eV}, \quad \gamma_1 = 0.38\,\text{eV}, \quad \gamma_2 = -0.021\,\text{eV}, \quad \gamma_3 = 0.315\,\text{eV}$$
$$\gamma_4 = 0.12\,\text{eV}, \quad \gamma_5 = -0.003\,\text{eV}, \quad \Delta = -0.002\,\text{eV}. \tag{128}$$

It should be noted that the resonance field values for transitions at the points K of the Brillouin zone are differently sensitive to the choice of the trigonal warping parameter γ_3, which determines the position of saddle points and leg levels. For this reason, the value of γ_3 is best determined by fitting the calculated field values to the experimental data.

The value for the trigonal warping parameter $\gamma_3 = (0.315 \pm 0.015)$ eV found in this way agrees well with the values of γ_3 determined from the data of oscillation, magneto-optical, and other measurements (see sect. 4.7. and table 5). The underrated value of 0.21 eV for γ_3, obtained by Ushio et al. (1972), can evidently be explained by their neglect of the complicated structure of the Landau levels associated with the saddle points in the electron spectrum located near the K-extrema. Another parameter whose upper and lower limits have been reliable determined is the value of the Fermi energy ε_F. This estimate, $-27\,\text{meV} < \varepsilon_F < -25\,\text{meV}$, is in good agreement with the value $\varepsilon_F = (-25.1 \pm 0.1)$ meV found by Dillon et al. (1978b) on the basis of oscillation measurement data.

We also note that, in addition to peaks due to the cyclotron resonance, caused by majority electrons, Doezema et al. (1979) have succeeded in observing a resonance peak due to holes, which permitted a determination of the hole cyclotron masses m_c^h to be made. This determination has yielded values of m_c^h equal to $0.039m_0$, $0.040m_0$, $0.045m_0$, and $0.044m_0$ at values of 3.68, 10.45, 14.70 and 15.80 meV, respectively. The hole cyclotron mass values thus obtained are close to the relevant values determined from oscillation measurements (see table 12) and appear to be due to the extremal cross section for majority holes.

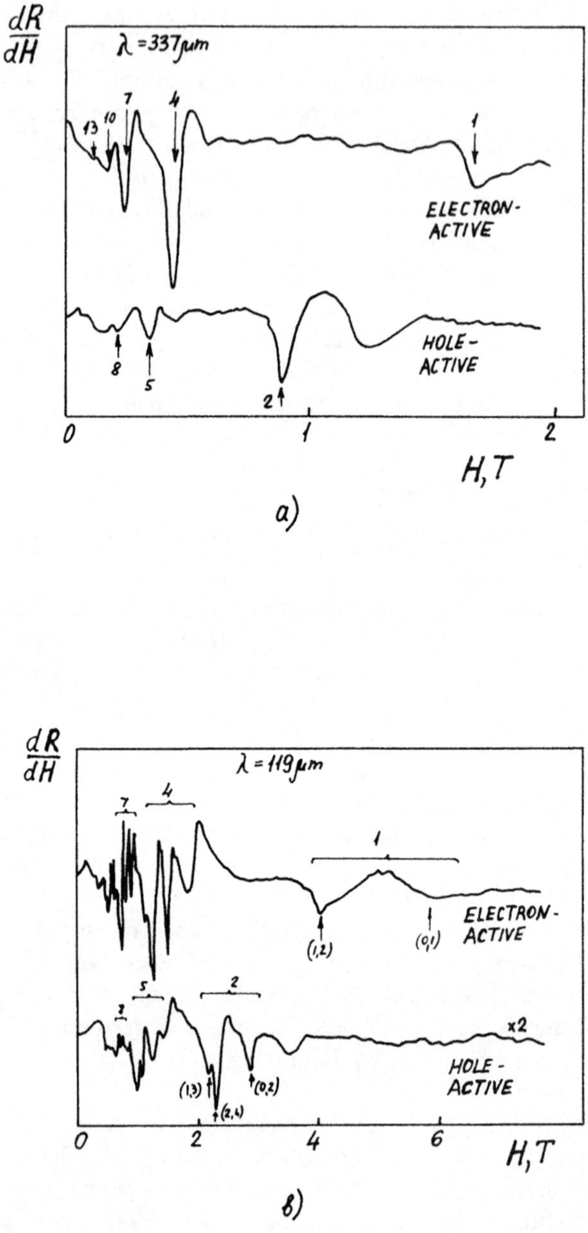

Fig. 64. Magnetoreflectance derivative spectra of graphite: (a) $\lambda = 337\,\mu m$; peaks are labelled by the number of the harmonic; (b) $\lambda = 119\,\mu m$ the bracketed ranges are labelled by harmonic number; (c) low-field detail of the 119-μm derivative spectra; the parenthesized number indicates the Landau levels in the valence band and in the conduction band which correspond to the transition; (d) $\lambda = 84\,\mu m$. (Doezema et al. 1979.)

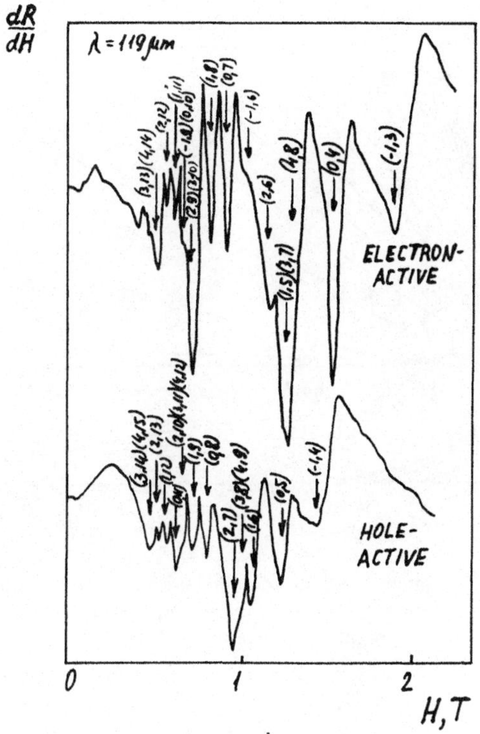
dR/dH
λ = 119 μm
ELECTRON-ACTIVE
HOLE-ACTIVE
0
1
2
H, T

c)

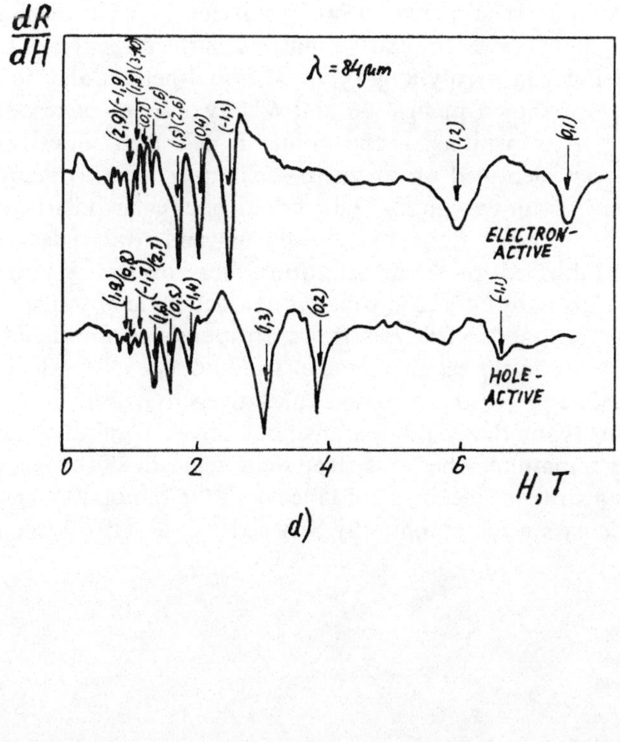
dR/dH
λ = 84 μm
ELECTRON-ACTIVE
HOLE-ACTIVE
0
2
4
6
H, T

d)

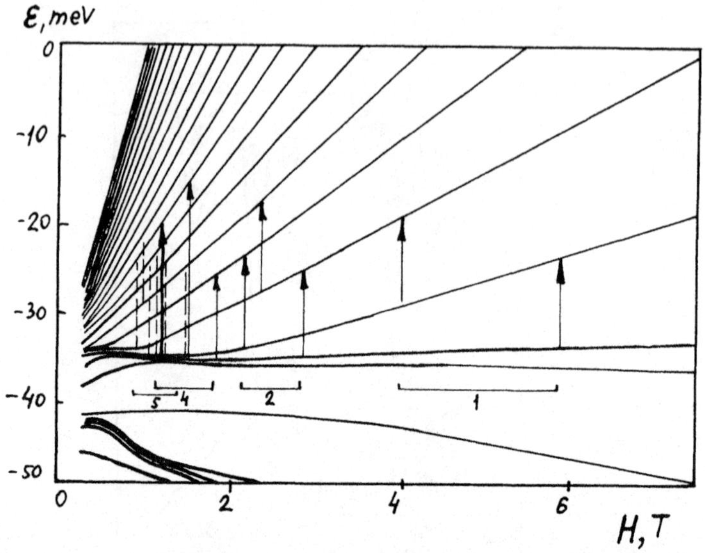

Fig. 65. Structure of the Landau levels in graphite, as calculated by Nakao (1976). Transitions corresponding to $\lambda = 119\,\mu\text{m}$ are indicated by arrows; bracketed ranges are labelled by harmonic number. (Doezema et al. 1979.)

7.4. Magnetoreflection in graphite

In measuring magneto-optical effects in graphite one usually observes the oscillations of the reflection coefficient R in a magnetic field for light waves reflected from the basal plane of a crystal for $\boldsymbol{H} \| c$. Typical oscillations, recorded using light waves circularly polarized in the direction of precession of electrons and holes in pyrolytic graphite at liquid-helium and liquid-nitrogen temperatures, are shown in figs. 66 and 67. Two series of resonance peaks, corresponding to transitions at the points K (electron series) and H (hole series), have been identified on magneto-oscillation curves, obtained with the use of light waves with two opposite directions of circular polarization, already in the early experiments done by Dresselhaus and Mavroides (1964, 1966). They have established that the selection rules for the K-point (electron) transitions are given by eq. (125), which is characteristic of transitions between trigonally warped bands. The H-series corresponds to interband transitions from Landau levels in the bands ε_1 and ε_2 to Landau levels in the band ε_3 near the points H. They obey the selection rules given by (120).

Resonant fields for the K-point transitions are read off at a point close to the reflectivity minimum, where the steep high-field side of the resonance curve deviates from a straight line, in accordance with the K-point magnetoreflection line-shape calculation for graphite by Schroeder et al. (1970) (see figs. 66 and

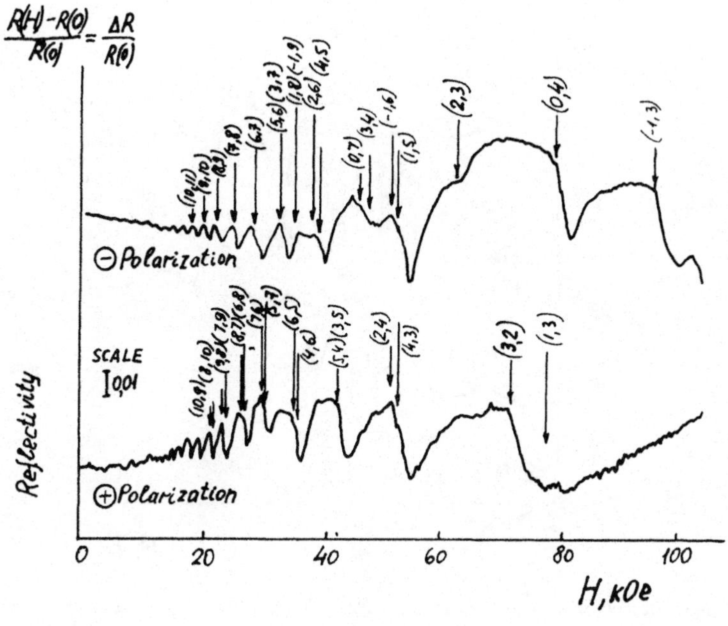

Fig. 66. Magnetoreflectance oscillations in pyrolytic graphite for $T = 4.2\,\mathrm{K}$ and $\hbar\omega = 0.069336\,\mathrm{eV}$. The curves correspond to circular polarization of the electromagnetic wave in the direction of the precession of electrons (⊕) and holes (⊖). The parenthesized numbers indicate the Landau levels in the valence and conduction bands which correspond to the transition. The resonance-field values calculated by the SWMC-model are shown by arrows. (Schroeder et al. 1970.)

67). The resonance point for the H-point transitions is taken at the reflectivity maximum, in accordance with the relevant H-point line-shape calculation by Toy et al. (1977).

Since Landau levels at the points H are defined by eq. (63) and are dependent solely on the SWMC-model parameters γ_0 and Δ, the study of H-point transition makes possible a determination of exactly these parameters. Figure 68 illustrates the experimental dependences which relate the photon energy to the field values of the corresponding resonance H-point transitions. The solid curves show the results of the calculation by eqs. (63) and (124) for a definite choice of γ_0 and Δ values.

The K-point resonance-transition field values, in turn, make it possible to find combinations of the SWMC-model parameters:

$$\frac{\gamma_0^2}{\gamma_1} = \frac{4}{3}\frac{\hbar^2}{a_0^2 m^*} \quad \text{and} \quad \beta = \frac{\gamma_4}{\gamma_0} - \frac{2\gamma_2 - 2\gamma_5 - \Delta}{8\gamma_1}$$

[see (70) in sect. 4.5]. Figure 69 presents the dependences of inverse resonance-field values $1/H_n$ on the number $n = n_c + n_v$ of the K-point transitions for three

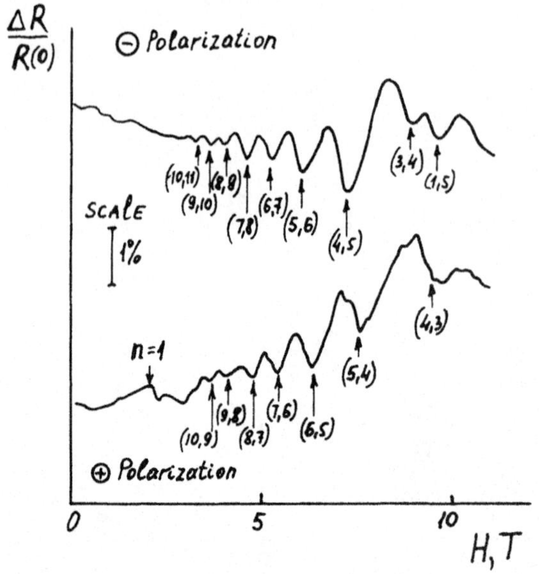

Fig. 67. Magnetoreflectance oscillations in pyrolytic graphite for $T = 77$ K and $\hbar\omega = 0.117$ eV. The curves correspond to the circular polarization of the electromagnetic wave in the direction of the precession of electrons (⊕) and holes (⊖). The figure shows the K-series transitions for the (⊕) and (⊖) circularly polarized light (arrows pointing upward) and one H-series transition for the (⊕) polarization (the arrow pointing downward). The parenthesized numbers indicate the Landau levels in the valence and conduction bands which correspond to the transition. (Mendez et al. 1980b.)

photon-energy values $\hbar\omega = 0.069$, 0.117, and 0.230 eV. The dip-angle tangent of the straight lines obtained and also the position of the point of their intersection with the abscissa yield the data necessary for finding the parameters $1/m^*$ and β. In this way, Mendez et al. (1980b) found that

$$m^* = (0.068 \pm 0.001)m_0 \quad \text{and} \quad \gamma_1/\gamma_0^2 = (4.0 \pm 0.1) \times 10^{-2}\,(\text{eV})^{-1}.$$

The latter is in good agreement with the value $(3.8 \pm 0.5) \times 10^{-2}\,(\text{eV})^{-1}$ obtained by Schroeder et al. (1970) in calculations which explicitly took into account trigonal warping ($\gamma_3 = 0.29$ eV). Attention is called to the fact that these authors have succeeded, owing to the high resolution of the peaks of the magneto-optical spectrum excited by a monochromatic light beam, in conclusively determining the sign and value for the parameter γ_4, viz., $\gamma_4 = +0.044$ eV.

Along with oscillation effects, a major role in studying the group of minority carriers in graphite is played by magneto-optical effects. H-point magneto-optical transitions are especially sensitive to the peculiarities of the electron spectrum. A detailed analysis of the data available in the literature on the H-

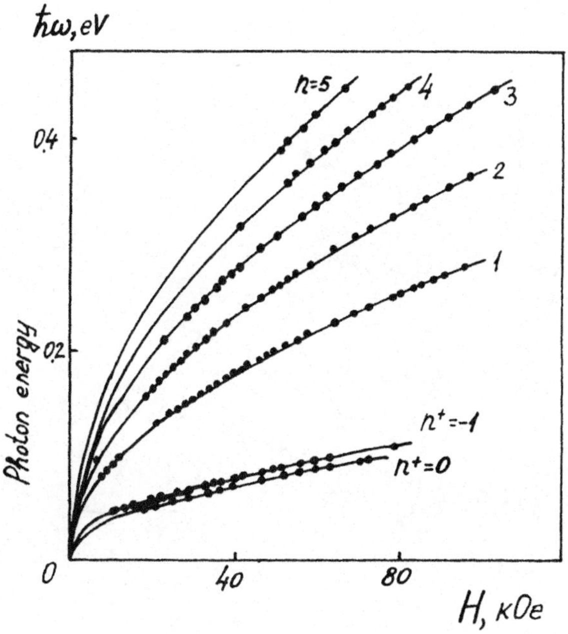

Fig. 68. Summary of H-point resonances covering a wide range of photon energies and magnetic fields. The curves passing through the experimental points represent a least-squares fit of these data to eqs. (63) and (124). (Toy et al. 1977.)

point magneto-optical transitions has been made by Toy et al. (1977) to determine the SWMC-model parameters responsible for the minority H-holes (primarily, γ_2 and Δ). They compared not only the resonance-field values and the structure of the Landau levels at the points H of the Brillouin zone of graphite, but also the shape of magneto-optical reflection curves and of individual resonance lines to the results of a calculation based on the SWMC-model at various values for its parameters γ_i. This analysis of magneto-optical data was combined with an analysis of oscillation data obtained in a study on long-period oscillations in graphite corresponding to the small cross sections of constant-energy surfaces $S_{m1} = 0.79 \times 10^{-42}$ and $S_{m2} = 0.35 \times 10^{-42}\ g^2\,cm^2/s^2$ (see sections 4.2 and 4.4). As a result of fitting the theoretical dependences to the experimental curves, the SWMC-model parameter Δ has been determined: $\Delta = (-0.008 \pm 0.004)$eV. As is known, both the condition of existence and the characteristics of minority holes at the points H depend on the sign and magnitude of Δ. The above value of Δ corresponds to the situation in which the extremal cross section for the H-point minority-hole Fermi surface coincides, for $\boldsymbol{H} \| c$, with the minimum value for

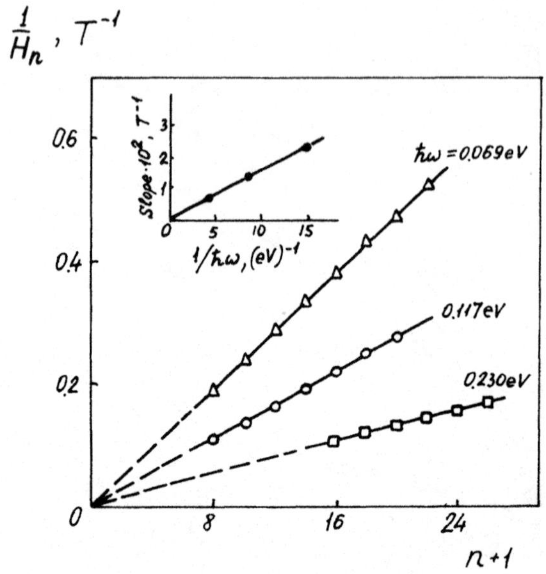

Fig. 69. Reciprocal of the resonant magnetic field H_n for the ⊕ polarization (see fig. 66) versus $(n+1)$, for various photon energies $\hbar\omega$. In the figure, n represents the sum of the quantum numbers corresponding to the two Landau levels involved in the interband transition: $n = n_c + n_v$. The resonant magnetic fields for $\hbar\omega = 0.117\,\text{eV}$ correspond to the data of Mendez et al. (1980b). The data for $\hbar\omega = 0.069$ and $0.230\,\text{eV}$ are taken from Schroeder et al. (1970). Magnetic fields are given in tesla. Inset shows that the slope of $1/H_n$ versus $(n+1)$ is inversely proportional to the photon energy $\hbar\omega$. (Mendez et al. 1980b.)

$S_{m2} = 0.35 \times 10^{-42}\,\text{g}^2\,\text{cm}^2/\text{s}^2$, while the cross section for the "neck" of the H-point majority-hole Fermi surface corresponds with $S_{m1} = 0.79 \times 10^{-42}\,\text{g}^2\,\text{cm}^2/\text{s}^2$.

In connection with this, we would like again to draw the reader's attention to the difficulties that arise if the cross section S_{m1} is ascribed to the neck of the majority-hole surface. These difficulties are associated with the fact that the dependence of S_{m1} on angle θ corresponds to the ellipsoidal, rather than distinctly dumbbell-shaped, surface. But it is the latter that is to be expected under the above assumption, as can be seen from fig. 51. This question has already been discussed in sect. 4.4.

The value of Δ determined by Toy et al. (1977) is in accordance with the values $\Delta = (-0.010 \pm 0.002)\,\text{eV}$ found by Sharma et al. (1974) for a similar situation on the basis of oscillation data alone. The other parameters obtained by Toy et al. (1977) are $\gamma_0 = 3.16\,\text{eV}$, $\gamma_1 = 0.39\,\text{eV}$, and $\varepsilon_F = -0.026\,\text{eV}$.

CHAPTER 8

The Effects of Hydrostatic Pressure on the Electronic Properties and Energy Spectrum of Graphite and Pyrocarbon Materials

From the results of the X-ray measurements made by Kabalkina and Vereshchagin (1960) in graphite under pressures up to 16 kbar it is known that the compressibility of graphite in the basal plane is negligibly small in comparison to the change in interlayer distance. Therefore, the action of hydrostatic pressure upon graphite, at least in the pressure range up to several tens of kbar, may be regarded, in a first approximation, as a uniaxial compression. On the basis of the data of Kabalkina and Vereshchagin (1960), as well as those of Green et al. (1973) and Gauster and Fritz (1974), the change in the interlayer distance in graphite under pressure up to 20 kbar has been found to obey the following empiric relation:

$$\frac{c_0(p) - c_0(0)}{c_0(0)} = \frac{\Delta c_0}{c_0} = -28 \times 10^{-7} p + 45 \times 10^{-12} p^2, \tag{129}$$

where $c_0(0) = 2d_0 = 6.7076$ Å is the initial distance between graphite layers at $p = 0$, p being expressed in bars. The empiric relation given by eq. (129) is commonly used to analyze the experimental results of measurements made in hydrostatically compressed graphite.

The uniform compression of a crystal does not change the structure and symmetry of its lattice. Accordingly, the SWMC-model of the electron spectrum of graphite, based on lattice symmetry, remains valid in describing its energy spectrum under pressure. It has already been mentioned (see sect. 4.2) that the SWMC-model parameters depend on the interactions between carbon atoms both in a layer and in neighbouring layers and are classified accordingly as in-plane (γ_0) and out-of-plane (γ_i, $i = 1, 2, \ldots, 6$) parameters.

If, because of the high anisotropy in graphite compressibility, amounting up to ~ 30, one neglects the change in the interatomic distance in a layer under pressure, then the consideration may be confined to an analysis of the change in the out-of-plane parameters and the quantities depending on them. Needless to say, this does not entail that the change in interatomic distance in a graphite layer under pressure may be neglected in all cases without exception. Taking this change into account is absolutely necessary, in particular, in analyzing those characteristics of graphite which essentially depend on the magnitude of the overlap between the wave functions for the neighbouring carbon atoms in a layer, as well as on the magnitude of the overlap of the wave functions for impurity states or defects. For instance, in the case of pyrocarbon materials, the magnitude of the negative magnetoresistance, associated with the delocalization of a definite number of carriers in a graphite layer, represents such a characteristic. The effect of pressure upon the electronic properties of graphite was investigated mainly by studying its optical and magneto-optical characteristics, as well as the transport and oscillation phenomena in it.

8.1. Optical properties of graphite under pressure

In studying the character of the change in the electronic properties of graphite under pressure, which are largely determined by the band structure in the region close to the Fermi level and about 1 eV in width, measurements of the optical spectra in the near infrared part are currently the most useful.

The reflection spectrum for graphite under a pressure to 2.8 kbar has been recorded by Misu et al. (1979) using HOPG specimens at room temperature in the photon energy range 0.655–1.125 eV. As the pressure-transmitting medium, gaseous helium was utilized.

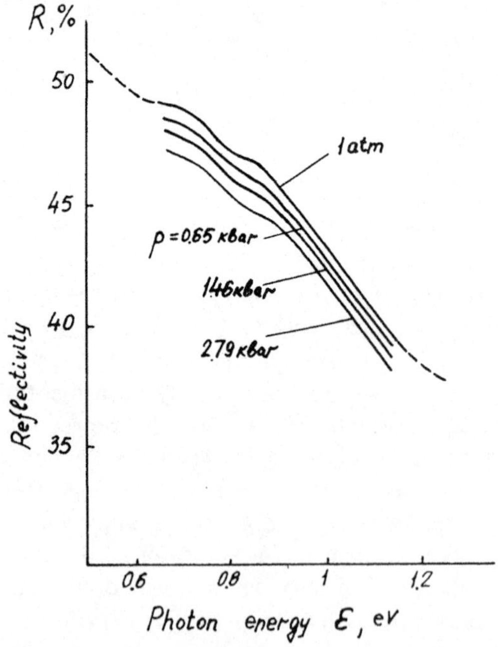

Fig. 70. Room-temperature basal-plane reflectivity spectra of graphite at 1 atm and under pressure. The dashed lines represent a smooth connection with the data of Taft and Philipp (1965). The structures on the low-energy and high-energy sides correspond to the A_1 and A_2 transitions, respectively (see text). (Misu et al. 1979.)

The reflection spectra $R(\hbar\omega)$ of graphite taken in the near IR region at $p = 1$ atm and p up to 2.79 kbar are shown in fig. 70. In these spectra, two low-intensity peaks are observable at $\hbar\omega \simeq 0.7$ and $\simeq 0.85$ eV, superimposed on the steady decrease of R with increasing energy which is typical of graphite. We note that two similar peaks have also been resolved in the thermal reflection spectrum of graphite at $p = 1$ atm, recorded by Bellodi et al. (1975). This structure of the IR-spectrum of graphite corresponds to two transitions at the points K of the Brillouin zone between the bands ε_2 and ε_3 (transition A_1) and ε_3 and ε_1 (transition A_2). A scheme for these transitions is presented in fig. 71. Within the SWMC-model, the transition energies ε_{A_2} and ε_{A_1} are given by

$$\varepsilon_{A_1} = 2(\gamma_1 - \gamma_2 - \gamma_5 + \varepsilon_F) - \Delta, \qquad \varepsilon_{A_2} = 2(\gamma_1 - \gamma_2 + \gamma_5) + \Delta \tag{130}$$

Taking advantage of the set of the parameters γ_i designated in table 5 by M, one can calculate the theoretical values for the transition energies: $\varepsilon^{\text{calc}} = 0.704 \pm 0.015$ eV and $\varepsilon_{A2}^{\text{calc}} = 0.888 \pm 0.015$ eV. Using a procedure developed by Philipp and Taft (1959) on the basis of the Kramers–Kronig relation, one may determine, from the reflection spectra of fig. 70, the spectra for the real κ_1 and imaginary κ_2 parts of the dielectric constant and their energy derivatives. The spectra of $\partial\kappa_1/\partial\varepsilon$ and $\partial\kappa_2/\partial\varepsilon$ at normal pressure, obtained in this fashion by Misu et al. (1979), are shown in fig. 72. The analytical calculation on the line shape of such spectra done by Nakao (1979) has enabled him to analyze in detail the features of the spectral curves

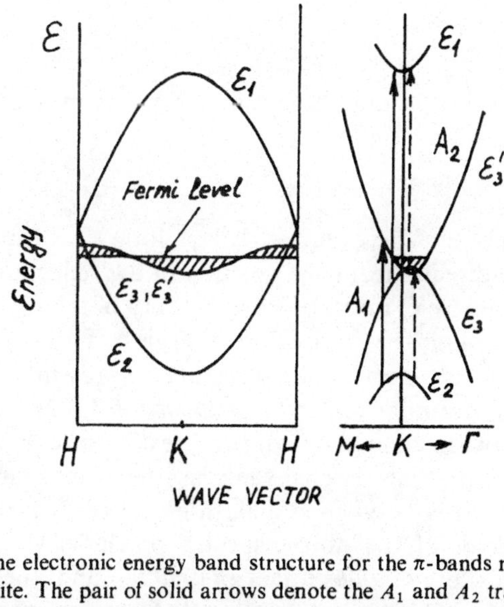

Fig. 71. A sketch for the electronic energy band structure for the π-bands near the K point in the Brillouin zone of graphite. The pair of solid arrows denote the A_1 and A_2 transitions (see text) that would be observed for $\gamma_3 = 0$. Additional important contributions to these transitions, indicated by the dashed arrows, are made since $\gamma_3 \neq 0$. (Misu et al. 1979.)

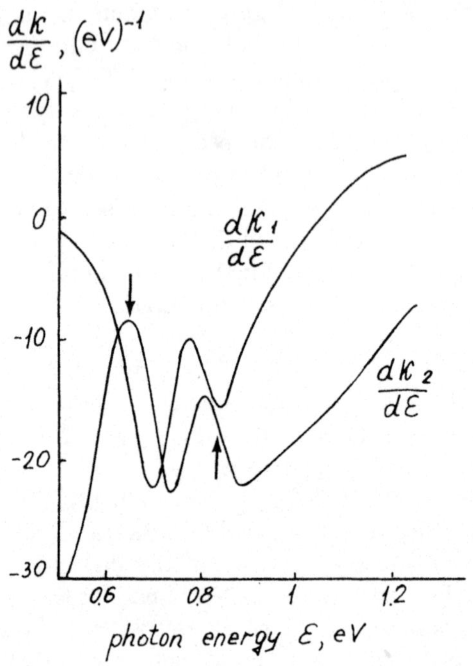

Fig. 72. Spectra of the photon-energy derivatives of the real part, $d\kappa_1/d\varepsilon$, and the imaginary part, $d\kappa_2/d\varepsilon$, of the in-plane dielectric constant of graphite at 1 atm and at room temperature. The arrows show the location of the A_1 and A_2 transitions. (Misu et al. 1979.)

obtained from experiment and determine from them the transition energy values at $p = 1$ atm:

$$\varepsilon_{A_1}^{\text{exp}} = 0.669\,\text{eV}, \qquad \varepsilon_{A_2}^{\text{exp}} = 0.847\,\text{eV}.$$

The difference between the obtained values for $\varepsilon_{A_1}^{\text{exp}}$, $\varepsilon_{A_2}^{\text{exp}}$ and those found by Bellodi et al. (1975) directly from the position of the reflection-coefficient peaks ($\varepsilon_{A_1} \simeq 0.7$ eV, $\varepsilon_{A_2} \simeq 0.85$ eV) indicates the necessity of taking the line-shape into account in determining the transition energies. This becomes still more significant in analyzing spectra obtained under pressure. The study of the reflection spectra of graphite under pressure made by Misu et al. (1979) and the comparison between their data and the results of calculations by Nakao (1979) have shown that an intense and asymmetric change in line shape normally occurs in graphite when singularities are shifted on the energy scale. This specific character of the above spectra precludes an accurate determination of transition energy values under pressure – and, therefore, of the baric coefficients $\partial\varepsilon_{A_i}/\partial p$ for these energies – directly from the energy dependences of reflectivities.

Misu et al. (1979) have proposed to determine the change in transition energy from reflection spectra obtained under pressure, not from the shift of peaks but by calculating the shift $\overline{\Delta\varepsilon_{A_i}}$ $(i = 1, 2)$ averaged over energy. This is given by the relation

$$\int \Delta\kappa_i \, \mathrm{d}\varepsilon = -\overline{\Delta\varepsilon_{A_j}} \int \frac{\mathrm{d}\kappa_i}{\mathrm{d}\varepsilon} \mathrm{d}\varepsilon \quad (i = 1, 2; \quad j = 1, 2). \tag{131}$$

Equation (131) contains the energy derivatives of the real and imaginary parts of the dielectric constant, as well as those of the increments of κ_i under pressure:

$$\Delta\kappa_i = \kappa_i(\varepsilon)|_p - \kappa_i(\varepsilon)|_{p=0}. \tag{132}$$

The spectra of $\Delta\kappa_i$ $(i = 1, 2)$, deduced from the reflection spectra of fig. 70, are illustrated in figs. 73a,b. Comparing the spectra in figs. 70 and 73a,b, one may

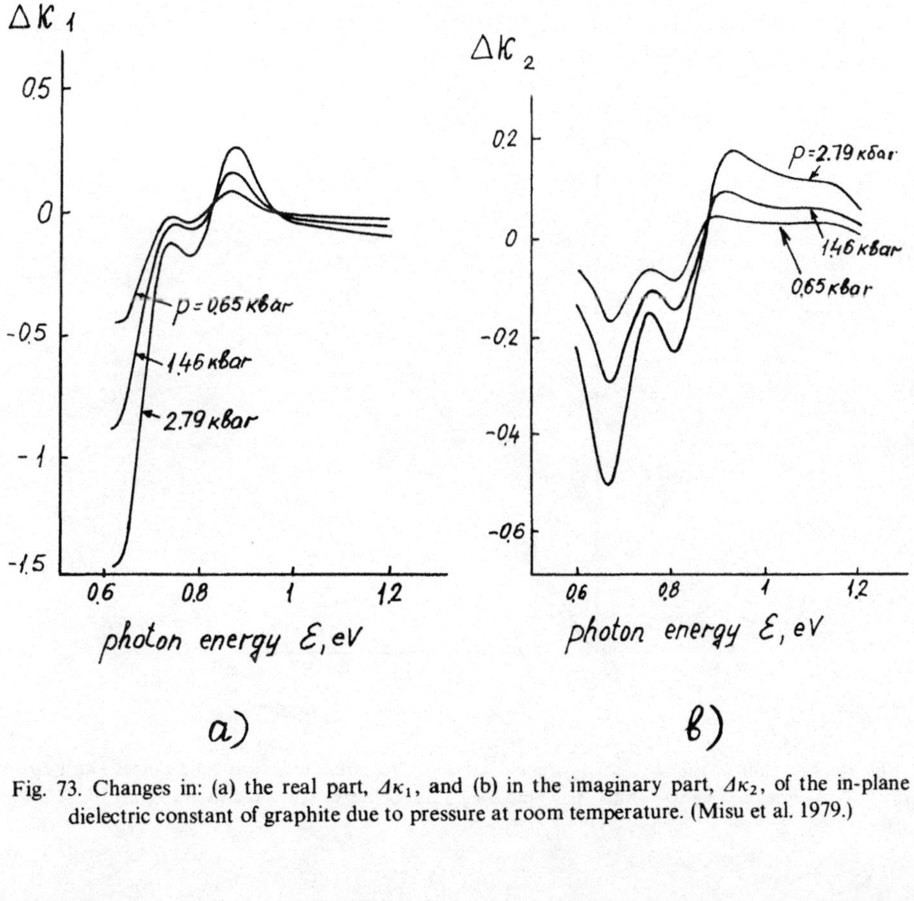

Fig. 73. Changes in: (a) the real part, $\Delta\kappa_1$, and (b) in the imaginary part, $\Delta\kappa_2$, of the in-plane dielectric constant of graphite due to pressure at room temperature. (Misu et al. 1979.)

notice that transitions A_1 and A_2 are shifted under pressure towards the higher energies.

Attention is drawn to the fact that the averaged shift $\overline{\Delta\varepsilon_{A_i}}$ differs from the true one only in the case when a resonance line alters asymmetrically as it shifts on the energy scale. If the spectral line $\kappa_i(\varepsilon)$ broadens symmetrically, i.e., it is an even energy function relative to the extremum point, whereas the line $d\kappa_i(\varepsilon)/d\varepsilon$ is an odd energy function, then $\overline{\Delta\varepsilon_{A_i}} = \Delta\varepsilon_{A_i}$. The results obtained by Misu et al. (1979) by processing the experimental data with the use of (131) indicate that the averaged shifts $\overline{\Delta\varepsilon_{A_i}}$ for transitions A_1 and A_2 are, within the experimental error, linear functions of pressure (fig. 74). The baric coefficients for these averaged shifts will then be equal to

$$\frac{\overline{\Delta\varepsilon_{A_1}}}{\Delta p} = 0.0176 \pm 0.0005\,\mathrm{eV/kbar},$$

$$\frac{\overline{\Delta\varepsilon_{A_2}}}{\Delta p} = 0.026 \pm 0.002\,\mathrm{eV/kbar}. \tag{133}$$

These values give the rates of the displacement of the bands ε_1, ε_2 and ε_3 relative to one another under pressure.

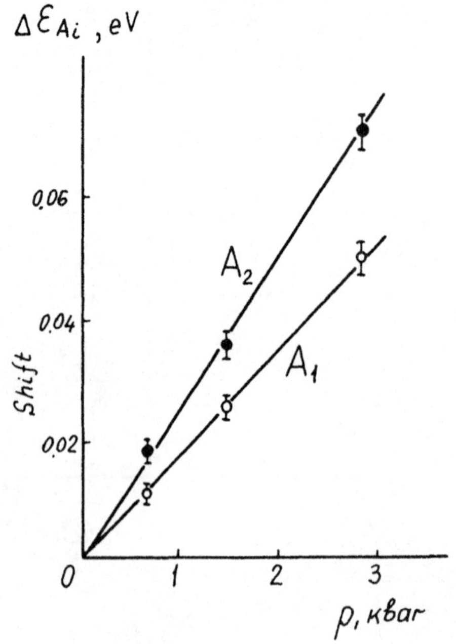

Fig. 74. The shift of the A_1 and A_2 transitions with pressure. Open and solid circles (with error bars) are the average shifts $\overline{\Delta\varepsilon_{A_1}}$ and $\overline{\Delta\varepsilon_{A_2}}$, respectively (see text). (Misu et al. 1979).

8.2. Magneto-optical measurements in graphite under pressure

Magneto-optical measurements yield much more complete and direct information about the change in band structure and the energy-spectrum parameters than do the measurements of optical spectra. In connection with this, we should note that the potentialities of the method are far from being exhausted by the results obtained to date.

Magneto-optical spectra of HOPG specimens at liquid-nitrogen temperatures and pressures up to 2 kbar have been measured by Mendez et al. (1980b), who utilized gaseous helium as the pressure-transmitting medium. The magnetoreflection spectra were taken with the help of a circularly polarized light beam from a CO_2-laser ($\lambda = 10.6\mu m$, $\hbar\omega = 117\,meV$). The magnetic field, as high as 10 T, was produced using a Bitter solenoid. Typical spectra of magnetoreflection in graphite at atmospheric pressure, $T = 77\,K$ and two directions of circular polarization for light waves were presented in fig. 67. In recording the spectra under a pressure to 2 kbar, no qualitative change in the shape of oscillation curves or that of individual resonance peaks has been noticed. This permits one to conclude that in analyzing the spectra at the K- and H-points of the Brillouin zone obtained under pressure one can apply the same criteria that have been used at atmospheric pressure (see sect. 7.4). The

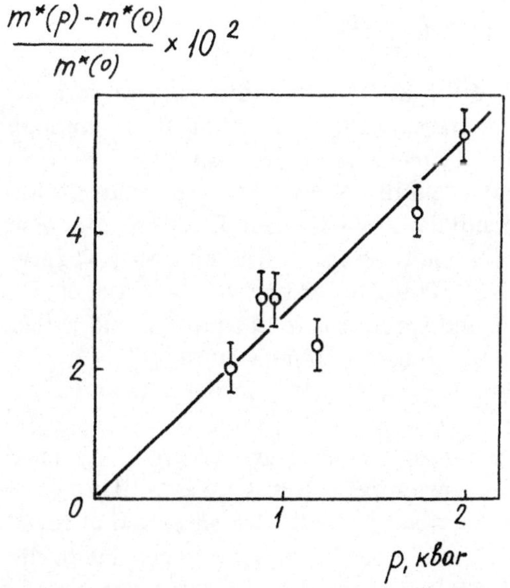

Fig. 75. The relative change in the parameter m^* as a function of pressure. The circles correspond to experimental determinations from plots of $1/H_n$ versus $(n + 1)$. The solid line represents a linear least-squares fit of the experimental points. (Misu et al. 1979.)

results of the measurements suggest that, in a first approximation, the resonance fields H_n, corresponding to H-point transitions are pressure-independent. This corresponds to a negligibly small pressure dependence of the parameter γ_0, which determines the values of H_n for the H-point transitions (63) and (124). The resonance fields corresponding to K-point transitions under pressure are shifted towards the higher magnetic fields. The increase in resonance-field values is reflected in the change of the slope of curves described by (121), due to the pressure dependence of the parameters m^* and β [see (70), sect. 4.5]. The experimental dependence of m^* on p, is shown in fig. 75 and is in good agreement with the data of Lynch and Drickamer (1966). The straight line in fig. 75 has resulted from a least-squares fit to the experimental data, and describes the baric coefficient:

$$\frac{\mathrm{d} \ln m^*}{\mathrm{d}p} = 0.027 \pm 0.004\,(\mathrm{kbar})^{-1}. \tag{134}$$

The insufficiently accurate experimental data preclude a quantitative determination of the baric coefficient $\mathrm{d}\beta/\mathrm{d}p$ for the parameter β. However, making use of the baric coefficients for resonance fields and eq. (122), one can conclude that $\mathrm{d}\beta/\mathrm{d}p > 0$.

8.3. Galvanomagnetic measurements in graphite and pyrocarbon materials under pressure

The measurement of the electrical conductivity of various kinds of graphite under uniform compression indicates that both components of electrical conductivity, σ_a and σ_c, increase with pressure over a wide temperature range extending from liquid-helium up to room temperatures. The baric coefficient of the electrical conductivity along the c-axis, $(\mathrm{d} \ln \sigma_c)/\mathrm{d}p$, reaches the value of 2 or 3% per kilobar at low temperatures and very slightly decreases as the temperature is raised. Thus, according to the data of Yeoman and Young (1969), who investigated specimens of hot-pressed and annealed graphite in the pressure range 0–0.4 kbar, at temperatures of 4.2, 77, and 300 K the coefficients $(\mathrm{d} \ln \sigma_c)/\mathrm{d}p$ are 2.82 ± 0.1, 2.76 ± 0.1, and 2.55 ± 0.1 (in units of $10^{-2}\,\mathrm{kbar}^{-1}$), respectively.

The baric coefficient of the electrical conductivity of graphite along the layer, $(\mathrm{d} \ln \sigma_a)/\mathrm{d}p$, at low temperatures is close to $(\mathrm{d} \ln \sigma_c)/\mathrm{d}p$ in value, but with increasing temperature it very rapidly decreases and at room temperature does not exceed 3% of its low-temperature value. According to the work by Yeoman and Young (1969) quoted above, for the same specimens of graphite at the same temperatures 4.2, 77, and 300 K, the values for the $(\mathrm{d} \ln \sigma_a)/\mathrm{d}p$ are 2.8, 1.2–1.4, and ± 0.1 (in units of $10^{-2}\,\mathrm{kbar}^{-1}$), respectively.

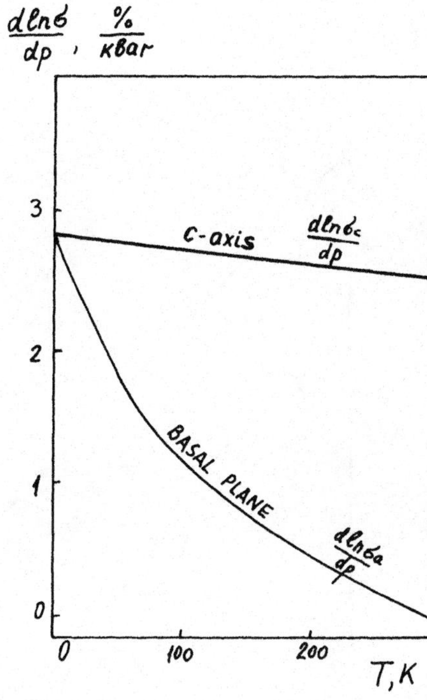

Fig. 76. Variation in the pressure coefficient of conductivity for current flow both parallel and perpendicular to the basal plane. (Yeoman and Young 1969.)

Typical temperature dependences of the baric coefficients $(\mathrm{d}\ln\sigma_a)/\mathrm{d}p$ and $(\mathrm{d}\ln\sigma_c)/\mathrm{d}p$ of graphite are illustrated in fig. 76.

As the electrical conductivity of a solid under pressure may change as a result of variation of both the density and the mobility of carriers, one should have additional information of the pressure dependence of either of these parameters in order to be able to analyze the data on the conductivity of uniformly compressed graphite. Such pressure dependences can be obtained from galvanomagnetic measurements.

The galvanomagnetic effects occurring in graphite under pressure have been explored by several workers. They have found that in the entire range of temperatures under study – from liquid-helium up to room – the transverse magnetoresistance $\Delta\rho_{a\perp}(H)/\rho_{a\perp}(0)$ $(\boldsymbol{H}\|c)$, which determines, to within the imbalance $n_h - n_e$, the squared average mobility $\bar{\mu}^2$ of carriers in the basal plane [see eq. (88), sect. 5.2]. According to Spain (1969), for single-crystal graphite at room temperature the parameter $\{\Delta\rho_{a\perp}(H)/[H^2\rho_{a\perp}(0)]\}^{1/2}$ decreases almost linearly and rather steeply ($\sim 4\%$ per kilobar) with increasing pressure (fig. 77).

Such a dependence is consistent with the idea that for $T \simeq 300\,\mathrm{K}$ the

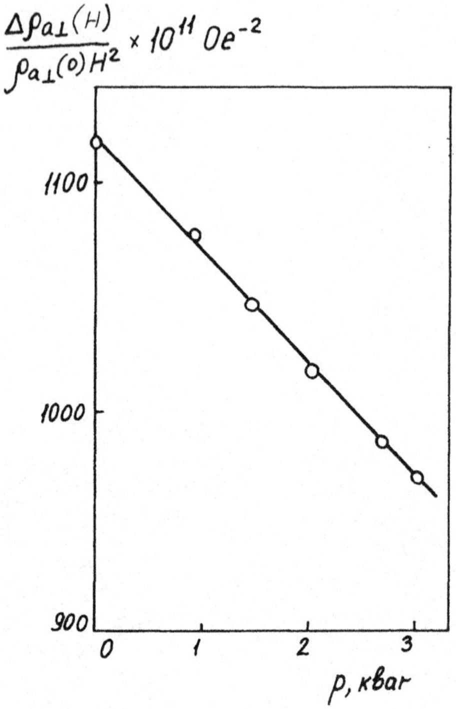

Fig. 77. Variation of the parameter $\Delta\rho_{a\perp}(H)/(\rho_{a\perp}(0)\ H^2)$ with pressure, showing that the quantity $(\bar{\mu})^2$ varies linearly with pressure. Measurement at 298 K. This result is consistent with a model in which electron–phonon scattering predominates. (Spain 1969.)

scattering of electrons by phonons is prevalent in single-crystal graphite. At $T = 77$ K, this dependence becomes more complicated and cannot be interpreted so unambiguously. However, at both $T = 77$ K and $T = 4.2$ K (see Kechin et al. 1968) the magnetoresistance in single-crystal graphite, and therefore the average mobility $\bar{\mu}$, decrease with increasing pressure.

The above data on the change in the average mobility of carriers moving along graphite layers enable one to conclude that the observed increase in the conductivity σ_a of uniformly compressed graphite must be the consequence of a rather rapid increase in electron and hole density under pressure. Such a conclusion is consistent with the results of the measurements of the total carrier density $n = n_e + n_h$ in graphite under pressure made by Arkhipov et al. (1963), Likhter and Kechin (1964), and Spain (1969). They have found that the baric coefficient $(d \ln n)/dp$ is positive and ranges from 3×10^{-2} to 4×10^{-2} kbar^{-1}, its value changing but slightly as the temperature rises from 4.2 to 300 K.

A comparison of the baric coefficients $(d \ln \bar{\mu})/dp$, $(d \ln n)/dp$, and $(d \ln \sigma_a)/dp$

in graphite at various temperatures shows that in the low-temperature range the conductivity σ_a increases due to a rapidly increasing carrier density which more than compensates for the decrease of average mobility. As Spain (1969) has demonstrated, at room temperature the rates of change of these two opposite effects become practically the same, as a result of which the baric coefficient of conductivity becomes abnormally small: the increase in carrier density under pressure is almost completely counterbalanced by the decrease of average mobility.

Attention should be called to fact that the decrease of carrier mobility under pressure may, to a certain extent, be associated with the increased carrier density which in turn results from the greater overlap of the π-bands and the increased Fermi energy for electrons and holes. The above parameters may affect carrier mobility because of the nonparabolicity of the electron spectrum of graphite, due to which an increase in Fermi energy causes the effective masses of electrons to increase as well.

The weak dependence of the baric coefficient $(\mathrm{d}\ln\sigma_c)/\mathrm{d}p$ on temperature and its closeness to the value of the $(\mathrm{d}\ln n)/\mathrm{d}p$ enables one to conclude that the effective mobility μ^c_{eff} of carriers moving along the c-axis weakly depends, at least at pressures of several kbar, on the magnitude of compression and the coefficient $(\mathrm{d}\ln\mu^c_{\mathrm{eff}})/\mathrm{d}p$ is small in comparison with the baric coefficient of total carrier density.

8.4. Oscillation effects in graphite under pressure

Oscillation effects in uniformly compressed graphite have been studied in sufficient detail. Thus, Itskevich and Fisher (1967), who used a liquid-medium chamber designed by Itskevich (1963), investigated the ShdH oscillations for $\boldsymbol{H}\|c$ in monocrystalline synthetic graphite under pressures up to 8 kbar. The oscillations of the field derivative of magnetoresistance $\partial\rho(H)/\partial H$ they observed (fig. 78) are currently believed to be due to the electron part of the Fermi surface. At $p = 8$ kbar, the increase in the extremal Fermi-surface cross section is $[S_e(p) - S_e(0)]/S_e(0) = \Delta S_e/S_e = 31\%$, the baric coefficient for the electron cross section having been estimated as $(\mathrm{d}\ln S_e)/\mathrm{d}p = 39 \times 10^{-3}\,\mathrm{kbar}^{-1}$.

Anderson et al. (1967) studied the dHvA oscillations in graphite for pressures up to 4 kbar transmitted by gaseous helium. Owing to this, they were able to observe oscillations due to majority electrons and holes, as well as the long-period oscillations which they ascribed to minority holes at the points H. These authors have found the following baric coefficients for the extremal cross sections (in kbar)$^{-1}$:

$$\frac{\mathrm{d}\ln S_e}{\mathrm{d}p} = 40 \times 10^{-3}, \qquad \frac{\mathrm{d}\ln S_h}{\mathrm{d}p} = 34 \times 10^{-3}, \quad \text{and}$$

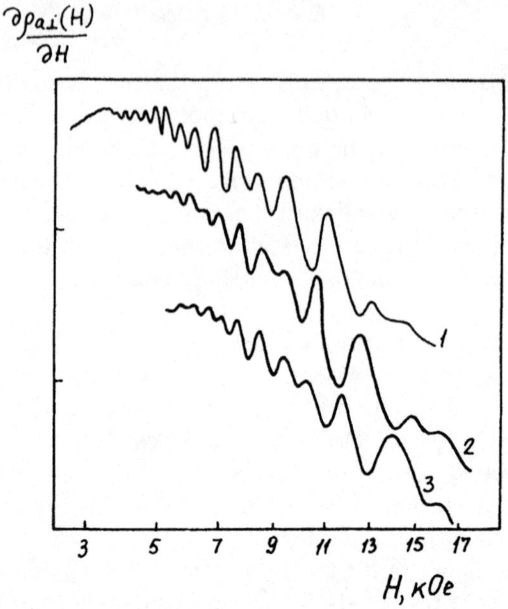

Fig. 78. ShdH oscillations in synthetic single-crystal graphite at $T = 1.5\,\mathrm{K}$, $\boldsymbol{H} \| c$ and various pressures p; 1, $p = 1$ bar; 2, $p = 4.8$ kbar; 3, $p = 8.0$ kbar. (Itskevich and Fisher 1967.)

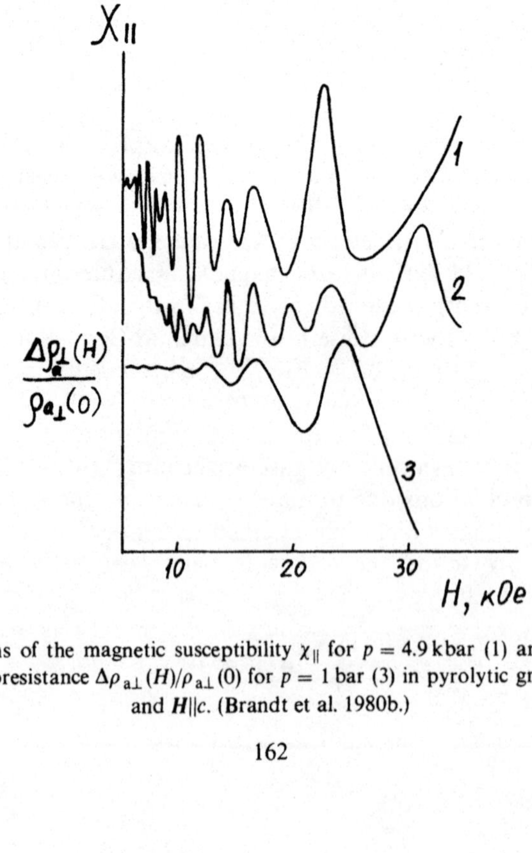

Fig. 79. Oscillations of the magnetic susceptibility $\chi_{\|}$ for $p = 4.9$ kbar (1) and $p = 17.3$ kbar (2), and of the magnetoresistance $\Delta\rho_{a\perp}(H)/\rho_{a\perp}(0)$ for $p = 1$ bar (3) in pyrolytic graphite for $T = 2.1\,\mathrm{K}$ and $\boldsymbol{H} \| c$. (Brandt et al. 1980b.)

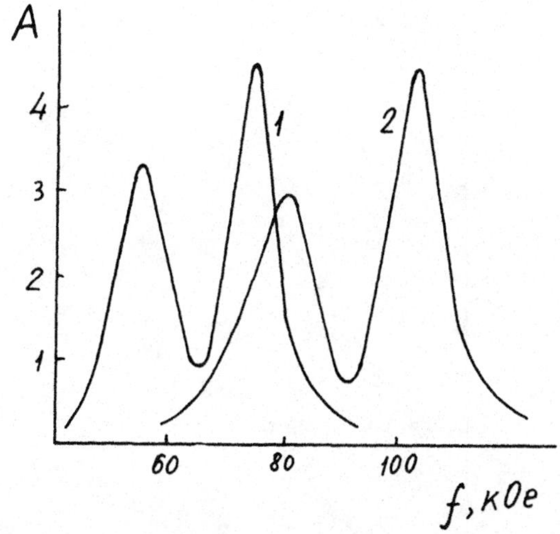

Fig. 80. Curves of the spectral density A for oscillations of the magnetic susceptibility $\chi_{\parallel}$ in pyrolytic graphite at pressures $p = 4.9$ kbar (1) and $p = 17.3$ kbar (2) (the oscillations are shown in fig. 79). "f" is the oscillation frequency. (Brandt et al. 1980b.)

$$\frac{\mathrm{d} \ln S_{\mathrm{m}}}{\mathrm{d}p} = 60 \times 10^{-3}.$$

Detailed studies of oscillation effects, both ShdH and dHvA, in single crystals of synthetic graphite under pressure have been carried out by Brandt et al. (1980a,b) and Kuvshinnikov (1980). They have determined, in addition to the pressure dependence of Fermi-surface cross sections, the baric dependences of the cyclotron masses m_{c} and of the Dingle temperature T_{D}. Typical curves for the observed oscillations of magnetoresistance and magnetic susceptibility are presented in fig. 79. The dHvA oscillations in fields higher than 4 kOe represent the superposition of two frequencies, corresponding to the electron and hole parts of the Fermi surface for majority carriers. Figure 80 shows the curves for the dHvA spectral density. The ShdH oscillations are monochromatic and due to the hole Fermi surface. The pressure dependences of S_{e} and S_{h} in synthetic (pyrolytic) graphite for $p > 10$ kbar have been found for the first time to deviate considerably from the linear behaviour (fig. 81). From the temperature and field dependences of the oscillation amplitude, the cyclotron masses $m_{\mathrm{c}}^{\mathrm{e}}$, $m_{\mathrm{c}}^{\mathrm{h}}$ and the Dingle temperatures $T_{\mathrm{D}}^{\mathrm{e}}$, $T_{\mathrm{D}}^{\mathrm{h}}$ for electrons and holes have been determined. The variation of the parameters $m_{\mathrm{c}}^{\mathrm{e}}$, $m_{\mathrm{c}}^{\mathrm{h}}$, and $T_{\mathrm{D}}^{\mathrm{e}}$ with pressure is illustrated in figs. 82 and 83. One can see that not only the Fermi-surface cross sections, but also the cyclotron masses $m_{\mathrm{c}}^{\mathrm{e}}$ and $m_{\mathrm{c}}^{\mathrm{h}}$ increase rather rapidly with pressure. The linear character of these dependences is broken down also in the pressure range $p > 10$ kbar. The observed increase in Dingle temperature with pressure generally agrees with the reversible decrease of the average mobility of carriers in graphite under pressure (see fig. 77). The

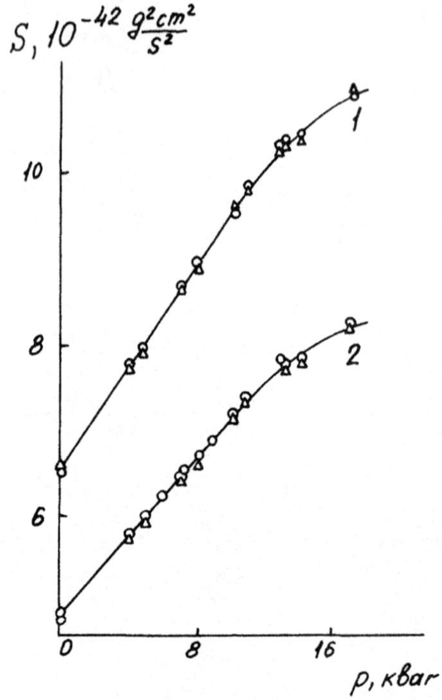

Fig. 81. Pressure dependences of the extremal cross-sections for electrons $S_e(1)$ and holes $S_h(2)$ in pyrolytic graphite. The dependences have been obtained using dHvA (○, 4.2 K; △, 2.1 K) and ShdH (□, 4.2K) effects; $\boldsymbol{H} \| c$. (Brandt et al. 1980b.)

increase in the cyclotron masses with pressure is also qualitatively compatible with the magnetoreflection data of Mendez et al. (1980b) (see fig. 75), while the increased cross sections for the majority-carrier Fermi surface is consistent with the galvanomagnetic data which are indicative of the growth of the electron (and hole) density.

The values for the Fermi-surface cross sections, cyclotron masses at $p = 1$ bar, and their baric coefficients in graphite are compared in table 13.

We note that Brandt et al. (1980a,b) and Kuvshinnikov (1980) investigated quasi-monocrystalline specimens of pyrolytic graphite precipitated from the gaseous phase at ~2100°C and subsequently heat-treated under pressure at ~3000°C. The thus prepared quasi-single crystals had, according to the results of an X-ray diffraction analysis, an interplanar distance $c_0 = 6.730$ Å., i.e., 0.0224 Å greater than in single-crystal graphite (6.7076 Å). For this reason, the extremal cross sections S_e and S_h in these quasi-monocrystalline specimens are slightly smaller in absolute value than in single-crystal graphite (see tables 11 and 13). The variation of interplanar distance Δc_0 in graphite under pressure is described by empiric relation (129), according to which the value $\Delta c_0 = -0.0224$ Å corresponds to a pressure $p \simeq 1.2$ kbar. Table 13 gives in

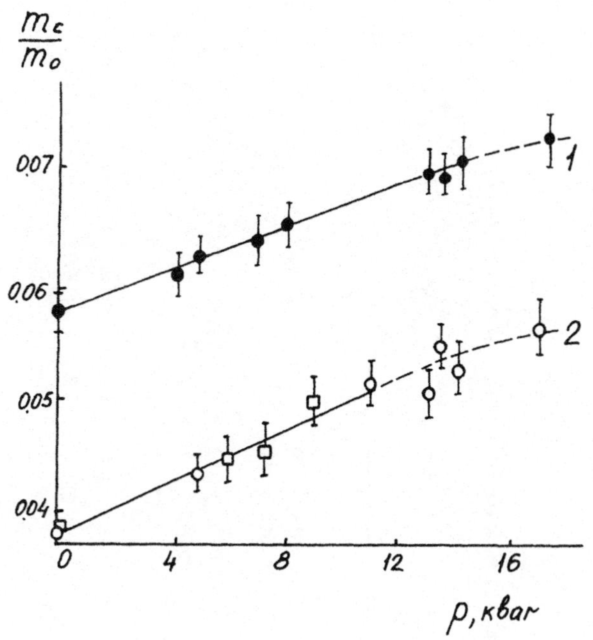

Fig. 82. Pressure dependences of the cyclotron masses $m_c^{e,h}$ of electrons (curve 1) and holes (curve 2); $\boldsymbol{H} \| c$. (Kuvshinnikov 1980.)

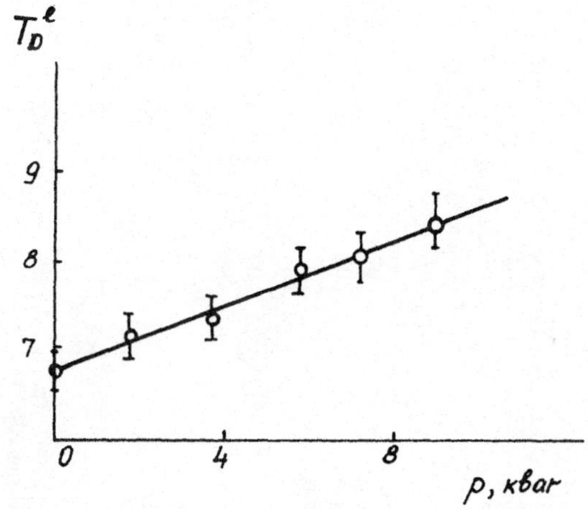

Fig. 83. Pressure dependence of the Dingle electron temperature T_D^e in pyrolytic graphite. (Kuvshinnikov 1980.)

parentheses the values for the cross sections S_e and S_h and for the cyclotron masses m_c^e and m_c^h (see Brandt et al. 1980b) recalculated to the pressure $p \simeq 1.2\,\text{kbar}$, at which $c_0 = 6.7076\,\text{Å}$. These values are in good agreement with those found for single-crystal graphite (see also tables 11 and 12).

Table 13
Comparison of extremal Fermi-surface cross sections, cyclotron masses and baric coefficients obtained by various authors.

Carriers	Extremal cross sections S $(10^{-42}\,g^2\,cm^2/s^2)$	$\frac{d \ln S}{dp}$ $(kbar^{-1})$	m_c/m_0	$d \ln m_c/dp$, $(kbar^{-1})$	Ref.*
Majority electrons	7.07	0.034 ± 0.006	—	—	[1]
(maximum cross secti-	7.03	0.039	—	—	[2]
on at points K of BZ)	6.52 ± 0.06	0.0468 ± 0.001	0.058 ± 0.003	0.017 ± 0.003	[3]
	(6.92 ± 0.06)		(0.059 ± 0.003)		[3]
Majority holes	5.09	0.040 ± 0.004	—	—	[1]
(maximum cross section)	4.80 ± 0.05	0.0485 ± 0.001	0.038 ± 0.02	0.024 ± 0.005	[3]
	(5.14 ± 0.05)		(0.040 ± 0.02)		[3]
Minority carriers	0.79	0.06 ± 0.015	—	—	[1]
	0.36 ± 0.01	—	0.0038 ± 0.003	—	[3]

*References
[1] Anderson et al. (1967).
[2] Itskevich and Fisher (1967).
[3] Brandt et al. (1980b).

8.5. *Transformation of the energy spectrum of graphite under pressure*

The optical, magneto-optical, galvanomagnetic and oscillation data obtained for specimens of natural and synthetic monocrystalline graphite (see the preceding four subsections) all suggest that under the influence of uniform compression the band overlap in graphite increases. This is accompanied by the growth of the **k**-space regions occupied by electrons and holes, and, therefore, by an increase in charge-carrier density. The reason for this change in electron spectrum is the diminished interplanar distance and the increased overlap between for electron wave functions for neighbouring carbon layers. The associated increase in the Fermi energy for electrons and holes in highly nonparabolic bands of graphite will, in turn, lead to an increase in effective carrier masses. The above-mentioned deviation of the pressure dependences of S_e and S_h for $p > 10\,\mathrm{kbar}$ from the linear behaviour (fig. 81) is due to the nonlinear dependence of the interplanar distance c_0 in graphite upon pressure. Empiric relation (129) between Δc_0 and p enables one to recalculate the dependences of S_e and S_h on p into their dependences on Δc_0. The latter, as shown by Kuvshinnikov (1980), may rather accurately be regarded as linear over a wide range of variation of interplanar distance (to $|\Delta c_0|/c_0 \simeq 3.5\%$), as

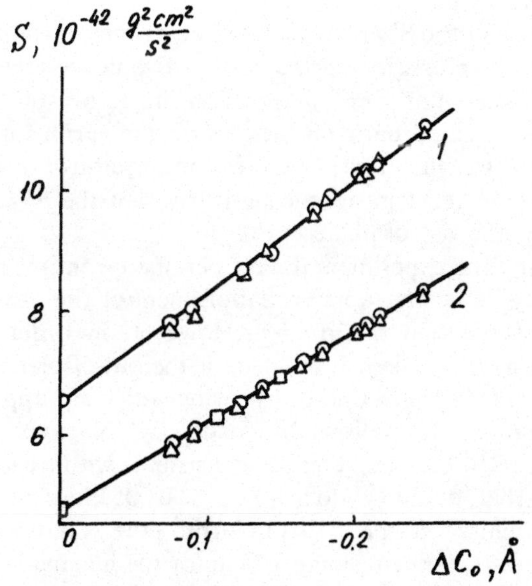

Fig. 84. Extremal cross-sections of the Fermi surface for electrons, S_e (curve 1), and holes, S_h (curve 2), in pyrolytic graphite versus interlayer distance Δc_0 under pressure. ○, △, □ see caption of fig. 81. (Kuvshinnikov 1980.)

shown in fig. 84. Because of the high anisotropy of graphite compressibility, $c_{11}/c_{33} \simeq 30$, this entails that the Fermi-surface cross sections S_e and S_h depend linearly on the change in crystal volume ΔV.

At the same time, one should mention that the baric coefficients $(\mathrm{d}\ln S)/\mathrm{d}p$ for the Fermi-surface cross sections in graphite exceed by more than an order of magnitude the value of the bulk compressibility, $(\mathrm{d}\ln V)/\mathrm{d}p$, for graphite, which equals 2.5×10^{-3} kbar (see table 13). For comparison, recall that in the case of isotropic compression of the lattice in the model of nearly free electrons the changes in the Fermi-surface cross sections S and in the volume V are related by:

$$\frac{\Delta S}{S} = -\frac{2}{3}\frac{\Delta V}{V}. \tag{135}$$

Within this model, the shape of individual parts of the Fermi surface remains unchanged as these parts grow, while remaining similar to themselves, due to the fact that the Fermi-sphere radius and the Brillouin-zone size both change inversely proportional to the size of a unit crystal cell. Comparing the above values for $(\mathrm{d}\ln S)/\mathrm{d}p$ and $(\mathrm{d}\ln V)/\mathrm{d}p$, one may conclude that in graphite the proportionality factor for a relation similar to (135) is much greater than $\frac{2}{3}$, reaching a value of ~ 15 for majority carriers and ~ 25 for minority ones. As ΔV is proportional to Δc_0, it follows that the interaction between the conduction electrons in neighbouring graphite layers is extremely sensitive to a change in interplanar distance.

The applicability of the SWMC-model in describing the electronic spectrum of graphite under pressure allows one to analyze the band-structure transformation under uniform compression in terms of this model. The experimentally determined baric coefficients for the various quantities (optical transition energies, Fermi-surface cross sections, cyclotron masses, etc.) offer a possibility of finding the logarithmic derivatives of the SWMC-model parameters γ_i (mostly, the out-of-plane ones).

When examining the experimental data obtained from measurements under pressure, one may, as a first approximation, neglect the change in the parameter γ_0 and assume that $\mathrm{d}\gamma_0/\mathrm{d}p = 0$. However, in order to estimate the accuracy of the baric coefficients obtained for the out-of-plane parameters, one should know the error introduced by using such an approximation. The analytical estimation of $(\mathrm{d}\ln\gamma_0)/\mathrm{d}p$ made by Nagayoshi (1977) yields $(\mathrm{d}\ln\gamma_0)/\mathrm{d}p = -0.0016\,\mathrm{kbar}^{-1}$. This is consistent with the estimate of the upper limit for that quantity $|(\mathrm{d}\ln\gamma_0)/\mathrm{d}p| < 0.005\,\mathrm{kbar}^{-1}$, which has been obtained by Mendez et al. (1980b) from the relation $(\mathrm{d}\ln H_n)/\mathrm{d}p \simeq 2(\mathrm{d}\ln\gamma_0)/\mathrm{d}p$, where H_n is the resonance field for the magneto-optical transition at the point H of the Brillouin zone [see eq. (122)]. The above estimate of the upper limit for $|(\mathrm{d}\ln\gamma_0)/\mathrm{d}p|$ has been made on the basis of an evaluation of the accuracy with which resonance fields are measured; within the experimental error, they are pressure-independent.

When examining the effect of pressure on the SWMC-model parameters determined from experimental data one should bear in mind that out-of-plane parameters heavily depend on temperature. This is a consequence of the strong dependence of γ_i upon the interplanar distance c_0, which varies as a result of the thermal expansion of the lattice. According to the data of Baskin and Meyer (1955), the interplanar distance in graphite decreases to 6.6753 Å and 6.6720 Å from its room-temperature values $c_0 = 6.7076$ Å as the lattice is cooled down to 77 K and 4.2 K, respectively. Such a decrease of c_0 as a result of lattice cooling is equivalent to a room-temperature compression of the lattice. Hence it follows that, in order to compare the parameters γ_i determined from experimental data obtained at different temperatures, one should make a correction associated with the change in the interplanar distance, At present, there are no data on the temperature dependence of the compressibility of graphite in the temperature range ⊕300 K. However, according to the estimate made by Mendez et al. (1980b), the change in compressibility in this temperature range does not exceed 3%. This value of the relative error for the compressibility does not exceed the error of determination of the baric coefficients $(\mathrm{d}\ln\gamma_i)/\mathrm{d}p$ from the experimental data obtained from measurements in graphite under pressure.

One may, therefore, conclude that, although the absolute values of the parameters significantly depend on temperature, the change in their baric coefficients may be neglected. The constancy of the baric coefficients makes it possible to recalculate the values of γ_i from one temperature to another by using $(\mathrm{d}\ln\gamma_i)/\mathrm{d}p$ and the dependence of the interplanar distance on p and T.

As a general remark concerning the determination of baric coefficients from experimental data, we should like to note that since the number of quantities measured under pressure is usually smaller than that of the parameters $(\mathrm{d}\ln\gamma_i)/\mathrm{d}p$ to be determined, it is necessary that some additional relations between baric coefficients be used. These are based on the assumption (Arkhipov et al. 1963) that the parameters γ_i are determined by the overlap of the wave functions, which decrease exponentially at large distances. In such a case, parameters describing the interaction between atoms in neighbouring layers (the first-order parameters $\gamma_1, \gamma_3, \gamma_4$) should vary as $\exp(-c_0/a^*)$, while those describing the interaction between atoms in alternate layers (second-order parameters, e.g. γ_2, γ_5), as $\exp(-2c_0/a^*)$, where a^* is a distance characterizing the localization of the wave functions. The above assumption about the dependence of γ_i on c_0 is not rigorously made, because not all parameters determine the interaction in directions strictly perpendicular to the graphite layers (see fig. 10).

In deriving the additional relations between the baric coefficients for the parameters γ_i it is assumed that the fundamental dependence of these parameters upon pressure is the consequence of the change in the interplanar distance c_0 caused by the pressure. In conjunction with the assumption about the character of the dependence of γ_i on c_0, made by Arkhipov et al. (1963),

Table 14
Baric coefficients of conductivity in graphite at various temperatures

Material	Pressure range (k bar)	Temperature (K)	$(d \ln \sigma_a)/dp$ $(10^{-6}\,bar^{-1})$	$(d \ln \sigma_c)/dp$ $(10^{-6}\,bar^{-1})$	Ref.*
Natural graphite	0–0.1	295	+5	–	[1]
Pyrolytic graphite	0–0.4	295	–	–	[2]
Natural graphite	0–0.6	77, 295	–	–	[3]
Natural graphite	0–9.0	290–420	+3.0	–	[4]
Pyrolytic graphite	0–8.0	4.2	—	–	[5]
Hot-pressed and	0–0.4	4.2	28	28.2 ± 1	[6]
annealed graphite		77	12–14	27.6 ± 1	[6]
		295	±1	25.5 ± 1.5	[6]
Hot-pressed pyrolytic	0–4.0	77	10–13	–	[7]
graphite		300	±1	–	[7]

*References
[1] Bridgman (1952).
[2] Samara and Drickamer (1962).
[3] Aust and Drickamer (1963).
[4] Likhter and Kechin (1964).
[5] Kechin et al. (1968).
[6] Yeoman and Young (1969).
[7] Spain (1969).

this enables one to conclude that the baric coefficients of the first-order parameters should either be equal or close to one another, the same being true of the baric coefficients for the second-order parameters; the former coefficients should be twice as high as the latter.

The idealized additional relations between the baric coefficients of the parameters γ_i thus have the following form

$$\frac{\mathrm{d}\ln|\gamma_2|}{\mathrm{d}p} = \frac{\mathrm{d}\ln|\gamma_5|}{\mathrm{d}p} = 2\frac{\mathrm{d}\ln\gamma_1}{p} = 2\frac{\mathrm{d}\ln\gamma_3}{\mathrm{d}p} = 2\frac{\mathrm{d}\ln\gamma_4}{\mathrm{d}p}. \tag{136}$$

In order to determine the $(\mathrm{d}\ln\gamma_i)/\mathrm{d}p$ from the experimental data, one can use, along with the relations given by eq. (136), the following, less rigorous relations, in which either the a priori relationship between the coefficients of first-order and second-order parameters or the exact equality of the coefficients for all first-order parameters is not assumed:

$$\frac{\mathrm{d}\ln|\gamma_2|}{\mathrm{d}p} = \frac{\mathrm{d}\ln|\gamma_5|}{\mathrm{d}p}; \qquad \frac{\mathrm{d}\ln\gamma_1}{\mathrm{d}p} = \frac{\mathrm{d}\ln\gamma_3}{\mathrm{d}p} = \frac{\mathrm{d}\ln\gamma_4}{\mathrm{d}p}; \tag{137}$$

$$\frac{\mathrm{d}\ln|\gamma_2|}{\mathrm{d}p} = \frac{\mathrm{d}\ln|\gamma_5|}{\mathrm{d}p} = 2\frac{\mathrm{d}\ln\gamma_1}{\mathrm{d}p}; \qquad \frac{\mathrm{d}\ln\gamma_3}{\mathrm{d}p} = \frac{\mathrm{d}\ln\gamma_4}{\mathrm{d}p}. \tag{138}$$

In accordance with the assumptions made above, let us consider what information about the baric coefficients for the parameters γ_i can be extracted from the measurements of the optical spectra of graphite under pressure taken in the near infrared region (see sect. 8.1).

Obviously, the experimentally determined values of $\mathrm{d}\varepsilon_{A_1}/\mathrm{d}p$ and $\mathrm{d}\varepsilon_{A_2}/\mathrm{d}p$ are not enough to enable one to find the baric coefficients for all the parameters contained in expression (130), viz., γ_1, γ_2, γ_5, Δ, and ε_F. Indeed, the sum of the shifts for the transitions A_1 and A_2 under pressure is

$$\Delta\varepsilon_{A_2} + \Delta\varepsilon_{A_1} = 4\gamma_1 - 4\gamma_2 + 2\varepsilon_F, \tag{139}$$

while their difference:

$$\Delta\varepsilon_{A_2} + \Delta\varepsilon_{A_1} = 4\gamma_5 + 2\Delta - 2\varepsilon_F, \tag{140}$$

and we have two equations to find five unknowns. Yet, as Misu et al. (1979) have shown on the basis of Nakao's calculation (Nakao 1979), the number of parameters necessary to determine the averaged shifts $\Delta\varepsilon_{A_1}$ and $\Delta\varepsilon_{A_2}$ under pressure can be reduced to only two: γ_1 and γ_5. Then the sum and difference of the averaged shifts have the form:

$$\overline{\Delta\varepsilon_{A_2}} + \overline{\Delta\varepsilon_{A_1}} = 4\Delta\gamma_1, \qquad \overline{\Delta\varepsilon_{A_2}} - \overline{\Delta\varepsilon_{A_1}} = 4\Delta\gamma_5. \tag{141}$$

Taking advantage of the baric coefficients for averaged shifts (133), relations (141) and the absolute values for the parameters γ_1 and γ_5, one can obtain the following values for the logarithmic pressure derivatives of γ_1 and γ_5:

$$\frac{\mathrm{d}\ln\gamma_1}{\mathrm{d}p} = 0.028 \pm 0.003\ \mathrm{k\,bar^{-1}}, \qquad \frac{\mathrm{d}\ln\gamma_5}{\mathrm{d}p} = 0.055 \pm 0.016\,\mathrm{k\,bar^{-1}}, \tag{142}$$

which satisfy the condition $(\mathrm{d}\ln\gamma_5)/\mathrm{d}p = 2(\mathrm{d}\ln\gamma_1)/\mathrm{d}p$ [see eqs. (136) and (138)]. Direct information about the value of the baric coefficient for the parameter γ_1 can be obtained from an analysis of the spectra of magneto-optical K-point transitions, from which the dependence of the parameter m^* upon pressure is obtained (see fig. 75).

Indeed, if trigonal warping is neglected, then, in accordance with (70), the parameter m^* is determined solely by the ratio $\gamma_0^{(2)}/\gamma_1$. The influence of trigonal warping upon m^* has been examined by Schroeder et al. (1970), Nakao (1976), and Doezema et al. (1979) who demonstrated that the contribution of the parameter γ_3 to the baric coefficient $(\mathrm{d}\ln m^*)/\mathrm{d}p$ is a second-order effect. Thus, in a first approximation, the relation

$$\frac{\mathrm{d}\ln m^*}{\mathrm{d}p} \simeq \frac{\mathrm{d}\ln\gamma_1}{\mathrm{d}p} - 2\frac{\mathrm{d}\ln\gamma_0}{\mathrm{d}p} \tag{143}$$

will be valid. Using the theoretical estimate of Nagayoshi (1977), according to which $(\mathrm{d}\ln\gamma_0)/\mathrm{d}p = 0.0016\,\mathrm{kbar^{-1}}$, Mendez et al. (1980b) obtained

$$\frac{\mathrm{d}\ln\gamma_1}{\mathrm{d}p} = 0.024 \pm 0.004\,\mathrm{kbar^{-1}}. \tag{144}$$

This value coincides, within the error of its determination, with the results of the calculation by eq. (142) on the basis of optical measurements. It is of interest to note that the above logarithmic pressure derivative of γ_1 given by (142) and (144) are more than twice as high as their values previously found by Kechin et al. (1968) and Kechin (1969) from galvanomagnetic data by using relations (138) and assuming that $(\mathrm{d}\ln\gamma_0)/\mathrm{d}p = 0$ (see table 15). In these papers the baric coefficients for the parameters γ_1 and γ_4 have been found in an analysis of the pressure dependence of the low-field Hall coefficient, using Kechin's approximate relation

$$-R_{\mathrm{H}}\bigg|_{H\to 0} = \frac{0.487a_0^2 c_0 \rho_T^2\,\gamma_0^2\gamma_2}{e(kT)^2\,|\gamma_1|}\left(1 - 17.02\frac{\gamma_4 kT}{\gamma_0\gamma_2} - 1.247\left|\frac{\gamma_4\gamma_2}{\gamma_0 kT}\right|\right), \tag{145}$$

where ρ_T is a parameter of the order of unity. This relation is valid for $T > 50\,\mathrm{K}$. According to Brandt et al. (1980b), the accuracy of determination of the logarithmic derivatives $(\mathrm{d}\ln\gamma_1)/\mathrm{d}p$ and $(\mathrm{d}\ln\gamma_4)/\mathrm{d}p$ by this approximate formula is rather low.

A detailed analysis of the baric coefficients for all the parameters of the SWMC-model, including ε_{F}, has been made by Dillon et al. (1977) and Brandt et al. (1980b) on the basis of oscillation and galvanomagnetic data and the results of the calculations by Nagayoshi et al. (1976b). Dillon and his co-

Table 15
Comparison of baric coefficients of the SWMC-model parameters (in $kbar^{-1}$) by various workers.

$\frac{d\ln\gamma_0}{dp}$	$\frac{d\ln\gamma_1}{dp}$	$\frac{d\ln\|\gamma_2\|}{dp}$	$\frac{d\ln\gamma_3}{dp}$	$\frac{d\ln\gamma_4}{dp}$	$\frac{d\ln\|\gamma_5\|}{dp}$	$\frac{d\ln\|\gamma_6\|}{dp}$	$\frac{d\ln\|\varepsilon_F\|}{dp}$	Ref.*
0	0.012	0.019	0.012	0.012	0.024	–	0.019	[1]
0	0.020	–	–	–	–	–	–	[2]
0	–	–	–	0.008	–	–	–	[3]
0	0.012	0.024	0.012	0.012	0.024	0.09	0.024	[4]
0	0.012	0.024	0.012	0.012	0.024	–	–	[5]
−0.0016	0.020	–	–	0.022	0.038	0.031	–	[6]
–	0.0281 ±0.003	–	–	–	0.055 ±0.016	–	–	[7]
−0.0016	0.024	0.011 ±0.008	–	–	–	–	0.011 ±0.008	[8]
0	0.021	0.042	0.075	0.075	0.042	−0.036	0.022	[9]
0	0.022	0.043	0.032	0.032	0.043	−0.048	0.023	[10]

*References:
[1] Itskevich and Fisher (1967).
[2] Arkhipov et al. (1963).
[3] Kechin (1969).
[4] Anderson et al. (1967).
[5] Kechin et al. (1968).
[6] Nagayoshi (1977).
[7] Misu et al. (1979).
[8] Mendez et al. (1980b).
[9] Dillon et al. (1977).
[10] Brandt et al. (1980b).

workers used the expressions of the four-parameter approximation of the SWMC-model for S_e, S_h, and S_m.

As the original experimental data, they have taken the baric coefficients for Fermi-surface cross sections from the papers by Itskevich and Fisher (1967) and Anderson et al. (1967) (see table 13). The logarithmic derivatives $(\mathrm{d} \ln \gamma_i)/\mathrm{d}p$ have been found using relations (138) between the desired coefficients and the set C of the values for the parameters γ_i from table 5. The values for $(\mathrm{d} \ln \gamma_i)/\mathrm{d}p$ or $(\mathrm{d} \ln |\gamma_i|)/\mathrm{d}p$ and $(\mathrm{d} \ln |\varepsilon_F|)/\mathrm{d}p$, found by various authors, are summarized in table 15.

CHAPTER 9

Magnetic Susceptibility of Graphite and Carbon Materials

9.1. *Magnetic properties of single-crystal and pyrolytic graphite in weak and quantizing magnetic fields*

The magnetic properties of monocrystalline graphite are described by two components of the magnetic susceptibility tensor, $\chi_\parallel$ and $\chi_\perp$, which correspond to the magnetic field directions, the one along the *c*-axis ($\boldsymbol{H} \parallel c$) and the other along the graphite layers ($\boldsymbol{H} \perp c$). Monocrystalline graphite is diamagnetic at any temperature and has a very high anisotropy of magnetic susceptibility: $\chi_\parallel \gg \chi_\perp$. Its diamagnetism is abnormally high in the direction perpendicular to the plane of the graphite layer. It can be maintained that graphite has the largest value of specific diamagnetism among all known materials. The specific magnetic susceptibilities of a typical single-crystal graphite at room temperature are $\chi_\parallel = -22 \times 10^{-6}$ and $\chi_\perp = -0.5 \times 10^{-6}$ emu/g (Krishnan and Ganguli 1939). The difference between these values increases with lowering temperature, tending to $\chi_\parallel - \chi_\perp \simeq -30 \times 10^{-6}$ emu/g for $T \lesssim 100$ K. The temperature dependences of $\chi_\perp$ and of the difference $\chi_\parallel - \chi_\perp$ are shown in fig. 85.

The smaller component of magnetic susceptibility, $\chi_\perp$, is of the same order of magnitude as the diamagnetic susceptibility of free carbon atoms. It weakly depends on temperature and steadily varies from -1.4×10^{-6} emu/g for $T = 100$ K to -0.4×10^{-6} emu/g for $T = 940$ K.

The difference of the susceptibilities, $\chi_\parallel - \chi_\perp$, which characterizes the anisotropy of graphite, also exhibits a steady temperature dependence, varying from -28×10^{-6} emu/g at $T = 140$ K to -7.8×10^{-6} emu/g at $T = 1270$ K.

The abnormally high diamagnetism of graphite in the direction perpendicular to the plane of the layer, and the large value of the anisotropy of its magnetic susceptibility have long remained unexplained. In this connection, we note that attempts have been made to treat graphite as a limiting case of the series of aromatic hydrocarbons, which corresponds to a very large number of benzene rings. According to Van Vleck (1932), the molecular magnetism of aromatic hydrocarbons is approximately additively constituted by the elementary magnetism of electronic orbits associated with the atoms of hydrogen and carbon. Therefore, the absolute value of magnetic susceptibility $|\chi_\parallel|$ in the direction perpendicular to the plane of an aromatic hydrocarbon molecule should increase with increasing number of benzene rings.

However, in the direction parallel to the plane of the molecule the diamagnetic susceptibility $\chi_\perp$ should differ only slightly from that of many nonaromatic hydrocarbons. These features of the magnetic properties of the sequence of aromatic hydrocarbons have actually been observed in experiment. According to Langevin's theory of molecular magnetism, diamagnetism is associated with closed electron shells, which represent current circuits precessing in a magnetic field. The total molecular susceptibility, χ_M, is obtained by summation over all the closed-shell orbits:

$$\chi_M = -\frac{ne^2}{6m_0c^2}\sum_i \overline{r_i^2}. \tag{146}$$

In this expression, the most significant part is played by the term that describes the projections of electron-orbit areas, $\Sigma_i \overline{r_i^2}$, upon the plane perpendicular to the direction of the magnetic field. The increase in the magnetic susceptibility of aromatic hydrocarbon molecules per molecule means that the aromatic (or π-) electrons produce current circuits whose area is, roughly speaking, proportional to the number of rings, i.e., to the area of the molecular frame, which is far greater than that of an electron orbit in an atom. Hence the large value of diamagnetism.

A quantum-mechanical calculation of magnetic susceptibility also yields an expression of the type of (146), the only difference being that, because of the presence of excited levels, there appear additional terms, associated with transitions between these levels.

However, an explanation of the magnetic properties of graphite proceeding from the above concepts would be incorrect, because the nature of the conduction-electron diamagnetism in metals differs from that of the molecular diamagnetism.

Indeed, whereas in aromatic hydrocarbons the magnitude of the diamagnetism $\chi_\parallel$ increases with increasing number of benzene rings [and this increase is decribed by the classical expression given by eq. (146)], in metals the classical precession of electrons in a magnetic field generally fails to result in diamagnetism, because the resulting diamagnetic current of the precessing electrons is completely counterbalanced by the oppositely directed current produced by the electrons moving near the metal surface. The diamagnetism of electrons has, as Landau first demonstrated in 1930, a purely quantum origin associated with the quantization of the electron energy in a magnetic field. Its magnitude is, therefore, determined by the structure of the carrier-energy spectrum, rather than by the number of unit cells making up a crystal.

Irrespective of the electron precession radius $\bar{r}_i$, in metals, semimetals, and doped semiconductors the magnetism of carriers depends on the value of their cyclotron mass and their effective g-factor. As a rule, the spin paramagnetic contribution exceeds the diamagnetic one, so that the gas of carriers is usually paramagnetic.

As for graphite, its diamagnetism is highly sensitive to the parameters of the the energy spectrum and, for certain values of these parameters and for certain Fermi energies of the carriers, it may be quite small, which is hardly compatible with the "molecular" hypothesis of its nature.

Physically much more meaningful is the attempt made by Mrozowski (1952) to calculate the magnetic susceptibility of graphite on the basis of the concept of a two-dimensional electron gas in a graphite layer. However, the values of $\chi_{\parallel}$ and $\chi_{\perp}$ calculated with the help of the known parameters for the band spectrum of graphite (see table 5) differ to a considerable extent from the experimental data. The attempts to exploit the Landau–Peierls electron diamagnetism theory for calculating the magnetic susceptibility of graphite, made by Smoluchowski (1953) and Hove (1955), have failed, too. Although they obtained a slope of the dependences of susceptibility on temperature and on impurity concentration which is qualitatively correct, the absolute value has turned out to be much lower than the experimental value.

The inapplicability of the Landau–Peierls theory of electron diamagnetism to calculations of magnetic susceptibility of substances whose spectra have several close bands at one and the same point of momentum space (as is the case with graphite) has been pointed out first by Adams (1953). He has demonstrated that in substances possessing such a spectrum the determining contribution to susceptibility comes from the interband electron transitions in a magnetic field.

A theory of diamagnetism of graphite, which took into account interband transitions on the basis of a simple two-dimensional model of the spectrum, has been advanced by McClure (1956), who used the formalism developed by Luttinger and Kohn (1955) to calculate the energy levels of graphite in a magnetic field. McClure has modified this formalism when considering degenerate bands. The contribution of the energy of atomic orbital momenta in a magnetic field and the contribution of the Pauli spin paramagnetism was neglected in his calculation. In this manner, he succeeded in obtaining a correct value for the magnetic susceptibility of graphite and in explaining its temperature dependence $\chi_{\parallel} \sim 1/T$ in the high-temperature range. This calculation has shown that within the two-dimensional approximation the entire electron diamagnetism is determined by interband transitions disregarded in the earlier works. The large value of diamagnetism can be explained by the contribution of electrons, whose effective masses are very small.

McClure (1956) has derived an expression for the susceptibility $\chi_{\parallel}$ of graphite, which is exact in the limit of weak magnetic fields and high temperatures for $kT \gg \gamma_0 a_0 (eH/c\hbar)^{1/2}$ and which has the form

$$\chi_{\parallel} = -0.044 \frac{4}{\pi c_0 \rho} (\gamma_0 a_0)^2 \left(\frac{e}{c\hbar}\right)^2 \frac{\operatorname{sech}^2 (\varepsilon_F/kT)}{kT} \quad \text{(in emu/g)}, \tag{147}$$

where $\rho = 2.265\ \text{g/cm}^3$ is the density of graphite. The above formula describes

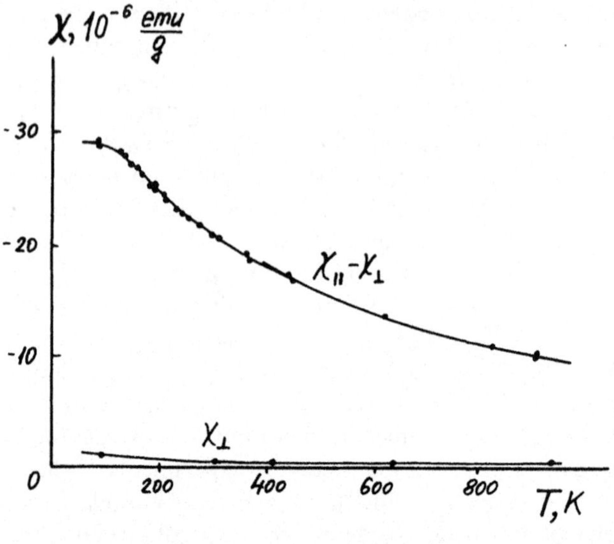

Fig. 85. Temperature dependences of the magnetic anisotropy in single-crystal graphite. (Krishnan and Ganguli 1939.)

satisfactorily the behaviour of $\chi_{\parallel}$ (see fig. 85) in the range of high temperatures at a reasonable value for $\gamma_0 \simeq 2.8\,\text{eV}$ (see table 5). In the region of quantizing magnetic fields, where the energy-level spacing exceeds the energy of thermal vibrations: $\gamma_0 a_0 (eH/c\hbar)^{1/2} > kT$, the dHvA effect occurs. In such fields as these, the results of the calculation by McClure (1956) are at variance with the experimental data. First of all, his calculation suggests that the average value for susceptibility in quantizing magnetic fields will be zero, and therefore dHvA oscillations in graphite should be about the zero level. This prediction has been refuted by the data of experimental studies of the dHvA effect in graphite and pyrocarbons which indicate that magnetic susceptibility oscillations occur against the background of a large diamagnetic component, the sign reversal (from negative to positive) being observed only for the highest peak at $H \sim 20\,\text{kOe}$ (see fig. 46). Moreover, McClure's expression for the oscillation amplitude

$$\Delta\chi_{\parallel} \simeq 0.01\gamma_0^2 \frac{kN}{\varepsilon_F}, \qquad (148)$$

where N is the number of an oscillation peak, yields values which are practically two orders of magnitude as high as the real values for the relevant amplitudes.

Subsequently, McClure (1960) has refined his calculation of the magnetic susceptibility of graphite by using, instead of the two-dimensional spectrum approximation, a simplified model of a three-dimensional spectrum, equivalent to the SWMC-model with $\gamma_3 = 0$. Taking into account the trigonal warping in the SWMC-model ($\gamma_3 \neq 0$) leads to a mixing of states with quantum numbers n, $n+3$, and $n-3$, thereby precluding the derivation of analytical expressions for energy levels in a magnetic field, necessary for calculating the magnetic susceptibility.

The calculation of the susceptibility of graphite, based on the simplified model of a three-dimensional spectrum, has proved to be sufficiently correct to describe the temperature dependence of the non-oscillating part of $\chi_\parallel$ over the whole temperature range from liquid-helium to room temperatures. Specifically, at low temperatures the estimated $\chi_\parallel$ is equal to $-3\gamma_0^2 \times 10^{-6}$ emu/g (γ_0 in eV), which, at a reasonable value for $\gamma_0 (\simeq 3\,\text{eV})$, is in good agreement with the experimental value of -30×10^{-6} emu/g. The agreement between the calculated temperature dependences of $\chi_\parallel - \chi_\perp$ is illustrated in fig. 86.

The reader's attention is drawn to the fact that McClure, in his 1960 calculation, used the following set of the SWMC-model parameters: $\gamma_0 = 2.8\,\text{eV}$, $\gamma_1 = 0.27\,\text{eV}$, $\gamma_2 = 0.02\,\text{eV}$, $\gamma_4 = -0.03\,\text{eV}$, $\gamma_5 = 0.02\,\text{eV}$ and

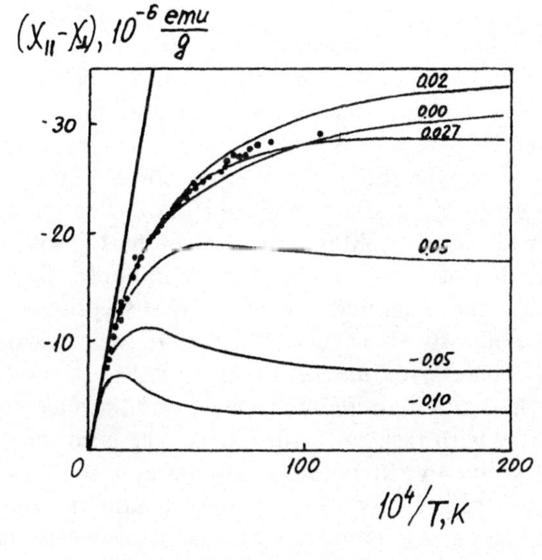

Fig. 86. The magnetic susceptibility of graphite. The solid circles present the data of Ganguli and Krishnan (1941). The curves are calculated using the following values for the parameters of the SWMC-model: $\gamma_0 = 2.8$, $\gamma_1 = 0.27$, $\gamma_2 = 0.02$, $\gamma_4 = -0.03$, $\gamma_5 = 0.02$, $\Delta \equiv \gamma_6 = -0.025$ (in eV). The straight line is the "two-dimensional" susceptibility for $\gamma_0 = 2.8\,\text{eV}$ and $\varepsilon_F = 0$. The numbers at the curves give the Fermi levels in eV. The $+0.027\,\text{eV}$ level corresponds to pure graphite. The 0.05 eV level corresponds to 10^{-4} more electrons per atom than holes per atom, while the levels 0.02, 0.0, -0.05 and $-0.10\,\text{eV}$ correspond, respectively, to 0.25×10^{-4}; 1.0×10^{-4}; 3.7×10^{-4} and 6×10^{-4} more holes per atom than electrons per atom. (McClure 1960.)

$\Delta = -0.025\,\text{eV}$. In this set the parameters γ_2 and γ_4 differ from the currently adopted ones not only in value but even in sign (see table 5).

For a positive γ_2, the value of ε_F for undoped graphite with equal densities of electrons and holes is also positive ($\varepsilon_F \simeq 0.027\,\text{eV}$). Accordingly, McClure (1960) compared the experimental values for $\chi_\parallel - \chi_\perp$ for pure graphite with the curve corresponding exactly to that value ε_F (see fig. 86). Therefore, the reasonably good agreement between the experimental data and the curve calculated for $\varepsilon_F = 0.027\,\text{eV}$ is somewhat surprising. The point is that in calculating the non-oscillating part of the magnetic susceptibility only some of the SWMC-model parameters play a leading role, while the influence of the others is practically negligible.

In the high-temperature range, magnetic susceptibility is largely determined by the parameter γ_0 alone and, at $T = 1000\,\text{K}$, the $\chi_\parallel$ closely (90%) agrees with the susceptibility calculated within the two-dimensional model. The only other parameter that appreciably affects the magnitude of $\chi_\parallel$ for $T \geqslant 300\,\text{K}$ is γ_1. In McClure's calculation (1960) the parameters γ_0 and γ_1 were chosen to obtain the best fit of the calculated curve to the experimental points. The agreement for $T \sim 100\,\text{K}$ was assured by a suitable choice of Δ. One may assume that the remaining parameters (among them γ_2 and γ_4) are really rather unimportant for determining the value of $\chi_\parallel$ and that their choice has not affected the agreement between theory and experiment. Interestingly, McClure's calculation (1960) has enabled him to predict a peak in the temperature dependence of the susceptibility of graphite with the Fermi level shifted downward. In practice, the Fermi level may be lowered by doping graphite with acceptor-type impurities (see the lower curves in fig. 86, which correspond to $\varepsilon_F = 0.05$, -0.05, and $-0.10\,\text{eV}$). The predicted peak has subsequently been detected by Marchand and Dupart (1967) who studied the magnetic susceptibility of pyrocarbons doped with boron (fig. 87). The effect of trigonal warping on the magnitude of magnetic susceptibility of graphite has been taken into account by Sharma et al. (1974). In their calculation, they have used the general formula for magnetic susceptibility derived by Fukuyama (1971), in which it is not required that the Fermi surface should possess a rotational symmetry with respect to the c-axis. The effect of trigonal warping was taken into account by expanding in the parameter $\gamma_3/\gamma_0 \sim 0.1$ to within the terms of second order. Thus, the expression for the magnetic susceptibility in the SWMC-model at $\boldsymbol{H} \parallel c$ is given by the sum of two terms,

$$\chi_{\text{SW}} = \chi_0 + \delta\chi, \tag{149}$$

of which only the second one is dependent on the parameter γ_3.

In the high-temperature limit, these terms are given by

$$\chi_0 = -\pi A_0 \frac{\gamma_0^2 - \frac{3}{5}\gamma_4^2}{6kT}, \tag{150}$$

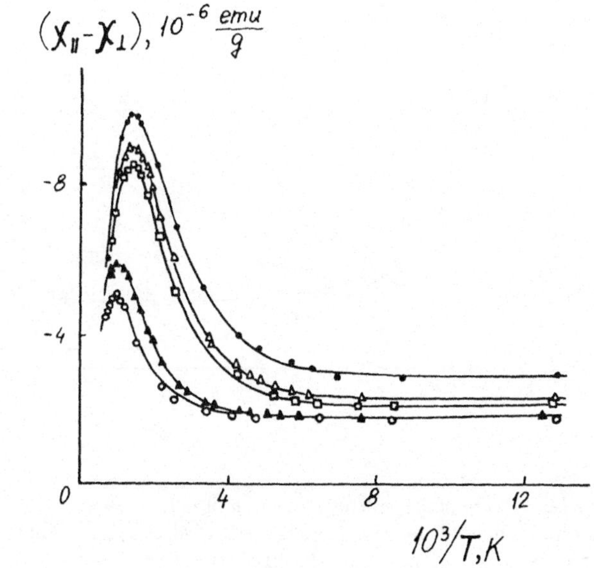

Fig. 87. Temperature dependences of the diamagnetic susceptibility of boron-doped pyrocarbons. The experimental data of Marchand and Dupart (1967) are represented by the symbols: ●, 40 AS specimen; △, 40 AP; □, 56 P; ▲, 100 P; ○, 150 S. The solid lines show the calculations of Levintovich and Kotosonov (1980).

$$\delta\chi = \pi A_0 \frac{\gamma_3^2\left[1 - \frac{24}{5}\left(\frac{\gamma_4}{\gamma_0}\right)^2\right]}{6kT}, \tag{151}$$

where

$$A_0 = \left(\frac{e}{\hbar c}\right)^2 \frac{3a_0^2}{2\pi^2 c_0}.$$

The correction introduced by taking into account trigonal warping at high temperatures is small (not greater than 1%).

At low temperatures, $\chi_0 \sim \gamma_0^2/\gamma_1$ and $\delta\chi \sim \gamma_3^2/\gamma_2$. In this case, the correction introduced by trigonal warping is $(\gamma_3/\gamma_0)^2(\gamma_1/\gamma_2) \sim 13\%$, which considerably exceeds the correction at high temperatures $[\sim(\gamma_3/\gamma_0)^2]$. The above estimates suggest that in magnetic susceptibility calculations the effect of trigonal warping should be taken into consideration only in the low-temperature range. The role of trigonal warping is especially significant in calculating the dHvA oscillation amplitude.

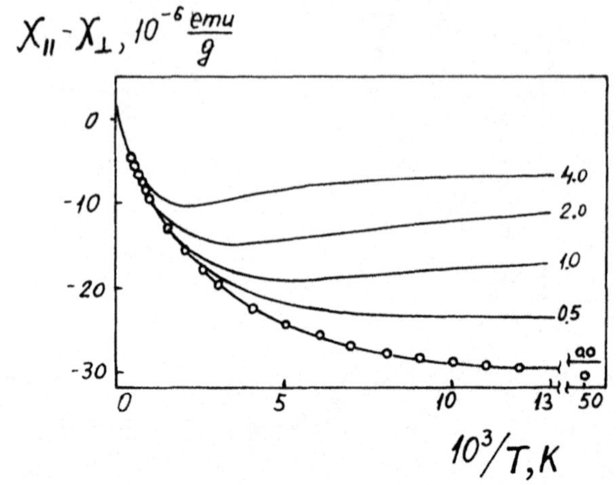

Fig. 88. Magnetic susceptibility of graphite as a function of inverse temperature. The open circles represent the experimental data for pure graphite (Poquet et al. 1960). The numbers at the curves represent $(n_h - n_e) \times 10^4$, where n_h and n_e are the numbers of free holes and electrons per carbon atom. The curves were calculated by Sharma et al. (1974) using the following parameter values: $\gamma_0 = 3.11$, $\gamma_1 = 0.376$, $\gamma_2 = -0.0208$, $\gamma_3 = 0.29$, $\gamma_4 = 0.197$, $\gamma_5 = -0.01$, $\Delta = \gamma_6 = -0.012$ (eV).

Sharma et al. (1974) have made a numerical calculation of the magnetic susceptibility of graphite as a function of temperature and position of the Fermi level. The value of ε_F in undoped graphite was taken to be equal to -0.024 eV, in accordance with the currently accepted signs and values for the spectrum parameters (see the data of table 5).

The magnetic susceptibility has been found to have a strongly pronounced diamagnetic character if the Fermi level is near the doubly degenerate band ε_3; it is weakly diamagnetic or paramagnetic if the Fermi level lies elsewhere in the band-overlap region for $|\varepsilon_F - \Delta - 2\gamma_5| > |2\gamma_1|$ and it is close to zero if the Fermi level is outside the overlap region. Sharma et al. (1974) have calculated the temperature dependence of the difference $\chi_{\parallel} - \chi_{\perp}$ in graphite for various values for the imbalance $n_h - n_e$ (and therefore, at various values for the Fermi level). The calculated curves are shown in fig. 88. The points on the curve for undoped graphite are the magnetic susceptibility data of Poquet et al. (1960). It can be seen that the curves agree rather well with the experimental data at all temperatures, except for the lowest ones ($T < 20$ K), where the discrepancy of theory with experiment is about $\sim 6.6\%$.

It may therefore be stated that on the basis of the SWMC-model for the electron spectrum graphite a sufficiently exact theory has been developed, which enables one to calculate the non-oscillating part of the low-field magnetic susceptibility of pure graphite over a wide range of temperature. As for the dependence of magnetic susceptibility on the position of the Fermi level, one should note that in comparing the calculated values of $\chi_{\parallel}$ with experiment difficulties appear which are associated with the determination of the degree of ionization of incorporated impurities.

In conclusion we note that, to our knowledge, at present there is no calculation whose results would be consistent with the experimental amplitude of magnetic-susceptibility quantum oscillations in graphite. One may except that the theory developed by Sharma et al. (1974) on the basis of the SWMC-model for the energy spectrum of graphite will permit a correct description of the dHvA effect in this substance. Yet, the mathematical difficulties involved are too formidable to allow one to obtain analytical expressions for the dependence of $\chi_{\parallel}$ on the magnetic field in the low-temperature range.

9.2. *Magnetic properties of graphite in the extreme quantum region of magnetic fields*

The above discussion of the magnetic properties of graphite has shown that in weak and quantizing magnetic fields this material is a highly anisotropic diamagnetic substance whose non-oscillating part of the magnetic susceptibility ($\chi_{\parallel}$) weakly depends on magnetic field.

In fields $H > 65\,\mathrm{kOe}$, electrons and holes in graphite are on the extreme Landau levels with $n = 0$. In that range of fields, which is called the extreme quantum limit (EQL), the magnetic properties of graphite have been studied less thoroughly. This is due to the difficulty of creating static magnetic fields $H \gtrsim 100\,\mathrm{kOe}$ and also due to the difficulty of making measurements in pulsed magnetic fields where the eddy currents may strongly affect the results of measurements. For these reasons, most of the measurements of the absolute values of the magnetic susceptibility or the magnetic moment of graphite have been performed in fields of 70–80 kOe. In higher magnetic fields (i.e., directly in the EQL fields), the magnetic properties of graphite have been investigated in very few experiments, although these studies are of great interest.

Chouteau and Briggs (1977) have measured the magnetic moment M of pyrolytic graphite at temperatures of 1.5 to 4.2 K in static fields up to 150 kOe produced by means of a water-cooled Bitter magnet. The measurements of M were made on two specimens of PG, the more perfect of which (number 1) was characterized by the resistivity ratio $\rho_a(300\,\mathrm{K}):\rho_a(4.2\,\mathrm{K}) \simeq 3$; the disorientation of the hexagonal axes of crystallites in this specimen did not exceed 0.3°. The obtained field dependence of the magnetic moment in PG speciments for $\boldsymbol{H}\parallel c$ is shown in fig. 89.

The differential magnetic susceptibility $\chi_{\parallel} = \mathrm{d}M/\mathrm{d}H$ for pyrolytic graphite at $T = 4.2\,\mathrm{K}$ in pulsed magnetic fields up to 400 kOe has been measured by Kapustin et al. (1981), who used specimens of quasi-single-crystal PG obtained by precipitating carbon from the gaseous phase at $T \sim 2100°\mathrm{C}$ and sub-

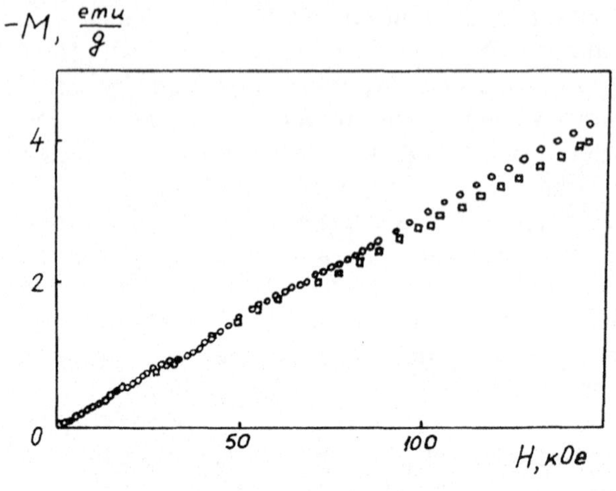

Fig. 89. Magnetisation of pyrolytic graphite at 4.2 K as function of the magnetic field. ○ specimen 1; □ specimen 2. (Chouteau and Briggs 1977.)

sequently heat treated under pressure at $T \sim 3000°C$. These workers have employed the improved induction method proposed by Shapira and Foner (1970), which permitted them to measure the differential susceptibility $\chi_{\parallel} = dM/dH$ which is of the order of 10^{-5} emu/g in pulsed magnetic fields. The obtained field dependence of $\chi_{\parallel}$ in PG is presented in fig. 90. The

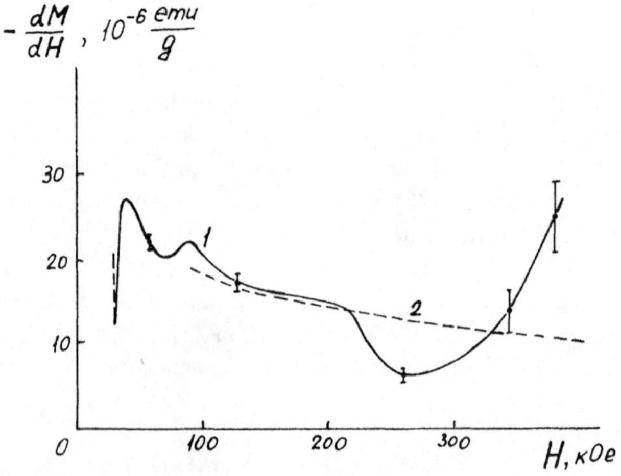

Fig. 90. The differential magnetic susceptibility in pyrolytic graphite as a function of the magnetic field $\boldsymbol{H} \| c$ at $T = 4.2$ K (solid curve 1). Dashed curve 2 shows the calculation by Jacobi and Jay-Gerin (1977). (Kapustin et al. 1981.)

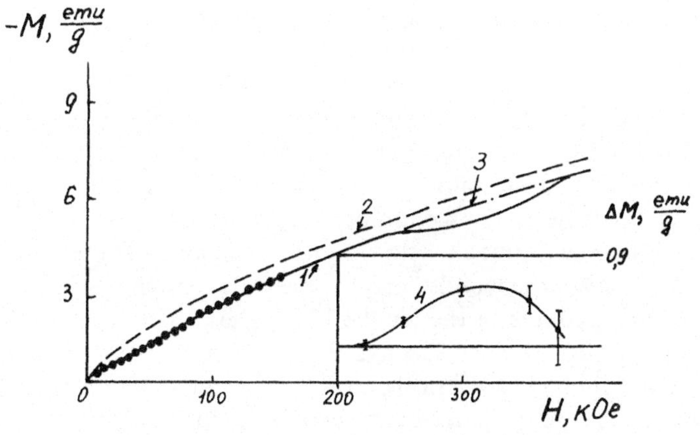

Fig. 91. Magnetic moment in pyrolytic graphite as a function of magnetic field H//c at $T = 4.2$ K (solid curves 1 and 4). Dashed curve 2 shows the calculation by Jacobi and Jay-Gerin (1977), curve 3 shows the same calculation now taking into account a constant paramagnetic addition; the solid circles represent the data of Dillon et al. (1978b). (Kapustin et al. 1981.)

dependences of $\chi_{\parallel}$ on H in the field range up to 150 kOe, which have been obtained in these studies are in good agreement with one another. In the range 100–150 kOe one can observe a slow but steady decrease of the absolute value of $\chi_{\parallel}$ by about 40%. A qualitatively different dependence of $\chi_{\parallel}$ on H is observed in fields $H > 220$ kOe. As can be seen in fig. 90, in this range the absolute value of $\chi_{\parallel}$ drastically decreases, but on passing a minimum at $H \simeq 260$ kOe it again begins rapidly to increase, so that in fields $H > 330$ kOe its value is greater than the values observed in fields from 100 to 200 kOe.

It is more convenient to discuss the magnetic properties of graphite in the EQL in terms of magnetization. Accordingly, fig. 91 shows the field dependence of the magnetic moment M of pyrolytic graphite taken from the work by Kapustin et al. (1981). This dependence has been obtained by integrating the experimental curve of fig. 90. In the field range 220kOe $\lesssim H \lesssim$ 380kOe there is a precipitous decrease on the curve for the H-dependence of M, which corresponds to a sharply nonmonotonic change in differential susceptibility in that range. This feature appears to be due to the phenomenon of magnetic freeze-out of impurity carriers in graphite (see sect. 5.4), and will be discussed at greater length later on.

9.2.1. *Magnetic susceptibility of graphite in EQL magnetic fields*

The study of the EQL behaviour of magnetic susceptiblity is of interest for the following reasons. As has been mentioned above, the diamagnetism of graphite in weak magnetic fields is due exclusively to the filled-band diamagnetic contribution, whereas that of charge carriers (the Landau–Peierls magnetism) is negligibly small.

The magnetization of the carriers is determined by the ordering in the magnetic field of the effective magnetic moments associated with the electron spins, which produces the paramagnetic effect, and by the Larmor electron precession (the orbital motion), which results in diamagnetism. How do these contributions to magnetic susceptibility change in EQL fields?

The magnitude of diamagnetism due to filled bands is determined by the energy gap between the bands and by the value of the effective masses of carriers near the extrema of the bands. The highest diamagnetism arises when the gap tends to zero and the effective masses are abnormally small. One may therefore expect that a change in the direct energy gap due to the displacement of the (conduction and valence) band boundaries in the magnetic field, as well as a change in the effective masses of carriers resulting from the nonparabolicity of the spectrum, may somehow change this component of diamagnetism in the EQL. Furthermore, the paramagnetic contribution from electrons should also change, because in the EQL at the last Landau levels with quantum number $n = 0$ all carriers have one and the same spin orientation.

The SWMC-model allows one to calculate the change in all the components of the magnetic susceptiblity condisered above for graphite in the range of very high magnetic fields. Let us analyze the experimental data available and compare them with theory.

The experimental data on the magnetic properties of pyrolytic graphite, obtained by Chouteau and Briggs (1977) and also by Kapustin et al. (1981), enable one to conclude that the transition to the EQL, which is accompanied by a considerable change in the density of states near the Fermi level, does not result in a radical change in the magnitude of diamagnetism (at least for $H < 200\,\text{kOe}$). As before, this magnitude is largely determined by the contribution from the filled bands, although this contribution appears to decrease steadily in a magnetic field.

The contribution provided by the filled bands to the magnetization of graphite in the EQL has been calculated by Jacobi and Jay-Gerin (1977). They have derived an analytical expression for the magnitude of magnetic moment M_{FB} due to filled bands

$$M_{\text{FB}}(H) = \frac{3}{4}\frac{\gamma_0^2 a_0^2}{\pi\gamma_1}\left(\frac{e}{ch}\right)^2 \int_0^{\pi/2} \frac{\mathrm{d}x}{\cos x} \ln\left[\frac{\alpha}{H} + \left(\frac{\alpha^2}{H^2} + 1\right)^{1/2}\right], \tag{152}$$

where

$$\alpha = \frac{8}{3}\frac{\gamma_1}{\gamma_0^2 a_0^2}\frac{ch}{e}\cos x\left[\gamma_1 \cos x + \left(\gamma_1^2 \cos^2 x + \tfrac{3}{4}\gamma_0^2 a_0^2 \frac{eH}{ch}\right)^{1/2}\right], \tag{153}$$

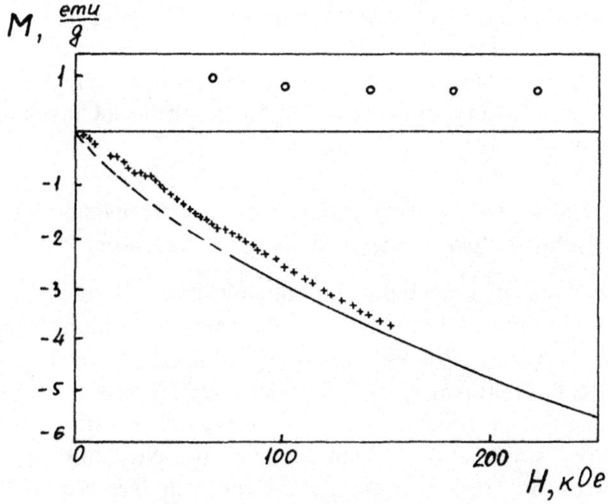

Fig. 92. Comparison of the experimental results of Chouteau and Briggs (1977) (+) with the Wallace and Gupta (1974) prediction (O) and with the calculation of Jacobi and Jay–Gerin (1977) (dashed line continued as solid curve) performed for $\gamma_0 = 3.12$eV and $\gamma_1 = 0.377$eV.

and

$$x = k_z c_0.$$

The dependence of M_{FB} on H given by (152) is shown by the solid curve in fig. 92. In the range $H \lesssim 70$kOe, the dependence is indicated by the dashed curve, because the calculation is valid only in the EQL. Figure 92 also shows the results of magnetic-moment measurements.

The contribution from the magnetization of carriers to the magnetism of graphite in EQL fields when only the Landau levels with $n = 0$ remain in the spectrum, has been calculated by Wallace and Gupta (1974). Their calculation indicates that, because of the magnetization of carriers at levels with $n = 0$, graphite in EQL fields should become a weak paramagnetic whose positive magnetic moment M slowly decreases with increasing magnetic field. This implies that in the EQL one should observe a positive value for the ratio M/H in the presence of a weak, negative differential magnetic susceptibility dM/dH, whose absolute value is, according to the calculation, an order of magnitude lower than $\chi_{\parallel}$ in weak magnetic fields. The results of this calculation are shown in fig. 92.

One can see that the diamagnetic moment due to the filled bands somewhat exceeds the experimental values. The discrepancy may be due to the paramagnetic contribution of carriers to susceptibility, although it is about twice as low as the calculated value of this contribution. Nevertheless, the main contribution to the magnetization of graphite in the EQL still comes from the filled bands. It is of interest to note that the H-dependence of M in the EQL is nonlinear. A similar behaviour of magnetic moment has been observed in bismuth by Brandt and Semyonov (1975) and by McClure and Shoenberg (1976).

9.2.2. The relation between the field dependence of the magnetic susceptibility of graphite and the phenomenon of magnetic freeze-out

The above analysis of the change in magnetic properties of graphite in the EQL does not permit one to explain the essentially nonmonotonic dependence of $\chi_{\parallel}$ on H in the range $220\,\mathrm{kOe} \lesssim H \lesssim 380\,\mathrm{kOe}$ and the precipitous decrease observed on the curve for magnetic moment $-M(H)$ (see fig. 91). The observed features may be understood, however, if one assumes that in the field range indicated above some of the conduction band electrons are localized on impurity centres as a result of magnetic freeze-out (see sect. 5.4). Indeed, the localization of electrons leads to a lowering of the total energy of an electronic system, thus giving rise to an additional (with respect to the contribution from the filled-band electrons) paramagnetic contribution to the magnetization of graphite. This contribution may be considered on the basis of phenomenological concepts about electron localization, which makes it possible to evaluate the binding energy of an electron on an impurity site in a magnetic field. The electron-system energy in a magnetic field may be written as

$$\varepsilon = \varepsilon_{\mathrm{FB}} + \varepsilon_{\mathrm{I}}, \tag{154}$$

where the first term on the right-hand side represents the contribution from the lower filled energy bands of graphite (see Jacobi and Jay-Gerin 1977) and the second is an addition to the energy, associated with impurities. If $N_{\mathrm{i}} = N_{\ell}(H)$ stands for the ionized-impurity concentration and $N_{\ell} = N_{\ell}(H)$, for the concentration of those impurities on which electrons have been localized, then the addition ε_{I} to the filled-band energy may be written as

$$\varepsilon_{\mathrm{I}} = \sum_k (\varepsilon_{\mathrm{i}} N_{\mathrm{i}} + \varepsilon_{\ell} N_{\ell})_k = \sum_k [\varepsilon_{\mathrm{i}} N_{\mathrm{i}}^{(0)} - N_{\ell}(\varepsilon_{\mathrm{i}} - \varepsilon_{\ell})]_k. \tag{155}$$

where ε_{i} and ε_{ℓ} are the energies of an impurity-atom electron in the delocalized and localized states, which are measured relative to the lowest Landau level, $N_{\mathrm{i}}^{(0)} = N_{\mathrm{i}}(H) + N_{\ell}(H) = \mathrm{const.}$ is the total concentration of impurities fully ionized at $H = 0$, the summation over k implies summation over the various types of impurities. The magnetization of a specimen is given by

$$M = -\frac{\partial \varepsilon}{\partial H} = -\frac{\partial \varepsilon_{\mathrm{FB}}}{\partial H} - \sum_k \left[\frac{\partial \varepsilon_{\mathrm{i}}}{\partial H} N_{\mathrm{i}}^{(0)} - \frac{\partial N_{\ell}}{\partial H}(\varepsilon_{\mathrm{i}} - \varepsilon_{\ell}) - N_{\ell}\frac{\partial(\varepsilon_{\mathrm{i}} - \varepsilon_{\ell})}{\partial H}\right]_k, \tag{156}$$

whence, neglecting the variation of ε_i in the field, we find

$$M = -\frac{\partial \varepsilon_{FB}}{\partial H} + \sum_k \left[\frac{\partial N_\ell}{\partial H}(\varepsilon_i - \varepsilon_\ell) - N_\ell \frac{\partial \varepsilon_\ell}{\partial H} \right]_k . \tag{157}$$

The first term on the right-hand side of (157) has been calculated by Jacobi and Jay-Gerin (1977) and its dependence on H is shown in fig. 91 (curve 2). The second term represents the contribution from the localized electrons to magnetization. It is positive, because the binding energy $-\varepsilon_\ell$ of a localized electron and the number of the localized electrons N_ℓ increase in a magnetic field. Thus the effect of electron localization on impurity sites provides a positive (paramagnetic) contribution to the susceptibility of graphite, which is precisely what is observed experimentally. The presence of several types of impurity centres with various field values at which electrons begin to be localized should lead to a stepwise field dependence of magnetic moment, each step corresponding to a peak of magnetic susceptibility. These features should be much better pronounced in magnetic characteristics of graphite than in characteristics proportional to the magnitude of the carrier imbalance $n_h - n_e$ (e.g., the Hall conductivity component $\sigma_{xy}H$), because the magnetic characteristics are determined by the field derivatives of N_ℓ and ε_ℓ, rather than by these quantities themselves.

That the experimental dependences of M or $\chi_\parallel$ upon H do not show a stepwise character enables one to conclude that in the PG specimens investigated by Kapustin et al. (1981) the magnetic freeze-out effect occurs on impurity centres of one type only (or on centres of different types, but with close field dependences of the effect). Therefore, the summation over k in eqs. (155)–(157) may be omitted. In the field range where the magnetic freeze-out takes place ($H > H_{FOE}$, where H_{FOE} is a characteristic value of the magnetic field at which electron localization on impurity centres begins), the change in electron energy on an impurity centre will be

$$\varepsilon_\ell(H) - \varepsilon_i(H_{FOE}) = \frac{1}{N_\ell} \int_{H_{FOE}}^{H} \Delta M \, dH, \tag{158}$$

where

$$\Delta M = M^{exp} - M_{FB}^{theor},$$

M_{FB}^{theor} is the calculated value for the magnetic moment of the filled-band electrons (see fig. 91, curve 3). The integral in (158) represents the area confined under curve 4 in fig. 91. Hence, assuming that for $H \simeq 380\,kOe$ the process of magnetic freeze-out is practically ended, one can estimate the change in the electron energy due to electron localization on impurities for $H \simeq 380\,kOe$, viz., $\varepsilon_i - \varepsilon_\ell \simeq 100\,meV$. This estimate is consistent with the estimated binding energy of the electron localized on a solitary impurity centre in a substance with the sharply anisotropic spectrum (97) (see sect. 5.4), which is $\varepsilon_B \simeq 50\,meV$

– for graphite in a field $H \simeq 380$ kOe, satisfying inequality (98). In a rather high magnetic field, when the localization of carriers on impurity sites comes to an end ($N_l = N_i^{(0)}$), the amount of the paramagnetic addition to magnetization will be determined by the rate of increase in the localized-electron binding energy in a magnetic field.

Magnetic measurements, therefore, afford, in principle, a means for checking the validity of formula (97) for the electron binding energy in graphite. To this end, data on the field dependence of M in fields exceeding 500 kOe are needed.

The experimentally observed paramagnetic contribution to the magnetization of graphite occurs in the same magnetic fields ($H > 200$ kOe) in which the galvanomagnetic measurements of Brandt et al. (1974) have revealed a drastic decrease of the unbalance $\Delta n \sim \sigma_{xy} H$.

Significantly, the peculiar character of the galvanomagnetic and magnetic characteristics was observed by Brandt et al. (1974) and Kapustin et al. (1981) in specimens of kish-graphite and pyrolytic graphite with very similar parameters and with a concentration of ionized impurities about $10^{18}\,\mathrm{cm}^{-3}$. Therefore, the fact that their data are in good agreement is an additional argument for the assumption about electron localization in graphite under the influence of a strong magnetic field.

9.3. Diamagnetism of quasi-two-dimensional graphite

The search for substances characterized by a high diamagnetism retained in a wide range of magnetic fields and temperature is of great scientific and practical value. From this point of view, quasi-two-dimensional pyrocarbons with an increased interplanar distance c_0, owing to which their energy spectrum becomes very similar to that of gapless semiconductors characterized by the linear dispersion relation given by (30) (see sect. 4.1), is of considerable interest. In such materials, a great increase in diamagnetism may be expected as the Fermi level approaches the point of band degeneracy. Materials of this type include pyrocarbons and quasi-monocrystalline pyrographite precipitated at ~2000–2200°C. Brandt et al. (1979) and Kuvshinnikov (1980) have found that all these materials have a hole-type conductivity and a large value of diamagnetic susceptibility $\chi_{\parallel}$, whose absolute value rapidly increases with decreasing carrier density n_{h} (i.e., as the Fermi level approaches the point of band degeneracy), reaching the value of $|\chi_{\parallel}| \simeq 140 \times 10^{-6}\,\mathrm{emu/cm^3}$ at liquid-helium temperatures, $n_{\mathrm{h}} \simeq 0.6 \times 10^{18}\,\mathrm{cm}^{-3}$ and $H \simeq 12$ kOe. The value of $|\chi_{\parallel}|$ for quasi-monocrystalline graphite at $T = 4.2$ K at first slightly increases with the field, then remains constant or passes through a gently sloping maximum and only then, beginning with a certain field value $H = H_k$, linearly (in a first approximation) decreases (fig. 93). The magnetic-field value H_k decreases with

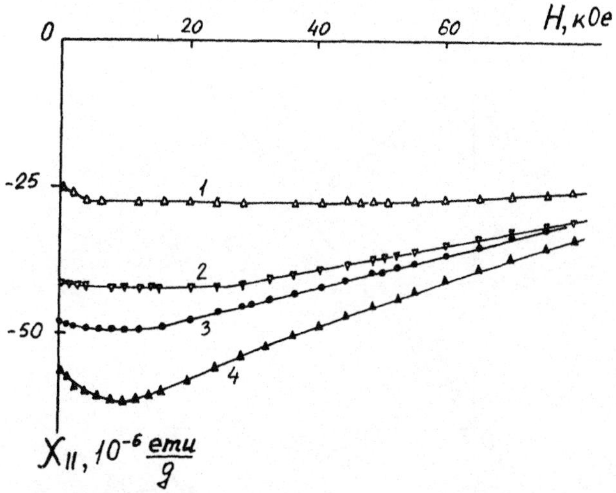

Fig. 93. The magnetic susceptibility $\chi_{\parallel}$ of quasi-single-crystal graphite (pyrocarbon) as a function of the magnetic field $\boldsymbol{H} \| c$ at $T = 4.2\,\mathrm{K}$ and for various hole concentration n_h in the specimens: curve 1, $n_h = 5.7$; 2, $n_h = 2.8$; 3, $n_h = 1.5$; 4, $n_h = 0.6$ (in $10^{18}\,\mathrm{cm}^{-3}$). (Brandt et al. 1979.)

decreasing hole density n_h, the maximum value of $|\chi_{\parallel}|$ simultaneously increasing (fig. 94). In order to elucidate the cause of the decrease of $|\chi_{\parallel}|$ for $H > H_k$, it is of interest to compare H_k with H_{EQL} which corresponds to the boundary of the EQL range of magnetic fields. The magnetic field H_{EQL} for two-dimensional graphite may then be estimated as

$$H_{EQL} = \frac{\pi h^2 c}{e} n_h d, \tag{159}$$

where d is the interplanar distance, which for pyrocarbon constitutes $\simeq 3.44\,\text{Å}$. The estimation of H_{EQL} for PG specimens yields values close to the value of H_k in fig. 94. Thus the linearly decreasing dependence of $|\chi_{\parallel}|$ on H in pyrocarbons (see fig. 93) seems to be associated with the transition to the EQL field range.

Since the contribution from charge carriers to susceptibility is small, one would naturally expect that, similarly as for three-dimensional graphite (see fig. 90), the decrease of the filled-band contribution should be the main reason for the decrease of $|\chi_{\parallel}|$ in the EQL. Brandt et al. (1979) and Kuvshinnikov (1980) observed a record value of magnetic susceptibility $\chi_{\parallel}$ in pyrocarbon specimens, constituting 0.2% of $-1/4\pi$, the theoretically ideal diamagnetism of superconductors. The susceptibility value they observed exceeds the known values for susceptibility of all other diamagnetics (except for superconductors). Figure 95 presents the temperature dependences of the differential magnetic susceptibility $\chi_{\parallel}$ of pyrocarbons and pyrographite in weak and in high magnetic fields. It should be noted that the $|\chi_{\parallel}|$ in these materials is

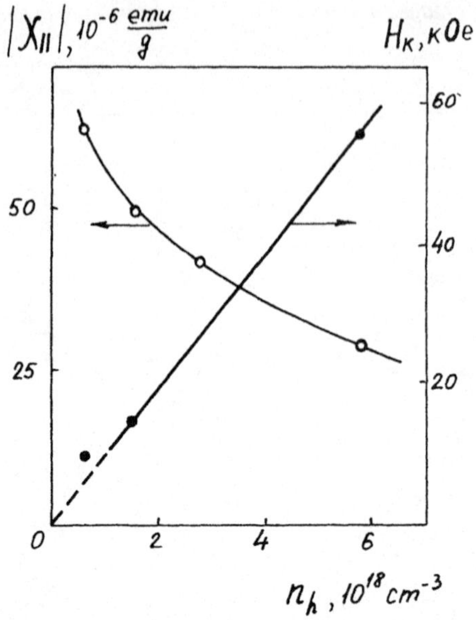

Fig. 94. Dependence of the maximum magnetic susceptibility $|\chi_{\parallel}|$ (left-hand scale) and magnetic field H_k (right-hand scale) in quasi-single-crystal graphite upon hole concentration n_{h} (see fig. 93). (Brandt et al. 1979.)

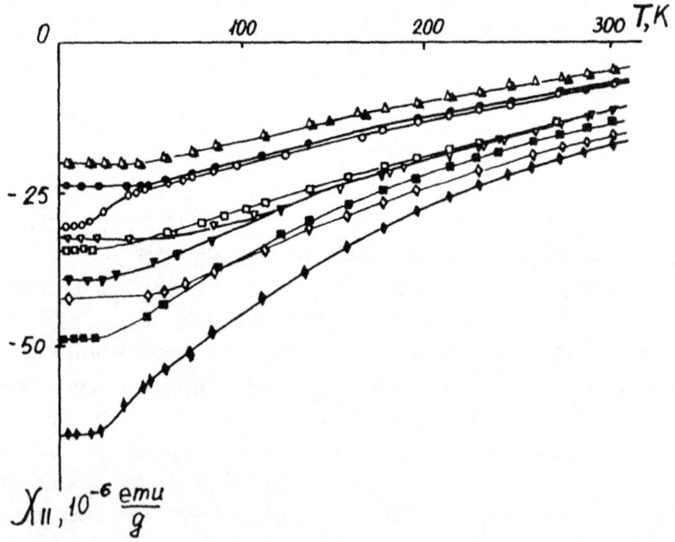

Fig. 95. Temperature dependences of the differential magnetic susceptibility in pyrocarbon and pyrolytic graphite for various magnetic fields. Pyrolytic graphite: ●, $H = 0.6\,\mathrm{kOe}$; ○, $H = 52\,\mathrm{kOe}$. Pyrocarbon: ▲ $H = 8\,\mathrm{kOe}$; △, $H = 50\,\mathrm{kOe}$ (1); ▼, $H = 6\,\mathrm{kOe}$; ▽, $H = 50\,\mathrm{kOe}$ (2); ■ $H = 6\,\mathrm{kOe}$; □, $H = 52\,\mathrm{kOe}$, (3); ◆, $H = 10\,\mathrm{kOe}$; ◇, $H = 50\,\mathrm{kOe}$ (4). Labels (1)–(4) indicate various specimens of pyrocarbon. (Kuvshinnikov 1980.)

larger than that in quasi-monocrystalline pyrographite and in monocrystalline graphite even at room temperature.

The increased diamagnetism of quasi-two-dimensional pyrocarbons, which is observed with decreasing carrier density, justifies one's expectation that in specimens with a still lower carrier density it will be possible to come still closer to the theoretical upper limit for the diamagnetic susceptibility $\chi_{\parallel}$ of two-dimensional graphite, which, according to the estimate of Levintovich and Kotosonov (1980), amounts to about -400×10^{-6} emu/g.

The maximum value for diamagnetism in quasi-two-dimensional graphite is attained in the absence of any carriers, when the Fermi level crosses the degeneracy (conical) point of the π-bands. In order to reach these values of $\chi_{\parallel}$, one should have defectless graphite-like layers which do not interact with one another.

9.4. *Magnetic properties of carbon materials*

As has already been pointed out, the high magnetic susceptibility of graphite is directly due to the structure of its electron energy spectrum, which is shaped as the crystal lattice of graphite forms. For carbon materials (cokes, synthetic pyrocarbons, soots, etc.), whose crystal structure arises in the course of heat treatment, a strong dependence of their magnetic susceptibility magnitude on the heat treatment temperature (HTT) is typical (fig. 96).

At an HTT $\lesssim 1000°$C, one usually observes a weak diamagnetism; however; sometimes a weak paramagnetism has been observed, whose origin may be

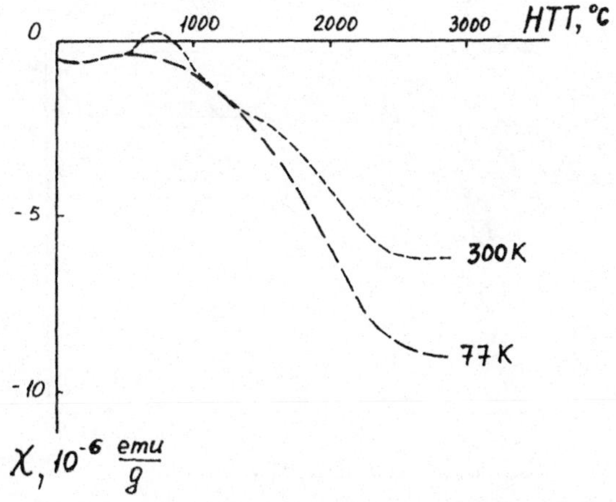

Fig. 96. Magnetic susceptibility of pyrocarbons as a function of heat-treatment temperature (HTT). (Mrozowski 1950.)

due to the presence, in the structure of these materials, of free radicals with unpaired spins. This has experimentally been confirmed by EPR-experiments (e.g., Honda 1959, Uebersfeld and Erb 1959).

At higher HTT values there is a range, 1600–2400°C, in which the diamagnetic susceptibility rapidly increases, exhibiting a characteristic temperature dependence.

Since the size of crystallites increases as the HTT is raised, it is of interest to compare the dependence shown in fig. 96 with the dependence of χ upon crystallite size.

If, instead of the HTT, one takes the effective crystallite size as the variable, then the increase in diamagnetism will correspond to the region where the diameter of crystallites occurring in the direction perpendicular to the hexagonal axis ranges from 50 to 150 Å. The dependence of the magnetic susceptibility of synthetic pyrocarbons upon the size of crystallites is illustrated in fig. 97. First of all, attention should be called to the fact that at room temperature soot powders with crystallites with a cross section of over 200 Å have a magnetic susceptibility $\chi \simeq -7 \times 10^{-6}$ emu/g closely approaching that of randomly oriented crystals of natural graphite, because for polycrystals the susceptibility is given by

$$\chi_{\text{polycr}} = \tfrac{1}{3}(\chi_{\parallel} + 2\chi_{\perp}) \simeq \tfrac{1}{3}\chi_{\parallel} = -7.3 \times 10^{-6}\,\text{emu/g} \tag{160}$$

(see sect. 9.1).

It would, therefore, be natural to assume that the electron energy spectrum characteristic of graphite is formed only when crystallites grow to a definite, critical size.

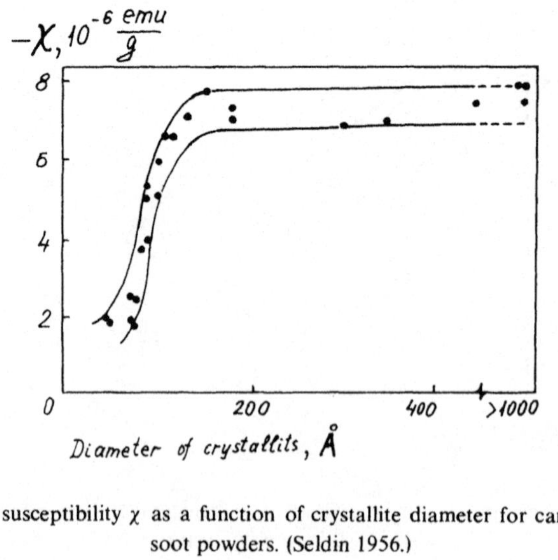

Fig. 97. Magnetic susceptibility χ as a function of crystallite diameter for carbon rods and gas-soot powders. (Seldin 1956.)

Unfortunately, the data available at present do not permit a distinct separation of the contribution due to the change in crystallite size from that due to the change (improvement, ordering) in the general structure of a speciment with rising HTT. Such a separation would, in certain measure, be aided by X-ray studies of powders with particles of various sizes.

At present the reason for which the susceptibility of carbons reaches a certain limiting value at crystallite sizes of about 150 Å remains obscure. In connection with this, it seems appropriate to discuss, though briefly, the very interesting question of how the magnetic properties of specimens are related to their size. The point is that in crystallites of rather small sizes the dominant role, at low temperatures, should be played by the effects of dimensional quantization. The associated abnormal magnetism of small metallic particles has been considered by Buzdin et al. (1984) who succeeded in demonstrating that, due to the effects of dimensional quantization in metallic clusters with an average size L placed in weak magnetic field, there should appear a high magnetic susceptibility, which may reach several percent of the ideal one. Particles whose size is rather small, but still much greater than the lattice constant a, i.e., their size satisfies the condition $L \gg a$, should possess a high magnetic susceptibility in the range of magnetic fields $\mu_B H \ll \varepsilon_F (a/L)^2$, where $\mu_B = e\hbar/2m_0c$ is the Bohr magneton and $\varepsilon_F(a/L)^2$ is a quantity determining the spacing between the dimensionally quantized levels of the electron energy.

At temperatures $T \ll (\varepsilon_F/k)(a/L)^2$ and with a completely filled upper level, the paramagnetic energy does not contribute to magnetic susceptibility. As a result, diamagnetism should increase by a factor of $(L/a)^2$ so that the magnetic susceptibility will be of the order of

$$\chi_d \simeq -\chi_0 (L/a)^2, \tag{161}$$

where $\chi_0 = \mu_B^2 \nu(\varepsilon_F)$ is the susceptibility of a massive specimen, $\nu(\varepsilon_F)$ is the density of states at the Fermi level. The best objects for observing the abnormal diamagnetism are small rings, spheres, and discs.

As the particle size increases, the ranges of magnetic fields and temperature in which the effect is observable become increasingly narrower, so that for rather large particles it will be impossible to measure the values of H and T required for the observation. Accordingly, in measuring χ_d at some given values for the magnetic field and temperature the dependence of the susceptibility on particle size should pass through a sharp peak.

In the investigations carried out thus far, the value for the magnetic field and the temperature were not those which are necessary to observe the anomalous magnetic properties in small particles. At the same time, graphite is a very conventient object for observation of the above effect. Therefore, it would be very interesting to try to discover if the effect of Buzdin et al. (1984) found for metallic clusters, also occurs in graphite powders in which the indispensable control of the quality and size of crystallites can be guaranteed.

It should, of course, be borne in mind that a positive result may be achieved only in the case when the quality of crystallites and the structure of their crystal lattice are independent of their size. In the opposite case, when the crystal structure of small-sized particles greatly differs from that of graphite, one should observe a steady increase in susceptibility associated with formation of the graphite lattice (an increase in the degree of two-dimensionality) with the increasing crystallite size, as can be seen from fig. 97.

CHAPTER 10

Graphite Intercalation Compounds (GIC)

10.1 Concepts and properties

The high anisotropy of the binding forces between carbon atoms in graphite, to which the graphite crystal lattice owes its lamellar structure, allows atoms and molecules of various impurities to be inserted into interlayer space. As a result, a new class of crystalline graphite compounds in whose lattice flat layers of the carbon atoms alternate with layers of the implanted matter arises. The incorporation of impurity atoms or molecules into the interlayer space of a graphite crystal is called intercalation, the impurities themselves are called intercalants. The lamellar compounds of graphite which result from intercalation have been given the name of graphite intercalation compounds (GIC). Unlike pure graphite, which is a compensated semimetal with an equal concentration of electrons and holes, GIC display typically metallic properties, in particular, a high electrical conductivity in the direction of the layers; some of the GIC (e.g., C_8K, C_4HgK, $C_4KTl_{1.5}$) even being superconductors.

At the present time, several hundreds of various substances, both atomic and chemical compounds, which can be implanted into the interlayer space of graphite, are known. Among them, one can mention alkali metals, inorganic acids and their salts, halogens, transition-metal dichalcogenides, certain silicates and chlorides of metals, etc.

The most characteristic feature of GIC is their peculiar crystal-lattice structure, in which intercalate layers periodically alternate with a definite number of carbon layers. The formation of a periodic structure as a result of graphite intercalation has been called the "staging phenomenon", the number, *n*, of graphite layers occurring between the adjacent layers made up by impurity atoms or molecules (intercalants) being referred to as "stage index". The concept of such a structure of the GIC lattice was first introduced by Rüdorff (1940) on the basis of X-ray studies, the concept of stage index, which has since become classical, is also due to him.

Another property which is typcial of GIC is that the intercalate atoms or molecules found in the interlayer space exhibit a high degree of structural order. The tendency towards ordering makes possible structural transitions of the "order–disorder" type as the temperature of intercalate layers is lowered.

It should be noted that the origination of a periodic structure by alternating layers of graphite and intercalate (staging phenomenon) is independent of the

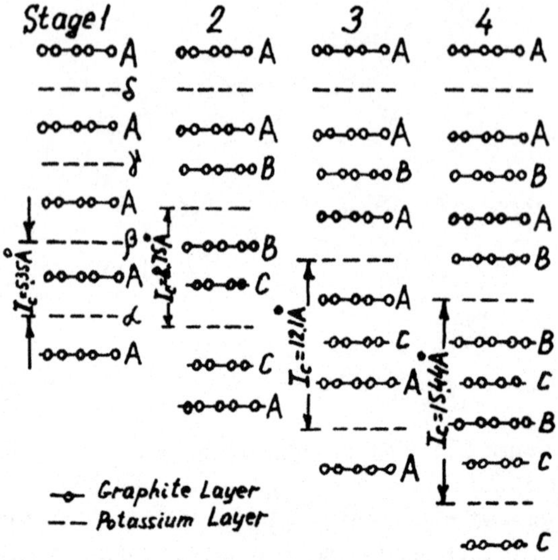

Fig. 98. Schematic diagram illustrating the staging phenomenon in graphite-potassium compounds for stages $1 \leqslant n \leqslant 4$. The potassium layers are indicated by dashed lines and the graphite layers by solid lines connecting open circles, and indicating schematically a projection of the carbon-atom positions. The ... ABAB... graphite-layer stacking for stages $n \geqslant 2$ is maintained between intercalate layers, although a rhombohedral stacking arrangement appears across them (see fig. 3a,b). For each stage, the distance I_c between adjacent intercalate layers is indicated. For first-stage C_8K, the unit cell includes intercalate layers with stacking indices α, β, γ, δ.

degree of ordering and of the structural type in an intercalate layer. The periodicity of structure in GIC in the direction perpendicular to the layers takes place for both commensurate and incommensurate structures in the intercalate layer, as compared to the structure of the adjacent graphite layers, as well as in the absence of any ordering in the intercalate layer. The periodicity of layers takes place also at a very low intercalant concentration (i.e., for $n \gg 1$). Up to now there have been produced GIC with $n \gtrsim 20$, in which the period of repetition of the structure in the direction perpendicular to the layers exceeds 70 Å. The staging phenomenon in GIC is illustrated in fig. 98, which shows several schemes for the arrangement of layers in graphite intercalated by potassium, with stage-index values ranging from 1 to 4. The model for the arrangement of carbon and potassium atoms in this compound for $n = 1$ (C_8K) proposed by Rüdorff and Schulze (1954) is shown in fig. 99.

As a result of graphite intercalation, there occurs a redistribution of electron density between carbon atoms in graphite layers and the atoms or molecules of the implanted substance in intercalate layers. According to the character of this redistribution, all GIC may be divided into two broad classes: donor-type GIC and acceptor-type GIC.

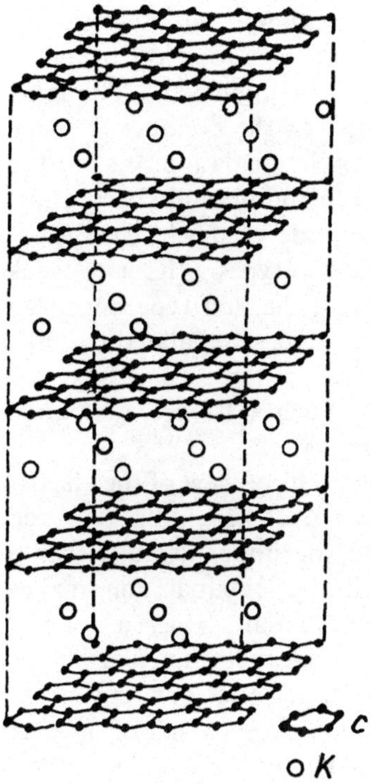

Fig. 99. Model for C_8K according to Rüdorff and Schulze (1954) showing the stacking of graphite layers (networks of small solid balls) and of potassium layers (networks of large hollow balls).

In donor-type GIC, the valence electrons of the intercalant become delocalized in the graphite layers, as a result of which the density of free electrons in them is significantly increased. The intercalant then appears as a donor which provides s-electrons to graphite layers. Compounds of this kind are produced when graphite is intercalated by alkali, alkaline-earth, and rare-earth metals. Donor-type GIC may be exemplified by C_6Li, C_8K, C_8Rb, C_8Cs, C_6Ca, C_6Ba, C_6Sr, C_6Eu, and other compounds.

In acceptor-type GIC, π-electrons withdraw out of the graphite layers and, subsequently, become localized in the intercalate layers. As a result, in carbon layers there appear free holes, whose density is considerably higher than their initial density in pure graphite. In this case, the intercalant acts as an acceptor.

Acceptor-type GIC include intercalation compounds with inorganic acids, halogens, interhaloids, etc.... Examples of such compounds are C_5HNO_3, $C_8H_2SO_4$, $C_{16}Br_2$, C_8AsF_5, C_9AlCl_3, C_8ICl, etc.

Important characteristics of GIC, which make it possible to describe charge redistribution due to intercalation, are the factors f_{chem} and f_{el}. The former determines the fraction of the electrons lost from or gained by an impurity per one molecule (or atom) of the intercalant. The latter describes the fraction of the free electrons supplied to the conduction band or the valence band of graphite in the course of intercalation. As a rule, $f_{chem} > f_{el}$ for all GIC, because some fraction of the additional charges turn out to be localized.

According to the magnitude of the factors f_{chem} and f_{el}, all acceptor-type GIC may be divided into two types: GIC with weak acceptors and GIC with strong acceptors. In GIC of the first type both these factors are small (both f_{chem} and f_{el} are $\lesssim 0.2$). Graphite intercalated by halogens serves as an example of such compounds. In GIC of the second type both these factors are rather large: $0.5 \lesssim f_{chem}$, $f_{el} \lesssim 1$. Such GIC are exemplified by the compounds of graphite with metal fluorides.

We note that, from the point of view of the electronic properties of graphite, the production of intercalated structures is a peculiar and highly specific method of implanting doping impurities of either the donor or the acceptor type into the initial material. Intercalation makes it possible to vary the physical properties of a material – over a wide range and in a controlled manner. This is due to the fact that intercalation occurs in a semimetal with a low concentration ($\sim 10^4$) of electrons per atom, into which donors or acceptors with concentration close to that of the carbon atoms are introduced. As a result, the electric, magnetic, thermal, and other characteristics of GIC are significantly different from those of pure graphite.

Of the properties common to GIC, we should mention that many compounds of graphite are unstable in air and must be kept in a hermetically sealed envelope. This is a consequence of the instability of intercalants: donor intercalants are readily oxidized, while acceptor ones are just as readily decomposed. Only few types of GIC are stable in air. They include, for instance, the compounds of graphite with $FeCl_3$, $SbCl_5$, etc. The decomposition of GIC is drastically slowed down as they are cooled down to liquid-nitrogen temperatures.

If, however, an unstable GIC is kept at room temperature outside a hermetic envelope, then the process of decomposition lasts only for a certain length of time, at the end of which it practically stops. The material thus produced has been called the residue compound. This compound is chemically stable and contains some concentration of intercalant (usually low), whose value depends on the temperature and the initial concentration of the intercalant in GIC.

10.2. Mechanisms of formation of a periodic structure in GIC

Let us consider which factors underlie the formation of the periodic structure of GIC in the direction perpendicular to the layers.

From general considerations it follows that the staging phenomenon in GIC may be due to two processes: the electrostatic interaction between the intercalate layers and the formation of a field of elastic stress in the graphite lattice in the course of intercalation.

The electrostatic interaction takes place when a certain amount of charge passes from graphite layers to an acceptor-type intercalate layer or, conversely, is drawn back from a donor-type intercalate layer to the nearest graphite layers. Naturally, the graphite layers that directly bound an intercalate layer (so-called bounding layers) should screen the electric field produced by the intercalate layer, thereby precluding the appearance of the long-range electrostatic interaction. However, as Pietronero et al. (1978, 1979) and Pietronero and Strässler (1979) have shown, the distribution of the screening-charge density ρ as a function of distance z from the intercalate layer is given by a power, and not by an exponential, law:

$$\rho(z) = e\frac{3z_0^3}{(z + z_0)^4}, \tag{162}$$

where the characteristic screening length z_0 depends on the two-dimensional charge density ρ_0 at $z = 0$, the Fermi velocity v_F, and the transverse dielectric constant $\kappa_\perp$:

$$z_0 = \left[\frac{e}{\pi\rho_0}\left(\frac{\kappa_\perp \hbar v_F}{6e^2}\right)^2\right]^{1/3}. \tag{163}$$

Pietronero et al. (1979) estimate z_0 as 3.0–4.0 Å if $\rho_0 \simeq 6.4 \times 10^{14}\,\text{cm}^{-2}$ and $\kappa_\perp \simeq 3.4$. This estimate has been confirmed subsequently by Safran and Hamman (1981). With such a dependence of ρ on z, the electrostatic interaction, although remaining long-range, is weak enough to account completely for the formation of the periodic structure of GIC. That the intercalate layers are rather effectively screened in such compounds is borne out by the data of Bok et al. (1978), independently obtained in measuring the electrical conductivity of GIC both in the plane of the layers and in the direction perpendicular to them, as well as the data of Dresselhaus et al. (1977) and Underhill et al. (1979) on the Raman and infrared spectra of GIC.

Another mechanism that ensures the long-range interaction between intercalate layers in the lattice of GIC is the appearance of an elastic stress as impurity atoms are being implanted into the lattice (the long-range lattice stress interaction). The significant role played by the elastic stress in the formation of the periodic structure of GIC is confirmed by the data of Nixon and Parry (1969), who measured lattice constants along the direction perpendicular to the layers of $C_{8n}K$ for various stage indices n, and by those of Underhill et al. (1979), who investigated the $1/n$-dependence of the shift of the graphite-lattice vibration frequency that arises from the intercalation of graphite by donor and acceptor impurities.

Practically the entire volume of data obtained in studying the kinetics of graphite intercalation by various impurities can be explained on the basis of the model considering the periodic field of lattice stress that corresponds to the minimum of deformation energy (the so-called stress model). In particular, this model explains well the fact that thin specimens of graphite are intercalated much more rapidly than thick ones.

A calculation of the stress-induced interaction force between two intercalant atoms placed between two graphite layers has indicated (see Kittel 1978) that the intercalant atoms are attracted to each other, tending to form an "island" of implanted matter.

At the same time, intercalant atoms inserted in different graphite layers repel each other. The appearance of "intercalate islands" leads to the decrease of elastic energy by about 0.1 eV per intercalant atom. The free energy of a lamellar compound is minimal when the completely filled and periodically arrayed intercalate layers are formed. In this case, the value of the energy turns out to be much lower than for randomly placed or underfilled intercalate layers.

Kittel's calculation (1978) is supplemented by the computations performed by Safran and Hamman (1979, 1980), which made it possible to demonstrate that in intercalating graphite by randomly placed atoms the elastic energy will be the lowest only in the case when the intercalant atoms form clusters in periodically arrayed intervals (layers). The minimum energy is then reached when the intercalant atoms completely fill each layer in which these clusters may occur. Although the atoms of the neighbouring graphite layers suffer the greatest displacement due to the insertion of an intercalate layer, the more distant layers of carbon atoms also undergo some displacement. Owing to this, elastic interactions become long-range and the resulting elastic energy turns out to be distributed over all graphite layers, otherwise the crystal would split.

We may thus assume that, regardless of the character of bonds between intercalate particles and carbon atoms (ionic, covalent or metallic), the elastic energy that should be expended on the expansion of the initial graphite specimen in the course of its intercalation appears to play a major role in the formation of the periodic structure of GIC.

What is the relationship between the mechanisms of the electrostatic and elastic interaction in graphite intercalation? This question has not yet been completely settled, but some preliminary results, derived from theoretical calculations, make it possible to elucidate the roles played by the respective mechanisms.

Taking into account the effect of both the elastic and electrostatic interactions on the process of intercalation, and calculating the overall configuration of the intercalate layers thus formed (see Safran and Hamman 1981) suggests that it is the elastic interaction that is dominant in the intercalation kinetics, while the electrostatic interaction is mainly responsible for the equilibrium state of the system to which the periodic structure corresponds.

An important contribution to elucidating the intercalation mechanism has come with the work of Hooley (1972), who discovered that the expansion of a graphite specimen being intercalated would stop if the basal planes (those perpendicular to the c-axis) are coated by some protecting lacquer. Under these conditions, intercalation stops even if the specimen is subjected to a tensile stress along the c-axis. These experiments suggest that, although impurity particles penetrate into the space between the graphite layers from the side of their facets, there can be no intercalation without the preliminary adsorption of the intercalant upon the basal planes of the crystal.

On the basis of these experimental results one may propose the following intercalation mechanism. The adsorption of the intercalant on the basal plane gives rise to a double electrical layer on the surface of the specimen. The excessive charge of the intercalate layer (e.g., the positive one for donor GIC) is screened by the free electrons in the graphite layers.

By analogy with the problem of charged-centre screening in metals and semiconductors, one may assume a damping and periodic distribution of the density of screening charges. (Equation (162) for the dependence of ρ on z describes the monotonic component of the screening-charge distribution.) The oscillation period of $\rho(z)$, whose nature is analogous to that of the Friedel oscillations of the charge density about charged centres in metals and semiconductors, is determined by the screening-electron density in the adjacent graphite layers, i.e., by the charge-transfer fraction, f, which depends on the intercalant type, pressure of its vapours, temperature gradient, etc.

The appearance of electron-density oscillations gives rise to a damping, periodic-in-z modulation of interplanar distance and, therefore, to an elastic stress in graphite. As a result, the interplanar bonds turn out to be weakened at some distance from the surface, which favours the insertion of the intercalant precisely between these planes.

Thus the above adsorption of intercalants through side facets assures, under certain, given, conditions, the opening of a definite interplanar interval in the graphite lattice, into which intercalant atoms start penetrating. When the entire interplanar interval has been filled, the intercalate layer begins to play the same role as the layer of adsorbed atoms on the surface, thereby creating favourable conditions for opening a new interplanar interval at a distance determined by the density-oscillation period for the charge that screens the excessive charge in the intercalate layer (see fig. 100). From this point of view the distance between the specimen's surface and the first intercalate layer should be smaller than that between the first intercalate layer and the next. This is due to the fact that the symmetry of the charge distribution between the intercalate layer and the bounding graphite layers will be different for the surface-adsorbed and for the implanted layers. Since the screening-charge density near the adsorbed layer should exceed the charge density near the

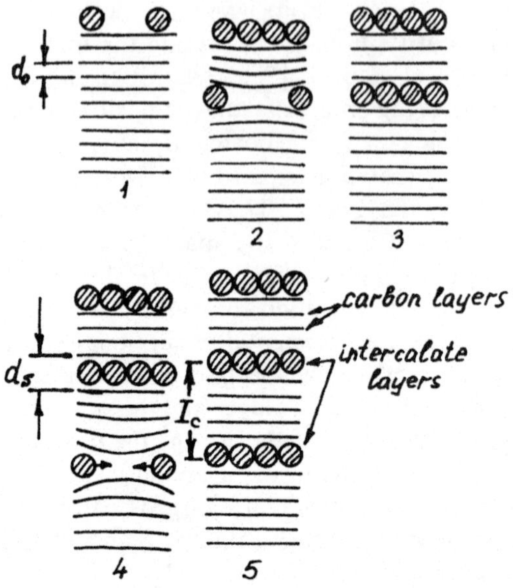

Fig. 100. Succession of formation of intercalation compounds in fifth-stage graphite.

implanted layers, the oscillation period will be shorter in the former case. It will be recalled that the Friedel screening-charge oscillations obey the law $\sim\cos k_F r$, where k_F is the Fermi electron wave vector, r the distance from the charged centre. Thus the oscillation period, which is equal to $2\pi/k_F$, decreases with increasing electron density.

In this representation, intercalation is a dynamical process of successive spreading of "cracks" from the periphery of a crystal and the subsequent filling of these cracks by diffusing atoms or molecules of the intercalant.

At the same time, many details of the origination of the periodic structure of GIC have not yet been fully understood. For instance, the very important question of the relation between the external conditions of intercalation (the intercalant-vapour pressure, the temperature of a graphite specimen, the temperature gradient near it, etc.) and the intercalation stage index.

10.3. Synthesis of GIC

The possibility of intercalating graphite was already known in the last century. The synthesis of a graphite intercalation compound was first reported by Schaffäult as far back as 1841. But the systematic studies of GIC were started only in the early 1930s as a result of the advances in X-ray diffraction techniques.

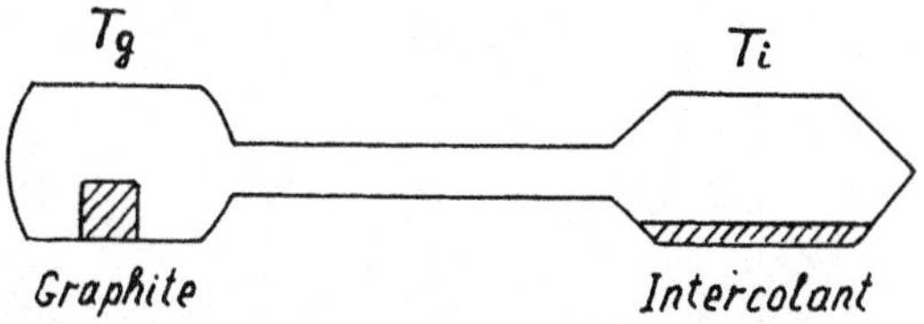

Fig. 101. A two-zone method of producing graphite intercalation compounds. Graphite is maintained at T_g and the intercalant, at T_i.

At the present time, graphite intercalation compounds are being produced by various methods, e.g., gas-transport reactions, intercalation from the liquid phase, co-intercalation, and electrochemical intercalation (see Hérold 1979).

Let us briefly describe the principal methods of intercalation. The most common technique (proposed by Hérold in 1955) is intercalation from the gaseous phase. In this way, an enormous amount of various substances may be introduced into graphite.

To run the process, pure graphite and an intercalant are placed in glass ampules connected by a thin tube, in which suitable temperatures, labelled T_g and T_i, are maintained (fig. 101). The vapours of the intercalant pass through the tube and enter into the ampule containing graphite.

To keep the process of intercalation going, it is necessary that there should always be a difference in temperature between the ampules, viz., $T_g > T_i$. Otherwise, intercalant vapours would precipitate on the graphite surface, which precludes intercalation.

The required stage index for GIC may be obtained by selecting the temperature difference $T_g - T_i$, which in turn is determined by the size and shape of the graphite specimen, the presence of defects, the amount of intercalant and the geometry of the glass ampules and the connecting tube. Figure 102 presents an isobar for GIC synthesis (graphite intercalated with potassium), which illustrates the dependence of potassium percentage in graphite upon the temperature difference $T_g - T_i$. The dashed curve indicates the segments corresponding to the stage indices 1, 2, and 3, respectively. In the synthesis of GIC of various stage indices, T_i is usually kept constant, while T_g is varied, which determines the value of n. The value of T_i is then determined by the threshold pressure, p_i, of intercalant vapours, below which the process comes to an end. Perhaps the main disadvantage of the method just described is that it is applicable only for introducing highly volatile substances, i.e., for intercalants with a high pressure of saturating vapours.

Another common technique is intercalation from the liquid phase, i.e., an intercalant is dissolved in a suitable solvent. The introduction of the intercalant occurs when a graphite crystal is put in the solution. The temperature at which intercalation takes place may be varied over a wide range up to the boiling point of the solution. This method is successfully used, for example,

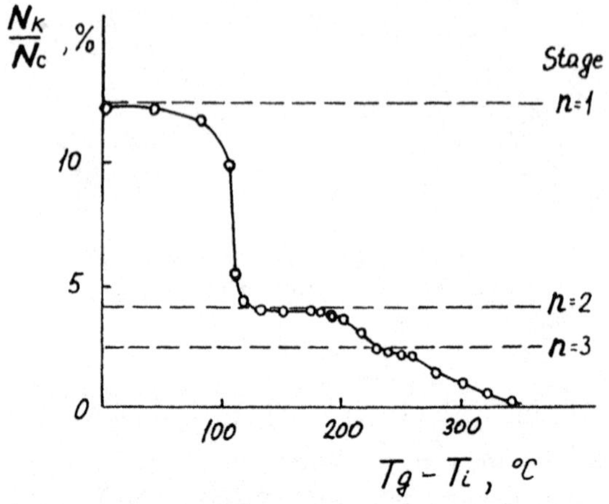

Fig. 102. Potassium percentage in graphite versus the temperature difference $T_g - T_i$ in intercalation from the gaseous phase. N_K, N_C are the concentrations of potassium and carbon, respectively. (Nixon and Parry 1968.)

for introducing into graphite bromides (with nitromethane as the solvent) or chlorides (with tetrachloromethane as the solvent).

Materials of low volatility and poor solubility may be incorporated into graphite directly from a melt. By this method, it is possible to obtain, for instance, compounds of graphite with lithium and other alkali metals. According to Zanini et al. (1978b), the process occurs when graphite is placed in a molten metal surrounded by an argon atmosphere.

In producing GIC by the co-intercalation technique, graphite is put into an intercalant powder and this system is subjected to compression (up to 10–20 kbar) in an inert atmosphere. In this way, it is possible to obtain, in particular, graphite compounds with lithium characterized by $n = 1$–4 (see Guérard and Hérold 1975).

The advantage of the co-intercalation method is that it allows a wider range of impurity substances to be incorporated in the host material. For instance, according to Billaud et al. (1980), it is possible to introduce into graphite Na simultaneously with K and Cs. The method also allows the synthesis of ternary lamellar graphite compounds although the components themselves may not serve as intercalants for graphite.

The electrochemical technique is customarily used for strong-acid intercalants. Two electrodes are placed in an acid solution, graphite being connected with the anode. Synthesis of the required stage index is ensured by a suitable choice of the voltage applied to the electrodes. According to Aronson et al. (1971), it is, in this way, possible to incorporate into graphite both donor and acceptor intercalants.

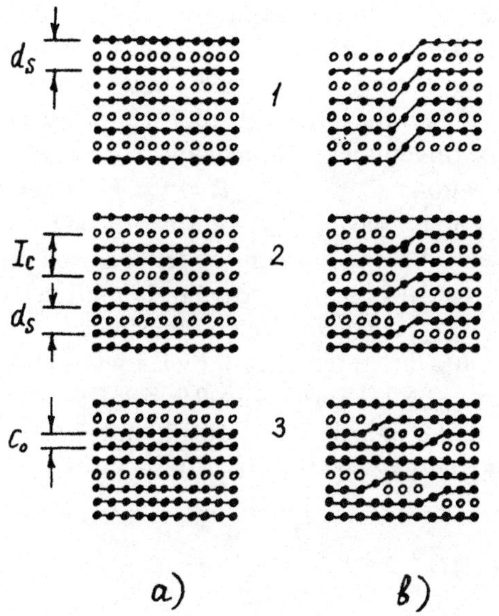

Fig. 103. Schematic diagram illustrating various stages of intercalation in graphite according to two models: (a) the classical model of Rüdorff (1940), (b) the folded (domain) model of Daumas and Hérold (1969).

The rate of intercalation is critically dependent on the sort of implanted matter and on the ambient parameters (pressure, the temperature of graphite and that of the intercalant, the size of the specimen, the concentration of defects, etc.). Intercalation begins only in the case when the intercalant-vapour pressure at the given temperature exceeds some threshold value characteristic of the formation of GIC of some stage number.

In order to obtain other stages of intercalation, it is necessary to vary the temperature and pressure of the intercalant vapour so that the expansion of the next interlayer interval would be thermodynamically advantageous.

Since the structure of pristine graphite is imperfect (of block-type), the filling of the layers is nonuniform. At first, there appear individual filled islands (domains), whose size is, as a rule, smaller than that of crystallites.

Owing to the fact that intercalation begins from all side facets, the filled planes are not always brought together inside the specimen. A model of domain-structure formation in the course of intercalation has been developed by Daumas and Hérold (1969). The formation of GIC structure in accordance with their model is schematically illustrated in fig. 103b. For comparison, fig. 103a represents a scheme for the GIC structure according to the classical model of Rüdorff (1940), in which the layers of the intercalant and those of graphite are homogeneous and parallel to one another throughout the specimen.

It should be noted that the domain model describes much better the structure of real GIC. In particular, the model enables one to explain transitions between reciprocally nonmultiple stages, which are observed when certain GIC of low stage index are decomposed in atmospheres of various gases. On the basis of this model, Daumas and Hérold (1969) have explained the sequence of transitions $C_8K \rightarrow C_{24}K \rightarrow C_{36}K$ observed in the decomposition of the C_8K compound of the first stage ($n = 1$).

Experimental evidence for the domain model has been supplied by Thomas et al. (1980), who directly observed domain boundaries in the third-stage compound $C_{35.54}FeCl_{2.12}$, using an electron microscope.

Nevertheless, it should be noted that the classical model of homogeneous parallel layers (see figs. 98, 99, 103a) due to Rüdorff (1940) still serves as a basis for both electron-structure calculations and the interpretation of experiments in which the electron energy spectrum of GIC is investigated.

10.4. Structure of GIC

X-ray studies have shown that the increase in the size of the initial graphite crystal in the direction of the *c*-axis during intercalation results mainly from the increase in interplanar distance, d_s, between the neighbouring carbon layers where an intercalant is inserted. The distance between the remaining carbon layers (which is $d_0 = 3.35$ Å for pure graphite) varies insignificantly so that, in a first approximation, it may, especially for $n > 2$–3, be regarded as unchanged. Table 16 presents values for interplanar distances d_s at $T = 300$ K for several types of GIC.

The structure of GIC is primarily characterized by the lattice period, I_c, along the *c*-axis (the identity period). If we assume, as a first approximation, that during intercalation the distance between the neighbouring graphite layers remains unchanged and equal to d_0, then the period I_c may be approximately expressed in terms of the stage index n and the distances d_s and d_0:

$$I_c \simeq (n-1)d_0 + d_s = nd_0 + d_i, \qquad (164)$$

where d_i stands for the quantity $d_s - d_0$ which determines the increase in the size of a crystal, originally consisting of n graphite layers, due to intercalation. The validity of the assumption about the weak dependence of d_0 and d_s on stage index n, and, therefore, the validity of the approximate relation given by (164) is confirmed when one compares the experimental values of I_c obtained by X-ray measurements with the calculated I_c.

In table 17 a comparison between the experimental and calculated values of I_c for $C_{8n}K$ compounds with $n = 1,2,3$ and 4, is given.

The data of table 17 suggest that relation (164) holds well to within several

Table 16
Interplanar distances d_s for some GIC at $T = 300\,K$.

Substance	Stage index n	Interplanar distance d_s (Å)	Ref.*
graphite	∞	3.35	[1]
C_8Br_2	2	7.05	[1]
$C_{12}Br_2$	3	7.05	[1]
$C_{16}Br_2$	4	7.05	[1]
C_9AlCl_3	1	9.53	[1]
$C_{18}AlCl_3$	2	9.53	[1]
$C_{27}AlCl_3$	3	9.53	[1]
$C_{36}AlCl_3$	4	9.53	[1]
C_8K	1	5.40	[1]
$C_{24}K$	2	5.40	[1]
$C_{36}K$	3	5.40	[1]
C_8Rb	1	5.65	[1]
$C_{24}Rb$	2	5.65	[1]
$C_{36}Rb$	3	5.65	[1]
C_5ICl	1	7.24	[2]
C_8FeCl_3	1, 2	9.37–9.45	[3]
$C_{24}HSO_4(H_2SO_4)_2$	2	7.98	[4]

*References:
[1] Rüdorff (1941).
[2] Rüdorff and Schulze (1940).
[3] Rüdorff and Schulze (1954).
[4] Rüdorff (1939).

hundredths of Å, i.e., to an accuracy of 1%, which is obviously insufficient for an analysis of the internal stresses arising during intercalation between the inner layers of graphite situated between the adjacent intercalate layers. For these layers, interplanar distances are dependent upon stage index and if n is fixed

Table 17
Comparison between the experimental and theoretical I_c for $C_{8n}K$ ($n = 1,2,3,4$). The theoretical values of I_c were calculated by using the formula $I_c = 5.41 + (n - 1)\,3.35$ (Å) [see (164)]. All values are given in Å.

	Stage index n				Ref.*
	1	2	3	4	
I_c (experimental)	5.41	8.77	12.12	15.49	[1]
	5.35	8.72	12.10	15.45	[2]
	5.32	8.74	12.07	15.44	[3]
I_c (theoretical)	5.41	8.76	12.11	15.46	

*References:
[1] Rüdorff and Schulze (1954).
[2] Dresselhaus and Dresselhaus (1981).
[3] Nixon and Parry (1968).

Table 18
Values of d_i in some common intercalates.

Intercalant	d_i (Å)
K	2.05
Rb	2.35
Cs	2.66
Li	0.35
HNO_3	4.44
$SbCl_5$	6.01
SbF_5	5.06
Br_2	3.69
$AlCl_3$	6.13
$FeCl_3$	6.10
$NiCl_3$	5.98
AsF_5	4.80

they depend on the relative position of the pair of graphite layers being considered with respect to the intercalate layer. The above dependences manifest themselves in the change of interplanar distances that occurs in the third decimal place; they are not described by a simple expression of the type of (164).

Equation (164) can be used for a crude determination of the period I_c of any stage index if the period of any one stage index is already known; it can also be used to determine the stage index from the change in specimen size during intercalation. As it follows from (164), the increase Δl in the size of a graphite crystal in the direction of the c-axis, due to intercalation, over the initial size l_0 of the crystal may be written as:

$$\frac{\Delta l}{l_0} = \frac{d_i}{n d_0}, \tag{165}$$

whence

$$n = \frac{d_i}{d_0} \frac{l_0}{\Delta l}. \tag{166}$$

In order to find n by formula (166), one should know not only the experimentally measured quantities l_0 and Δl, but also d_i. The latter quantity depends on the kind of the intercalant and represents a parameter which is characteristic of the implanted substance, and which is determined from additional measurements. The values for d_i in some more common intercalants are given in table 18. We note that the inaccuracy contained in eq. (166) is unimportant in a determination of stage index because the value thus found is rounded (up or down) to the nearest integer.

Table 19
Variations in the value of d_s for various intercalates.

Intercalant	d_s (Å)		
	$n = 1$	$n = 2$	$n = 3$
K	5.35	5.39	5.40
Rb	5.65	5.70	5.70
Cs	5.94	6.01	—
Li	3.706	3.71	3.70
HNO_3	7.84	7.79	7.79
$HClO_4$	7.96	7.90	7.90
Cl_2O_7	7.98	7.78	7.70
$SbCl_5$	9.42	9.36	9.36
SbF_5	8.46	8.41	8.41
Br_2	—	7.04	7.04
$AlCl_3$	9.54	9.48	—

Let us now consider the weaker changes that occur in the GIC lattice with various stage indices, viz., the change in distance between the graphite layers and the change in spacing between carbon atoms in the graphite layers. Although these changes are small, they play a major role in the creation of elastic stress fields in GIC.

The greatest changes in interplanar distance occur in the bounding graphite layers.

It has been established that in going over from compounds with $n = 2$ the distance d_s between the graphite layers that bound the intercalate layer (d_s is called the intercalate sandwich thickness) increases for donor type intercalants and decreases for acceptor-type ones (the change in d_s expressed in Å is observed in the second decimal place). The variations in the value of d_s are given in table 19. The dependence of d_s on stage index practically disappears for $n > 3$.

The experimental data on the variation of d_s in passing from compounds with $n = 1$ to those with $n = 2$ are in qualitative agreement with the calculated charge distribution in the layers. Indeed, the calculations indicate that the charge distribution on the graphite layers that bound a donor-intercalant layer changes with the increase in stage number from 1 to 2 in such a way that the bonds between the graphite layers weaken. For the acceptor-type intercalants, as Caswell et al. (1980) have shown, the reverse takes place.

During intercalation, the distance d_{c-c} between carbon atoms in the graphite layers also undergoes slight changes. The precision X-ray measurements recently performed by several groups have shown that in donor-type GIC this distance increases, while in acceptor-type GIC it decreases, the symmetry of the graphite layer remaining unchanged. Pietronero and Strässler (1981) have found that the change, $\Delta d_{c\text{-}c} = d_{c\text{-}c} - 1.4203$ (Å), in the distance between the carbon atoms in the graphite layer during intercalation can be described as follows:

$$\Delta d_{c\text{-}c}|_{n=1} = 0.157f_c + 0.146|f_c|^{3/2} + 0.236f_c^2, \qquad (167)$$

for compounds with stage index one and

$$\Delta d_{c\text{-}c}|_n = \frac{2}{n}[0.157f_c + 0.146|f_c|^{3/2} + 0.236f_c^2], \qquad (168)$$

for compounds with stage index greater than one. In these formulae, f_c is the fraction of the delocalized charge carriers in the given layer per carbon atom, its sign being plus for donor-type GIC and minus for the acceptor-type ones.

The results of the calculations by Pietronero and Strässler (1981) are in good agreement with those of Nixon and Parry (1969), who have found that $\Delta d_{c\text{-}c}$ in compounds $C_{8n}K$ is described satisfactorily by the expression $\Delta d_{c\text{-}c} = 0.0113/n$ (Å) (changes in the second and third decimal places).

Along with the dependence of $\Delta d_{c\text{-}c}$ on n, it is possible to observe in experiment the dependence of $\Delta d_{c\text{-}c}$ on the magnitude of the ionic radius, r_i, and the valence, Z, of metallic intercalants ($Z = 1, 2, 3$ for alkali and alkaline-earth metals and lanthanides, respectively). The empiric relations between $d_{c\text{-}c}$ and these parameters, established by Guérard et al. (1976), have the form:

$$\Delta d_{c\text{-}c} = 0.03266\frac{Z}{r_i}\,(\text{Å}) \qquad \text{for } C_8X,$$

$$\Delta d_{c\text{-}c} = 0.01958\frac{Z}{r_i}\,(\text{Å}) \qquad \text{for } C_6X, \qquad (169)$$

where r_i is expressed in Å, X is a metal. The change in $\Delta d_{c\text{-}c}$ occurs also in the second and third decimal places.

The above dependences of the $d_{c\text{-}c}$ on $1/n$, Z, and r_i reflect qualitatively the distribution of stress between the graphite layers in intercalation.

Let us now consider the arrangement of the atoms or molecules of implanted matter in the intercalate layers. It should at once be noted that, although thicknesses d_s of the filled layers and identity periods I_c have been determined for all synthesized GIC, there have been rather few crystal analyses of the intercalate layer. Information on the structure of this layer can be obtained by studying the X-ray reflections ($hk0$) in GIC. However, in such investigations one has to surmount considerable difficulties associated with the chemical and thermal instabilities of the compounds used, the lack of sufficiently large single crystals of GIC, the emergence of fresh defects in the course of intercalation, etc. Because of all these adverse factors, the structure of the intercalate layer has only been identified in relatively few GIC. The intercalate-layer structure

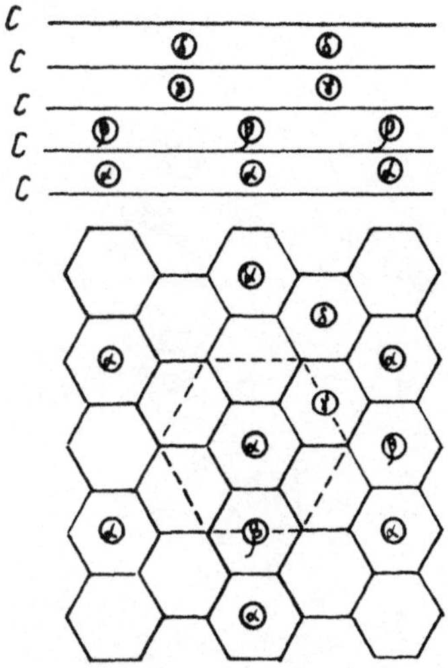

Fig. 104. Schematic arrangement of potassium atoms in the intercalate layers in the lattice of C_8K. The dashed line shows the projection of the unit cell on the graphite-layer plane. (Rüdorff 1959.)

has been studied in most detail in donor-type GIC with potassium and lithium as the intercalate. These compounds are illustrative of the case when the structure of the intercalate layer is commensurate to that of the graphite layer.

Let us consider the character of the commensurate intercalate-layer structure in the particular case of the compound C_8K of stage index one, whose lattice is schematically depicted in fig. 99. According to the X-ray data of Rüdorff (1959), potassium atoms in the intercalate layer occur in front of the centres of the hexagonal cells in the carbon layers. In fig. 104 the projections of these atoms in one of the intercalate layers upon the carbon-layer plane are denoted by α. In the neighbouring intercalate layer, potassium atoms occupy other equivalent positions opposite to the hexagonal-cell centres, the projections of these atoms being designated as β in fig. 104. In the next two intercalate layers, these atoms occupy positions designated as γ and δ. The original arrangement of the K atoms is repeated in every fifth intercalate layer. In other words, in the lattice of C_8K the intercalate layers are stacked as follows: $\alpha, \beta, \gamma, \delta, \alpha, \ldots$ (see fig. 104, top). Accordingly, the unit cell in the C_8K lattice contains 16 carbon and 2 potassium atoms.

According to Inoshita et al. (1977), such a cell can be described by the space group of symmetry D_{2h}^{24} in an orthorhombic system. We should note that in compounds with alkali metals the intercalant does not always crystallize in a structure commensurate to that of the graphite layer. This can be exemplified by the $C_{24}Cs$ compound (see Clarke et al. 1979a,b).

Returning to the general problem of the arrangement of atoms in the

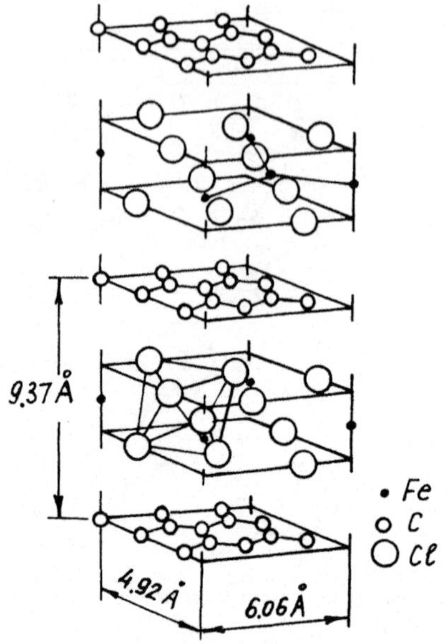

Fig. 105. Schematic diagram of stage-one graphite–$FeCl_3$ showing intercalate layers with the bulk $FeCl_3$ structure sandwiched between graphite planes. The intercalate layers are incommensurate with those of graphite, the pertinent lattice constants being indicated. The octahedral coordination of the ferric ion is shown in the lower sandwich and the two inequivalent ferric ions are indicated on the upper sandwich. (Cowley and Ibers 1956.)

intercalate layer, we note that the lattice constants in this layer are typically much greater than those in the graphite layers. This is a consequence of the larger ionic radii of the intercalant particles and the greater length of interatomic bonds characteristic of an intercalant. However, the symmetry of the graphite layer imposes a definite symmetry also upon the arrangement of intercalant atoms or molecules. Intercalation proceeds at a more rapid rate for those substances in which the intralayer ordering arises for the minimum distortion of the bonds typical of the intercalant itself.

For molecular intercalants, the layers of implanted matter may have both the commensurate structure (e.g., for C_xBr_2 compounds, see Eeles and Turnbull 1965) and the incommensurate one (e.g., for C_6CrCl_3, see Vangelisti et al. 1974, for C_xFeCl_3 compounds, see Cowley and Ibers 1956). The C_xFeCl_3 compound typifies the situation when the intercalant forms three, and not one, atomic layers placed between the carbon layers [in the case of C_xFeCl_3, these are Fe and Cl atoms, see fig. 105 which schematically shows the lattice of compounds containing $FeCl_3$ with stage index one]. In such cases, intercalate layers have a sandwich structure along the *c*-axis of a lamellar compound. There are numerous examples of sandwich-type GIC, in particular, the compounds of C_xAlCl_3 (Leung et al. 1980), C_xCrCl_3 (Vangelisti and Hérold 1976),

C_xMoCl_5 (Syme-Johnson 1967), C_xSbCl_5 (Melin and Hérold 1975), $C_xH_2SO_4$ (Rüdorf and Hoffmann 1938), C_xHNO_3 (Rüdorf 1939, Fuzellier et al. 1977), etc.

Let us now consider the character of graphite-layer stacking in GIC. X-ray studies have shown that in compounds with $n = 1$ there occur two main types of stacking:

(1) type-I stacking, for which the structure of the graphite layers is one and the same on both sides of the intercalate layer (A/A/A...), and
(2) type-II stacking, for which the structure of these layers (A/B/A/B...) is different.

In the notation used, the slashes denote the intercalate layer and the letters A and B denote the layers of the carbon atoms in the hexagonal graphite lattice (see sect. 2.3 and fig. 3a). Sometimes, the type of stacking is temperature dependent. There are also compounds of stage index one in which the different types of stacking may co-exist at one and the same temperature.

Type I of graphite layer stacking is realized in compounds of stage index one with alkali metals K, Rb, Cs. For compounds of stage index two with the same metals, there is, according to Nixon and Parry (1968), another order of alternation, viz. AB/BC/CA/AB/BC, where A, B and C denote the graphite layers in a rhombohedral system (see sect. 2.3 and fig. 3b).

For compounds with $n > 2$, the graphite layers are placed between two intercalate layers in accordance with the ordinary layer stacking in the hexagonal system of initial graphite: ABAB..., ACAC..., or BCBC.... In certain cases, the rhombohedral stacking of graphite layers, ABCABC..., arises.

Compounds of stage index two with acceptor intercalants exhibit the stacking of the AB/BA/AB... type (characteristic of C_xSO_3 compounds) and of the AB/BC/CA... type (characteristic of C_xHNO_3 compounds). A second type of graphite-layer stacking (A/B) arises in many compounds of stage index two with acceptor intercalants. For the same compounds with a higher stage index, the A/B stacking is preserved for the graphite layers which are adjacent to the intercalate layer. Thus, according to Hérold (1979), compounds with $n = 2$ with SbF_5 have the hexagonal type of stacking, i.e. AB/AB/AB, whereas in compounds of stage index four with $MoCl_5$ both hexagonal, A/BABA/BABA/B..., and rhombohedral, A/BCAB/ABCA/B..., stackings are realized (Syme-Johnson 1967).

For certain kinds of intercalant, both types (I and II) of stacking of the carbon layers bounding the intercalate layer are observable. Among such intercalants, one should mention H_2SO_4. In this case, for compounds of stage index two the stacking A/AB/BA/AB... or A/BA/BA/BA... is observed (Hérold 1979).

We should note that the hexagonal stacking is more stable and is realized in the majority of cases. Furthermore, it is known that the process of intercalation in rhombohedral crystals nearly always results in a transition to a hexagonal system.

10.5. Phase transitions in GIC

Lamellar compounds of graphite possess a great variety of differently ordered structures or phases, some of which have been discussed in the preceding section. Between these phases there may occur various phase transitions, including structural transitions associated with a change in stage index, in the intralayer ordering of the intercalant, and in the stacking order of layers. It is also possible to observe transitions due to the change in magnetic ordering and transitions to the superconducting state.

Structural phase transitions may be induced by changes in temperature, pressure and, sometimes, variations in the pressure of the intercalant vapours. At the present time, only some of the observed phases have been given adequate interpretation and only some transitions between them have been studied in sufficient detail. No complete phase diagram in coordinates of temperature–pressure–intercalant density has yet been constructed even for simplest compounds.

Below we will examine the most common types of phase transitions

10.5.1. Transitions accompanied by a change in the stage index, i.e., transitions of the form $n \leftrightarrows n-1$

These transitions can be observed in compounds with alkali metals and also with some acceptors (e.g., H_2SO_4, ICl). These transitions may be induced by variations of temperature (see, e.g., Aronson et al. 1968, 1971, Sasa et al. 1971, Falardeau et al. 1978) or pressure (see, e.g., Brandt et al. 1983).

10.5.2. Transitions accompanied by a change in the order of succession of the intercalate layers (intercalate layer stacking)

Such transitions, induced by temperature variations, can be observed, for instance, in compounds with alkali metals at low stage indices. A reversible structural transition with a change in intercalate-layer stacking from $\alpha\beta\gamma\alpha\ldots$ in the low-temperature phase to $\alpha\alpha\alpha\ldots$ in the high-temperature phase was observed by Guérard and Hérold (1975) at $T \simeq 220\,K$ in C_6Li. A similar reversible transition was observed in an electron diffraction analysis by Kambe et al. (1980) in C_8Rb, whose low-temperature phase corresponds to the order of succession of Rb layers $\alpha\beta\gamma\delta\ldots$ (cf. fig. 104).

10.5.3. Structural transitions associated with a change in the ordering of atoms or molecules in the intercalate layer (from commensurate to incommensurate relative to the graphite-layer structure)

Such structural transitions are observed, for example, in graphite intercalated with inert gases and also with some acceptor intercalants (HNO_3, etc.) when the temperature of the system is varied. In particular, according to Nixon et al. (1966), for $T < 252$ K one may observe some ordering in the intercalate layer of the C_xHNO_3 system, the intercalate structure becoming commensurate to that of the graphite layers. Also, in this temperature range, some correlation between the mutual arrangement of the intercalate layers is observed. For $T > 252$ K both the correlation and the ordering in the intercalate layer vanish. At those temperatures, the intercalate resembles by its properties a liquid; therefore, the phase transition at $T \simeq 252$ K may be regarded as an order–disorder transition.

Similar order–disorder transitions have been observed by Chung and Dresselhaus (1977), Chung et al. (1977), and Baiker et al. (1981) in compounds with Br_2, IBr, ICl, $AlCl_3$, etc. They have established, in particular, that phase transitions are accompanied by heat evolution or absorption, and should be classed as first-order transitions. Another example of a transition from commensurate to incommensurate structure in the intercalate layers is a phase transition accompanied by the appearance of a charge density wave (CDW) – a periodic charge-density distribution in the plane of the intercalate layer. In the presence of several equivalent directions in the plane, the charge density distribution in the CDW is described by the superposition of standing waves in these directions.

As a result of the electron–phonon interaction, the appearance of CDW leads to a modulation of the intercalant atom (or molecule) distribution density, i.e., to the establishment of a superlattice, whose period is determined by the Fermi-electron wave vector. As Inoshita et al. (1977) have shown, the formation of CDW is the most probable in $C_{8n}K$ compounds with low stage indices. However, until now there have been no conclusive experimental evidence for the presence of CDW in any GIC, although Chung (1977) attributes the phase transitions in C_xBr_2 and C_xCs, observed by electron-diffraction techniques, to the formation of CDW. An attempt has been made to relate some additional frequencies of dHvA oscillations in C_xBr_2 observed by Rosenmann et al. (1979) to the formation of CDW.

In some GIC (e.g., in donor compounds with $n = 2$), there have been detected successive structural transitions induced by temperature variations. These transitions are accompanied by a change in the type of ordering in the intercalate layers. Such transitions are observed, in particular, in compounds with alkali metals (see, e.g., Berker et al. 1980, Kambe et al. 1980). For alkali-metal compounds, two transition temperatures, T_ℓ and T_u ($T_\ell < T_u$), have been determined. These temperatures allow one to specify three temperature ranges, $T < T_\ell$, $T_\ell < T < T_u$, and $T > T_u$ with different intercalate-layer structures. The temperature of structural transitions in GIC may be determined by the appearance of anomalies in transport or other properties. As an illustration, fig. 106 shows the temperature dependences of electrical resistance and magnetoresistance in a C_xK compound with stage index three. The discontinuities in

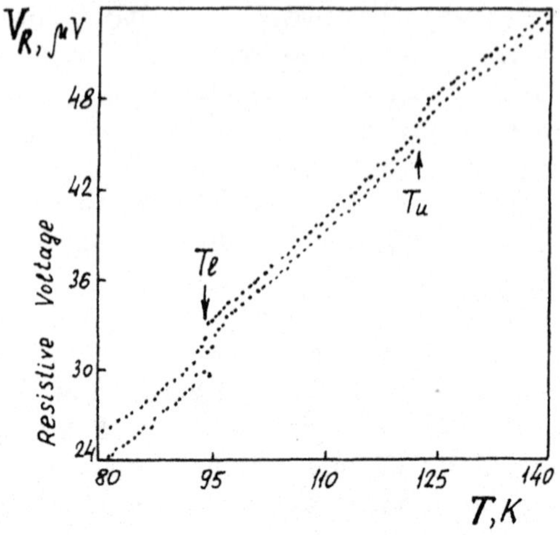

Fig. 106. Measurement of Onn et al. (1977) of resistive voltage V_R versus temperature for $C_{24}K$ for zero magnetic field (lower curve) and for $H = 1.5\,T$ (upper curve). The transition temperatures $T_l = 95\,K$ and $T_u = 124\,K$ are indicated. Similar resistance anomalies are observed for compounds with intercalants Rb and Cs.

the monotonous curves at $T = T_l$ and $T = T_u$ are associated with structural transitions in the intercalate layer.

10.5.4. Transitions to the superconducting state

The superconductivity of GIC was first discovered by Henning and Mayer (1952). More recently, Hannay et al. (1965) and Koike et al. (1978) have found superconductivity in compounds with alkali metals (K, Rb, Cs) with stage index one for $T \lesssim 1\,K$. Recently, the class of superconducting GIC has been broadened by inclusion of ternary compounds C_8KHg, C_8RbHg (Yeh et al. 1982, Iye and Tanuma 1982) and $C_4KTl_{1.5}$ (Wachnik et al. 1982), as well as quaternary compounds $C_8KHg_{0.9}Pb_{0.1}$ and $C_4KTl_{1.35}Hg_{0.15}$ (Iye and Tanuma 1983), etc. The character of superconducting transitions in GIC has been studied in detail by Koike et al. (1980). They have found that, depending on the orientation of the applied magnetic field relative to the c-axis, a transition to the superconducting state may be of the first or second order. A theory of superconductivity in GIC has been developed by Shimizu and Kamimura (1983). The question of the superconductivity of GIC is dealt with in more detail in sect. 10.13.

10.5.5. Transitions accompanied by magnetic ordering

Such transitions occur in GIC in which the intercalate layer is composed of transition-metal atoms. These transitions were observed in compounds with stage indices one and two with $FeCl_3$, $FeCl_2$, $CoCl_2$, and $NiCl_2$. According to Ohhashi and Tsujikawa (1974a,b), in compounds with $FeCl_3$ there takes place an antiferromagnetic ordering for temperatures below 3.6 K for $n = 1$ and a ferromagnetic ordering below 8.5 K for $n = 2$.

In both cases, the magnetic moments are ordered in the intercalate layer. The interaction between the magnetic moments in the neighbouring intercalate layers is quite small, except for the stage-one compound with $FeCl_3$, in which the intralayer interaction is antiferromagnetic in character, the interlayer interaction also resulting in an antiferromagnetic alternation of magnetic moments along the c-axis. In compounds with $NiCl_2$ and $CoCl_2$ (see Hohlwein et al. 1974), transitions to the ferromagnetic state occur. The corresponding Curie temperatures are 20.3 K in the former case and 9.05 K in the latter. Thus graphite intercalation compounds offer an interesting possibility of studying the properties of two-dimensional magnetic systems.

10.6. The energy spectrum of electrons and holes in GIC

10.6.1. Peculiarities of the energy spectrum of carriers in GIC

The energy spectrum of GIC is much more complicated than that of pure graphite. The complexity of its structure and the difficulties involved in its study are primarily due to the spatial inhomogeneity of GIC, which leads, firstly, to a nonuniform distribution of carriers between the graphite layers and the intercalate layers; secondly, to carriers in the graphite layers being nonuniformly distributed along the c-axis for $n \geqslant 3$; thirdly, to the appearance of an additional periodicity of the lattice potential both in the direction of the c-axis and in the plane of the graphite and intercalate layers. Let us examine these aspects in more detail.

(A) Because of the spatial inhomogeneity of graphite intercalation compounds, the conditions for electrons in the intercalate layer and in the adjacent and interior layers of graphite will be quite different.

Donor intercalants lose the valence electrons to the adjacent graphite layers, thereby significantly increasing the concentration of free carriers (electrons) in the graphite layers. By contrast, acceptor intercalants gain and localize the electrons from the adjacent graphite layers, thereby increasing the concentration of free holes there.

In studying the charge transport along the layers in GIC, it has been found that electrons and holes can freely move only in the graphite layers, whereas in the intercalate layers charges are transported mainly by hopping processes with an effective mobility many orders of magnitude lower than that of electrons and holes in the graphite layers. Therefore, it may be assumed that there are practically no free charge carriers in the intercalate layers, i.e., these represent monatomic (or monomolecular) dielectric layers separated by the graphite layers.

Thus whenever we refer to the energy spectrum of quasi-particles in GIC we mean the spectrum of electrons and holes in the graphite layers.

Charge transport in the direction perpendicular to the layers and the bonding of free carriers in the neighbouring stacks of graphite layers separated by the intercalate layer is determined by the value of the resonance overlap integral of the electron wave functions in the graphite layers adjacent to the intercalate layer. As is suggested by the results of measurements of the electrical conductivity in GIC along the layers (σ_a) and across them (σ_c) (see sect. 10.7), the anisotropy of the conductivity, $\eta = \sigma_a/\sigma_c$, which varies from 10^3 to 10^4 in pure graphite, increases to 10^5–10^6 in the presence of acceptor intercalants, but decreases to 10^1–10^2 if the intercalant is of the donor type. The very high anisotropy of electrical conductivity in acceptor-type GIC implies that, in these compounds, there are practically no bonds between the electrons in the neighbouring stacks of the graphite layers separated by the intercalate layer. Hence it follows that in acceptor-type GIC the stacks of the graphite layers themselves may be treated as independent systems, each of them having a completely developed energy spectrum of the carriers it contains. Basically, acceptor GIC represent sets of identical unbound crystals in the form of stacks of graphite layers, the exchange of electrons between which may, in a first approximation, be completely neglected. Such a notion of independent graphite layer stacks is, however, less satisfactory when the spectrum of the donor GIC is to be described. This is due to the much lower anisotropy of electrical conductivity and to the ensuing much tighter bonding between the electrons in stacks separated by the intercalate layer. Nevertheless, in the case of the donor GIC the above-mentioned anisotropy is still much greater than unity, which justifies the use of the independent-stack approximation for a crude description of the electron spectrum of these compounds. In a more detailed description of the energy spectrum of carriers in GIC one should take into account the possibility of electron (hole) transitions between the graphite-layer stacks. The influence of these transitions on the formation of the spectrum in GIC is significant when the mean free path of carriers along the c-axis exceeds the identity period, I_c, taking account of the motion of carriers in the direction perpendicular to the layers being evidently more important for the donor GIC.

In considering the transitions of electrons and holes from one stack of graphite layers to another there arise effects associated with the new, resulting from intercalation, periodicity along the c-axis with identity period I_c. Here the presence of the intercalate layers in the lattice of pure graphite may be regarded as a superlattice structure. The long-period potential component in such a superlattice (with period I_c) may give rise to a splitting of the Fermi surface of the initial material into a number of separate constant-energy surfaces ("sheets"). The possibility of such an effect was first pointed out by Brandt et al. (1972).

Thus in considering the energy spectrum of electrons and holes in GIC one should bear in mind that, in a first approximation, this spectrum is formed by the motion of carriers in graphite-layer stacks and, in the next approximation, by the interaction between the carriers and the intercalate layers.

One should also recall that such a straightforward approach to the consideration of the spectrum of carriers in GIC is unsatisfactory for low-stage index compounds (with $n = 1, 2, 3$). The point is that the spectrum of charge carriers in the graphite bounding layers is largely formed due to the hybridization of the electronic states of the carbon atoms and the atoms (or molecules) of the intercalant. And at low stage indices the behaviour of carriers in the graphite layers is in fact fully determined by the bounding-layer structure. The above peculiarity of the low-stage index compounds shows that a sufficiently rigorous consideration of the energy spectrum of the carriers in them requires a special approach (see below, concerning the calculation of the energy spectrum of GIC from first principles).

(B) In considering the energy spectrum of the electrons and holes delocalized in a stack of the graphite layers it is necessary to take into account the nonuniform distribution of the carriers along the c-axis within the stack, this distribution arising in compounds with $n \geqslant 3$ as a result of intercalant charge screening. As a consequence of the nonuniform carrier distribution within the graphite layers, the individual graphite layers in a stack do not play the same role. We will now discuss this question at greater length, comparing GIC of the first, second, and higher stage indices.

In GIC with stage index one there is a single graphite layer placed between the neighbouring intercalate layers. The concentration of electrons or holes in this layer significantly exceeds the concentrations of these carriers in pure graphite, which constitute about 10^{-4} quasi-particles per carbon atom. So a GIC of stage index one essentially represents a highly doped semimetal in which the concentration of donor or acceptor impurities is close (in order of magnitude) to that of the parent atoms. Each carbon layer in such a GIC is practically equivalent to a two-dimensional metallic layer.

In GIC of stage index two there are two graphite layers, which are placed between the neighbouring intercalate layers and which are completely equivalent to each other. From this point of view the system of the graphite layers in GIC of stage index two is homogeneous and the concentrations of electrons (or holes) transferred from the intercalate layer to them are equal to each other. The homogeneity of the system of the graphite layers within a single stack in GIC of stage indices one or two makes it possible to speak sufficiently strictly of the electron (hole) energy spectrum of these compounds.

As the stage index further increases the situation radically changes. Already for $n = 3$, when three graphite layers are placed between the neighbouring intercalate layers, only the extreme graphite layers of a stack, i.e., the ones adjacent to the intercalate layer, remain equivalent. Because of the screening, the transferred-charge (electron or hole) density in the central, nonequivalent graphite layer is lower than in the extreme ones. Now it is no longer possible to speak of the single electronic structure of carriers in a graphite-layer stack.

As has already been mentioned in sect. 10.2, in moving away from the intercalate layer the monotonic charge-density component $\rho(z)$ in the graphite layers is described by the function given by (162). Estimates of the screening length z_0 vary from 3 to 4 Å, which is close to the value of interplanar distance in a stack of graphite layers. Hence it follows that in GIC with $n = 3$ the charge density in the graphite interior layer may be 10 to 15 times lower than that in the equivalent graphite layers adjacent to the intercalate layer.

Still greater is the difference in charge density between the adjacent and interior graphite layers in GIC with $n > 3$. Owing to the screening, the graphite interior layers remain, for sufficiently large n, practically undisturbed by the charged intercalate layers, the concentrations of electrons and holes in the graphite interior layer being close to those observed in pure graphite. This conclusion is in good agreement with the data of oscillation measurements in compounds with high stage indices, where one can observe quantum oscillations whose frequencies coincide with those due to electrons and holes in pure graphite.

Thus, owing to the screening, the interior part of a graphite-layer stack for $n \gg 1$ may again be treated, in the limit of low intercalant concentration (the so-called dilute limit), as a homogeneous system, whose electron energy spectrum in the basal plane coincides with the carrier spectrum in pure graphite. The change in carrier density occurs then practically only in one or two graphite layers adjacent to the intercalate layer. In other words, in order to describe a stack of graphite layers in GIC of a high stage index, one may use the approximate model assuming a homogeneous system of graphite interior layers (close in character to pure graphite), which are situated between rather thin and also (in a first approximation) homogeneous "blankets" [$(1–2)z_0$ in thickness] made up by layers with a high carrier concentration.

In the next, more exact, approximation, one may independently consider the electron energy spectrum in each of the layers constituting the "blanket" of high carrier concentration.

(C) The next peculiarity of the GIC electron spectrum is associated with new periodicities in the lattice, which result from intercalation and which may alter the shape and size of the Brillouin zone of pure graphite and, as has already been mentioned above, give rise to a Fermi-surface splitting into a number of constant-energy surfaces. We have already briefly discussed [see (A)] the role of the new period I_c (identity period) along the c-axis. Let us discuss this point in more detail.

We shall start from the simplest case of compounds with $n = 1$ or 2, in which the charge density distribution in a graphite-layer stack is homogeneous. In these compounds, the identity period is the only periodicity along the c-axis, which determines the size of the Brillouin zone along the k_z-axis. If the motion of electrons and holes along the c-axis were free (the charge-carrier gas is then three-dimensional), then their scattering governed by a periodic potential (with period I_c) in this direction would give, in momentum space at $k_z = \pm(\pi/I_c)i$ (i is an integer), rise to planes of discontinuity in energy, perpendicular to the k_z-axis. The two planes that are closest to the origin of the coordinate system determine, at $k_z = \pm\pi/I_c$, the boundary of the first Brillouin zone, whose size along the k_z-axis is then equal to $2\pi/I_c$.

The case we have just considered is not very different from the situation in donor GIC with $n = 1$ or 2, in which the bonding between graphite-layer stacks and the electron motion along the c-axis are so important that the electron system might be regarded as three-dimensional.

In acceptor-type compounds with stage index one or two, stacks of graphite layers are practically independent of one another, charge transport being largely two-dimensional in character. The graphite layers in acceptor GIC are in fact two-dimensional neighbour crystals, whose hole spectrum is entirely determined by their internal structure. The presence of the periodicity I_c does not introduce here any changes in the carrier energy spectrum. However, although in these compounds the hole motion along the c-axis may be neglected, the size of the Brillouin zone along the k_z-axis (equal to $2\pi/I_c$) should be taken into account in a calculation of, for example, the bulk carrier concentration, because the carrier gas is then assumed to be three-dimentional.

Thus in compounds with $n = 1$ or 2, the Brillouin zone of pure graphite, which has the shape of a hexagonal prism with a height of $2\pi/c_0$ along the k_z axis (see fig. 9), reduces, due to taking into account the periodicity with period I_c, to a prism of the same shape, though less high, $2\pi/I_c$. [We keep in mind that the identity period I_c in GIC always exceeds the interplanar distance d_0 in pure graphite, see tables 16 and 17 and eq. (164)]. The change in the Fermi surface due to the reduction of the Brillouin zone can be determined by taking into account the new periodicity in momentum space, which has the shorter period $2\pi/I_c$, as has been done by Brandt et al. (1972) on an example of an initial spherical surface.

It should be noted that if the initial Fermi surface is close in shape to a cylinder (a strongly prolate ellipsoid), then the coaxial constant-energy surfaces resulting from the Brillouin-zone reduction may turn out to be so close to each other that, because of the uncertainty in the carrier energy, $\Delta\varepsilon \sim \hbar/\tau$ (τ being the relaxation time), it would be impossible to detect this splitting in experiment. Nor can the splitting be detected if the energy discontinuity at the boundary of the resulting Brillouin zone is lower than the thermal vibration energy, kT.

Let us now turn to GIC with stage index greater than two. As the stage index increases, the effect exerted by the superlattice structure with period I_c upon the carrier energy spectrum weakens. This occurs due to the following reasons.

First, with increasing stage index the size of a graphite-layer stack also increases; simultaneously, the isolation of the interior layers in a stack becomes greater as a result of screening.

Secondly, with the increasing number of graphite layers in a stack, the periodicity along the c-axis with the period equal to the distance d_0 between the graphite layers will play an increasingly important role. At the same time, the periods d_0 and I_c are incommensurate [see eq. (164) and the data of tables 16 and 17]. As a result, the planes of energy discontinuity at $k_z = \pm(\pi/I_c)i$, where $i = 1, 2, 3, \ldots$, are shifted relative to the boundaries of the Brillouin zone with the size $2\pi/c_0$ in going over from one zone to another in the scheme of expanded zones. This leads to the smearing of features in momentum space, which occur with the shorter period $2\pi/I_c$ and which completely lose their significance for $I_c \gg c_0$ (i.e., for $n \gg 1$). This means that in GIC with $n \gg 1$ the periodicity with period c_0 along the c-axis is alone vital for the formation of the electron spectrum in the graphite interior layers and the Fermi surface due to carriers in these layers should be considered as if it were in the Brillouin zone for initial pure graphite (fig. 9). Owing to the strong screening of the graphite interior layers, this approach proves adequate already for $n \gtrsim 5$ or 6.

In compounds of the intermediate stages (with $3 \leqslant n \lesssim 5$ or 6) a description of the energy spectrum turns out to be a rather arduous task. In this case, the spectrum in the graphite interior layers is affected by the periodic potential components with periods both I_c and c_0. The experimental data available at present indicate that in compounds with $n = 3$ or 4 the superlattice structure with period I_c gives rise to the Brillouin-zone reduction and Fermi-surface splitting due to energy discontinuities at $k_z = \pm(\pi/I_c)i$ $(i = 1, 2, 3, \ldots)$. It should be noted that in compounds with $n > 2$ the Fermi surfaces due to carriers in the graphite bounding layers considerably differ from those in the graphite interior layers. The Brillouin zone reduction, corresponding to the superlattice-structure period I_c, evidently affects in a different manner the electron spectrum in these different groups of graphite layers.

Let us now examine how the electron energy spectrum in GIC is influenced by the new periodicity associated with the ordering of atoms or molecules in the intercalate layer. If a periodic structure emerges in this layer, then the periodic potential associated with the structure leads to the appearance of new planes of energy discontinuity in momentum space which will be parallel to the k_z-axis. If the structure in the intercalate layer is commensurate with that in the graphite layer, then, since the translation periods in the intercalate layer are always greater than the analogous periods $a_0 = 2.46\,\text{Å}$ in the graphite layer, these new energy-discontinuity planes cut the Brillouin zone of initial graphite (see fig. 9) into smaller, though equal, parts. This phenomenon is called the "zone-folding effect". The change in the Brillouin zone due to this effect may be illustrated by the example of the simplest compounds with $n = 1$, C_8X, which are produced by intercalating graphite with alkali metals. In such compounds, the projection of a unit cell of the intercalant upon the graphite-layer plane is shown by the dashed line in fig. 104. The area of this projection is four times as large as that of a unit cell of the graphite layer; accordingly, the area of the two-dimensional Brillouin zone corresponding to the intercalate layer structure constitutes $\frac{1}{4}$ of that of the two-dimensional Brillouin zone corresponding to the structure of a single graphite layer. In fig. 107, the outer hexagon represents the Brillouin zone for the graphite layer, while the inner hexagon represents the Brillouin zone for the intercalate layer. The dashed lines indicate the traces of the new energy-discontinuity planes parallel to the k_z-axis, which are generated by the zone-folding effect. One can see that, as a result of this, the middles of the Brillouin-zone edges in graphite (points M and M') becomes physically equivalent to the Brillouin-zone centre (point Γ).

Thus the lattice potential, whose symmetry is determined by the intercalate layer, will strengthen the interaction between electrons at the points M, M', and Γ.

Owing to the zone-folding effect, the Fermi surface in initial GIC with disordered intercalate layers turns out to be cut into separate pieces by the new energy-discontinuity planes. Thus the zone-folding effect should lead to the appearance of complicated electron and hole orbits in the magnetic field (see Batallan et al. 1978, Simon et al. 1981). Attention of the reader is called to

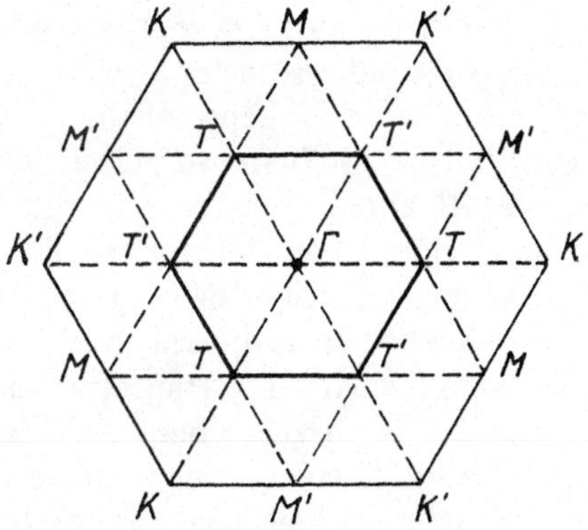

Fig. 107 Brillouin zone for the graphite layer (outer hexagon) and the projection of the BZ for C_8X on the plane perpendicular to the k_z-axis (inner hezagon). Dashed lines indicate the traces of the new planes of energy discontinuity, due to intercalation, which are parallel to the k_z-axis.

the fact that the zone-folding effect is obviously best pronounced in GIC with small n. The influence exerted by zone folding upon the electron energy spectrum in the graphite interior layers in GIC with $n \gg 1$ should evidently vanish because of the screening of the intercalate layer. For $n \gg 1$ new orbits may become available only for the electrons in the graphite bounding layers. If the structure in the intercalate layer is incommensurate with that in the graphite layers, then there will appear no new orbits, because in the expanded-zone scheme the new energy-discontinuity planes are shifted as one moves from the Brillouin zone of a given number to that of the larger (or smaller) number. This means that in the carrier energy spectrum there will not appear a new periodic disturbance, which would lead to the appearance of new Fermi-surface pieces and new electron and hole orbits (the effect of new energy-discontinuity planes is smeared). Thus, in regard to its influence upon the electron spectrum of GIC, the situation of the incommensurate ordering in the intercalate layer is equivalent to the situation with a random arrangement of the intercalant particles in a layer.

(D) Finally, among the peculiarities of the energy spectrum of GIC, one should also mention that, unlike the spectrum of pure graphite, which is a compensated semimetal with equal concentrations of electrons and holes, the spectrum of GIC is that of a metal, in which carriers of either one or the other sign prevail. In this respect, the energy spectrum of GIC is rather similar to the spectrum of a semimetal highly doped with donor or acceptor impurities. As in the case of polyvalent metals, the presence of carriers of opposite sign in GIC is associated with the complicated band structure of their spectrum.

10.6.2. Models of the carrier energy spectrum in GIC of low state indices in the tight-binding approximation

As has been mentioned above, the electron energy spectrum may correctly be treated only for GIC with stage indices one or two, in which the graphite-layer system is spatially homogeneous in the sense of the layer equivalence. Similar to the Wallace model (Wallace 1947) for the electron spectrum of a single graphite layer or the SWMC-model for the spectrum of three-dimensional graphite, the currently most popular models for the electron spectrum in GIC of low stage index are based on the tight-binding approximation (TBA).

Models of this kind have been proposed by Blinowski et al. (1980), Blinowski and Rigaux (1980), Nguyen Hy Hau et al. (1981). Their phenomenological models are applicable in describing the electron spectrum in a subsystem (a stack) of the graphite layers, which are assumed to be equivalent to each other. Therefore, the phenomenological models may rigorously be applied only to GIC of stage index one or two. From the above considerations we can state that these models give a much better description for the spectrum of acceptor-type GIC, which have a high anisotropy of electrical conductivity and negligibly small values of the resonance overlap integrals for the graphite layers separated by the intercalate layer.

10.6.2.1. A TBA model for the spectrum of GIC with stage index one. In constructing a model for the energy spectrum of GIC with stage index one, in a first approximation, a system of equivalent noninteracting graphite layers is considered. A unit cell contains two carbon atoms and represents a regular hexagon (fig. 2a). In this approximation, the electron spectrum in GIC with $n=1$ is identical with the electron spectrum of a single graphite layer obtained by Wallace (1947). The dependence of the energy upon the wave vector $\boldsymbol{k}$ in the graphite plane is written as

$$\varepsilon_{c,v} = \pm\tfrac{3}{2}\gamma_0 a_0 k, \tag{170}$$

which coincides completely with eq. (30) and describes the conduction band (index c, sign +) and the valence band (index v, sign −) of a gapless semiconductor with a conical band-degeneracy point. In (170), just as in (30), γ_0 represents the resonance overlap integral of the π-electron wave functions for two nearest-neighbour carbon atoms in the graphite layer, $\boldsymbol{k}$ is the wave vector in the graphite plane, measured from the Brillouin-zone corners. The general appearance of the line spectrum given by (170) for GIC of stage index one is shown in fig. 108.

The Fermi surface for GIC of stage index one represents the totality of

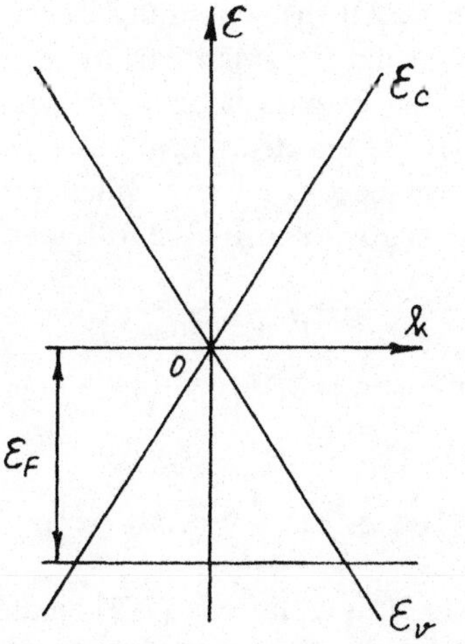

Fig. 108. Schematic diagram showing the energy spectrum of first-stage acceptor-type GIC. (Blinowski et al. 1980.)

circular cylinders whose geometrical axes coincide with the Brillouin-zone edges parallel to the k_z-axis. The height of the cylinders, which equals that of the Brillouin zone, is $2\pi/I_c$.

Hence one may conclude that the bulk electron (hole) concentration in these compounds is related to the Fermi wave vector $\boldsymbol{k}_F$ in the plane perpendicular to the k_z-axis and with the identity period I_c, by the following expression,

$$n_{e,h} = \frac{4\pi k_F^2}{(2\pi)^2 I_c} = \frac{4S_{extr}}{(2\pi\hbar)^2 I_c}, \tag{171}$$

where S_{extr} is the extremal cross section of the cylinder formed by a plane perpendicular to the k_z-axis. The cyclotron carrier mass ($m_c^{e,h}$) is then expressed in terms of the extremal cross section as follows:

$$m_c^{e,h} = \frac{1}{2\pi}\frac{\partial S_{extr}}{\partial \varepsilon} = \frac{2}{3}\frac{\hbar}{\gamma_0 a_0}\left(\frac{S_{extr}}{3\pi}\right)^{1/2}. \tag{172}$$

10.6.2.2. A TBA-model for the spectrum of GIC with $n = 2$. In these compounds, pairs of graphite layers are separated by the intercalate layer. Therefore, if one uses the tight-binding approximation one should take into account, apart from the interaction between the nearest neighbours in a layer, which is described by the γ_0 parameter, also the interaction between the nearest-neighbour carbon atoms in two neighbouring layers, which is described by the resonance overlap integral γ_1. All the remaining interactions may approximately be taken into account by using the small parameter Δ. The interaction between two layers described by the γ_1 parameter leads to the removal of the double degeneracy of the bands in the spectrum given by (170). As a result, there appear two conduction bands (index c) and two valence bands (index v), in which the dispersion of carriers is governed by the following laws:

$$\begin{aligned}
\varepsilon_{v_1}(k) &= \tfrac{1}{2}\{\gamma_1 + \Delta - [(\gamma_1 + \Delta)^2 + 9\gamma_0^2 a_0^2 k^2]^{1/2}\},\\
\varepsilon_{v_2}(k) &= \tfrac{1}{2}\{-\gamma_1 + \Delta - [(\gamma_1 - \Delta)^2 + 9\gamma_0^2 a_0^2 k^2]^{1/2}\},\\
\varepsilon_{c_1}(k) &= \tfrac{1}{2}\{-\gamma_1 + \Delta + [(\gamma_1 - \Delta)^2 + 9\gamma_0^2 a_0^2 k^2]^{1/2}\},\\
\varepsilon_{c_2}(k) &= \tfrac{1}{2}\{\gamma_1 + \Delta + [(\gamma_1 + \Delta)^2 + 9\gamma_0^2 a_0^2 k^2]^{1/2}\}.
\end{aligned} \tag{173}$$

The additional parameter Δ in equations (173) creates a small overlap between the valence band and the conduction band in the energy spectrum of GIC with $n = 2$. Its role is, however, insignificant and, in approximate descriptions of the spectrum of these compounds with real Fermi-level values, the parameter Δ may be assumed to be equal to zero. The resulting band-

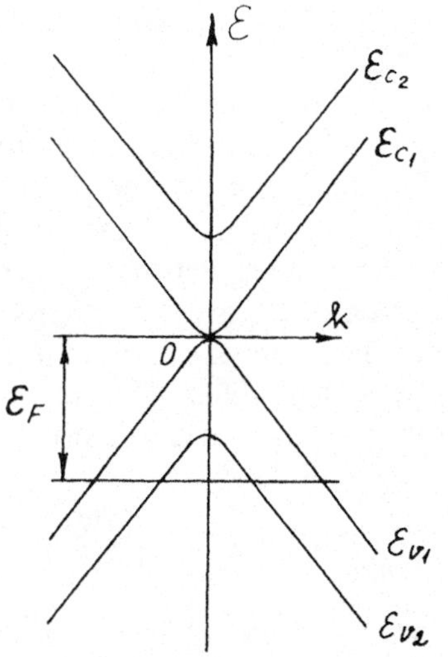

Fig. 109. Schematic diagram illustrating the energy spectrum second-stage acceptor-type GIC for $\Delta = 0$. (Blinowski et al. 1980.)

structure is described by

$$-\varepsilon_{v_1} = \varepsilon_{c_1} = \tfrac{1}{2}\{-\gamma_1 + [\gamma_1^2 + 9\gamma_0^2 a_0^2 k^2]^{1/2}\},$$
$$-\varepsilon_{v_2} = \varepsilon_{c_2} = \tfrac{1}{2}\{\gamma_1 + [\gamma_1^2 + 9\gamma_0^2 a_0^2 k^2]^{1/2}\}, \qquad (174)$$

and is illustrated in fig. 109. As can be seen from the above equations and fig. 109, for $\Delta = 0$ the spectrum of GIC with $n = 2$ is inverted with respect to the degeneracy point of the bands ε_{c_1} and ε_{v_1}. The difference between this spectrum in the vicinity of the degeneracy point and the spectrum of GIC with $n = 1$ [see (170)] is that, although at $k = 0$ the direct forbidden band is zero, this does not lead to the linearization of the spectrum and also does not lead to abnormally small effective masses of electrons and holes at zone edges. This is due to the prohibition imposed on the virtual transitions between the bands ε_{v_1} and ε_{c_1}, as a result of which the effective carrier mass at zone edges is determined by the distance between the respective bands ε_{v_1}, ε_{c_2} and ε_{v_2}, ε_{c_1}, which is close to the value of γ_1, rather than by the gap between them.

Depending on the position of the Fermi level, in GIC with $n = 2$ there may exist one or two groups of carriers. As an illustration, fig. 109 shows the position of the Fermi level in acceptor-type GIC with stage index two, which corresponds to the presence of two groups of holes (the Fermi level crosses both valence bands, ε_{v_1} and ε_{v_2}). The position of the Fermi level in GIC with $n = 2$ is in its turn determined by the extent of the electron density redistribution between the intercalate layer and the graphite layers, i.e., by the quantity f_{el}. It should also be noted that if the intercalant species is changed, one can observe a small change in the parameters γ_0 and γ_1, which results in a change in the shape of the energy bands and their relative shift in energy.

The size of the Brillouin zone along the k_z-axis in compounds with $n = 2$ is determined, as for GIC with $n = 1$, by the identity period I_c. In the case of only one group of carriers (at $|\varepsilon_F| < \gamma_1$), the Fermi surface of GIC with $n = 2$ consists of circular cylinders, whose geometrical axes coincide with the Brillouin zone edges. The carrier density is then described by the same expression, (eq. (171), as that used for GIC with $n = 1$. Since, however, for GIC with $n = 2$ the relation between the extremal cross section, S_{extr}, of the cylinders and the energy will be different from that for GIC with $n = 1$, eq. (172) for the cyclotron mass should be replaced by the following expression:

$$m_c^{e,h} = \frac{1}{2\pi}\frac{\partial S_{extr}}{\partial \varepsilon} = \frac{2\hbar^2}{9\gamma_0^2 a_0^2}\left[(\gamma_1 \pm \Delta)^2 + \frac{9\gamma_0^2 a_0^2 S_{extr}}{\pi\hbar^2}\right]^{1/2}, \tag{175}$$

where the symbol "−" refers to electrons and "+" refers to holes.

In the case when, in GIC with $n = 2$, the Fermi energy exceeds the γ_1 parameter (this is true of acceptor-type compounds with strong acceptors or of donor-type compounds with alkali metals), the Fermi level crosses two conduction bands or two valence bands and, therefore, two types of carriers may appear. The Fermi surface then represents the totality of pairs of coaxial cylinders placed around the side edges of the Brillouin zone.

Let us designate as $S_{extr}^{(2)}$ and $S_{extr}^{(1)}$ the extremal cross sections of the inner and outer cylinders. The total carrier concentration, $n_{e,h}$, which equals the sum of the carrier concentrations for the first and second groups ($n_{e,h}^{(1)}$ and $n_{e,h}^{(2)}$) may then be written in the form:

$$n_{e,h} = n_{e,h}^{(1)} + n_{e,h}^{(2)} = \frac{4(S_{extr}^{(1)} + S_{extr}^{(2)})}{(2\pi\hbar)^2 I_c}. \tag{176}$$

The cyclotron masses of the carriers of the first group are given by eq. (175), in which for S_{extr} one should read $S_{extr}^{(1)}$. Similarly to this, the cyclotron masses of the carriers belonging to the second group, which occur on the inner coaxial cylinder, are given by

$$m_c^{e,h} = \frac{2\hbar^2}{9\gamma_0^2 a_0^2}\left[(\gamma_1 \pm \Delta)^2 + \frac{9\gamma_0^2 a_0^2 S_{extr}^{(2)}}{\pi\hbar^2}\right]^{1/2}, \tag{177}$$

where the sign "+" refers to electrons and "−" refers to holes.

10.6.2.3. A TBA-model for the spectrum of GIC with n = 3. As has been mentioned above, GIC with $n = 1$ or 2 are the only types of GIC for which the notion of the "electron energy spectrum" may be used correctly for the whole system of graphite layers. A theoretical treatment of the carrier energy spectrum in graphite layers of GIC with higher n inevitably involves making some simplifying assumptions. One of the TBA-models for the energy spectrum in GIC with $n = 3$, which was developed by Blinowski et al. (1980), is based, in particular, on the assumption that all three graphite layers in a stack are physically equivalent. The dispersion law has then six "branches":

$$\begin{aligned} \varepsilon^{(1)}_{c,v} &= \delta \pm |x|, \\ \varepsilon^{(2)}_{c,v} &= \pm\{\delta^2 + \gamma_1^2 + |x|^2 - [\gamma_1^4 + (4\delta^2 + 2\gamma_1^2)|x|^2]^{1/2}\}^{1/2}, \\ \varepsilon^{(3)}_{c,v} &= \pm\{\delta^2 + \gamma_1^2 + |x|^2 + [\gamma_1^4 + (4\delta^2 + 2\gamma_1^2)|x|^2]^{1/2}\}^{1/2}, \end{aligned} \tag{178}$$

where $x = \frac{3}{2}\gamma_0 a_0 k$, the indices c and v referring to the conduction band and the valence band, the signs "+" and "−", to electrons and holes, respectively.

The structure of a spectrum of the type given by (178) is illustrated in fig. 110. In GIC of stage index 3 with strong acceptors or with donors (alkali metals) the Fermi surface may consist of three coaxial cylinders around each Brillouin-zone edge, because in this case the Fermi level crosses three valence bands or three conduction bands.

The Blinowski model for GIC of $n = 3$ is, of course, a rather crude one. Nevertheless, it reflects the main features of the electron spectrum in a system consisting of three graphite layers, viz., the presence of three different conduction (or valence) bands, the presence of the band-degeneracy point, and the presence of a group of charge carriers of abnormally small masses near the degeneracy point.

From the data of Blinowski et al. (1980) on the optical reflection in the IR

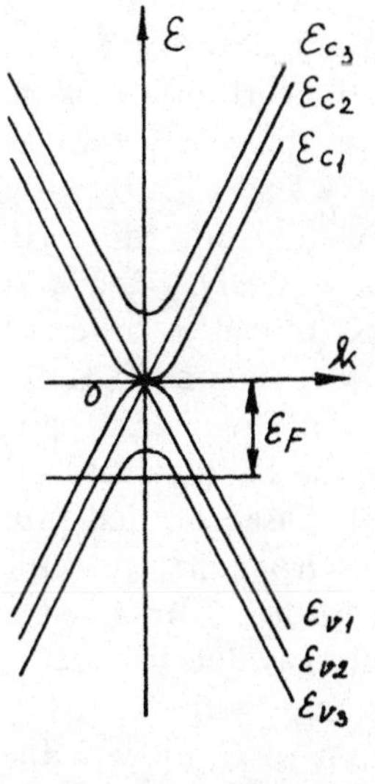

Fig. 110. Schematic diagram showing the energy spectrum of third-stage acceptor-type GIC. (Blinowski et al. 1980.)

range of GIC with $n = 2$ and 3 it follows that the γ_1 parameter contained in eqs. (173) and (178) is rather close in value to the corresponding parameter of the SWMC-model for the spectrum of three-dimensional single-crystal graphite. At the same time, the results of oscillational measurements indicate that the value of γ_1 in GIC with $n = 2$ and 3 depends on the intercalant species. An increase in carrier concentration in the graphite bounding layers leads to a decrease of γ_1 and, conversely, a decrease of carrier concentration results in an increase in γ_1. This regularity has been observed by Brandt et al. (1984a) in studying GIC with $n = 2$ with different intercalants, which affected each in a different way the concentration of carriers in the adjacent layers.

10.6.3. The calculation of the energy spectrum of GIC of low stage indices from first principles

The above-mentioned phenomenological TBA-models for the energy spectrum of GIC of low stage indices are capable of describing the structure of only a small number of bands lying in the immediate vicinity of the Fermi level. Such models may be used mainly for approximate descriptions of the transport and oscillation properties of GIC; they are unsuitable for an analysis of the GIC band structure in a wide energy region, which should be necessary, for instance, in the interpretation of optical or magneto-optical measurements.

The main disadvantage of the above phenomenological models is associated with the fact that, in these models, the electron spectrum is constructed for a system of graphite layers on the sole basis of interaction between the carbon atoms and without taking into account the contribution from the electronic states of the intercalant. At the same time, the states that result from the hybridization of the intercalant electron wave functions with the π- or σ-electron wave functions for the carbon atoms may occur sufficiently close to the Fermi level of GIC. In this case, the neglect of such states in the phenomenological models may lead to substantial errors and the model band structure near the Fermi energy will differ from the real one not only quantitatively, but even qualitatively, Such a situation arises, for example, when one uses the Blinowski model in order to describe the spectrum of a donor-type compound of $n = 1$, such as C_8K (see below).

In order to obtain a sufficiently realistic spectrum of carriers in GIC of low stage index and to calculate the band structure in a wide energy region, one usually resorts to calculations based on first principles.

Band-structure calculations from first principles have thus far been made only for the simplest GIC with $n = 1$ obtained in graphite intercalation with alkali metals. Such a limitation is due to the fact that computations become increasingly more cumbersome with increasing stage index. Indeed, in calculating the band spectrum of C_6Li with the simplest stacking sequence $A\alpha A\alpha A\alpha\ldots$, whose unit cell contains six carbon atoms and one lithium atom, it is necessary to take into account seventy orbitals pertaining to the $1s^22s^22p^6$ electronic states in C and Li atoms. The unit cell of the same compound with $n = 2$ contains 36 carbon and 3 lithium atoms (it is assumed that the stacking sequence of the layers is CAαABαBCαCA...). Accordingly, the $1s^22s^22p^6$ electronic states of these atoms yield already 390 orbitals, and 225 of them are filled.

The necessity of a rigorous consideration of so great a number of orbitals renders a calculation extremely laborious. Nevertheless, in spite of the great computational difficulties, such calculations from first principles are highly important and interesting because of the detailed character of the obtained dependences of energy on momentum for carriers and the more realistic shape of the Fermi surfaces constructed in this way, as compared to what may come from the various phenomenological models. The most detailed calculations on the energy spectrum of GIC from first principles have been made for C_6Li by Holzwarth et al. (1978b) and for C_8K by Ohno et al. (1979). As an illustration, we shall consider the results of the calculations of the spectrum for the compounds just indicated.

The calculations made for C_6Li show that the bands originating from the s-electron band in Li lie higher than the Fermi level in GIC throughout the Brillouin zone. This means that the lithium atoms in this compound are fully ionized and that the fraction, f_{el}, of the charge transferred to the graphite layers from a single Li atom is equal to unity. At the same time, the bands originating from carbon π-electrons suffer but an insignificant perturbation due to intercalation. Thus the calculation from first principles made for C_6Li indicates that, in the given case, the Blinowski phenomenological model for GIC with $n = 1$ obeying the dispersion relation of the type of (170) is applicable as a first approximation. The position of the energy levels in the spectrum of C_6Li, calculated by the Slater–Koster LCAO-interpolation expansion technique (1954) in a paper by Holzwarth et al. (1978b), is in good quantitative agreement with the data obtained by Eberhardt et al. (1980), who measured the angular dependences of the photoemission spectra in C_6Li. Table 20 compares the calculated with the experimental values for the energy of the levels in the spectrum of C_6Li, occurring at the Γ-point in the Brillouin zone ($k = 0$). The calculation has taken into account the presence of a commensurate periodic structure in the intercalate layer (the zone-folding effect). The additional σ- and π-bands in the spectrum which are due to this phenomenon are marked with an asterisk. The energies given in table 20 are measured relative to the Fermi level in GIC.

Table 20 shows an excellent agreement between theory and experiment (a noticeable discrepancy is seen only for the π_1^* band), thus lending support for the validity of the method used for the calculation.

The band structure of another simple compound with stage index one, C_8K, has been determined in a self-consistent band calculation by Ohno et al.

Table 20
Comparison between the calculated and the experimental energy levels in the spectrum of C_6Li. All values are given in eV.

Band	Calculation (Holzwarth et al. 1978b)	Experiment (Eberhardt et al. 1980)
π_1^*	−1.3	-0.5 ± 0.3
$\sigma_{2,3}$	−5.9	-5.5 ± 0.4
π_1	−9.3	-9.3 ± 0.3
$\sigma_{2,3}^*$	−13.3	-13.0 ± 0.5
σ_1^*	−15.4	-15.2 ± 0.5
σ_1	−21.8	-22.5 ± 0.4

(1979), who also took into account the zone-folding effect. In the calculations for C and K, use was made of atomic pseudo-potentials and an ionic Madelung potential was introduced to account for charge transfer. The Madelung potential was determined self-consistently, yielding a fractional charge transfer of $f_{el} = 0.6$, which is to be contrasted with $f_{el} = 1$ for the fully ionized intercalant obtained theoretically by Holzwarth et al. (1978b) in C_6Li and confirmed by the angle-resolved photoemission experiment of Eberhardt et al. (1980).

The general appearance of the energy levels for C_8K in the energy range ~20 eV is illustrated in fig. 111. Without going into the details of the band structure obtained we wish only to note that the bands which have their origin in the π-bands of graphite retain their properties, as do the bands originating from the s-band in potassium.

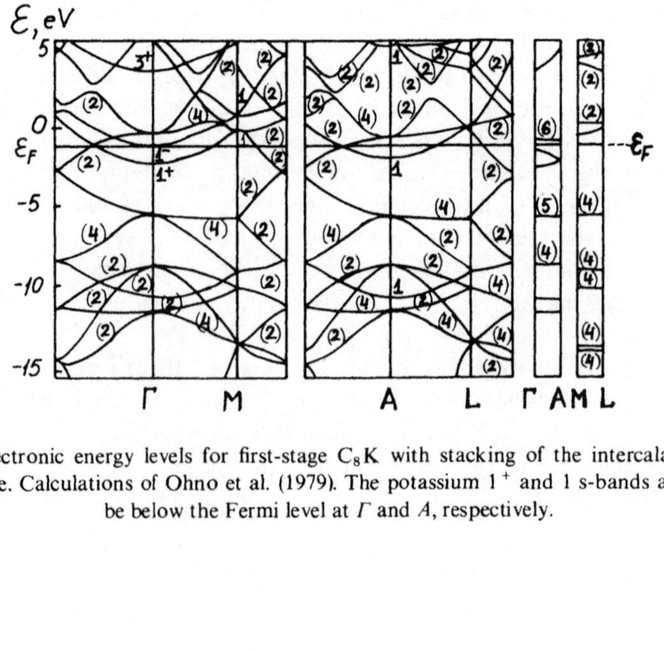

Fig. 111. Electronic energy levels for first-stage C_8K with stacking of the intercalate layers of $\alpha\beta\gamma\delta\alpha\ldots$-type. Calculations of Ohno et al. (1979). The potassium 1^+ and 1 s-bands are shown to be below the Fermi level at Γ and A, respectively.

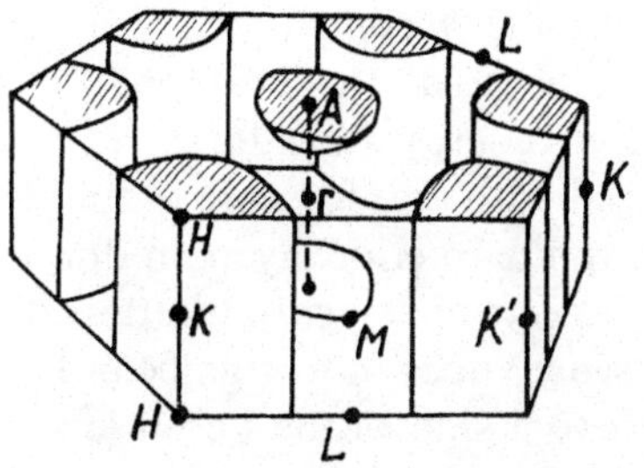

Fig. 112. The Fermi surface in C_8K calculated from first principles by Inoshita et al. (1977), neglecting the effects of zone folding. The open surfaces give rise to the two-dimensional transport behaviour and the closed surfaces correspond to the three-dimensional behaviour. The high-symmetry points in the BZ for C_8K are designated as those in the BZ for graphite (see fig. 9).

The Fermi surface for C_8K has been obtained theoretically by Inoshita et al. (1977). These authors also found that the value of f_{el} in this compound is about 0.6. The overall appearance of the Fermi surface for C_8K, with the zone folding effect ignored, is shown in fig. 112. As one would expect, the main constant-energy surfaces, calculated on the basis of the phenomenological model without the zone-folding effect, represent a system of open cylinders placed around the vertical Brillouin-zone edges. Measured across, each cylinder is about the same size as the Brillouin zone itself. It is of interest to note that the first-principles calculation suggests the existence, in the spectrum of C_8K below the Fermi level, of a second conduction band, which is not disclosed by the approximate phenomenological model of the spectrum of GIC with stage index one, discussed in the preceding section (see fig. 108). The extrema of the second conduction band are in the centres of the upper and lower edges of the Brillouin zone (points A and A'). Because of this, near these points there appear two closed, lens-shaped electronic Fermi surfaces (fig. 112). The second conduction band thus obtained (as well as a second group of electrons associated with it) owes its origin to the hybridization of the potassium s-electrons with the carbon π-states. Being fundamentally associated with the electronic states of the intercalant, it could not, naturally, emerge within a simple phenomenological model for the spectrum of $n = 1$ compounds, based merely on the consideration of the interaction between the carbon atoms in the graphite layers.

10.6.4. The rigid-band model for GIC with high n

It has already been mentioned that the energy spectrum of such compounds cannot be calculated from first principles. At the same time, if n is sufficiently large the electron (hole) spectrum for the graphite interior layers turns out to be rather close to that of pure graphite. This has motivated the use of the latter in describing the band structure of a quasi-homogeneous system of the graphite interior layers in high-n compounds. The specific character of GIC is then manifested by the fact that the position of the Fermi level is shifted upward or downward in energy (for donor or acceptor-type compounds, respectively), as compared with that for pure graphite. With such an approach, the spectrum for the graphite interior layers in donor or acceptor compounds may be treated as the spectrum of graphite doped with donor or acceptor impurities, whose presence does not significantly change band-parameter values, but merely leads to a shift of the Fermi level. This treatment has been referred to as the rigid-band model.

For semiconductors and semimetals, the rigid-band model is commonly used in the case when the dopant concentration is many orders of magnitude lower than the concentration of the host atoms, which is always true for any real situation because of the limited solubility of impurities. For graphite intercalation compounds the concentration of impurities (the intercalant) may be as high as that of the carbon atoms. Evidently, under these conditions, the rigid-band model is no longer applicable for semimetals. Actually, however, the extent of applicability of this model for describing GIC is much larger than for doped semiconductors or semimetals. This is due to the following circumstances. First of all, in GIC, impurities are not randomly distributed in the host material, but fill completely periodically repeated layers. Secondly, the intercalate layers are strongly screened by the graphite layers adjacent to them, so that the disturbance caused by them ceases to be felt at a distance of one or two graphite layers. Hence it follows that the spectrum peculiar to pure graphite remains practically undisturbed already in the centre of a system of four or five graphite layers.

From the experimental data obtained thus far, one can conclude that the rigid-band model may successfully be applied for GIC with $n = 4$ and higher. It is of interest to mention that this model has even been used for a qualitative interpretation of the results of oscillational measurements in $C_{24}AsF_5$, a compound with $n = 3$ (see Iye et al. 1980). However, we should again emphasize that the rigid-band model is valid only in describing the carrier energy spectrum in the graphite interior layers, which may be approximately treated as equivalent to each other. The model cannot, of course, be extended to the description of the spectrum in the graphite bounding layers, in which both the carrier concentration and the band structure may significantly differ from those in the interior layers. In the bounding layers, it is primarily the hybridization of the carbon-atom and intercalant states that plays the major role. The electrons that occur in these layers provide their own, independent contribution to the transport, oscillational, optical, and other effects. The relative weight of this contribution naturally decreases with the increasing stage index as $\sim 1/n$, so that, at sufficiently large n, the carrier spectrum in the graphite interior layers yeilds an increasingly accurate description of the results of measurements of various characteristics of GIC.

The rigid-band model, based on the conceptions of the SWMC-model for the graphite spectrum, was first used by McClure (1960) to describe acceptor-type compounds C_xBr_2. We note that eq. (35), which gives the dispersion relation adopted in the SWMC-model for pure graphite, is valid in a narrow region of momentum space near the Brillouin-zone side edges and may really be applied to GIC for $n \geqslant 6$, when the shift of the Fermi level, ε_F, relative to the initial Fermi level, ε_F^0, in graphite, is small. One then usually adopts the four parameter model relations [see sect. 4.3 eqs. (36) and (37)] to make calculations. However, the use of the SWMC-model for the graphite spectrum as a basis for the rigid-band model in describing the GIC spectrum at not very high n values gives rise to difficulties associated with the fact that the Fermi-surface size in these compounds is comparable with the Brillouin-zone size. In order to overcome this difficulty, Johnson and Dresselhaus (1973) have extended the dispersion relations of the SWMC-model to cover the entire Brillouin zone of graphite. Using symmetry requirements to specify the form of the Hamiltonian, they utilized a three-dimensional Fourier-expansion to serve as the basis functions. The band parameters of the model were evaluated using:

(1) Fermi-surface data in the vicinity of the Brillouin-zone edges *HKH* and *H'K'H'*;

(2) fits to the optical data below 6 eV; and

(3) the requirement that the dispersion relations should reduce to those of the SWMC-model in the vicinity of the Brillouin-zone edges.

The resulting dispersion relations for the π-bands along several high-symmetry directions are shown in fig. 113. The extension of this three-dimensional

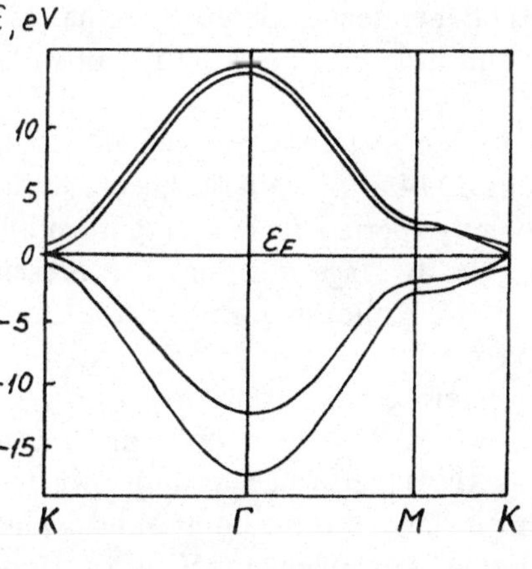

Fig. 113. Graphite π-bands along several high-symmetry directions calculated by Johnson and Dresselhaus (1973) on the basis of a three-dimensional Fourier expansion of $\varepsilon(k)$. The Fourier expansion coefficients are evaluated by fitting $\varepsilon(k)$ to the SWMC-model along the HK-axis and by comparing the calculated frequency-dependent dielectric function $\kappa(\omega)$ with optical reflectivity measurements.

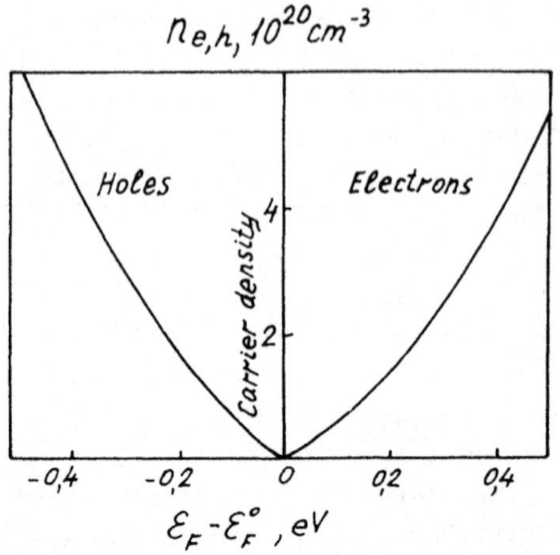

Fig. 114. Carrier density obtained from the SWMC-band model for various Fermi-level shifts relative to the graphite Fermi level ε_F^0 where the electron and hole carrier densities are equal. In the band overlap region, both electron and hole carriers are present. (Dresselhaus and Dresselhaus 1981.)

Fourier-expansion model to describe dispersion relations in graphite intercalation compounds was first made by Dresselhaus and Leung (1980). The results of the calculation by Johnson and Dresselhaus (1973) enable one to calculate the concentration of electrons and holes in the graphite interior layers of GIC as a function of Fermi energy ε_F measured from its value in pure graphite, ε_F^0. Such a dependence of $n_{e,h}$ on $(\varepsilon_F - \varepsilon_F^0)$ is illustrated in fig. 114, while the corresponding $(\varepsilon_F - \varepsilon_F^0)$-dependences of the extremal cross sections S_{extr} of the electron (at points K) and hole (at points H) Fermi surfaces in GIC are shown in fig. 115.

Attention is called to the fact that the dependences of $n_{e,h}$ and $S_{extr}^{e,h}$ upon $(\varepsilon_F - \varepsilon_F^0)$ presented in figs. 114 and 115 have been obtained without taking into account the new periodicity (period I_c) resulting from intercalation, because, as already mentioned above, at large n values the presence of the superlattice structure does not alter the Brillouin zone and the carrier energy spectrum in the graphite interior layers.

Let us dwell at greater length upon the energy spectrum for GIC of intermediate stage index, i.e., $3 \leqslant n \leqslant 5$ or 6, as described by the rigid-band model and find out how the superlattice structure with period I_c affects the size and shape of the Fermi surface in these compounds. This influence will now be illustrated on the particular examples of GIC of the donor and acceptor types.

As a first example, we shall examine, following Tanuma et al. (1978), the Fermi surface in a GIC with potassium for $n = 4$. According to the rigid-band model, the electron Fermi surface for the graphite interior layers in such a compound is a quasi-ellipsoid (fig. 116a) if the superlattice structure is ignored.

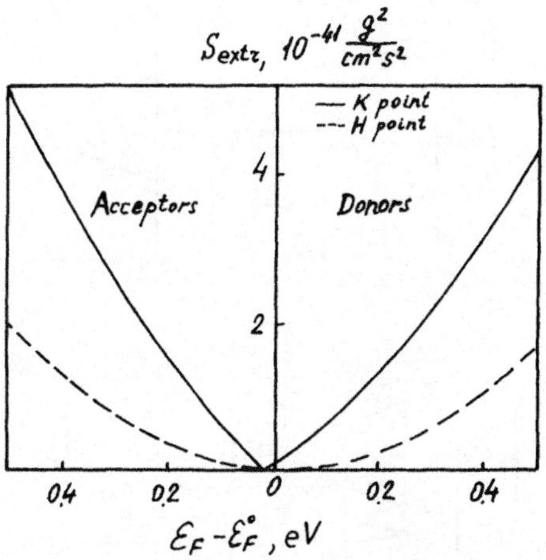

Fig. 115. Extremal Fermi-surface cross sections S_{extr} in GIC at the K and H points of the Brillouin zone calculated on the basis of the SWMC-band model for various Fermi-level shifts $\varepsilon_F - \varepsilon_F^0$. The effects of the superlattice with period $2\pi I_c$ and zone folding are neglected. (Dresselhaus and Dresselhaus 1981.)

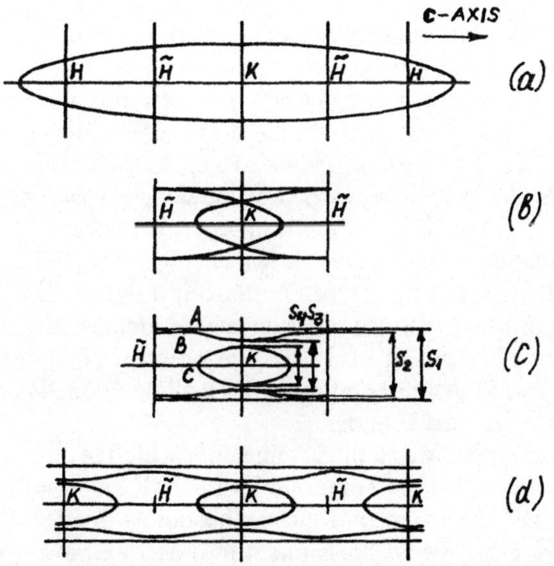

Fig. 116. The Fermi surfaces constructed by Tanuma et al. (1978) for a fourth-stage graphite–K compound: (a) Cross section of the Fermi surface is shown assuming a rigid-band model for graphite. (b) Fermi surface is folded back to the new Brillouin zone $\tilde{H}K\tilde{H}$, which is introduced by the superlattice in the c-axis direction. (c) Schematic Fermi surfaces deduced from (b) by introducing energy gaps due to the periodic potential of the superlattice. (d) Fermi surfaces constructed in (c) are shown in the extended-zone scheme.

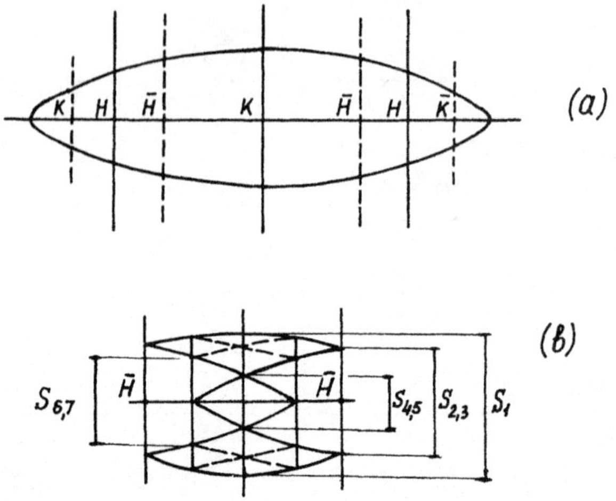

Fig. 117. Construction of the hole Fermi surface of a third-stage $C_{24}AsF_5$ acceptor-type compound assuming a rigid-band model for graphite: (a) The Fermi surface for holes constructed neglecting the superlattice along the *c*-axis which has a period I_c, the Fermi energy $\varepsilon_F = 0.611$ eV (b) The Fermi surface and the extremal cross section for $H \| c$ which arise when the effect of the superlattice along the *c*-axis, with period I_c, is taken into account. (Iye et al. 1980.)

The energy-discontinuity planes, associated with the periodicity with period I_c, cross the Fermi surface at $k_z = \pm\pi/I_c$ and $\pm 2\pi/I_c$. The surface of constant energy constructed with the new periodicity taken into account is shown in fig. 116b. In the next stage, one should take into account the energy discontinuity at the edge of the new Brillouin zone, which leads to the appearance of an open surface of the type of a warped cylinder and a closed, lens-shaped surface (see fig. 116c). The new constant-energy surfaces, constructed using the extended-band scheme, are shown in fig. 116d. These surfaces are characterized by five extremal cross sections S_i ($i = 1, 2, 3, 4, 5$) (see fig. 116c), which have been observed in oscillation measurements carried out in C_xK of stage index four (Tanuma et al. 1978). The only extremal cross section of the initial quasi-ellipsoid at the *K* point of the Brillouin zone is equal to the cross section S_1 of the warped cylinder.

As a second example, we shall examine, following Iye et al. (1980), the Fermi surface of the acceptor-type compound with AsF_5 at $n = 3$. The initial hole surface, which has the shape of a quasi-ellipsoid, is divided, if the superlattice structure with identity period I_c is taken into account, into two open surfaces of the "warped cylinder" type and one closed lens-shaped surface (see fig. 117a,b). The new constant-energy surfaces, constructed with the energy discontinuity at the Brillouin-zone edges taken into account, are characterised by seven different extremal cross sections (these cross sections obtained without the energy-discontinuity are shown in fig. 117b).

10.7. Electrical conductivity of GIC

10.7.1. Peculiarities of the electrical conductivity in GIC

As is known, the electrical conductivity of a normal metal is proportional to the concentration of free carriers and their mean free path; its limiting theoretical value, at room temperature, determined by the greatest possible concentration of electrons and their mean free path in scattering by phonons, being of the order of $10^6 \Omega^{-1} cm^{-1}$. The electrical conductivities of copper, σ_{Cu}, and silver, σ_{Ag}, which are close, at $T = 300 K$, to the limiting value, constitute 0.59×10^6 and $0.62 \times 10^6 \Omega^{-1} cm^{-1}$, respectively. Graphite intercalation leads to a considerable increase in the concentrations of electrons or holes in the graphite layers, but it relatively little affects their mobilities. Therefore, intercalation may significantly change the electrical conductivity of the host graphite material both in the direction of its layers (σ_a) and along the c-axis (σ_c). And while the intercalation of graphite with donor impurities results in an increase in σ_a by a factor of 4–10 (at $T = 300$ K), the intercalation with acceptor impurities leads to an increase in σ_a by 4–40 times at the same temperature and the σ_a of certain acceptor GIC may be as high as that of copper and silver, and even higher. The σ_c of graphite increases due to intercalation with donor impurities, but decreases when acceptor impurities are introduced. Thus, as far as their conductivity is concerned, GIC may be classified as good metals with a high anisotropy of conductivity, $\eta = \sigma_a/\sigma_c \gg 1$, which in acceptor-type GIC is typically much greater than that for pure graphite. In contrast, the anisotropy of the electrical conductivity in GIC of the donor type is considerably smaller than that for pure graphite.

Table 21 compares the conductivities (σ_a and σ_c) and their anisotropy $\eta = \sigma_a/\sigma_c$ at $T = 300$ K in certain donor and acceptor GIC with the analogous characteristics of pure graphite and with the conductivities of normal metals. The high values of the electrical conductivity σ_a at $T = 300$ K in some acceptor-type GIC (e.g., C_xAsF_5, C_xSbF_5, $C_8HNO_3 \cdot 3HNO_3$), which reach or even exceed the electrical conductivity of the best metallic conductors, has led to the introduction of the term "supermetallic conductivity" of GIC.

For a purposive search for compounds with high σ_a values in the basal plane it is important to know which properties of the intercalant are responsible for this. In the early investigations to attain increasingly higher σ_a in GIC, the investigators usually proceeded from the hypothesis advanced by Vogel in 1976. According to this hypothesis, the increase in electrical conductivity in the basal plane in graphite intercalation compounds is associated with the intercalant oxidation properties, i.e., in fact, it is ensured by the increase in carrier concentration in the graphite layers, which increases with increasing f_{el}. This hypothesis was in principle consistent with the experimental data available at that time, suggesting that in acceptor GIC the conductivity σ_a

Table 21
Conductivities (σ_a and σ_c) and their anisotropy η ($=\sigma_a/\sigma_c$) at $T = 300$ K in donor and acceptor GIC.

Material	Stage index n	Conductivity σ or σ_a ($10^6\ \Omega^{-1}\ cm^{-1}$)	Conductivity σ_c ($\Omega^{-1}\ cm^{-1}$)	$\eta = \sigma_a/\sigma_c$	Ref.*
Normal metals					
silver	–	0.62	–	–	
copper	–	0.59	–	–	
gold	–	0.40	–	–	
aluminum	–	0.30	–	–	
tin	–	0.085	–	–	
Graphite	∞	0.025	1–10	$(25–2.5) \times 10^3)$	[1, 2]
Donor GIC					
C_8K	1	0.1	3×10^3	~33	[3]
C_8K	1	0.11	1.94×10^3	~57	[4]
$C_{16}K$	2	0.17	1.97×10^2	~860	[4]
C_6Li	1	0.25	1.8×10^4	~14	[5]
C_8Rb	1	0.1	–	–	[6]
C_6Li	1	0.252	2.0×10^4	~12	[3]
C_8Cs	1	0.1	–	–	[7]
Acceptor GIC					
C_8AsF_5	1	0.50	0.20	2.5×10^6	[8]
$C_{16}AsF_5$	2	0.63	0.24	2.72×10^6	[8]
$C_{24}AsF_5$	3	0.58	–	–	[8]
C_8SbF_5	1	0.35	–	–	[9, 10]

$C_{16}SbF_5$	2	0.40	–	–	[9, 10]
$C_{24}SbF_5$	3	1.0	–	–	[9, 10]
$C_{48}SbF_5$	4	0.58	–	–	[9, 10]
C_8HNO_3	1	0.3	1.8	1.7×10^5	[11]
$C_{16}HNO_3$	2	0.35	–	–	[11]
$C_{24}HNO_3$	3	0.29	–	–	[11]
C_8FeCl_3	1	0.11	10	1.1×10^4	[10]
$C_{16}FeCl_3$	2	0.25	–	–	[10]
$C_3HNO_3 \cdot 3HNO_3$	1	1.19	–	–	[9, 10]
$C_8H_2SO_4 \cdot 2H_2SO_4$	1	0.61	–	–	[9, 10]
C_8AlCl_3	1	0.16	6.1	2.6×10^4	[11]
$C_8H_2SO_4$	1	0.16	0.9	1.8×10^5	[11]
$C_{16}Br_2$	2	0.22	1.6	1.4×10^5	[11]
$C_{8.3 \pm 0.2}ICl_{1.08 \pm 0.03}$	1	0.09	–	–	[12]
$C_{16.3 \pm 0.2}ICl_{1.1 \pm 0.03}$	2	0.28 ± 0.06	–	–	[12]
$C_{24.5 \pm 0.5}ICl_{1.1 \pm 0.03}$	3	0.44 ± 0.06	–	–	[12]
$C_{33.0 \pm 0.5}ICl_{1.1 \pm 0.03}$	4	0.13 ± 0.03	–	–	[12]
$C_{9.5 \pm 0.2}AlCl_3Br_{0.6 \pm 0.05}$	1	0.27 ± 0.02	–	–	[12]
$C_{9.8 \pm 0.1}CuCl_{2.05 \pm 0.02}$	2	0.12 ± 0.01	–	–	[12]

*References
[1] Spain (1973).
[2] Tsuzuku (1979).
[3] Fischer (1980).
[4] Murray and Ubbelohde (1969).
[5] Basu et al. (1979).
[6] Guérard et al. (1977).
[7] McRae et al. (1980).
[8] Foley et al. (1977).
[9] Ubbelohde (1976).
[10] Vogel (1979/1980).
[11] Ubbelohde (1972).
[12] Brandt et al. (1982).

would substantially increase if a weak-acceptor type intercalant is replaced by a powerful oxidizer such as AsF_5, SbF_5 + HF (see table 21). However, the subsequent studies by Bartlett et al. (1980) have shown that the application of Lewis acids, which are still stronger oxidizers, fails to yield the desired result. They have found that when very strong oxidizers are used there arise covalent bonds between the graphite and the intercalant, which leads to a distortion of the planar graphite-layer structure, resulting, in turn, in an appreciable decrease of carrier mobility. Evidently, in order to obtain maximum σ_a-values, one should optimize the product of carrier concentration and carrier mobility, which is ensured by the use of the intercalant, whose f_{el}-values, though very high, yet do not give rise to any change in graphite-layer plane structure. The investigations carried out to date suggest that the metal fluorides (AsF_5, SbF_5, etc.) may best serve this purpose. When considering the possibility of eventually replacing metallic conductors by graphite intercalation compounds we should note that as the temperature is lowered the electrical conductivity of GIC increases not so rapidly as that of normal metals (as an illustration, fig. 118 shows the temperature dependences of σ_a and $\rho_a = 1/\sigma_a$ in the acceptor-type compound C_{16} ICl) and, therefore, normal metals compare favourably in this respect with GIC at low temperatures. But at room temperature the competition between these two classes of materials is rather tough. One should, of course, bear in mind that in order to utilize GIC as electrical current conductors it is necessary to be able to weld them into thin, flexible and strong wires.

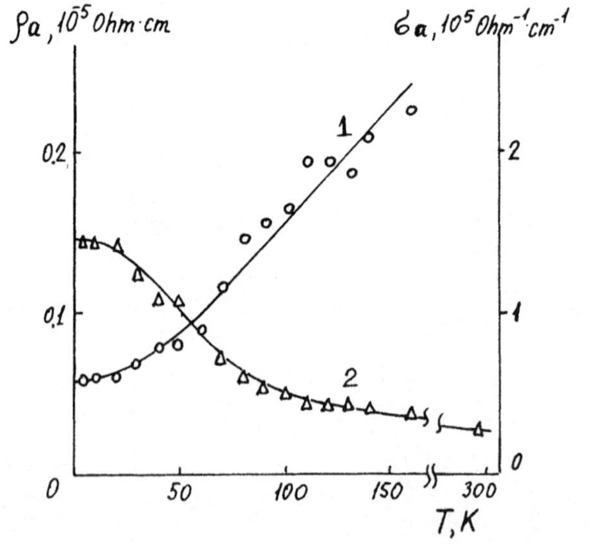

Fig. 118. Temperature dependence of the electrical resistivity ρ_a (curve 1) and the electrical conductivity σ_a (curve 2) of the $C_{16}ICl$ acceptor-type compound. The values of ρ_a and σ_a are measured using a DC four-probe method; the value of σ_a at $T = 300\,K$ is measured by a contactless induction method at a frequency of 100 kHz. (Brandt et al. 1982.)

This obstacle in the road to such a utilization of GIC has not yet been completely surmounted. There are two ways of solving the problem at hand. The first is to create composite wires out of good conductors (e.g., copper) and GIC. Vogel (1979/1980) has developed a wire composed of copper and C_xAsF_5 coated with copper. Its conductivity at room temperature is about $0.6 \times 10^5 \Omega^{-1} cm^{-1}$, its specific gravity being lower than that of copper.

The second way involves intercalation of carbon fibres, which are characterized by a high mechanical strength. As such well-graphitized fibres are intercalated, their strength does not change, whereas their electrical conductivity considerably increases. However, because of numerous structural defects, which limit the mobility of charge carriers, the high potentialities of fibres in this respect have not yet been realized.

10.7.2. The electrical conductivity of GIC in the basal plane

Let us discuss at greater length the conductivity σ_a in GIC along the layers. As has already been mentioned, as a result of intercalation, the σ_a increases by more than an order of magnitude in comparison with that of pure graphite, which in what follows will be referred to as $\sigma_a^{(g)}$.

The increase in σ_a in intercalation depends both on intercalant species and on stage index (see table 21). Large σ_a values for GIC may qualitatively be explained as follows. Pure graphite has a high average carrier mobility, which in perfect specimens at room temperature may be as high as $\bar{\mu} \sim (1–2) \times 10^4 cm^2/(V s)$, considerably exceeding the electron mobility in copper $[\sim 3.5 \times 10^1 cm^2/(V s)]$ under the same conditions. The concentration of electrons, n_e, and holes, n_h, in graphite at $T = 300 K$ constitutes $\sim 7 \times 10^{18} cm^{-3}$. Graphite intercalation leads to an increase in carrier concentration in the graphite layers up to $(2-6) \times 10^{20} cm^{-3}$, i.e., practically an increase by two orders of magnitude, whereas the mobility of electrons or holes decreases much more slowly (only 2 to 5 times). As a result, the increase in carrier concentration turns out to be the dominant effect leading to an increase in conductivity. Figure 119a,b shows the dependences of the absolute (σ_a) conductivities in GIC upon the inverse stage index, $1/n$, for various donor- and acceptor-type compounds. These dependences have been obtained by Dresselhaus and Dresselhaus (1981), Foley et al. (1977), Wu et al. (1981). Fischer (1979), and Brandt et al. (1982, 1983) and by other authors. The dependences of σ_a on $1/n$ have been obtained by the authors mentioned above. Figure 119 also shows the mean free path Λ and the concentrations n_h as a function of $1/n$.

These data indicate that in donor-type intercalants σ_a reaches a peak at $n \simeq 3$, while in acceptor-type intercalants this peak is reached at $n \simeq 2$.

The character of the dependence of σ_a on $1/n$ can be understood on the basis of a simple model for the electrical conductivity in GIC (see Dresselhaus and Leung 1979), which treats the graphite and intercalate layers as connected in

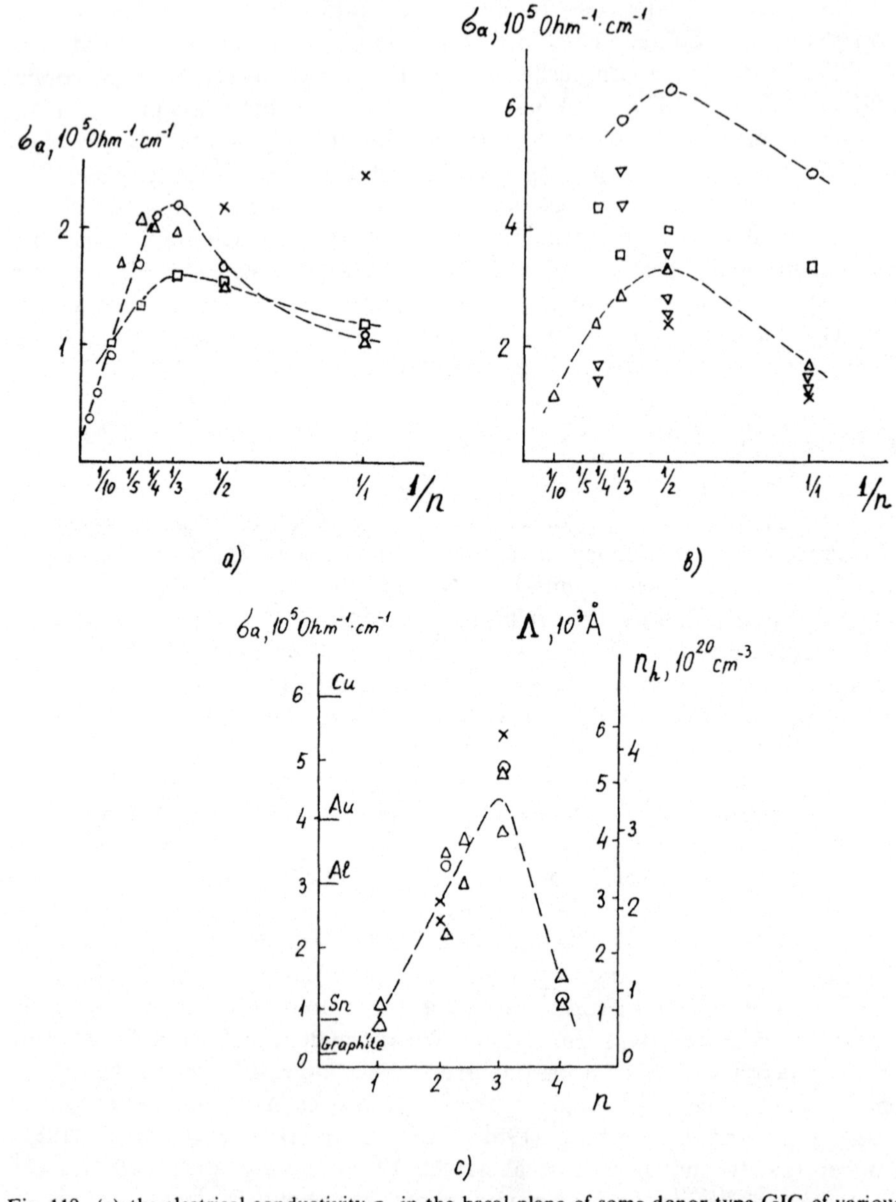

Fig. 119. (a) the electrical conductivity σ_a in the basal plane of some donor-type GIC of various stages at room temperature: ○, C_xK; △, C_xRb; □, C_xCs; ×, C_xLi (data of Fischer 1979, Potter et al. 1981, Basu et al. 1979). (b) σ_a in acceptor-type GIC of various stages at room temperature: ○, C_xAsF_5; □, C_xSbF_5; △, C_xHNO_3; ×, C_xFeCl_3; ▽, C_xICl (data of Vogel et al. 1977, Wu et al. 1981, Vogel 1979, Brandt et al. 1982). (c) σ_a(△), the hole density n_h(○), and the mean-free path $\Lambda(x)$ for holes in C_xICl of various stages at room temperature (noninteger n corresponding to a mixture of two stages) (Brandt et al. 1982). For comparison, the conductivities of graphite, Sn, Al, Au, and Cu at 300 K are marked by lines on the abscissa in the c-figure. In all these figures, the dashed lines are a guide for the eye.

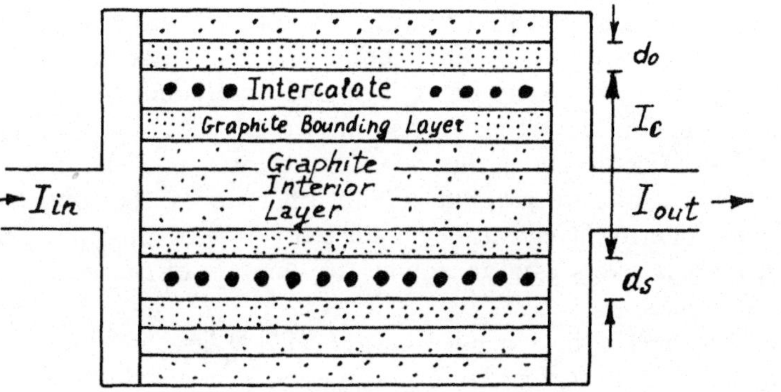

Fig. 120. Geometrical view of a conductance model; I_{in} and I_{out} are the incoming and outgoing currents. (Dresselhaus and Leung 1979.)

parallel (fig. 120). According to this model, the averaged relative conductivity $\sigma_a/\sigma_a^{(g)}$ along the layers for $n \geqslant 2$ may be described as

$$\frac{\sigma_a}{\sigma_a^{(g)}} I_c = \frac{\sigma_i}{\sigma_a^{(g)}} d_s + \frac{\sigma_{gb}}{\sigma_a^{(g)}} 2d'_0 + \frac{\sigma_{gi}}{\sigma_a^{(g)}} (n-2) d_0, \tag{179}$$

where the σ_i, σ_{gb} and σ_{gi} are the conductivities of the intercalate layer, of the graphite bounding layer, and of the graphite interior layer, respectively, $\sigma^{(g)}$ is the conductivity of the pure graphite in the basal plane. The symbol d_s has been explained in sect. 10.4 [see (164)] and the primed d_0 (i.e., d'_0) indicates that the distance between the graphite bounding layer and the graphite interior layer may be somewhat different from the distance between two adjacent graphite interior layers. For $n = 1$, eq. (179) may be rewritten as

$$\frac{\sigma_a}{\sigma_a^{(g)}} I_c = \frac{\sigma_i}{\sigma_a^{(g)}} d_s + \frac{\sigma_{gb}}{\sigma_a^{(g)}} d''_0, \tag{180}$$

where, for the reasons given above, d''_0 may somewhat differ from d_0 in pure graphite. Attention is called to the fact that in acceptor-type GIC the equality $d''_0 = d'_0 = d_0$ holds well, which is not the case with donor GIC.

Since the maximum carrier concentration occurs in the graphite bounding layers, it is these layers that have the highest electrical conductivity, σ_{gb}. For not too large n, the conductivity, σ_{gi}, of the graphite interior layers is also sufficiently high, because an increase in carrier concentration in them does not give practically rise to any change in the character of carrier scattering. As regards the conductivity σ_i, one should bear in mind that, as has already been mentioned above, the effective mobility in the intercalate layer is rather low and the conductivity is mainly of a hopping type. This is due to the fact that the Fermi level usually occurs considerably below the lower boundary of the intercalant conduction band (in donor GIC) and above the upper boundary of the intercalant valence band (in acceptor GIC).

Therefore, one may conclude that the highest electrical conductivity is to be observed in GIC with $n = 2$ and 3, because it is in these intercalation stages that the optimal conditions for a high electrical conductivity (namely, the highest conduction-layer density per unit cross section of the material) are realized. With increasing n this density decreases and the conductivity σ_a approaches that of pure graphite $\sigma_a^{(g)}$. In other words, the contribution from the graphite interior layers to the conductivity σ_a is of importance only in the case when these layers gain an appreciable amount of charge, as it occurs at low stage indices.

In order to obtain the dependece of σ_a on $1/n$ at large n, let us substitute the expression $I_c = d_i + nd_0$ for the identity period I_c in eq. (179) and rewrite this equation in the form

$$\frac{\sigma_a}{\sigma_a^{(g)}} = \frac{\sigma_i}{\sigma_a^{(g)}} \frac{d_s}{d_i + nd_0} + \frac{\sigma_{gb}}{\sigma_a^{(g)}} \frac{2d'_0}{d_i + nd_0} + \frac{\sigma_{gi}}{\sigma_a^{(g)}} \frac{(n-2)d_0}{d_i + nd_0}.$$

From this relation it follows that for $n \gg 1$ the conductivity σ_a in GIC may approximately be described as

$$\frac{\sigma_a}{\sigma_a^{(g)}} \simeq 2\frac{d'_0}{d_0} \frac{\sigma_{gb}}{\sigma_a^{(g)}} \frac{1}{n} + \frac{\sigma_{gi}}{\sigma_a^{(g)}} \tag{181}$$

and is linearly dependent on $1/n$. This linear dependence of σ_a on $1/n$ means that the electrical conductivity is proportional to the fraction of the high-conductivity layers in the periodic structure of GIC. This inference is in good agreement with the experimental data of McRae et al. (1980).

As the stage index drops below the optimal value, the part of the area, $d_s/(d_i + nd_0)$, occupied in the structure cross section by the intercalant increases, leading to the saturation and subsequent decrease of σ_a values as the stage index further decreases. This is confirmed by the experimental data (see fig. 119). The decrease of σ_a at low stage indices is best pronounced for the intercalants for which large d_s values are typical (e.g., HNO_3), the effect being less pronounced for the intercalants with typically low d_s values (e.g. Li). One should also take into consideration that in going over from $n = 2$ to $n = 1$ there should inevitably increase the probability of scattering of the graphite-layer electrons by charged centres in the adjacent intercalate layers, which results in a decreased carrier mobility.

Thus the above model of layers connected in parallel (fig. 120) predicts that the highest conductivity σ_a should be observed in the case when the charge density transferred from the intercalate layer to the graphite layers is maximal, the carrier mobility in these layers remains high, and the intercalate layer thickness is minimal. In the case of graphite intercalation with alkali metals (typical donors), the fraction of the charge transferred to the graphite layers is greater than that for intercalation with typical acceptors. However, the electron mobility in the graphite bounding layers is lower for the donor-type intercalants, because of the strong interaction between the graphite layers and the intercalate layer and, therefore, the intenser scattering of carrier.

The above analysis of the behaviour of σ_a in GIC is borne out by the results obtained from the experimental studies of the electrical conductivity, as well as galvanomagnetic and oscillation effects in acceptor compounds with ICl of various stage indices, carried out by Brandt et al. (1983). The dependences of σ_a, hole concentration n_h, hole mobility μ_h, and hole mean free path Λ_h upon stage index n obtained by these authors were illustrated in fig. 119c.

Let us briefly discuss the problem of the theoretical calculation of σ_a in GIC.

When calculating σ_a in donor GIC with relatively large n the rigid-band model and the concept of the two-dimensional electron (phonon) spectrum in the graphite layers are commonly used. It is assumed also that the s-bands of the donor intercalant lie much higher than the Fermi level in GIC. This was the basis on which Kamimura et al. (1980) calculated, in particular, the σ_a in graphite intercalated with alkali metals. Assuming that the dominant contribution to carrier relaxation comes from the processes of intravalley electron scattering, they have obtained a linear increase in resistivity with increasing temperature in the range below 200 K. For $T > 200$ K, the processes of intervalley scattering begin to play a significant role, which leads to deviations from the linear dependence of ρ_a on the temperature. Kamimura et al. (1980) have succeeded in calculating numerical values for electron mobilities in $C_{8n}K$ compound with $n = 3$ and 5 at room temperature, which are equal to ~400 and ~700 cm²/(V s), respectively. The obtained values are 20–25 times as low as the corresponding value for pure graphite under the same conditions. The estimated σ_a in the GIC $C_{8n}K$ with $n = 3$ and 5 agree well with the experimental data.

The calculations of σ_a in acceptor GIC based on the same assumptions (the two-dimensional electron spectrum of graphite and rigid bands) show a poorer agreement with experiment. Pietronero and Strässler (1979, 1980, 1981) and Pietronero et al. (1980a,b) have investigated GIC in which the Fermi level lies much lower than the region of band overlap in graphite, and holes are the only carriers available. In order to calculate the conductivity of a single graphite layer, they have used the following formula

$$\sigma = e^2 \nu(\varepsilon_F)\langle v_x^2(\varepsilon_F)\rangle \tau, \qquad (182)$$

where $\nu(\varepsilon_F)$ is the density of states at the Fermi level in the layer being considered, $\langle v_x^2(\varepsilon_F)\rangle$ is the mean-square Fermi velocity in electric field $\boldsymbol{E}$ directed along the x-axis, τ the carrier relaxation time. The value of ε_F for each layer is determined by the electronic density redistribution due to intercalation with the screening effect taken into account.

If we assume, as a first approximation, a linear law of carrier dispersion in a layer in the form given by (170), then the $\langle v_x^2(\varepsilon_F)\rangle$ can be given by

$$\langle v_x^2(\varepsilon_F)\rangle = \tfrac{9}{8}(\gamma_0 a_0)^2 \frac{1}{\hbar^2}. \tag{183}$$

The dependence of τ on the Fermi energy ε_F, temperature, and electron–phonon interaction constant $\langle g^2\rangle$ has the form

$$\tau = \frac{2}{\pi}\frac{\hbar}{\nu(\varepsilon_F)kT}\frac{(\gamma_0 a_0)^2}{\langle g^2\rangle}. \tag{184}$$

The conductivity of a graphite layer is then given by

$$\sigma = \frac{9}{4}\frac{e^2}{\pi\hbar kT}\frac{(\gamma_0^2 a_0^2)}{\langle g^2\rangle}. \tag{185}$$

Pietronero and Strässler (1980) have calculated, using TBA perturbation-theory techniques, the value of $\langle g^2\rangle$ for acceptor GIC and have estimated σ_a as $\sim 0.9 \times 10^5$ and $\sim 2.3 \times 10^5 \Omega^{-1}\,cm^{-1}$ for $n = 1$ and 2, respectively. Their estimates are much (5 to 7 times) lower than the experimentally observed σ_a in acceptor GIC with $n = 1$ and 2 (see table 21). This means that, under the assumptions they made, the effectiveness of hole scattering turned out to be higher than in reality. The largest discrepancy between the calculated and experimental σ_a values takes place for GIC in which the intercalant is a strong oxidizer (SbF_5, AsF_5, HNO_3). Thus the currently held theory of electrical conductivity σ_a in GIC only permits one to give a qualitative description of the linear part of the temperature dependence of the specific electrical resistivity $\rho_a = 1/\sigma_a$, but it is still very far from reaching the quantitative agreement with the experimental data for the all the graphite intercalation compounds investigated to date.

10.7.3. Electrical conductivity of GIC along the c-axis

From the conception of periodically occurring layers in a perfect GIC structure (see fig. 120) a simple phenomenological model follows which makes it possible to give a description of the charge transport along the *c*-axis. This model will be called the model of series-connected resistances. The electrical resistivity ρ_c of the structure along its *c*-axis must be equal to the sum of resistivities of individual layers within the period I_c

$$\rho_c = \sum_i \frac{\rho_{c_i} l_i}{I_c}, \tag{186}$$

where ρ_{c_i} is the resistivity of the *i*th layer, l_i being its thickness; the summation is over the number of the different layers of the system (the intercalate layer, the graphite bounding layer, the nearest-neighbour graphite layer, the graphite interior layer, etc.), which depends on the stage index of intercalation.

The experimental data suggest that the electrical conductivity $\sigma_c^{(g)}$ of graphite considerably increases when graphite is intercalated with donor impurities, the effect being best pronounced in GIC with $n = 1$. Thus in donor GIC C_8K and C_6Li the conductivity σ_c is ~230 and ~2170 times as high as $\sigma_c^{(g)}$ in pure graphite. Such a spectacular increase in σ_c suggests a very strong overlap between the p_z-orbitals in graphite and s-orbitals in alkali metals, as a result of which the donor impurities of the intercalate layer donate practically all their valence electrons to the graphite layers. Moreover, in donor GIC with $n = 1$ the direct overlap between the electron wave functions for the neighbouring intercalate layers appears to play an important role as well. This means that for donor GIC the effective resistivities ρ_{c_i}, of the graphite interior layers, as described by the phenomenological model of series-connected resistances given by (186), are much lower than for pure graphite. This may be due not only to the increased electron concentration in the layers, but also to the facilitated electron transport along the *c*-axis resulting from the larger value for the wave-function overlap integrals for the neighbouring layers.

On the other hand, in acceptor GIC one can observe a significant decrease of σ_c due to intercalation in comparison with $\sigma_c^{(g)}$ (see table 21). This decrease is associated with a low conductivity of the intercalate layer owing to the small overlap of the p_z-orbitals in graphite with the molecular orbitals of the intercalant. The intercalate layers then become only partially ionized, gaining electrons mainly from the adjacent graphite layers. As a result, a significant anisotropy of conductivity arises, reaching for typical strong acceptors values of 10^5–10^6.

10.7.4. Temperature dependence of the conductivities σ_a and σ_c in GIC

The high carrier concentration in GIC (~$10^{20}\,\mathrm{cm}^{-3}$) and the ensuing high temperature of the electron (hole) gas degeneracy, enable one to suppose that the major change in conductivity of GIC with varying temperature is caused by the change in mobility. In this respect, GIC differ from parent graphite, in which, because of the low Fermi energy value, the carrier concentration decreases by about a factor of 5 as the temperature is lowered from room to liquid-helium values.

As is suggested by the results of the measurements made by Zeller et al. (1979a,b) and by Brandt et al. (1982), for acceptor GIC the conductivity σ_a along the layers increases 3 to 4 times as the temperature is lowered from 300 K to 77 K, the increase being very slight as the temperature further decreases to the liquid-helium value (see, e.g., fig. 118). Much greater is the increase in σ_a in those donor GIC which were made by intercalating graphite with alkali metals. According to Potter et al. (1979), the σ_a in C_8K increases by an order of magnitude as the temperature falls from 300 to 77 K, but it further increases by about a factor of 20 with further cooling down to 4.2 K.

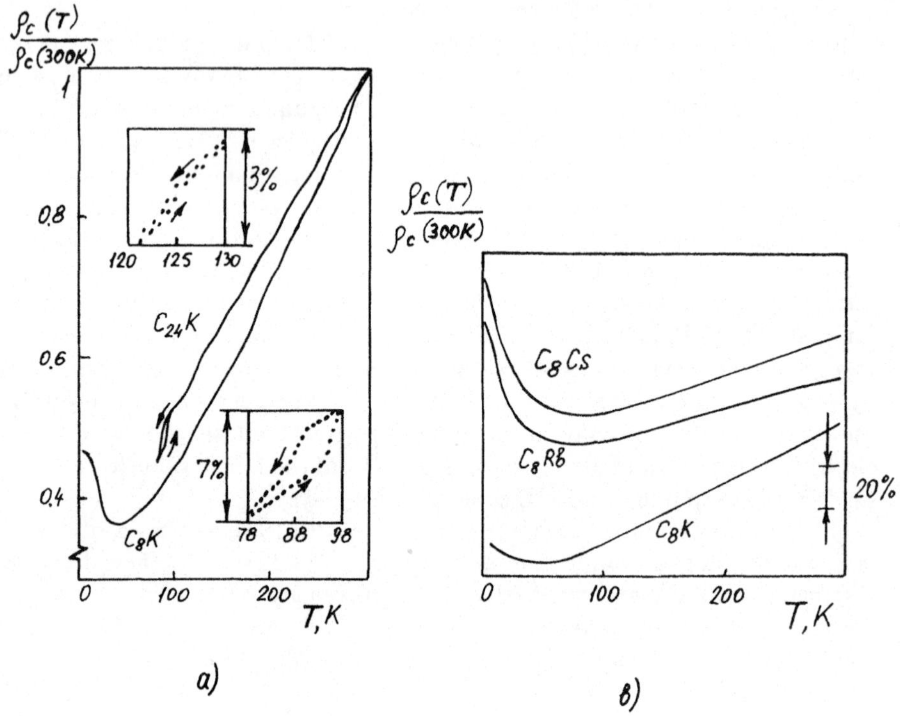

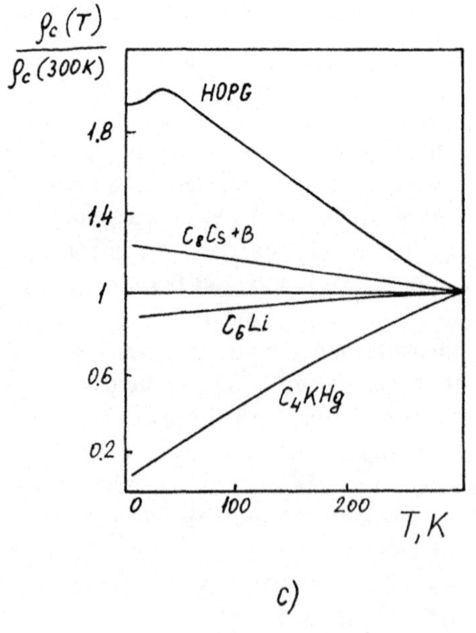

Fig. 121. (a) Normalized *c*-axis resistivity ρ_c versus temperature for C_8K and $C_{24}K$; the insets show the $C_{24}K$ anomalies on expanded scales; (b) Similar data for C_8Cs, C_8Rb and C_8K represent the range of results for HOPG-based samples; (c) Similar data for HOPG, C_8Cs prepared from HOPG containing 0.5%B, C_6Li, and C_4KHg. (Fischer et al. 1982.)

The temperature dependence of σ_a in C_8K in a wide temperature range may be described by the relation $\sigma_a^{-1} = A + BT + CT^2$, from which it follows that the dominant relaxation mechanisms are the electron–phonon scattering (the linear term) and the electron–electron scattering (the quadratic term). According to Zeller et al. (1979a), a similar dependence, though with much smaller coefficients, enables one to describe the weak change in σ_a with varying temperature observed in acceptor GIC.

Thus both donor and acceptor GIC show a typically metallic character of conductivity (σ_a) variation with temperature.

Temperature dependences of the conductivity σ_c in GIC are more complicated and, as a rule, qualitatively different from the similar dependences for pure single-crystal graphite or highly oriented pyrolytic graphite (see sect. 5.1). According to Fischer et al. (1982), the temperature dependences of σ_c in donor-type GIC, such as C_6Li, C_8K, C_8Rb, C_8Cs, obtained from both pure and boron-doped graphite, as well as in C_4KHg, all exhibit a metallic character, which for compounds of this kind at $T < 50 - 100$K is replaced by the dependence with a positive coefficient $d\sigma_c/dT$ (see fig. 121a,b,c).

The dependence with $d\sigma_c/dT > 0$ at low temperatures may in principle be associated with activation processes, and it indicates the possibility that in moving along the *c*-axis electrons may become localized. However, in donor GIC a semiconductor character of conductivity occurs at absolute values of $\sigma_c \sim 10^3 \Omega^{-1} cm^{-1}$ that are much greater than the value of the minimum metallic conductivity $\sigma_{min} = 0.026 e^2/\hbar a \sim 2 \times 10^2 \Omega^{-1} cm^{-1}$, which, according to Mott (1972), should determine the conductivity of a substance near the threshold of localization.

In acceptor GIC, the metallic character of the temperature dependence of σ_c in the range from liquid-helium to room temperatures is usually observed only in low-*n* compounds (see fig. 122 which shows the temperature curves for ρ_c for compounds with $SbCl_5$ at $n = 1$, 2, and 3). In high-*n* GIC at low temperatures, the *T*-dependences of σ_c display some parts which have an activation character (fig. 123). The salient points in the curves presented in figs. 122 and 123 at $T \simeq 210$ K correspond to a structural phase transition i.e., an "order–disorder" transition, in the intercalate layer (Clarke et al. 1982). The most striking peculiarity of the $\rho_c(T)$ dependences presented in fig. 122 is that they all have a metallic character over the entire temperature range under study, whereas the values of σ_c range from 1 to $10 \Omega^{-1}$ cm^{-1} which is one or two orders of magnitude lower than the minimum metallic conductivity in such systems. According to the currently held views, such low σ_c values would rather be suggestive of an activation character of the temperature dependence, indicating the effect of carrier localization. The co-existence of the metallic-type temperature dependence of σ_c with so low absolute values of σ_c ($\sigma_c \ll \sigma_{min}$) in acceptor GIC with small *n* would seem inexplicable if it is assumed that such a system is perfectly homogeneous in the layer plane. This co-existence may,

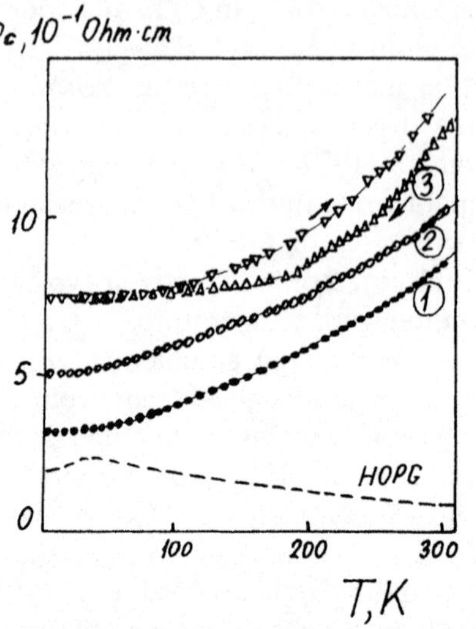

Fig. 122. The c-axis resistivity ρ_c of C_xSbCl_5 prepared from HOPG. Stage index is given by the encircled number. Thermal cycling of a third-stage compound is indicated by arrows. The c-axis resistivity of pure HOPG is shown by the dashed curve. (Uher and Morelli 1982.)

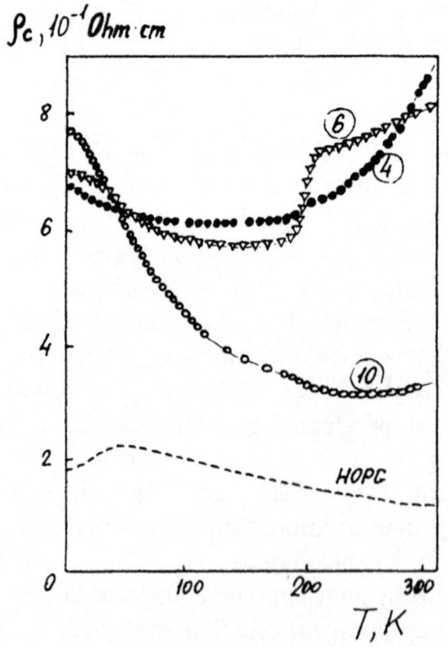

Fig. 123. The c-axis resistivity ρ_c of higher-stage C_xSbCl_5 prepared from HOPG. The dashed curve shows the behaviour of the resistivity of pure HOPG. Stage index is given by the encircled number. (Uher and Morelli 1982.)

however, be explained assuming that the motion of carriers along the *c*-axis takes place through rather narrow conducting channels formed by the defects introduced in the host material during intercalation. In the presence of such channels with a metallic-type conductivity the small value of σ_c, which is determined as the average conductivity for the whole layer, can be explained by the small total area of the cross section of the conducting channels. Similar explanations have been advanced by Uher and Morelli (1982). The possibility that such conducting channels along the *c*-axis may indeed be formed is suggested by the electron-microscope data on the structure of low-*n* GIC, which have been obtained by Daumas and Hérold (1969) and by Hardcastle et al. (1982). The hypothesis that there appear metallic-type conducting channels through which charges are transported along the *c*-axis is also supported by the experimentally observed increase in conductivity σ_c with decreasing stage index. Such an increase in σ_c may be associated with the increased number of defects with increasing intercalant concentration, which in turn leads to an increase in the number (and total area) of the conducting channels (see Fischer 1980).

Of an entirely different character are the temperature dependences of σ_c in acceptor-type high-*n* GIC, which are presented in fig. 123. Below a definite temperature (rising with increasing *n*) these dependences change from a metallic type to an activational type. This change in the character of conductivity appears to be brought about by a metal–dielectric transition due to the Anderson electron localization taking place as the Fermi level crosses the mobility threshold (see Uher and Morelli 1982). To conclude, at present there are some indications that in moving along the *c*-axis in acceptor GIC electrons may become localized and, therefore, at low temperatures, the electrical conductivity σ_c may have a hopping character.

We should stress, however, that the relatively scarce data on the temperature dependences of σ_c in various GIC obtained so far preclude the final conclusion about the character of conductivity along the *c*-axis. It is also still unclear to what extent the conductivity σ_c in acceptor GIC may be described by the diffusion relaxation mechanism proposed by Khanna et al. (1978) and Lauginie et al. (1980).

10.8. Galvanomagnetic effects in GIC

Unlike pure graphite, which is a compensated semimetal with equal concentrations of electrons and holes ($n_e = n_h$), graphite intercalation compounds represent metals with a high concentration of charge carriers.

As has been mentioned above, in GIC with $n \gg 1$ and with any intercalant, the Fermi level in the graphite interior layers never goes beyond the region of overlap of the π-bands in pure graphite, which entails the presence of both positive and negative charge carriers. According to the number of carriers and the type of their group, GIC with large n should be essentially similar to graphite doped with donor or acceptor impurities, whose spectrum can rather well be described by the rigid-band model. With decreasing n the Fermi level in GIC goes beyond the π-band overlap region in the spectrum of the host material. Under these conditions, one or several groups of electrons, in donor GIC, or holes, in acceptor GIC, may serve as carriers.

The Fermi surface in low-n GIC may, however, have a very intricate structure, in the first place because the Fermi level crosses several bands and, secondly, because there arise new periodicities along the c-axis (superlattice structure) and in the basal plane (zone-folding effect), which give rise to the splitting of the original surface into a number of sheets. Both these effects are, in turn, responsible for the emergence of several groups of charge carriers. Furthermore, the GIC energy spectrum is greatly complicated due to the nonuniformity of the charge distribution along the c-axis. From the above considerations one would expect that the galvanomagnetic effects in GIC would be rather complex in character, and will depend both on intercalant and stage index.

But in spite of the fact that in GIC there are several groups of carriers and some part of the surfaces of constant energy may be classified as open (see sect. 10.6), in most cases, the behaviour of galvanomagnetic effects in weak magnetic fields may adequately be described on a qualitative level by a very simple model based on the conception of a single dominant group of carriers.

The Hall effect and the transverse magnetoresistance in both donor and acceptor GIC were measured in a number of experiments. It has been found that in all GIC the sign of the Hall coefficient strictly corresponds to the intercalant being of the donor or acceptor type. This is true even of GIC with very large n, in which the increase in total carrier concentration does not exceed 20–30%, as compared with that for pure graphite. Another feature of the Hall effect in GIC is a very weak temperature dependence of R_H in a wide (from liquid-helium to room) temperature range. This is naturally associated with a high carrier concentration in GIC, which at low n reaches $\sim 10^{22}\,\mathrm{cm}^{-3}$.

The transverse magnetoresistance, $\Delta\rho_\perp(H)/\rho(0)$, in GIC in the weak-field range mostly shows a quadratic dependence upon magnetic field both for $\boldsymbol{H}\|c$ and for $\boldsymbol{H}\perp c$. This enables one to determine the average mobility, $\bar{\mu}$, which is related to the magnetoresistance coefficient $\Delta\rho_\perp(H)/[\rho(0)H^2]$ by the following expression:

$$\frac{\bar{\mu}}{c} = \left(\frac{\Delta\rho_\perp(H)}{\rho(0)\,H^2}\right)^{1/2}, \tag{187}$$

and which is, along with the Hall carrier concentration $n_H = (ec/R_H)^{-1}$, a rough characteristic for the system of carriers in graphite intercalation compounds. Using the thus found value of $\bar{\mu}$ and the measured value of σ_a at $H = 0$, one can estimate the concentration, $n_{e,h}$, of the dominant carrier group

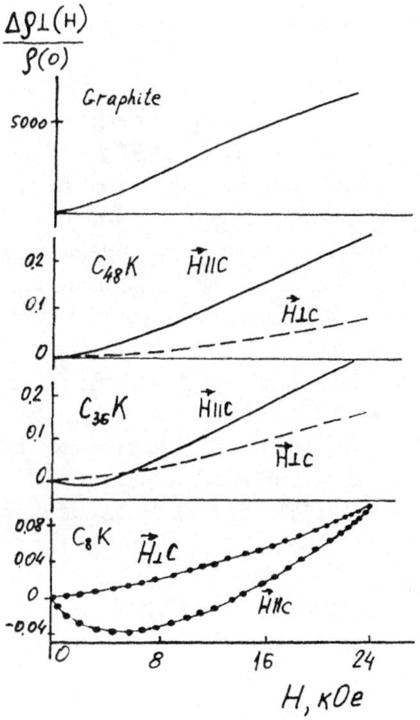

Fig. 124. Transverse magnetoresistance of graphite–potassium compounds at $T = 1.2\,\mathrm{K}$. (Tanuma et al. 1978.)

in GIC

$$n_{e,h} = \frac{\sigma_a}{e\bar{\mu}} . \tag{188}$$

It is of interest to note that in some donor-type GIC one can observe a negative magnetoresistance (NM) at low temperatures in the initial part of the field dependence of magnetoresistance (in fields of several kOe). This effect is most pronounced in compounds with K for $n = 1$. The character of magnetoresistance and the NM-effect in potassium-intercalated compounds with various n are illustrated in fig. 124. The appearance of the NM in single crystals with a high carrier concentration and with open surfaces of constant energy cannot be explained from the classical point of view. It seems, therefore, that the most likely cause of the appearance of the NM in GIC is the size-effect, which is best pronounced in the case of the dominant mechanism of carrier scattering by the boundaries of the constituent crystallites, whose size turns out to be comparable with that of the orbits of several groups of carriers in the magnetic field. Similar effects were previously observed in synthetic carbons (see Zorin 1979).

In sufficiently high magnetic fields ($H > 20$–$30\,\mathrm{kOe}$) and in the low-temperature range the transverse magnetoresistance in GIC depends linearly on the magnetic field both for $\boldsymbol{H} \| c$ and $\boldsymbol{H} \perp c$. The character of change of $\Delta\rho_{\perp}(H)/\rho(0)$ in high magnetic fields is illustrated by the field dependence of the transverse magnetoresistance in C_8K for H up to 15T, which is shown in fig. 125. As n increases, the transition to the linear dependence of $\Delta\rho_{\perp}(H)/\rho(0)$ takes place in increasingly lower fields, so that at sufficiently large n one observes a linear dependence typical of intercalated graphite in the field range from 5 to 150kOe (see fig. 35, sect. 5.4). The linear increase in resistance of graphite in a transverse magnetic field has been explained by McClure and Spry (1968) on the basis of the assumption that the principal mechanism of carrier relaxation is the scattering by charged impurity centres, whose screening radius is a function of the magnetic field. Evidently, a similar situation may take place in GIC in high magnetic fields. However, one should bear in mind that, because of the large volume of the Fermi surface in GIC, the field range $H \lesssim 150\,\mathrm{kOe}$ does not correspond to the extreme quantum limit, as is the case with pure graphite. Accordingly, the calculation of McClure and Spry (1968) may not be directly used for a quantitative interpretation of the H-dependence of $\Delta\rho_{\perp}(H)/\rho(0)$ in GIC.

The magnetoresistance coefficient $\Delta\rho_{\perp}(H)/[\rho(0)H^2]$ and the Hall coefficient

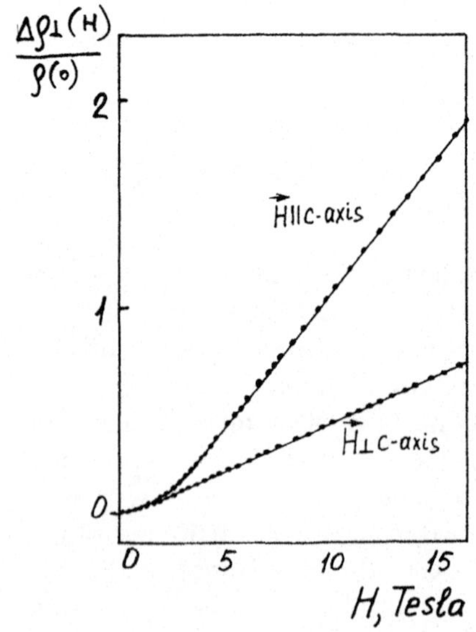

Fig. 125. Transverse magnetoresistance of C_8K at $T = 1.26\,\mathrm{K}$ for high-magnetic fields up to 15T. (Tanuma et al. 1978.)

Table 22
Parameters of donor and acceptor GIC found from galvanomagnetic measurements.

Compound	T (K)	$\bar{\mu}$ [$cm^2/(V\,s)$]	$n_{e,h}$ (cm^{-3})	$n_H^{e,h}$ (cm^{-3})	Ref.*
$C_{24}K$	4.2	7500	3.3×10^{21}	3.4×10^{21}	[1]
	77	1800	3.9×10^{21}	5.1×10^{21}	[1]
	300	250	4.3×10^{21}	7.1×10^{21}	[1]
$C_{24}K$	4.2	4980	5.12×10^{20}	9.86×10^{20}	[2]
	77	1780	6.48×10^{20}	1.27×10^{21}	[2]
	268	460	5.98×10^{20}	2.24×10^{21}	[2]
$C_{36}K$	4.2	7600	5.5×10^{21}	3.4×10^{21}	[1]
	77	2500	6.5×10^{21}	5.1×10^{21}	[1]
	300	466	4.5×10^{21}	7.1×10^{21}	[1]
$C_{48}K$	4.2	7700	2×10^{21}	1.1×10^{21}	[1]
	77	3700	1.5×10^{21}	1.1×10^{21}	[1]
	300	2300	6.0×10^{20}	7.8×10^{20}	[1]
$C_{24}Cs$	4.2	3770	6.2×10^{21}	3.3×10^{21}	[1]
	77	1300	4.9×10^{21}	3.3×10^{21}	[1]
	300	268	5.1×10^{21}	7.4×10^{21}	[1]
$C_{36}Cs$	4.2	3300	1.8×10^{21}	1.8×10^{21}	[3]
	77	1800	1.6×10^{21}	1.7×10^{21}	[3]
	300	1380	5.3×10^{21}	1.2×10^{21}	[3]
C_8AsF_5	300	1980	1.26×10^{21}	–	[4]
$C_{16}AsF_5$	300	3200	1.22×10^{21}	–	[4]
C_6HNO_3	300	1750	1.02×10^{21}	–	[4]
$C_{6.6}FeCl_3$	300	920	7.1×10^{20}	–	[4]
$C_{13.2}FeCl_3$	300	1630	8.9×10^{20}		[4]

*References
[1] Onn et al. (1979).
[2] Suematsu et al. (1980)
[3] Guérard et al. (1977).
[4] Vogel (1979).

R_H for donor and acceptor GIC with low n in weak magnetic fields have been calculated by Sugihara (1982) who used the approximation of the isotropic relaxation time τ. In his calculation, Sugihara assumes that the electron (hole) Fermi energy in GIC obeys the inequality $\varepsilon_F^2 \gg \frac{1}{4}\gamma_1^2$, where γ_1 is the SWMC-model parameter.

The expressions for $\Delta\rho_\perp(H)/[\rho(0)H^2]$ and R_H, which have been obtained in this way with the trigonal warping taken into account, have the form

$$\lim_{H\to 0} \frac{\Delta\rho_\perp(H)}{\rho(0)H^2} = \left(\frac{\bar{\mu}}{c}\right)^2 = \left(\frac{e^2}{\hbar^2 c_0}\right)^2 \left(\frac{\tau}{\varepsilon_F}\right)\left(\tfrac{3}{2}\gamma_0 a_0\right)^4 \left[\left(\frac{\gamma_1}{2\varepsilon_F}\right)^2 + \frac{57}{2}\left(\frac{\gamma_3}{\gamma_0}\right)^2\right], \quad (189)$$

$$\lim_{H\to 0} R_{\rm H} = \pi I_{\rm c} \frac{(\frac{3}{2}\gamma_0 a_0)^2}{2ec\varepsilon_{\rm F}[1 - \frac{21}{2}(\gamma_3/\gamma_0)^2]}. \tag{190}$$

The estimation made by formulae (189) and (190) using the γ_1 parameter value corresponding to pure graphite and $|\varepsilon_{\rm F}| \sim 1\,{\rm eV}$, which is typical of low-n GIC, shows that the trigonal warping affects relatively weakly the Hall-coefficient values (the contribution of the term with γ_3 does not exceed 5%), whereas its influence on the magnetoresistance value is decisive. We should also note that expression (189) provides a fairly reasonable estimation of the average mobility, which is quite consistent with the values for $\bar{\mu}$ in low-n GIC observed in galvanomagnetic measurements made in weak magnetic fields.

The values of the parameters $\bar{\mu}$, $n_{\rm e,h}$, $n_{\rm H}^{\rm e,h}$ for some donor- and acceptor-type GIC calculated by eqs. (187) and (188) at various temperatures are given in table 22.

10.9. Oscillation effects in GIC

10.9.1. General remarks on oscillation phenomena in GIC

Considerable advances in GIC technology, made over the last decade or so, have led to much longer lifetimes τ for carriers in these materials at low temperatures, thus offering the possibility of using quantum oscillation effects as tools for investigating the GIC energy spectrum.

The quantum oscillations of magnetoresistance (the ShdH effect) in GIC were first observed by Bender and Young (1971, 1972) in the C_xBr_2 system. Their observations have shown that it is possible to synthesize GIC perfect enough to ensure that the condition for observation of quantum oscillations ($\omega_{\rm c}\tau > 1$) be fulfilled in the liquid-helium temperature range, thereby stimulating studies of GIC using quantum oscillation effects. Among them, the de Haas–van Alphen effect (dHvA) is the most convenient for observation, because its detection does not require that specimens be provided with electrical contacts. Figure 126 shows, as an example, dHvA oscillations for some donor GIC with potassium. ShdH oscillations in acceptor GIC are illustrated in figs. 127, 128.

A feature that quantum oscillations in GIC for $\boldsymbol{H}\|c$ have in common is the presence of frequencies, F, which are more than an order of magnitude higher than those corresponding to majority carriers in pure graphite. Irrespective of the intercalant species used, these high oscillation frequencies typical of GIC for $\boldsymbol{H}\|c$ occur in the range from several tenths of an Oersted to several tens of millions of Oe, the highest frequencies (to 30×10^6 Oe) being observed in GIC with a low stage index ($n = 1, 2, 3$).

Along with the high oscillation frequencies, representative of GIC of high

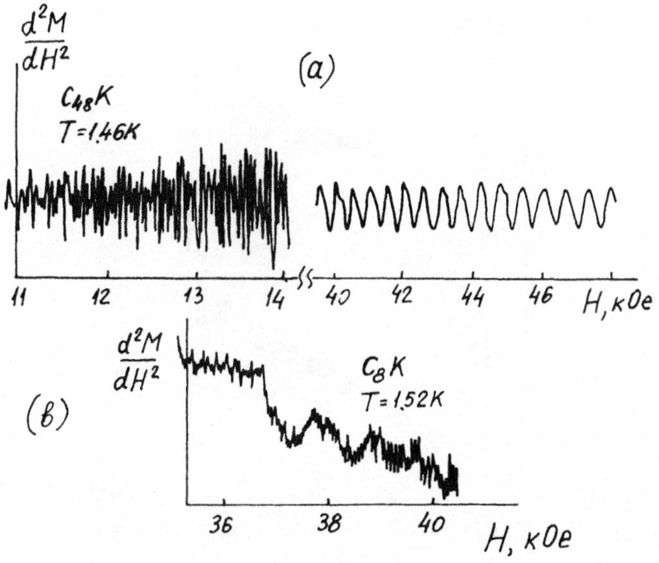

Fig. 126. De Haas–van Alphen oscillation curves for: (a) $C_{48}K$, and (b) C_8K ($H \| c$). For $C_{48}K$, the oscillations in the lower and intermediate fields are "abbreviated". (Tanuma et al. 1978.)

stage index are the oscillations of 0.046×10^6 and 0.064×10^6 Oe, which coincide with those for majority holes and electrons in pure graphite.

The high oscillation frequencies in GIC correspond to large cross-sectional values for the Fermi surface for electrons or holes arising due to the charge

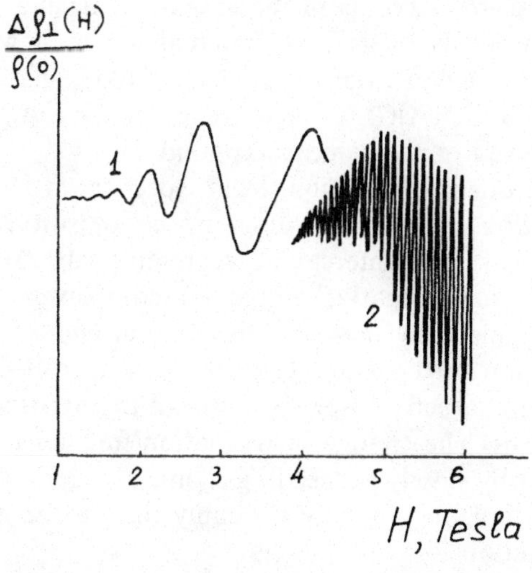

Fig. 127. The oscillating part of the transverse magnetoresistance versus magnetic field for GIC samples at $T = 4.2$ K (1, $C_{9.3}AlCl_{3.4}$; 2, $C_{9.5}AlCl_3Br_{0.6}$). (Brandt et al. 1984a.)

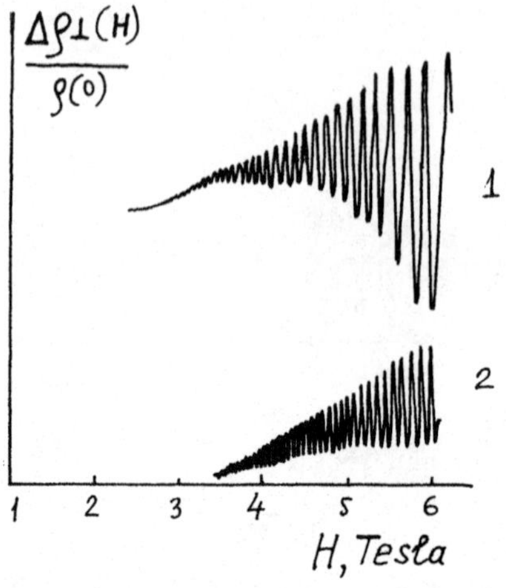

Fig. 128. The oscillating part of the transverse magnetoresistance versus magnetic field for $C_{9.8}CuCl_2$ (1) and $C_{27.5}ICl_3$ (2) at $T = 4.2$ K. (Brandt et al. 1984a.)

transfer from the intercalate layers to the graphite bounding layers.

On the other hand, the existence of the low oscillation frequencies coinciding with those due to electrons and holes in pure graphite may be ascribed to carriers found in the graphite layers sufficiently remote from the intercalate layers and, therefore, screened by the graphite bounding layers.

Another common feature of oscillations in GIC is a marked change in frequency from specimen to specimen and also a change in frequency with time. These phenomena have been observed by Brandt et al. (1984b) in studying oscillation effects in graphite intercalated with weak acceptors. In particular, a great spread of frequencies (up to 400%) has been observed in various specimens of $C_{9.3}AlCl_{3.3}$. A decrease of oscillation frequency with time has been observed in the same compound.

The observed frequency variation from specimen to specimen may be attributed to variations of the stoichiometric composition of the intercalate layers. In $C_{9.3}AlCl_{3.3}$, the intercalant represents an assembly of neutral molecules of $AlCl_3$ and negatively charged complexes $AlCl_4^-$, which are responsible for the π-electron density redistribution and which thus determine the hole concentration in the graphite layers. Uncontrolled variations in the concentration of the charged complexes give rise to variations in the observed oscillation frequencies. The change in oscillation frequencies with time seems to be brought about by slowly proceeding chemical reactions in the intercalate layer. Thus, for example, it is highly probably that in the intercalate layer of $C_{9.3}AlCl_{3.3}$ the reaction

$$2\ AlCl_4^- \rightarrow 2\ AlCl_3 + Cl_2 + 2e^-,$$

takes place resulting in excessive chlorine escaping from the compound, which has been confirmed by chemical assays. As a consequence, there appear additional electrons and, because of this, the hole number decreases. Another process which may lead to a decrease of the number of charged centres is a partial hydrolysis of GIC containing $AlCl_3$ in air.

Similar processes leading to the loss of excessive charges in the graphite layers may occur also in other GIC.

One should mention, however, that the mechanism of the donor or acceptor action of the intercalant inserted into graphite has not yet been completely elucidated. In particular, the experimentally observed significant change in carrier concentration in GIC at a comparatively insignificant change in intercalant composition still remains unexplained. This effect is typical of acceptor compounds. Thus, in substituting bromine for excessive chlorine in $C_{9.3}AlCl_{3.4}$ (resulting, in particular, in the formation of $C_{9.5}AlCl_3Br_{0.6}$) the frequency of the ShdH oscillations (see fig. 127) and, therefore, the hole concentration increase by more than a factor of 30, in spite of the fact that the oxidizing power of free bromine is lower than that of chlorine. In order to account for the observed change in hole concentration in $C_{9.3}AlCl_{3.4}$, there has been advanced the hypothesis that the complexes of the $AlCl_3Cl^-$ and $AlCl_3Br^-$ type, which form the molecules of the implanted matter and excessive halogen, take part in the charge redistribution between the carbon atoms and the intercalant.

10.9.2. Fermi-surface structure in GIC

Let us find out just to what extent the Fermi surface in GIC calculated from first principles or in a TBA model (e.g., Blinkowski's), or in the rigid-band model (see sect. 10.6) agrees with the experimentally observed oscillation frequencies. Such a comparison is possible only for some compounds, for which, on the one hand, detailed calculations have been made and, on the other hand, rather detailed data derived from oscillation-effect measurements are available. One of these compounds is C_8K, a donor GIC of stage index one, whose Fermi surface has been calculated from first principles by Inoshita et al. (1977) (see fig. 112). According to the experimental data of Tanuma et al. (1978, 1983), for $\boldsymbol{H} \| c$ in this compound one can observe monochromatic dHvA oscillations with a frequency $F = 28.75 \times 10^6$ Oe, which is more than two orders of magnitude higher than the oscillation frequencies corresponding to majority carriers in pure graphite (0.046×10^6 and 0.064×10^6 Oe). The value of the observed oscillation frequency in C_8K is in good agreement with the calculated frequency $F = 30.3 \times 10^6$ Oe, which corresponds to the cross section of a cylindrical Fermi surface surrounding the vertical Brillouin-zone edges (see fig. 112). Such a good agreement implies that in the specimens investigated by Tanuma et al. (1978, 1983) the intercalant atoms do not form an ordered structure, which would give rise to an additional periodicity in the basal plane and to the splitting of the Fermi surface into separate sheets (or zone-folding).

At the same time, the calculation of the energy spectrum of C_8K made by Inoshita et al. (1977) predicts that there should be Fermi-surface parts near the points A and A' (closed lens-shaped surfaces near the centres of the upper and lower-Brillouin-zone edges shown in fig. 112). The oscillation frequencies much lower than $F = 28.75 \times 10^6$ Oe which correspond to these parts of the Fermi surface have not yet been observed in experiment, so that it is still obscure whether this is due to incorrectly made calculations or to insufficiently perfect C_8K specimens, precluding observation of oscillations from the groups of carriers of a relatively low concentration.

The extent of correspondence between the oscillation data and the results obtained within the rigid-band model for donor compounds may be ascertained on the example of $C_{8n}K$ (a GIC with $n = 4$), whose Fermi-surface structure has already been discussed (see sect. 10.6), with the new periodicity (I_c) taken into account. It will be recalled that the rigid-band model taking account of the superlattice structure along the c-axis predicts for this compound five different extremal cross sections for the Fermi surface consisting of open surfaces of the "warped cylinder" type and of closed lens-shaped surfaces (see fig. 116). Tanuma et al. (1978) and Higuchi et al. (1980), who studied dHvA oscillations in $C_{8n}K$ with $n = 4$, actually observed five different frequencies F_i ($i = 1, 2, \ldots, 5$) of oscillations, whose values, in units of 10^6 Oe, are 1.21, 1.44, 2.38, 2.64, and 3.38, respectively. However, these values coincide with the results of rigid-band model calculations only in order of magnitude, which may indicate that the agreement between theory and experiment is merely qualitative. Tanuma et al. (1978) and Higuchi et al. (1980) have also investigated the character of dHvA oscillations in $C_{8n}K$ with $n = 3$ and have found that there are two oscillation frequencies present, viz., $F_1 = 1.46 \times 10^6$ Oe and $F_2 = 2.6 \times 10^6$ Oe. For the given case, the rigid-band model predicts three extermal cross sections, to which there correspond three oscillation frequencies. The experimentally observed frequencies are of the same order of magnitude as two out of the three theoretical frequencies. Oscillations with the third calculated frequency have not yet been observed. Here, as in the case of C_8K with $n = 1$, there is no complete clarity regarding the failure to observe these oscillations. The absence of a third frequency in the experimental curves may be due either to the insufficiently perfect specimens used or (which is very likely) to the inadequacy of the rigid-band model for calculating the spectrum of donor GIC with $n = 3$. One may attempt to use the equations of Blinowksi's TBA model to describe the spectrum. If one refers the experimentally observed oscillation frequencies to the extremal cross sections of the constant-energy surfaces associated with carriers in the two lower conduction bands (see fig. 110) and assumes the values for $\gamma_1 = 0.337$ eV and $\gamma_0 \sim 3.0\,\text{eV}$ (see Blinowski and Rigaux 1980), then one can show that the Fermi energy ε_F is about 0.71 eV and the parameter $\delta \ll \gamma_1, \gamma_0, \varepsilon_F$. The bottom of a third (upper) conduction band is at an energy $\varepsilon_3^c|_{k=0} \simeq \sqrt{2}\gamma_1 \simeq 0.533\,\text{eV}$ and therefore, this band also contains charge carriers, to which there corresponds the extremal cross section $S_3 \simeq 38 \times 10^{-42}\,\text{g}^2\,\text{cm}^2/\text{s}^2$ and the oscillation frequency $F_3 \simeq 0.357 \times \text{Oe}$. One can see that F_3 is approximately four times as low as the lowest frequency observed ($F_1 = 1.46 \times 10^6\,\text{Oe}$). It would be plausible to attribute the nonobservation of oscillations with so low a frequency as F_3 to the imperfection of the explored specimens of $C_{8n}K$ with $n = 3$.

Let us now consider oscillation effects in acceptor GIC. By now, these effects have been studied experimentally in a fairly large number of compounds of various stage indices with both weak and strong acceptors.

In acceptor compounds of high n one can, as a rule, observe oscillations representing the superposition of several frequencies. Thus, according to the data of Woollam et al. (1979), Batallan et al. (1980), and Iye et al. (1980), in compounds with the weak acceptors $FeCl_3$ and $PbCl_2$ with $n = 2, 4, 6$, and 11, one observes three oscillation frequencies, whose magnitude depends on the stage index. In compounds with the strong acceptor AsF_5, which were investigated by Iye et al. (1980), the spectrum of the observed frequencies is broader: whereas for $n = 2$ one observes, as the Blinowski model predicts, two-frequency oscillations, for $n = 3$ and 4 the number of the frequencies increases to 7 and 8, respectively. The number of the frequencies observed in compounds with AsF_5 for $n = 3$ agrees with the Fermi surface, as constructed by the rigid-band model (see sect. 10.6).

A detailed study of the energy spectrum and band parameters of acceptor GIC has been made by Brandt et al. (1984a,b, 1986) on the basis of oscillation measurements. These authors have explored oscillations in magnetic fields up to 65 kOe in weak-acceptor compounds with n from 1 to 4, such as $C_{9.3}AlCl_{3.3}$; $C_{9.3}AlCl_3Br_{0.6}$; $C_{16.6}ICl$; $C_{28}ICl$; $C_{9.8}CuCl_2$; $C_{10}CuCl_3$; $C_{24.5}SbCl_5$; $C_{16.2}Br_2$; $C_{32.8}ICl_{1.06}$; $C_{48.2}ICl_{1.06}$ and have determined the extremal cross sections S^h_{extr} for the hole surface and the cyclotron masses m^h_c. These data are given for several compounds with $n = 1$ and 2 in table 23, which also presents the Hall-coefficient values R_H, the Hall concentrations, $n^h_H = 1/ecR_H$, and Fermi-hole energies ε^h_F calculated using the Blinowski model (see below).

Let us examine in more detail the character of the oscillation dependences in acceptor GIC and establish the correspondence between these data and the models for the energy spectrum.

In acceptor compounds with $n = 1$, such as $C_{9.3}AlCl_{3.3}$ and $C_{9.5}AlCl_3Br_{0.6}$, oscillations are monochromatic, their amplitude in fields $H \simeq 60\,\text{kOe}$ being comparable to the magnitude of the steady part of the transverse magnetoresistance (see fig. 127). The presence of only one fre-

Table 23
Parameters of GIC ($n = 1$ and 2) found from oscillation and galvanomagnetic measurements.

Compound	Stage index n	S^{h}_{extr} (10^{-42} g^2 cm^2/s^2)	m^{h}_{c}/m_0	ε^{h}_{F} (eV)	R_H (10^{-1} cm^3/Coulomb)	n^{h}_{H} (10^{20} cm^{-3})
$C_{9.3}AlCl_{3.3}$	1	11.6 ± 0.2	0.056 ± 0.003	−0.12	5.0 ± 0.5	0.11
$C_{9.5}AlCl_3Br_{0.6}$	1	363 ± 3	—	−0.69	0.17 ± 0.01	3.6
$C_{16.6}ICl_{1.1}$	2	302 ± 3	0.132 ± 0.007	−0.36	0.24 ± 0.02	2.6
$C_{27.5}ICl_3$	2	319 ± 3	0.148 ± 0.008	−0.34	0.22 ± 0.02	2.8
$C_{9.8}CuCl_2$	2	180 ± 3	0.091 ± 0.005	−0.25	0.47 ± 0.05	1.3

quency in these compounds is in accordance with the Blinowski model for the energy spectrum of GIC with $n = 1$, which describes their Fermi surface as consisting of six identical cylinders surrounding the side Brillouin-zone edges. In the investigated compounds $C_{9.3}AlCl_{3.3}$ and $C_{9.5}AlCl_3Br_{0.6}$ with $n=1$, the carrier concentration $n_h^{(osc)}$ found from the measured oscillation frequency by eq. (171) is in excellent agreement with the concentration n_H^h determined from Hall measurements. Thus, for $C_{9.3}AlCl_{3.3}$, these values are as follows:

$$n_h^{(osc)}=0.11 \times 10^{20}\,\mathrm{cm}^{-3} \quad \text{and} \quad n_H^h = 0.12 \times 10^{20}\,\mathrm{cm}^{-3}.$$

The Fermi-energy values listed in table 23 were calculated using the expression

$$\varepsilon_F^h = -\tfrac{3}{2}\gamma_0 a_0 \left(\frac{S_{extr}^h}{\pi\hbar^2}\right)^{1/2}. \tag{191}$$

which follows from the dispersion relation given by (170) for GIC with $n = 1$. In these calculations, use was made of the value for pure graphite, which was taken to be equal to 3.2 eV [see Blinowski and Rigaux (1980), and also table 5 in ch. 4].

We note that the Blinowski model does not allow a qualitatively accurate description of all carrier parameters in acceptor GIC with $n = 1$. Thus, the cyclotron hole masses calculated within this model by formula (172) at reasonable γ_0 values are more than twice as high as the experimental values for m_c^h given in table 23.

In acceptor compounds with $n = 2$, such as $C_{16.6}ICl_{1.1}$, $C_{9.8}CuCl_2$, $C_{27.5}ICl$, ShdH oscillations are monochromatic (see, e.g., fig. 128). The values for the extremal hole cross sections S_{extr}^h and the cyclotron hole masses m_c^h found from the oscillation measurements are listed in table 23. The single-frequency oscillations in weak-acceptor compounds with $n = 2$ are indicative of the filling of only one valence band (see fig. 109). This is confirmed by the Fermi hole energy $\varepsilon_{F(v_1)}^h$ in the upper valence band ε_{v_1} estimated within the Blinowski model by formula

$$\varepsilon_{F(v_1)}^h = -\frac{1}{2}\left[-\gamma_1 + \left(\gamma_1^2 + 9a_0^2\gamma_0^2\frac{S_{extr(v_1)}^h}{\pi\hbar^2}\right)^{1/2}\right], \tag{192}$$

which follows from eq. (173) for the energy spectrum. In the above estimation, the parameter values $\gamma_1 = 0.377$eV and $\gamma_0 = 3.2$eV were used (see Blinowski and Rigaux 1980). The estimated absolute values for $\varepsilon_{F(v_1)}^h$ (see table 23) are smaller than the energy gap between the upper edges of the first and second valence bands, which (at $\Delta = 0$) is about equal to the γ_1 value, i.e., 0.377 eV. In this case, the second valence band is unfilled. At least it may be stated that if the above estimate of $\varepsilon_{F(v_1)}^h$ is not accurate enough to enable one to conclude that $|\varepsilon_{F(v_1)}^h| < \gamma_1$ for certain, then from the coincidence of $n_h^{(osc)}$ with n_H^h it follows that the hole concentration in the second valence band must be much lower

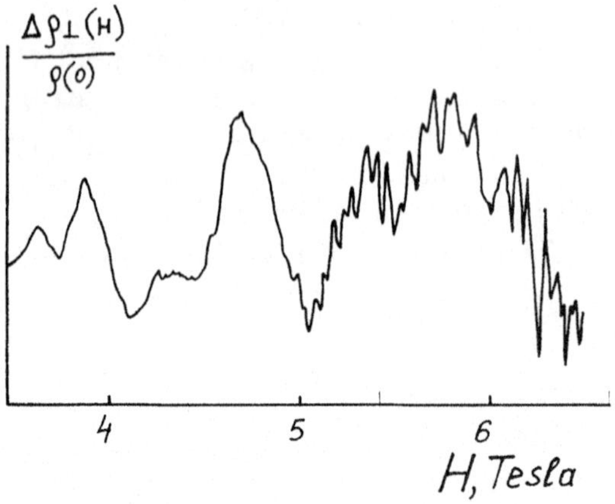

Fig. 129. The oscillating part of the transverse magnetoresistance versus magnetic field for $C_{24.5}SbCl_5$ at $T = 4.2$ K. (Brandt et al. 1984a.)

than in the first, so that the filling of the second band may be neglected.

The presence of single-frequency oscillations in the above-mentioned acceptor compounds with $n = 1$ and 2 makes it possible also to suppose that there is no zone folding in them, and therefore, at liquid-helium temperatures the intercalant molecules are in a disordered state.

Much more complicated oscillation curves, representing the superposition of several frequencies (see, e.g., fig. 129), are observed in the acceptor compounds $C_{24.5}SbCl_5$ and $C_{16.2}Br_2$ ($n = 2$), which have also been investigated by Brandt et al. (1984a). From the X-ray crystal analysis of these compounds made by Eeles and Turnbull (1965) and by Melin and Hérold (1975) it is known that at low temperatures the molecules of Br_2 and $SbCl_2$ in the intercalate layer form a flat lattice commensurate with the graphite lattice. The ordering of the intercalant leads to zone folding, which, in turn, makes possible the crossing of various parts of the Fermi surface and the emergence of complicated combinative orbits of carriers in the magnetic field (see Batallan et al. 1978, Simon et al. 1981). One may suppose that the complex multifrequency spectrum of oscillation curves in the compounds $C_{24.5}SbCl_5$ and $C_{16.2}Br_2$ is associated with the appearance of a new periodicity in the layer plane resulting from the ordering of the intercalant molecules.

The multifrequency oscillation spectrum due to zone folding was observed also in GIC with strong acceptors. Thus, Batallan et al. (1978), Rosemann et al. (1980), and Dresselhaus et al. (1980) observed combinative frequencies in $C_{8n}HNO_3$ of $n = 2$, 3, and 4. The values of the observed frequencies are given in table 24.

It has already been mentioned that in acceptor GIC with ICl for $n = 2$ the intercalant fails to form an ordered structure at liquid-helium temperature and

Table 24
Oscillation frequencies observed in $C_{8n}HNO_3$ with $n = 2, 3, 4$.

Oscillation frequency (10^6 Oe)	Frequency symbol	Oscillation frequency (10^6 Oe)	Frequency symbol
$n = 2$		$n = 4$	
2.35	α	1.04	α
3.35	$\beta - 3\alpha$	3.00	β
5.70	$\beta - 2\alpha$	3.51	γ
8.05	$\beta - \alpha$	4.11	$\beta + \alpha$
10.04	β	4.60	$\gamma + \alpha$
		5.21	$\beta + 2\alpha$
		5.90	$\gamma + 2\alpha$
		6.32	$\beta + 3\alpha$
$n = 3$			
0.40	γ		
2.35	$\beta - 2\alpha$		
4.21	α		
6.68	$\beta - \alpha$		
10.09	β		

the zone-folding effect is not observed. One would expect that in compounds of the higher n with this acceptor there will be no ordering in the intercalate layer, either. Therefore, the nonmonochromaticity of oscillations in such compounds for $n > 2$ appears to be due to several hole groups. As an example, fig. 130 shows the oscillation curve for the compound $C_{32.8}\,ICl_{1.06}$ with $n = 3$, which represents the superposition of two frequencies. The presence of several carrier groups is also suggested by the field dependence of the Hall-coefficient R_H, which is observed in GIC of the type under consideration (see, e.g., fig. 131). In compounds with ICl at $n = 4$, one also observes two-frequency oscillations. It is interesting to note that in such compounds one frequency exceeds the other by more than an order of magnitude, the lower frequency being of the same order as the oscillation frequency due to the majority holes in pure graphite. Hence, one may conclude that a highly effective screening of the charge of the intercalate layer occurs already at a distance equal to the thickness of a single graphite bounding layer (the charge density decreases by an order of magnitude at this distance), so that in compounds with weak acceptors already at $n = 4$ the hole energy spectrum in two equivalent graphite interior layers will be close to the spectrum of pure graphite with the Fermi level slightly shifted downward (by a value of the order of the band-overlap energy in graphite). In accordance with this conception, the high-frequency oscillations in compounds with ICl at $n = 4$ are associated with holes in the graphite bounding layers.

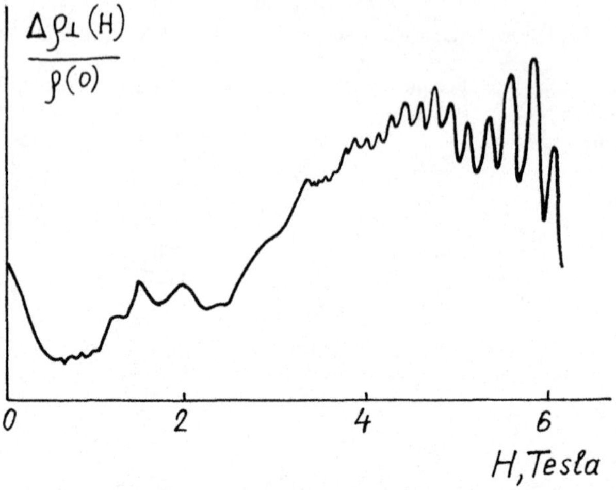

Fig. 130. The oscillating part of the transverse magnetoresistance versus magnetic field for a $C_{32.8}ICl_{1.1}$ sample at $T = 4.2$K. (Brandt et al. 1984a.)

10.9.3. Angular dependences of quantum oscillations in GIC

The angular dependencies of the frequency F and the amplitude A of oscillations were measured for a large number of GIC, both donor and acceptor, including compounds with K, Rb, $AlCl_3$, ICl, $CuCl_2$, etc.

The common characteristic of the dependence of the oscillation amplitude A

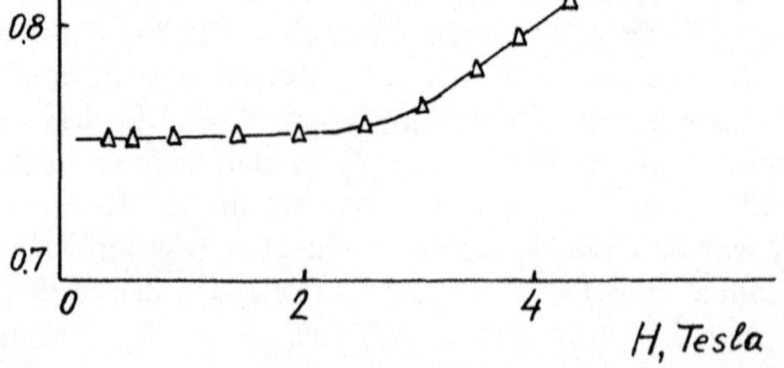

Fig. 131. Hall coefficient R_H versus magnetic field for $C_{32.8}ICl_{1.1}$ at $T = 4.2$K. (Brandt et al. 1984a.)

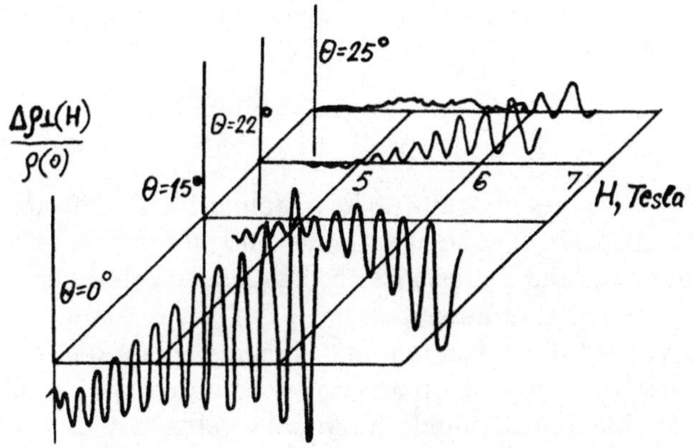

Fig. 132. Field dependence of the transverse magnetoresistance of $C_{9.8}\,CuCl_2$ for various angles θ between the magnetic field direction and the c-axis of the sample; $T = 4.2\,K$. (Brandt et al. 1984a,b.)

on the angle θ between $\boldsymbol{H}$ and the c-axis in GIC is that the amplitude drastically decreases with increasing θ, which greatly impedes oscillation measurements in the range $\theta > 40–60°$. This decrease is illustrated in fig. 132 which shows, as an example, oscillation curves for $C_{10}CuCl_2$, recorded under the same conditions at four θ values. The relative variations of the amplitude, $A(\theta)/A(0)$, and the frequency, $F(\theta)/F(0)$, of oscillations with varying θ in the compounds $C_{9.3}AlCl_{3.3}$, $C_{10}CuCl_2$, $C_{16}ICl$ and in pure graphite are shown in fig. 133. In this figure, the solid curve (1) corresponds to the angular

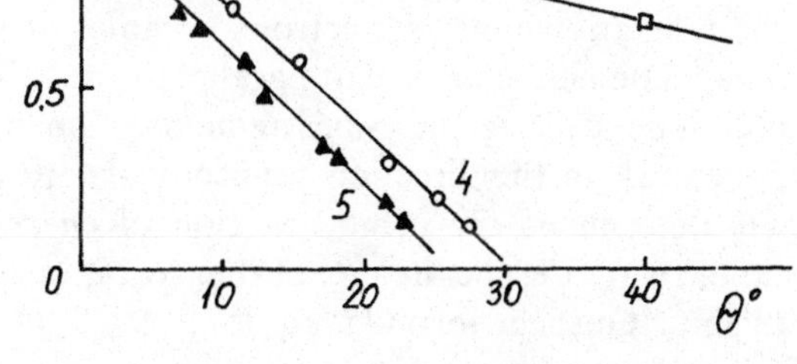

Fig. 133. Relative variation of oscillation frequency $F(\theta)/F(0)$ as a function of θ for graphite and GIC (curve 1), and relative change in oscillation amplitude $A(\theta)/A(0)$: 2, graphite (●); 3, $C_{9.3}AlCl_{3.4}$(□); 4, $C_{9.8}CuCl_2$(○); 5, $C_{16.6}ICl_{1.1}$(▲). (Brandt et al. 1984a,b.)

dependence of the relative oscillation frequency for a cylindrical Fermi surface, which is described by the expression

$$\frac{F(\theta)}{F(0)} = \cos^{-1}\theta. \tag{193}$$

The experimental values of $F(\theta)/F(0)$ for graphite and for the above-mentioned acceptor GIC are quite consistent with this curve for $\theta \leqslant 40^\circ$, which testifies to the Fermi surface of these substances being close to cylindrical, at least in the vicinity of the extremal cross section.

Attention is called to the fact that in the angular range 0°–40° oscillations in GIC completely disappear or undergo a drastic decrease of their amplitude, whereas the oscillation amplitude in graphite remains practically unchanged (curves 3, 4, and 5 in fig. 133).

The drastic decrease of oscillation amplitude in GIC with increasing angle θ may be explained as follows.

In intercalation compounds, systems of conducting graphite layers are separated by the intercalate layers in which carriers are in a certain measure localized. The conception of carriers being strongly localized in the intercalate layers is best applied to acceptor compounds. Owing to the fact that GIC consist of layers of greatly different conductivities, the orbits of carriers in a magnetic field are confined within the graphite layers and cannot cross the boundaries of the intercalate layers.

When the magnetic field $\boldsymbol{H}$ deviates from the direction of the c-axis at an angle θ, the component $H_\perp = H\sin\theta$, which lies in the layer plane, appears. The Lorentz force produced by this component, which acts on electrons or holes moving in the layer with the Fermi velocity v_F, is equal to $(e/c)H_\perp v_F$ and is directed perpendicular to the layer plane. This force tends to take the orbit out of the graphite plane, which increases the probability of carrier scattering by the charged ions of the intercalant and, therefore, reduces their relaxation time. This effect (and the oscillation amplitude determined by it) is enhanced with increasing angle θ, which also leads to the further increase in $H_\perp$. We should stress that such an effect is not observed in pure graphite even at $\theta \sim 60^\circ$–70°.

Thus in GIC the relaxation time of electrons or holes at the Landau level decreases when a magnetic field component parallel to the layer plane appears, even if the field component along the c-axis remains unchanged.

The field component $H_\parallel$ in the direction perpendicular to the layers, which is equal to $H\cos\theta$, determines the orbital motion of charge carriers in the conducting layer and, therefore, the area S_p of their orbit in momentum space, as given by the Lifshitz–Onsager formula, eq. (57):

$$S_p = \frac{2\pi e\hbar}{c} H_\parallel (N+\delta) = \frac{2\pi e\hbar}{c}(N+\delta) H\cos\theta, \tag{194}$$

where $N = 0, 1, 2, \ldots$, and δ is a constant of the order of unity.

Hence it follows that the oscillation period in the inverse magentic field is given by

$$\Delta(H^{-1}) = \frac{2\pi e\hbar}{cS_{\text{extr}}} \cos\theta$$

and the oscillation frequency, by

$$F(\theta) = \frac{cS_{\text{extr}}}{2\pi e\hbar} \cos^{-1}\theta = F(0)\cos^{-1}\theta, \tag{195}$$

where S_{extr} is the extremal cross section of the Fermi surface by a plane perpendicular to the k_z-axis. Equation (195) coincides with eq. (193) which describes the relative change in cross-sectional area and the relative change in oscillation frequency for a metal with a cylindrical Fermi surface.

Equation (195) suggests the following interpretation of the angular dependence of oscillation frequencies. Since on a cylindrical Fermi surface carriers are forbidden to move along the geometrical axis of the cylinder, the magnetic field inclination does not result in the inclination of the orbital plane, the orbit remaining in the plane perpendicular to the cylinder axis. However, the field component $H_{\parallel} = H\cos\theta$ perpendicular to the orbital plane decreases, which leads to an increase in the area described by the orbit in real space according to the law $\sim \cos^{-1}\theta$. This, in turn, leads to an increase in the oscillation frequencies, as referred to the absolute value of the total magnetic field vector $\boldsymbol{H} = \boldsymbol{H}_{\parallel} + \boldsymbol{H}_{\perp}$.

10.9.4. Trigonal warping of the Fermi surface and magnetic breakdown oscillations in GIC

Along with the ordinary quantum oscillation in GIC, such as the ShdH or dHvA effects, in acceptor compounds with AsF_5 with stage index one or two there have recently been observed oscillations resulting from the interference of electron waves scattered by interband-breakdown centres. This phenomenon is similar to the interference of light rays obtained by splitting the primary light beam by means of a semitransparent mirror. In the case of electron waves, it is the crystal itself placed in a sufficiently high magnetic field that serves as an interferometer. The centres of the interband magnetic breakdown, which are included in the system of the electronic orbits of this interferometer, act as a device that "splits" electron waves like semitransparent mirrors do in optics. Electron-wave interference arises from the uncertainty of the path along which an electron may travel from one point of the system of magnetic breakdown orbits to another point.

Oscillations of an interference nature were first observed in magnesium (see Stark and Friedberg 1971, Friedberg and Stark 1972) and subsequently in a number of other metals, such as niobium (Karim et al. 1978), aluminum (Gostishev et al. 1978), chromium (Reifenberger et al. 1980), and zinc (Fletcher 1981). Oscillations of this type have been investigated in detail in the theoretical papers by Slutskin (1967, 1970, 1973).

These oscillations have several features which distinguish them from the ordinary quantum oscillations, thus making it possible to substantiate their relation with the phenomenon of magnetic breakdown. The magnetic breakdown oscillations observed in GIC have furnished a fresh material for studying this interesting effect. Moreover, as has been established analytically, magnetic breakdown orbits in GIC are related to the trigonal warping of cylindrical sections of the Fermi surface and therefore, the detection of oscillations of this kind appears, first of all, as experimental evidence for the trigonal warping and, secondly, as a possibility of ascertaining its magnitude. This aspect of the study of magnetic breakdown oscillations is the more interesting because in the theoretical models for the GIC spectrum, as well as on the basis of first-principle calculations, the Fermi-surface sections surrounding the side Brillouin-zone edges are considered to be unwarped circular cylinders.

Because of the considerable importance of this problem, we will discuss the magnetic breakdown oscillations in GIC at greater length both from the theoretical and from the experimental viewpoint.

Let us first analyze magnetic breakdown oscillations as such. In order to understand the physical meaning of this phenomenon, we shall examine parts of two electronic orbits pertaining to different energy bands which come rather close to each other in the vicinity of the points J_1, and J_2 (segments of orbits 1 and 2 shown in fig. 134a). As is known, the probability p that an electron will pass from one orbit to another at the points of their closest approach is an exponential function of the magnetic field H:

$$p = \mathrm{e}^{-H_{\mathrm{B}}/H}, \tag{196}$$

where the quantity H_{B}, which has the dimensions of magnetic field, is called the magnetic breakdown field and is expressed in terms of the energy barrier $\Delta\varepsilon$ between the orbits, the Fermi electron energy ε_{F} on one of the orbits, and the Bohr magneton μ_{B}:

$$H_{\mathrm{B}} = \frac{(\Delta\varepsilon)^2}{\mu_{\mathrm{B}}\varepsilon_{\mathrm{F}}} \tag{197}$$

(see Lifshitz et al. 1971). It should be noted that the interband breakdown field H_{B} is much smaller than the intraband breakdown field, which is given by $H_{\mathrm{B}} = \Delta\varepsilon/\mu_{\mathrm{B}}$. The probability, $p_{J_1J_2}$, that an electron will travel via a system of orbits through the magnetic breakdown centres J_1 and J_2 may be calculated as follows: The wave functions for the transmitted, $|\mathrm{t}\rangle$, and reflected, $|\mathrm{r}\rangle$, electrons are related to the breakdown probability p by the expressions (see

Friedberg and Stark 1972)

$$|\mathrm{t}\rangle = p^{1/2}, \quad |\mathrm{r}\rangle = (1-p)^{1/2}\,\mathrm{e}^{\mathrm{i}\pi/2}. \tag{198}$$

Here, in determining the phase shift of the reflected wave, use was made of the analogy with the light beam splitting by a semitransparent mirror, resulting in the loss of a half-wave. As the electron moves in a magnetic field, the electron-wave phase changes by a quantity $\Delta\varphi$ given by the integral along the electron path:

$$\Delta\varphi = \frac{e}{c\hbar}\int \boldsymbol{A}\,\mathrm{d}\boldsymbol{r}, \tag{199}$$

where $\boldsymbol{A}$ is the vector potential of the magnetic field.

Let $\Delta\varphi_1$ and $\Delta\varphi_2$ be the phase shifts that result from an electron traveling along paths 1 and 2, between the magnetic breakdown centres J_1 and J_2. The probability $p_{J_1J_2}$ is then given by:

$$p_{J_1J_2} = |p^{1/2}\mathrm{e}^{\mathrm{i}\Delta\varphi_2}p^{1/2} + (1-p)^{1/2}\mathrm{e}^{\mathrm{i}\pi/2}\mathrm{e}^{\mathrm{i}\Delta\varphi_1}(1-p)^{1/2}\mathrm{e}^{\mathrm{i}\pi/2}|^2, \tag{200}$$

or, after some simple manipulations,

$$p_{J_1J_2} = p^2 + (1-p)^2 - 2p(1-p)\cos(\Delta\varphi_2 - \Delta\varphi_1). \tag{201}$$

The phase difference is, evidently, given by:

$$\begin{aligned}\Delta\varphi_2 - \Delta\varphi_1 &= \frac{e}{c\hbar}\left(\int_2 \boldsymbol{A}\,\mathrm{d}\boldsymbol{r} - \int_1 \boldsymbol{A}\,\mathrm{d}\boldsymbol{r}\right) = \frac{e}{c\hbar}\oint \boldsymbol{A}\,\mathrm{d}\boldsymbol{r} \\ &= \frac{e}{c\hbar}\iint \operatorname{rot}\boldsymbol{A}\,\mathrm{d}S_r = \frac{e}{c\hbar}HS_r = \frac{c}{e\hbar H}S_p,\end{aligned} \tag{202}$$

where S_r and S_p are the areas described by orbits 1 and 2 in the coordinate space and in momentum space, respectively (in fig. 134a the area S_p is shaded).

Thus the electron flow along paths 1 and 2 directed from J_1 to J_2 contains the noninterference fraction $I_0 \sim [p^2 + (1-p)^2]$ and the interference fraction $I_1 \sim 2p(1-p)\cos(\Delta\varphi_2 - \Delta\varphi_1)$. The latter component is an oscillating function, periodic in the inverse magnetic field. The frequency of the associated oscillations is equal to that of the ordinary quantum oscillations, which is determined by the cross-sectional area of the closed part of the system of orbits shaded in fig. 134a. It should be stressed that, unlike the ordinary quantum oscillations, the magnetic breakdown oscillations are unrelated to the quantization of the electron-orbit areas, because here there are no closed orbits at all.

Let us now examine how the amplitude of magnetic breakdown oscillations is affected by a finite temperature. We shall assume that there is no quasi-

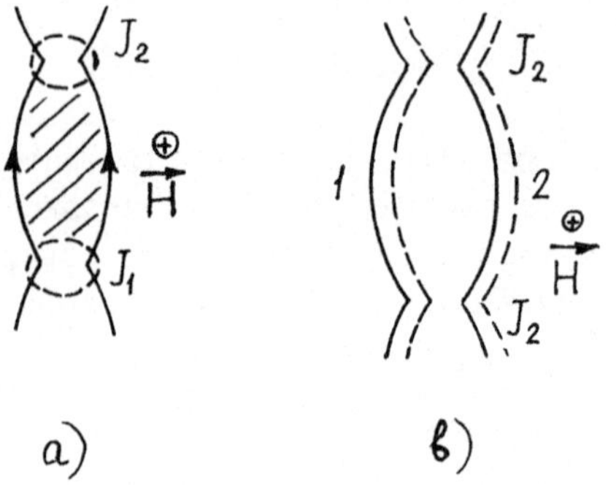

Fig. 134. (a) Electron trajectories 1 and 2 in a magnetic field referring to different energy bands. J_1 and J_2 are the magnetic breakdown centres (points, in whose vicinity the trajectories come rather close to each other); (b) Trajectories 1 and 2 at the initial (solid curves) and changed (dashed curves) electron energy values (see text).

particle scattering, i.e., the relaxation time τ is assumed to be infinite.

In passing along the interference part of the path shown in fig. 134a, each quasi-particle possesses a definite energy. Because the temperature is finite, the electron state distribution is changed in accordance with the Fermi–Dirac function $f_0(\varepsilon) = \{\exp[(\varepsilon - \varepsilon_F)/kT] + 1\}^{-1}$. The interference contribution to the current I_1, is evidently proportional to the integral

$$I_1(T) \sim \int \left(-\frac{\partial f_0}{\partial \varepsilon}\right) \cos[\Delta\varphi_2(\varepsilon) - \Delta\varphi_1(\varepsilon)]\, d\varepsilon, \tag{203}$$

where the phase difference $\Delta\varphi_2 - \Delta\varphi_1$ is given by eq. (202). The same expression gives the temperature dependence of the ordinary quantum oscillations as well, but, in that case, the S_p in eq. (202) is the area of a real closed electron orbit. For energies rather close to ε_F, the phase difference may be written in the form

$$\Delta\varphi_2 - \Delta\varphi_1 = \frac{c}{e\hbar H}\left[S_p(\varepsilon_F) + \left.\frac{\partial S_p}{\partial \varepsilon}\right|_{\varepsilon=\varepsilon_F} (\varepsilon - \varepsilon_F)\right].$$

Substituting this expression in eq. (203) yields

$$I_1(T) = \text{const.}\, \frac{x}{\sinh x} \cos\left[\frac{cS_p(\varepsilon_F)}{e\hbar H}\right], \tag{204}$$

where

$$x = \frac{2\pi^2 ck\, T m_c}{e\hbar H}, \quad m_c = \frac{1}{2\pi} \left.\frac{\partial S_p}{\partial \varepsilon}\right|_{\varepsilon=\varepsilon_F}.$$

Thus the temperature factor $x/\sinh x$ is determined by the magnitude of the cyclotronic electron mass at the Fermi level. This statement is valid, however, only for the ordinary quantum oscillations when S_p represents the area of a closed orbit in p-space, which varies with electron energy so that $\partial S_p/\partial \varepsilon = 2\pi m_c(\varepsilon_F)$. On the other hand, for the interference-type oscillations, the result may be entirely different, because the parts of paths 1 and 2 that bound the area shaded in fig. 134a belong to different orbits and as the energy is varied they may shift, quite independently of each other, either in the same or in the opposite directions. The former case is shown in fig. 134b, where the closed-orbit area remains practically unchanged and the derivative $\partial S_p/\partial \varepsilon$ is close to zero. Contrasting this situation with that of ordinary quantum oscillations, one may conclude that the former would correspond to the complete absence of any temperature dependence of the oscillation amplitude, and, therefore, to an abnormally low cyclotronic electron mass. We note, however, that even in the case when the above parts of paths 1 and 2 are shifted in the opposite directions with varying energy and when, formally, there arises a finite derivative $\partial S_p/\partial \varepsilon$, this shift takes place quite independently and the value of $(1/2\pi)\partial S_p/\partial \varepsilon$ does not correspond to a real cyclotronic mass of any carrier group, but represents only a factor determining the temperature dependence of the oscillation amplitude. The absence of any dependence of the oscillation amplitude on temperature is an important indication of the magnetic-breakdown (interference) nature of these oscillations. Furthermore, Stark and Friedberg (1971) have shown that magnetic breakdown oscillations should manifest themselves only as oscillations of transport properties (e.g., in the ShdH effect), but be absent from oscillations of thermodynamic parameters (e.g., in the dHvA effect).

Let us now turn to the discussion of some peculiarities in the oscillation pattern experimentally observed in acceptor compounds with AsF_5 for $n = 1$ and 2, in which magnetic-breakdown oscillations have first been detected in GIC by Markiewicz and Zahopoulos (1983). The Fermi surfaces for these compounds represent, respectively, one or two coaxial cylinders surrounding the side Brillouin-zone edges. Along with the ordinary quantum oscillations at $\boldsymbol{H} \| c$, whose frequencies F_i correspond to real cross sections S_i of the Fermi surfaces, Markiewicz and Zahopoulos (1983) have also observed long-period oscillations with frequencies multiple to some value of F_β (where $F_\beta = 1.25 \times 10^6$ Oe and 0.21×10^6 Oe for compounds with $n = 1$ and 2, respectively).

The low-frequency oscillations are characterized by the following features:

(1) These oscillations are observed only as oscillations of the magnetoresistance and are unobserved in the dHvA effect.

(2) Their amplitude is practically independent of temperature.

(3) The low-frequency oscillations appear as the modulation of the basic frequency F_i corresponding to a real Fermi-surface cross section with frequencies nF_β, where the n are integers.

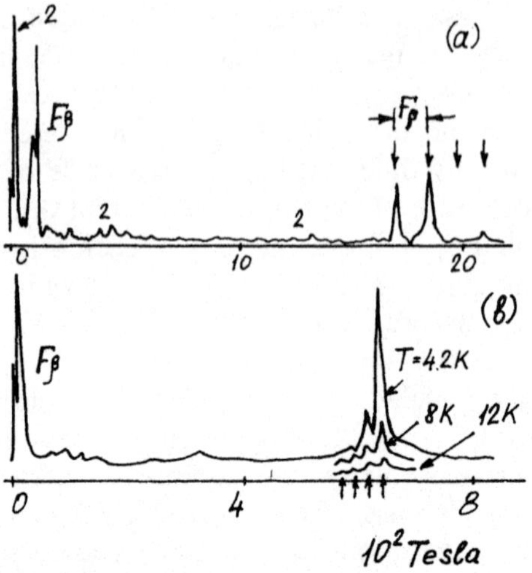

Fig. 135. (a) Fourier transform of the Shubnikov–de Haas spectrum for stage-1 AsF_5–graphite showing the F_β-frequency and the resulting modulation of the principal frequency (vertical arrows are separated by a distance equal to the F_β-frequency, $F_\beta = 1.25 \times 10^2$ T). Peaks marked 2 are probably due to regions of stage-2 in the sample; (b) Similar spectrum for a stage-2 compound; $F_\beta = 0.21 \times 10^2$ T. Note that as the temperature is raised, the modulations become more prominent. (Markiewicz and Zahopoulos 1983.)

Figure 135 illustrates the Fourier spectrum of ShdH oscillations in GIC with AsF_5 for $n = 1$ and 2. In this figure, the peaks corresponding to multiple modulation frequencies are indicated by arrows. The appearance of the multiple modulation frequencies may in general be ascribed to the emergence of a new periodic structure in the intercalate layer, i.e., to zone folding. However, the first two of the above-mentioned features of the long-period oscillations rule out this possibility and favour their interference nature. Markiewicz and Zahopoulos (1983) advanced an explanation of the possibility of the emergence of magnetic breakdown centres and electron interferometer structure in GIC with AsF_5 for $n = 1$ and 2. Their explanation is that in the course of intercalate-layer ordering ionized AsF_5 molecules form a hexagonal 3×3 superlattice (fig. 136a). For this particular lattice, the K point is folded into the Γ-point, and each trigonally warped cylindrical Fermi surface is rearranged to give two cylinders with hexagonal symmetry. As an illustration, fig. 136b shows the two new surfaces, B_1 and B_2 (B_1 bounded by the dashed line and B_2 bounded by the solid line), produced from the smaller, stage-two Fermi surface. The small superlattice gaps produced at the points where these two cylinders nearly touch are easily broken down in a field of about 10kOe, leading to the observed magnetic interferences. As is shown in fig. 136b, an electron may travel from point A to point B either by staying on the B_1 orbit

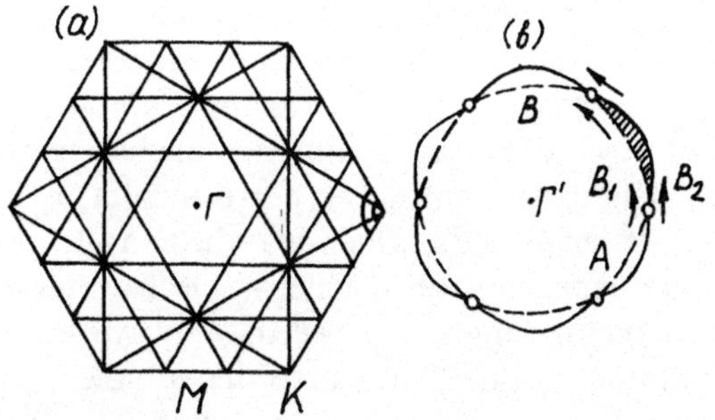

Fig. 136. (a) Two-dimensional Brillouin zone of graphite showing stage-2 AsF_5–graphite Fermi surfaces at one K-point and the new zone boundary introduced by a 3 × 3 superlattice. (b) Solid and dashed curves show the two Fermi surfaces generated by the superlattice potential from the smaller stage-2 Fermi surface. Magnetic breakdown occurs at the points of closest approach of the two surfaces leading to the interference of electrons moving along different paths. The area between these orbits is shaded. Arrows show the direction in which the electrons move. (Markiewicz and Zahopoulos 1983.)

or, via a double magnetic breakdown, by following a section of the B_2 orbit. The interference between these two electron beams leads to oscillations with frequency F_β, corresponding to the shaded area between the two orbits in fig. 136b. The frequencies that are a multiple of F_β appear as a result of interference between the beams that have suffered several splittings in successive magnetic breakdown centres. As follows from fig. 136b, the maximum value of the modulating frequency may be as high as $6F_\beta$, which corresponds to six consecutive areas between the solid and the dashed curves, each of the areas being equivalent to the shaded area in fig. 136b. Admittedly, magnetic breakdown centres cannot arise in the absence of trigonal warping, because the B_1 orbit would then coincide with the B_2 orbit.

Recall that in the SWMC-model for graphite the trigonal warping of the Fermi surface is ascribed to the interaction between the carbon atoms in the neighbouring graphite layers and is determined by the γ_3 parameter (see sect. 4.4.) This carbon-atom interaction should be greatly weakened in GIC with $n = 1$, and one may expect that in these GIC there will be almost no trigonal warping (small values of γ_3). The values of γ_3 in GIC with $n = 2$ should also be lower than in pure graphite, because in these compounds trigonal warping is largely due to the interaction between the carbon atoms only within a system of two graphite layers. The experimentally observed frequencies F_β of magnetic-breakdown oscillations in GIC with $n = 1$ and 2 (see fig. 135a, b) are at variance with the small value for the γ_3 parameter that can be estimated proceeding from the carbon-atom interaction between the neighbouring graphite layers in these compounds. This discrepancy can be eliminated by extending the SWMC-model dispersion relation, which is valid only in the immediate vicinity of the side Brillouin-zone edges, to the whole Brillouin zone, as has been done by Johnson and Dresselhaus. (1973) and by Holzwarth (1980). They have demonstrated that trigonal warping arises even in the case when one considers the carbon-atom interaction within a single graphite layer provided one goes beyond the approximation including the interaction between the nearest neighbours only. The correction of γ_0 for the interaction with the more remote neighbours, which may be considerable, leads to the γ_0 parameter becoming dependent on the Fermi electron wave vector k_F in the graphite plane and on the angle α between the wave vector $\boldsymbol{k}$ and the $K\Gamma$ direction in the Brillouin zone. The correction of the γ_0 parameter, which originally represents merely the resonance integral of the π-electron wave function overlap for the nearest neighbours, is equivalent to its replacement by a parameter γ'_0, given by the following expression,

$$\gamma'_0 = \gamma_0 \sqrt{1 + \delta \frac{k_F a_0}{2\sqrt{3}} \cos 3\alpha}, \tag{205}$$

where δ is a parameter of the order of unity.

Equation (205) contains the term $\cos 3\alpha$ whose multiplying factor increases with increasing Fermi energy. It describes the trigonal warping of the Fermi surface, the associated correction being the most significant for low-n GIC. The angular dependence of the γ'_0 parameter makes it possible to evaluate the shaded area in fig. 136b, and, therefore, the frequency F_β. The estimated F_β for GIC with AsF_5 for $n = 1$ and 2 (the value of k_F was determined from the Fermi-surface cross section using the experimental oscillation-frequency values) agrees well with the experimental F_β values.

Summarizing, the appearance of multiple modulation frequencies having the above-mentioned peculiarities in ShdH oscillations observed in GIC with AsF_5 for $n = 1$ and 2 enables one to establish not only the existence of an electron interferometer and magnetic breakdown oscillations, but also the presence of the trigonal warping of the Fermi surface in these compounds, which cannot be ascertained from conventional oscillation measurements. These regularities are of a rather general character and may be expected in other GIC as well.

10.10. Magneto-optical effects in GIC

10.10.1. General remarks

The study of magneto-optical transitions between Landau levels in GIC enables one to extract information on the band structure, Fermi-level position, and parameters of the energy spectrum in these substances. Moreover, the change in selection rules for interband transitions which is brought about by graphite intercalation allows one to infer that there has emerged a superlattice structure associated with a new periodicity both along the c-axis and in the intercalate-layer plane.

General conditions for the observation of magneto-optical transitions in GIC are, in principle, the same as those in pure graphite, a detailed discussion of which is contained in sect. 7.7.2. At the same time, graphite intercalation gives rise to some specific features of magneto-optical phenomena, characteristic of GIC only. Among them one should first of all mention the so-called Fermi-level cutoff phenomenon, associated with a Fermi-level shift and with the transition from the compensated-metal regime in graphite to the uncompensated-metal regime in GIC. Another feature of magneto-optical transitions in GIC is the change in selection rules resulting from the appearance of a new periodicity along the c-axis. Both these effects will be discussed in more detail in what follows.

Magneto-optical effects in GIC were first investigated by Chung and Dresselhaus (1976, 1977) in highly dilute acceptor compounds with halogens as the intercalant. Subsequently, the magnetoreflection spectra were observed in perfect specimens of GIC, both acceptor and donor, of various stage indices: C_xBr_2 ($n = 2, 3, 5$), C_xAlCl_3 ($n = 4, 8$), C_xFeCl_3 ($n = 5, 7, 11$), C_xRb ($n = 3, 6$).

By now, the magneto-optical spectra have been studied in most detail in high-n GIC. This is primarily due to the higher perfection of crystals of such compounds. It has been found that in high-n GIC with any intercalant the magneto-optical spectra have many common features, and are rather close to the spectrum in pure graphite. Below we shall discuss to a larger extent the data obtained for such compounds and analyze their implications.

10.10.2. Magneto-optical spectra in GIC with high n

Let us first of all discuss the structure of such spectra. Figures 137 and 138 present, as an illustration, the magneto-reflection spectra for C_xFeCl_3 ($n = 7$), C_xRb ($n = 6$), and C_xAlCl_3 ($n = 4, 8$) and compare them with those for pure graphite. In the figures, the arrows indicate resonance magnetic fields, the quantum numbers for the initial and final states being given at the arrows.

The magneto-optical spectra obtained at various photon energies can be conveniently represented as diagrams showing the photon energies and the corresponding resonance fields. An example of such a diagram, obtained by Mendez et al. (1980a) for $FeCl_3$ – containing GIC with $n = 7$ – is shown in fig. 139, which corresponds to transitions at the K points of the Brillouin zone for graphite due to circularly polarized photons, whose selection rules have the form $N_v - N_c = +1$. In this diagram, the open dots on the solid curves correspond to analogous transitions in pure graphite.

The comparison between the spectra in GIC and pure graphite indicates that, being similar in structure, these spectra differ mainly in the amplitude of their features, which generally decreases as a result of intercalation, and in the (slightly shifted) position of the equivalent resonance peaks on the magnetic-

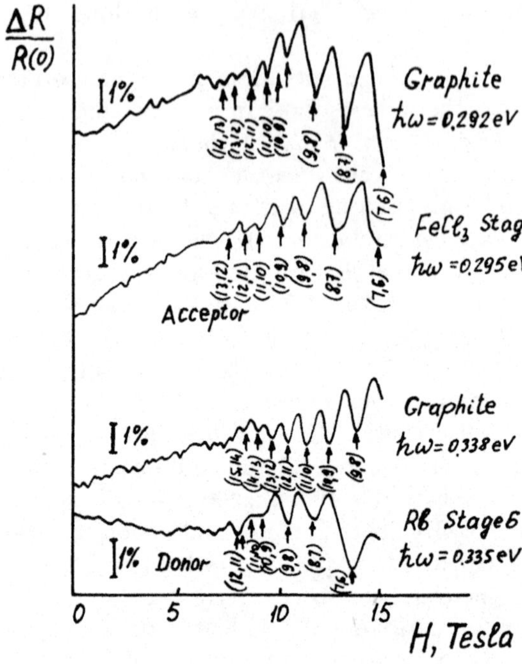

Fig. 137. Magnetoreflection spectrum $\Delta R(H)/R(0)$ using circular polarization for an acceptor compound ($FeCl_3$, stage-7) for a photon energy $\hbar\omega = 0.295$ eV and for a donor compound (Rb, stage-6) for $\hbar\omega = 0.335$ eV. For comparison, traces for graphite are shown at comparable photon energies. The resonances are specified by the quantum numbers for the initial and final states. (Mendez et al. 1980a.)

field scale. This, in turn, implies that the SWMC-model for the spectrum in pure graphite is capable of describing magneto-optical transitions in GIC with high n, at least in the energy region $\varepsilon_F^0 \pm 0.2$eV (ε_F^0 is the Fermi level in graphite) with relatively small changes in the γ_i parameters. An analysis of the experimental data for GIC, made within the SWMC-model, reveals that for GIC with $n \geqslant 4$ the γ_i parameters are relatively insensitive to both the value of n and the intercalant species used.

Thus magneto-optical measurements in high-n GIC essentially provide an experimental substantiation of the validity of the rigid-band model for these compounds, allowing the electron spectrum in the graphite interior layers to be treated as the one for pure graphite, which, in a first approximation, differs only in having the Fermi level shifted upward or downward in energy.

That the magneto-optical spectra in GIC are somewhat complicated in comparison with the spectrum in pure graphite is due to the nonuniform charge distribution in the graphite-layer system as a result of the charge transfer from the adjacent intercalate layers to the graphite bounding layers. A

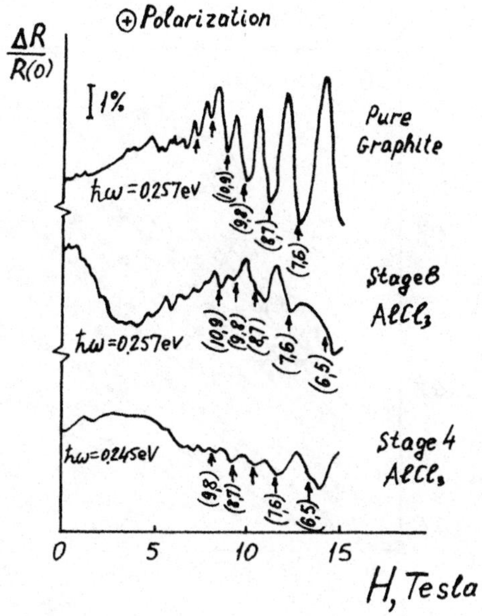

Fig. 138. Magnetoreflection traces for two graphite–$AlCl_3$ compounds (stage-8 and stage-4) and for graphite at approximately the same photon energy. On the ordinate is plotted the change in reflectivity, $\Delta R(H)/R(0)$, with magnetic field normalized to the zero-field reflectivity, and the reflectivity scale at 1% is indicated. The resonances are specified by the quantum numbers for the initial and final states. (Mendez et al. 1980a.)

high charge density in these layers gives rise to a strong distortion of band edges and magnetic levels near the boundaries of the graphite-layer systems. As a consequence, in the graphite bounding layers the initial and final states for typical transitions between Landau levels turn out to be either both filled or both empty which, according to the Pauli principle, makes these transitions forbidden.

Thus the experimentally observed magneto-optical transitions in GIC represent mainly transitions between the Landau levels corresponding to the graphite interior layers. This is specifically evidenced by the fact that the resonance line width in GIC is practically the same as that in pure graphite (see figs. 137 and 138). However, the general shift of all resonance peaks in GIC towards the lower magnetic fields, as compared to those in pure graphite, suggests that in all GIC, irrespective of their intercalants, the SWMC-model parameter γ_1, which describes the energy of the carbon-atom interaction between the neighbouring graphite layers, decreases or, in other words, the distance between the Landau levels increases.

Just as in pure graphite, in GIC the magneto-optical spectra measured for two directions of the circular polarization of light allow one to find the

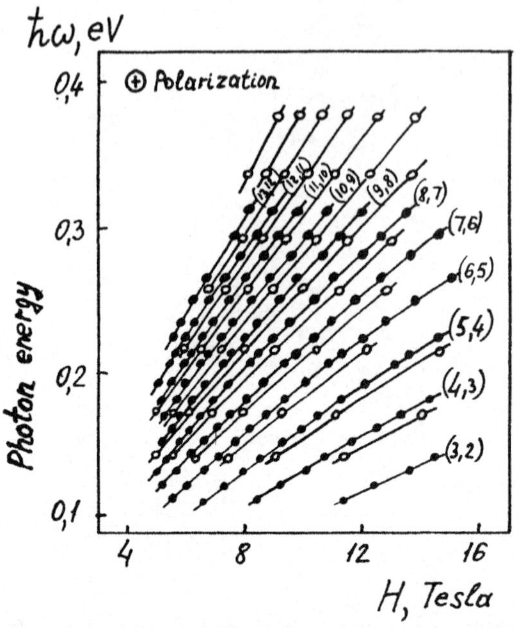

Fig. 139. Summary of resonant magnetic-fields "fan chart" for photon energies in the range $0.110 < \hbar\omega < 0.320\,eV$ for a $FeCl_3$ stage-7 sample using ⊕ circularly polarized radiation (●). To emphasize similar, but measurable, differences, a comparison between the results for this compound and those for pristine graphite (○) is presented in the same photon energy range. The numbers in parentheses give the quantum numbers for the initial and final states characterizing these transitions. (Mendez et al. 1980a.)

parameter m^*, introduced in sect. 4.4.5, the combination of the parameters $\gamma_0^2/\gamma_1 \sim 1/m^*$, and the cyclotronic masses in the conduction (m_c^*) and valence (m_v^*) bands.

The values of these parameters found from magneto-optical measurements by Mendez et al. (1980a) in donor and acceptor compounds, with $AlCl_3$, $FeCl_3$, Rb, and K with various stage indices are listed in table 25. For comparison this table also presents the corresponding parameters in pure graphite. The data of table 25 indicate that, in donor and acceptor GIC with high n, the parameters m^*, m_c^*, m_v^* and γ_0^2/γ_1 are not much (not more than 25%) different from the corresponding parameters in pure graphite. And the main change in the values of m^*, m_c^*, and m_v^* seems to be due to a change in the γ_1 parameter describing the interlayer overlap energy, whereas the γ_0 parameter describing the interaction between the carbon atoms in a layer is much less affected by intercalation.

10.10.3. Peculiarities of magneto-optical effects in GIC

One of the effects peculiar to GIC, which manifests itself in magneto-optical

Table 25

The parameters m^*, m_c^*, m_v^* *and* γ_0^2/γ_1 in donor and acceptor GIC with high stage indices and in pure graphite.

Parameter	Pure graphite	GIC with						
		C_xAlCl_3		C_xFeCl_3		C_xRb		C_xK
		$n = 8$	$n = 6$	$n = 7$	$n = 5$	$n = 6$	$n = 8$	$n = 9$
m^*/m_0	0.067 ±0.001	0.065 ±0.02	0.063 ±0.001	0.064 ±0.002	0.063 ±0.003	0.063 ±0.001	0.055 ±0.002	0.055 ±0.001
m_c^*/m_0	0.056 ±0.003	0.056 ±0.003	0.054 ±0.003	0.055 ±0.005	0.054 ±0.006	0.045 ±0.003	0.046 ±0.004	0.045 ±0.003
m_v^*/m_0	0.084 ±0.005	0.076 ±0.005	0.076 ±0.005	0.079 ±0.008	0.075 ±0.009	0.065 ±0.004	0.070 ±0.007	0.072 ±0.006
γ_0^2/γ_1 (eV)	25.1 ±0.5	26.0 ±0.9	26.6 ±0.3	26.1 ±0.7	26.6 ±1.3	31.7 ±0.8	30.6 ±1.1	30.3 ±0.8

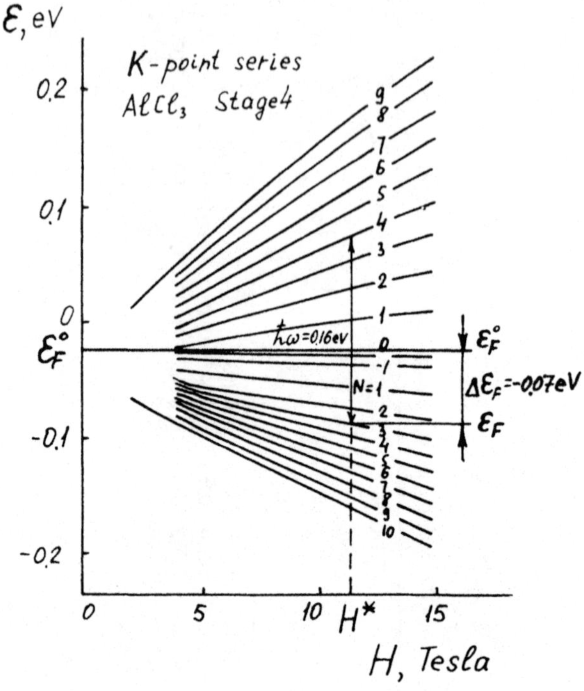

Fig. 140. Magnetic field dependence of various K-point Landau-level extrema for the valence and conduction bands. The Fermi level ε_F^0 for graphite is indicated. The interband transition for a photon energy of $\hbar\omega = 0.16\,\text{eV}$ between $N_v = 3 \rightarrow N_c = 4$ is cut off when ε_F in the intercalation compound falls below the arrow indicating the initial state. (Dresselhaus and Dresselhaus 1981.)

oscillations, is the disappearance of certain transition types observable in pure graphite. This effect is caused by the upward (or downward) shift of the Fermi level in energy resulting from intercalation and is called the Fermi-level cutoff phenomenon. Its essence may be conveniently illustrated by a scheme for the Landau levels at the K points of the Brillouin zone, shown in fig. 140. The scheme describes a situation corresponding to acceptor GIC whose Fermi level ε_F is shifted downward by 0.07 eV relative to the Fermi level ε_F^0 in pure graphite.

From the scheme shown in fig. 140 it follows that for the value of $\hbar\omega$ indicated by the arrow, the resonance transition from the level N_v to the level N_c is possible only in the case when the Fermi level ε_F lies between the initial and final states. For instance, for $\hbar\omega = 0.16\,\text{eV}$ (see fig. 140) the $N_v = 3 \rightarrow N_c = 4$ transition may be observed only in magnetic fields higher than the point at which the $N_v = 3$ level crosses the Fermi level ε_F, i.e., for $H > H^*$, because for $H < H^*$ the $N_v = 3$ level is unfilled. Below that crossing point the above-mentioned transition can be observed only when the quantum

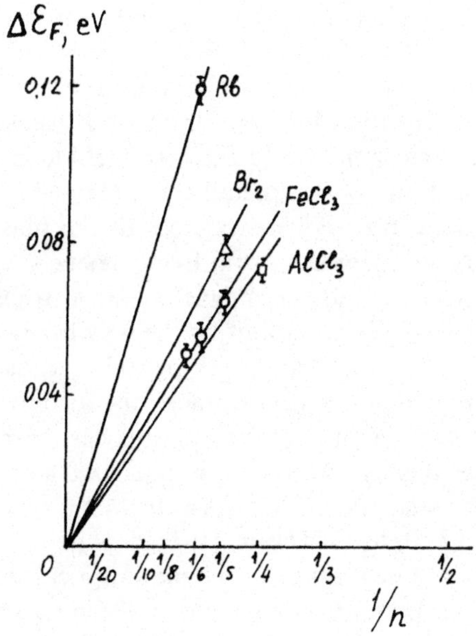

Fig. 141. Plot of the Fermi level ε_F in intercalation compounds relative to ε_F^0 versus reciprocal stage for several intercalants. The magnetoreflection experiment measures the passage of the initial K-point Landau level through the Fermi level. Because of the spatial dependence of the charge distribution on distance from the intercalate layer, the cut-off phenomena allow one to measure most sensitively the disappearance of the signal due to interband transitions from the innermost graphite layer(s) that are farthest from the intercalate layer. (Dresselhaus and Dresselhaus 1981.)

energy $\hbar\omega$ increases. Obviously, the cutoff phenomenon may not occur in the initial compensated semimetal with the unshifted Fermi level. Knowing the value of the cutoff field $H = H^*$ for a given photon energy enables one to determine the shift $\Delta\varepsilon_F = \varepsilon_F - \varepsilon_F^0$ of the Fermi level ε_F relative to its position ε_F^0 in pure graphite, provided that one has at his disposal the diagram of position of the Landau-levels in a magnetic field.

In the limit of strongly dilute GIC for $n \to \infty$ the shift $\Delta\varepsilon_F$ tends to zero. The dependence of $\Delta\varepsilon_F$ upon $1/n$ for various donor and acceptor intercalants is shown in fig. 141. At one and the same value of n, the shift $\Delta\varepsilon_F$ is greater for the donor intercalants. This conclusion, drawn from the results of magneto-optical measurements, is consistent with the results of IR-studies of lattice vibrations in GIC and also with the results of measurements of the conductivity σ_a in the basal plane. In the case when the Fermi level is shifted upward above the upper edge of the valence band of graphite at the H point in the Brillouin zone or downward below the lower edge of the conduction band of graphite at the K point, a semimetal, which has both electrons and holes, changes to a metal with carriers of only one sign. A transition of this type leads, in particular, to a change in dielectric constant κ. Indeed, as long as the Fermi level occurs in the region of overlap between the conduction and valence bands in the spectrum of graphite, the dielectric constant κ is mainly determined by the contribution from interband transitions (the so-called semimetallic regime corresponding to the zero forbidden band). When, due to intercalation, a transition to the metallic regime takes place, the dielectric constant begins (at least for frequencies close to the plasma frequency ω_p) to be determined by the contribution from free carriers.

Another distinguishing feature of magneto-optical transitions in GIC is the change in selection rules brought about by the new periodicity along the c-axis due to intercalation. The superlattice with period I_c leads, as has already been mentioned, to the appearance of new surfaces of constant energy, the system of Landau levels is complicated and new sequences of resonance transitions become possible. Specifically, for circularly polarized light waves in GIC there appear, along with those transitions for which $N_c - N_v = 1$, also the transitions with $N_c - N_v = 0$, forbidden in pure graphite.

As an illustration of a new series of magneto-optical transitions (those with $N_c - N_v = 0$), fig. 142 presents the magnetoreflection spectrum and the dependence of the reciprocal values for resonance fields $1/H_{res}$ upon $N_c + N_v + 1$ for C_xAlCl_3 with $n = 4$. The values of resonance field are indicated (in tesla) at each arrow, the values of quantum numbers N_v and N_c for a given transition

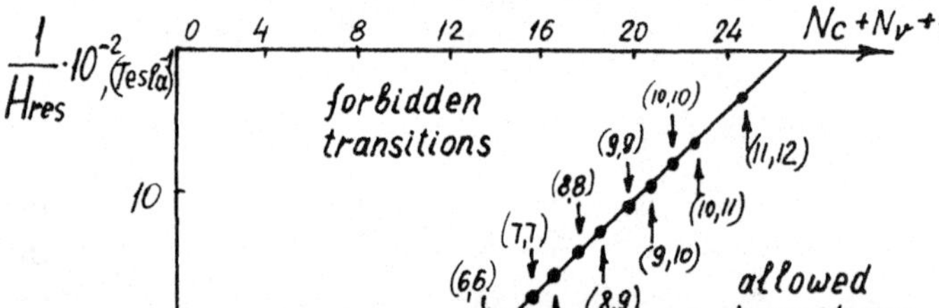

Fig. 142. In the lower portion is shown a magnetoreflection spectrum $[R(H) - R(0)]/R(0)$ for a stage-4 graphite–$AlCl_3$ compound for ⊖ circular polarization and a photon energy of 278.9 meV. The upper portion is a plot of resonant magnetic fields versus $(N_v + N_c + 1)$ for allowed and forbidden interband Landau-level transitions. (Dresselhaus and Dresselhaus 1981.)

being indicated in parentheses. The arrows above the curve for the dependence of $1/H_{res}$ upon $N_v + N_c + 1$ indicate the transitions forbidden in pure graphite.

10.11. The effect of pressure on the structure and energy spectrum of GIC

10.11.1. General remarks

Like the host material, graphite intercalation compounds have the greatest compressibility along the *c*-axis. Therefore, the action of hydrostatic pressure upon GIC single crystals is practically equivalent to axial compression along the *c*-axis. It should be noted that the compressibility of the graphite lattice does not radically change in intercalation, so that the compressibility factors along the *c*-axis ($I_c^{-1}\,dI_c/dp$) and in the graphite plane ($a_0^{-1}da_0/dp$) vary not very widely in going over from one kind of GIC to another or in varying the stage index. The value of the factor $I_c^{-1}dI_c/dp$ in GIC depends on the intercalant species and may be somewhat higher or somewhat lower than the value of $c_0^{-1}\,dc_0/dp = -2.9 \times 10^{-3}\,\text{kbar}^{-1}$, typical of graphite. Yet, basically, the value of $I_c^{-1}\,dI_c/dp$ in GIC varies as a function of the intercalant–graphite and graphite–graphite binding energies. Thus, for example, according to the data of Iye et al. (1981a,b) in GIC with $SbCl_5$, in which the intercalate layer is densely packed and rigidly bound with the graphite bounding layers, the value of $|I_c^{-1}\,dI_c/dp|$ decreases to $2 \times 10^{-3}\,\text{kbar}^{-1}$. In contrast to this, in GIC with HNO_3, in which the intercalate layers are loosely packed and weakly bound with the adjacent graphite layers, this value increases to $5 \times 10^{-3}\,\text{kbar}^{-1}$. At the same time, the compressibility factors of the above compounds in the basal plane, $a_0^{-1}\,da_0/dp$, are equal to zero within the experimental error, constituting 0.0 ± 0.05 and 0.0 ± 0.03 in units of $10^{-3}\,\text{kbar}^{-1}$. Recall that the analogous quantity in graphite equals 0.1×10^{-3} kbar^{-1}. In discussing the effect of pressure on the electronic properties and the energy spectrum of carriers in graphite (see ch. 8) it has been noted that uniform compression does not change the symmetry of the graphite lattice, the main change in the spectrum of graphite under pressure being associated with a change in the γ_i parameters determining the energy of the carbon-atom interaction between the neighbouring layers. In contrast to pure graphite, the intercalate layers in GIC give rise to some more complex effects which may be caused by pressure. Such pressure-induced effects involve a possibility of structural transitions in the intercalate layer. Thus, in particular, pressure may induce an "order–disorder" transition in the intercalate layer, may alter the character of ordering of intercalant particles and, finally, may bring about a change in stage index.

Until now, the experimental study of the structure and energy spectrum of

GIC under pressure has been carried out only for relatively few substances and the overall picture of change in GIC properties under pressure is not known in all details. The most complete data on the pressure effect upon GIC are available largely for acceptor compounds (as a rule, those with weak acceptors, such as ICl, $CuCl_2$, $SbCl_5$). One of the highly interesting effects discovered in the pressure studies is the structural phase transition between mutually nonmultiple intercalation stages (e.g., second and third). The transition is partially reversible and is observed in both donor and in acceptor compounds.

10.11.2. The effect of pressure on the energy spectrum of GIC with ICl and $CuCl_2$ for $n = 2$

As an illustration of the pressure-induced transformation of the energy spectrum of GIC, it is of interest to examine how the pressure affects the properties of compounds of the same stage index and with the same type of intercalant if these compounds have very similar energy spectra under normal conditions. Among such compounds one may mention, for example, GIC with weak acceptors such as $C_{8n}ICl$ and $C_{5n}CuCl_2$, the transformation of the spectra of which has been studied by Brandt et al. (1983, 1984a,b, 1986).

Consider, for example, the second-stage compounds $C_{16.6}ICl_{1.1}$ and $C_{9.8}CuCl_2$ which under normal pressure have very similar Fermi surfaces in the form of cylinders surrounding the Brillouin zone side edges. With such Fermi-surface structure, quantum oscillations for $\boldsymbol{H} \| c$ are monochromatic. The oscillation frequency determines the corresponding extremal cross sections, which we here designate for convenience as $S^{(2)}_{extr}$, the symbol (2) denoting the stage index. The values of $S^{(2)}_{extr}$, the cyclotronic masses m_c and hole concentrations n_h in the GIC under study were listed in table 23, which also gave the Fermi-energy values (ε_F) calculated for GIC with $n = 2$ on the basis of the Blinowski model.

Although the energy spectra of $C_{9.8}CuCl_2$ and $C_{16.6}ICl$ under normal conditions are of the same type, the character of change in oscillation curves for these compounds for $\boldsymbol{H} \| c$ under uniform compression is different.

In $C_{9.8}CuCl_2$, oscillations remain monochromatic at all the pressures investigated up to 12 kbar. Figure 143 shows oscillation curves for this compound at three pressure values: $p = 1$ bar, 3.9 kbar, and 10.1 kbar, respectively. The extremal cross section almost linearly decreases in $C_{9.8}CuCl_2$ under pressure at the rate

$$\frac{\partial \ln S^{(2)}_{extr}}{\partial p} = -0.017\,\text{kbar}^{-1} \tag{206}$$

(the lower curve in fig. 144).

On the other hand, in $C_{16.6}ICl_{1.1}$ at pressures $p \geqslant 1$ kbar there appear,

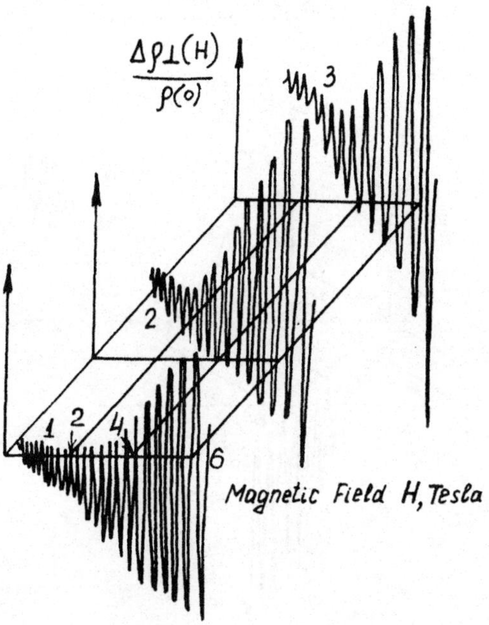

Fig. 143. Magnetic field dependence of the oscillating part of the transverse magnetoresistance $\Delta\rho_{\perp}(H)/\rho(0)$ for a $C_{9.8}CuCl_2$ sample at various pressures: (1) $p = 1$ bar, (2) $p = 3.9$ kbar, (3) $p = 10.1$ kbar. (Brandt et al. 1983.)

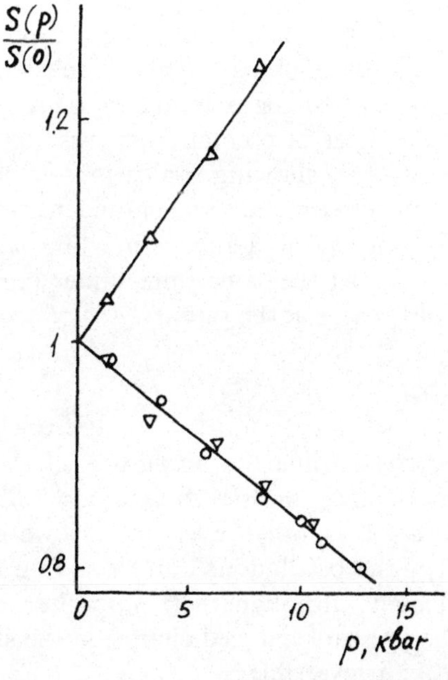

Fig. 144. Pressure dependence of the relative change in extremal Fermi-surface cross sections for $C_{16.6}ICl_{1.1}$ ($S^{(3)}_{extr}$, ▽; S', △) and $C_{9.8}CuCl_2$ ($S^{(2)}_{extr}$, ○). (Brandt et al. 1983.)

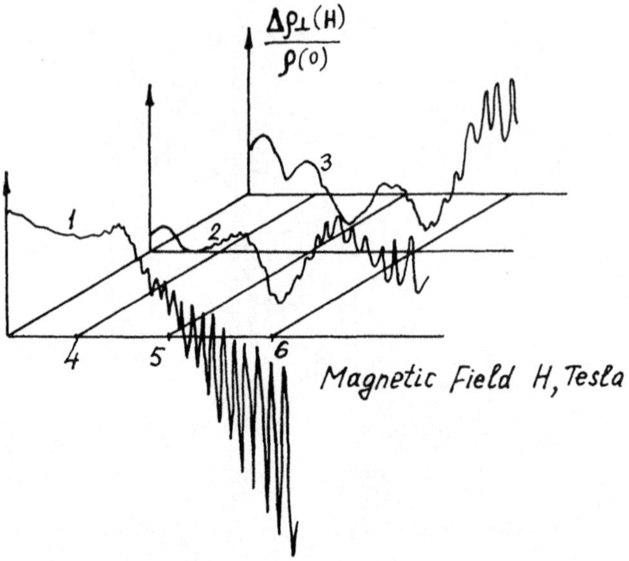

Fig. 145. Magnetic field dependence of the oscillating part of the transverse magnetoresistance for a $C_{16.5}ICl_{1.1}$ sample at various pressures: (1) $p = 1$ bar, (2) $p = 6.2$ kbar, (3) $p = 11$ kbar. (Brandt et al. 1983.)

along with the oscillations at the frequency corresponding to the extremal cross section $S^{(2)}_{extr}$, also oscillations with a considerably lower frequency corresponding at $p = 1$ kbar to the extremal cross section $S' = (25.4 \pm 0.6) \times 10^{-42}\,g\,cm^2/s^2$. The character of oscillation curves in $C_{16.6}ICl_{1.1}$ under pressure is illustrated in fig. 145, showing oscillations at $p = 1$ bar, 6.2. kbar and 11 kbar, respectively. In this compound when under pressure, the cross section $S^{(2)}_{extr}$ decreases almost linearly at a rate coinciding with that in $C_{9.8}CuCl_2$ (fig. 144, the lower curve). At the same time, under pressure the cross section S' practically linearly increases at the rate

$$\frac{\partial \ln S'}{\partial p} = 0.027\,\text{kbar}^{-1} \qquad (207)$$

(fig. 144, the upper curve), while the amplitude of the corresponding oscillations increases. The change in the cross section $S^{(2)}_{extr}$ in $C_{16.6}ICl_{1.1}$ and $C_{9.8}CuCl_2$ under pressure is fully reversible. However, as the pressure is released, in $C_{16.6}ICl_{1.1}$ the oscillations corresponding to the cross section S' do not vanish completely, although their amplitude significantly decreases. This indicates that the pressure-induced change in the structure of $C_{16.6}ICl_{1.1}$ may, to some extent, be irreversible.

The presence of hysteresis in the changing oscillation picture of $C_{16.6}\ ICl_{1.1}$ under pressure and the closeness of the value of the cross section S' at low pressures to that of the cross section $S^{(3)}_{\text{extr}}$ of the inner coaxial cylinder in the compound $C_{24.8}ICl_{1.1}$ with $n = 3$ has enabled Brandt et al. (1983) to conclude that the uniform compression of $C_{16.6}ICl_{1.1}$ produces regions where n equals three. The possibility of a transition with a change in n in this compound under pressure and its supposed character will be discussed in the next section. Now we will analyze the pressure-induced change in the energy spectra of $C_{9.8}CuCl_2$ and $C_{16.6}ICl_{1.1}$ with $n = 2$ proceeding from the baric dependences of their cross sections $S^{(2)}_{\text{extr}}$.

The pressure-induced change in hole concentration in acceptor GIC with $n = 2$ may be estimated by the following formula

$$\frac{\partial \ln n_{\text{h}}}{\partial p} = \frac{\partial \ln S^{(2)}_{\text{extr}}}{\partial p} - \frac{\partial \ln (d_0 + d_{\text{s}})}{\partial p}, \tag{208}$$

where the first term takes into account the change in the area of the Fermi-surface extremal cross sections while the second term takes into account the change in the Brillouin-zone size alond the k_z-axis.

From the compressibility data available at present in the literature on acceptor-type GIC (see, e.g., Iye et al. 1981a,b, 1982a) it follows that their compressibility is rather close to that of pure graphite, and does not exceed $5 \times 10^{-3}\,\text{kbar}^{-1}$. Hence one may conclude that [if (206) is taken into account] the first term in (208) should be much greater than the second and the magnitude and sign of the rate of change in hole concentration are practically entirely determined by the magnitude and sign of $\partial \ln S^{(2)}_{\text{extr}}/\partial p$. This means that, due to pressure, the hole concentration in $C_{16.6}ICl_{1.1}$ and $C_{9.8}CuCl_2$ decreases – in contrast to the pressure dependence of the carrier concentration in pure graphite, where the number of electrons and holes increases at the rate $\partial \ln n_e/\partial p = \partial \ln n_{\text{h}}/\partial p \simeq 0.045\,\text{kbar}^{-1}$ (see ch. 8). The extrapolation of the pressure dependence of $S^{(2)}_{\text{extr}}$ shows (fig. 144) that at $p \simeq 65$ or 70 kbar the cross section should go to zero, provided, of course, that the lattice structure and the energy spectrum of the GIC under consideration will be retained.

It is interesting to note that the decrease of hole concentration in $C_{16.6}ICl_{1.1}$ and $C_{9.8}CuCl_2$ with $n = 2$ occurs under pressure in spite of the decrease of interplanar distance, which results in the increased overlap between the wave functions for the carbon atoms in the neighbouring graphite layers and which would seem also to lead to the stronger overlap between the electron wave functions for the intercalate layer and the graphite bounding layers. The latter circumstance should in principle favour an increase in the relative amount of charge transferred to the intercalate layer, i.e., in the value of the factor f_{el} and, therefore, the hole concentration in the graphite layers. For this reason, the observed decrease of n_{h} in acceptor GIC with $n = 2$ can be explained only if it is assumed that the intercalate-layer structure changes under pressure such that the effective number of acceptors per carbon atom in a graphite layer decreases.

The shift of the Fermi level in $C_{16.6}ICl_{1.1}$ and $C_{9.8}CuCl_2$ under pressure may be found by differentiating eq. (192) for the Fermi energy ε_F^h in GIC with $n = 2$ in which for S_{extr}^h one should read $S_{extr}^{(2)}$. Taking account of the pressure-induced change in $S_{extr}^{(2)}$ and γ_1 and neglecting the baric dependence of γ_0 yields

$$\frac{\partial \varepsilon_F^h}{\partial p} = -\frac{1}{2}\left\{-\frac{\partial \gamma_1}{\partial p} + \frac{\gamma_1 \dfrac{\partial \gamma_1}{\partial p} + \dfrac{9}{2\pi\hbar^2}\gamma_0^2 a_0^2 \dfrac{\partial S_{extr}^{(2)}}{\partial p}}{\sqrt{\gamma_1^2 + 9\gamma_0^2 a_0^2 (S_{extr}^{(2)}/\pi\hbar^2)}}\right\}. \tag{209}$$

One may assume, as a first approximation, that the value of the baric factor of the γ_1 parameter in GIC coincides with that in pure graphite, i.e., $\partial\gamma_1/\partial p = 0.083\,\mathrm{eV\,kbar^{-1}}$. Making use of the experimental values of $S_{extr}^{(2)}$ and $\partial S_{extr}^{(2)}/\partial p$ for $C_{16.6}ICl_{1.1}$ and $C_{9.8}CuCl_2$ [see table 23 and eq. (206)] and the values of the parameters $\gamma_1 = 0.377\,\mathrm{eV}$ and $\gamma_0 \simeq 3.0\,\mathrm{eV}$, one may then find the values for $\partial\varepsilon_F^h/\partial p$ and $(\partial \ln|\varepsilon_F^h|)/\partial p$, which are

$$\frac{\partial \varepsilon_F^h}{\partial p} = +6.5\frac{\mathrm{meV}}{\mathrm{kbar}} \quad \text{and} \quad \frac{\partial \ln|\varepsilon_F^h|}{\partial p} = -0.018\,\mathrm{kbar^{-1}} \quad \text{for } C_{16.6}ICl_{1.1};$$

and

$$\frac{\partial \varepsilon_F^h}{\partial p} = +5.3\frac{\mathrm{meV}}{\mathrm{kbar}} \quad \text{and} \quad \frac{\partial \ln|\varepsilon_F^h|}{\partial p} = -0.021\,\mathrm{kbar^{-1}} \quad \text{for } C_{9.8}CuCl_2.$$

Thus one may conclude that in the acceptor compounds $C_{16.6}ICl_{1.1}$ and $C_{9.8}CuCl_2$ with $n = 2$ the application of pressure results in a practically identical transformation of the energy spectra of these compounds, leading to a decrease of the absolute value for the negative Fermi energy and to a decrease of the hole concentration.

It would be of great interest to look at the change of these spectra in the high-pressure range, where the Fermi energy in GIC approaches in absolute value the one in pure graphite. In the high-pressure range the competition between the processes leading to a decrease of carrier concentration as a result of the transformation of intercalate-layer structure and those leading to a decrease of interplanar distance may result in an essentially nonmonotonic change in carrier concentration in GIC being compressed, provided, of course, the picture is not distorted by phase transitions in the intercalate layers.

10.11.3. Pressure-induced staging transitions

Pressure-induced transitions were first observed by Wada et al. (1980a,b) and Clarke et al. (1980) in donor compounds with K. Their X-ray phase analysis has revealed that at pressures above 2 kbar in $C_{16}K$ with $n = 2$ there appear X-ray diffraction peaks characteristic of $C_{24}K$ with $n = 3$. Beginning with $p \simeq 7$ kbar the diffraction pattern of $C_{16}K$ corresponds completely to that of $C_{24}K$. As the pressure is released, the structure of this GIC again changes to that characteristic of GIC with $n = 2$. This phenomenon has been interpreted by the above-mentioned authors as a structural phase transition between two mutually nonmultiple stages (second to third). Similar changes in stage index under compression were observed in the pressure range from 4 to 7 kbar in GIC with potassium and other alkali metals (C_8K, $C_{24}K$, $C_{36}K$, $C_{48}K$, $C_{60}K$, $C_{24}Cs$, $C_{36}Cs$, $C_{24}Rb$) by Solin (1981), Wachnik et al. (1982), Fuerst et al. (1983). In acceptor GIC, the pressure-induced staging transitions were first observed in $C_{6n}HNO_3$ and $C_{5n}HNO_3$ by Iye et al. (1982a).

The common feature of the staging transitions observed in GIC is the simultaneous change in the stage number n and the in-plane density of intercalant atoms without changing the total number of these atoms in a sample. In compounds with alkali metals the transitions with the nonmultiple change from $n = 2$ to $n = 3$ are the commonest. Here the initial stoichiometry is $MC_{12\times 2}$ (M = K, Rb, Cs). In order that the total number of intercalant atoms be kept constant throughout the staging transition there must occur a simultaneous change in the in-plane density of intercalant atoms, i.e., $MC_{12\times 2} \rightarrow MC_{8\times 3}$. The results of the measurements of the c-axis resistivity in K-containing GIC also suggest that at the higher hydrostatic pressure ($\gtrsim$19 kbar) the intercalate layers become even denser (see Fuerst et al. 1983).

The mechanism of pressure-induced staging transitions in acceptor GIC has been investigated by Brandt et al. (1983, 1984a,b, 1986), who examined and compared oscillation effects in the similarly structured compounds $C_{16.6}ICl_{1.1}$ and $C_{9.8}CuCl_2$ ($n = 2$). The choice of these compounds has been dictated by the fact that, having an initial spectrum and structure of essentially the same type under normal conditions, they are significantly different in the temperature of their synthesis and, consequently, in the mobility of molecules in the intercalate layer at room temperature, at which the pressure is applied. $C_{16.6}ICl_{1.1}$ and $C_{9.8}CuCl_2$ are synthesized at ~40°C and ~600°C, respectively. Accordingly, for $C_{9.8}CuCl_2$ room temperature is sufficiently low in comparison with the temperature of intercalant crystallization and $CuCl_2$ molecules in $C_{9.8}CuCl_2$ at room temperature are in the solid-crystal phase. In contrast to this, in $C_{16.6}ICl_{1.1}$ room temperature is rather close to the temperature of intercalate-layer melting, so that intercalant molecules may comparatively easily move along the layers. The difference in the mobility of intercalant molecules at room temperature in the substances under consideration causes the different behaviour of these substances with increasing pressure. Recall that the difference in the character of change in oscillation curves for $C_{16.6}ICl_{1.1}$ and $C_{9.8}CuCl_2$ under pressure has already been discussed in the preceding section. This difference resides in the fact that in the former compound the original monochromatic oscillations, corresponding to the cylindrical Fermi-surface cross section $S^{(2)}_{extr} = (302 \pm 3) \times 10^{-42} g^2cm^2/s^2$ for $p \geqslant 1$ kbar are enriched in oscillations with a second, much lower, frequency, corresponding to the cross section S', which at low pressures constitutes $25.4 \times 10^{-42}\,g^2\,cm^2/s^2$, and is close to the cross section $S^{(3)}_{1\,extr}$ of the inner coaxial cylinder for the $n = 3$ compound $C_{24.8}ICl_{1.1}$. In $C_{9.8}CuCl_2$, on the other hand, the intercalate layers are characterized by a low mobility of molecules and there appears no additional frequency. As has already been mentioned in the preceding section, the appearance of a new cross section S' in $C_{16.6}ICl_{1.1}$ under pressure, which is close to $S^{(3)}_{1\,extr}$ in $C_{24.8}ICl_{1.1}$ has been attributed by Brandt et al. (1983) to the formation of regions with $n = 3$, in which the carrier concentration increases under compression. In contrast to this, in $C_{9.8}CuCl_2$ with low-mobility intercalant molecules no such regions emerge. The suppositions made by Brandt et al. (1983) to account for the rearrangement of the oscillation picture under pressure are sufficiently plausible; however, the character of oscillations observed in $C_{16.6}ICl_{1.1}$ under pressure does not fully correspond to the one that might be expected from a compound with a mixture of stages with $n = 2$ and 3. The point is that in $C_{24.8}ICl_{1.1}$ with $n = 3$ the Fermi surface consists of pairs of coaxial cylinders surrounding the Brillouin-zone side edges, of which the inner one has $S^{(3)}_{1\,extr} \sim S'$ and the outer one has a significantly larger cross section $S^{(3)}_{2\,extr} = (530 \pm 2) \times 10^{-42}\,g^2cm^2/s^2$. The high frequency oscillations corresponding to the outer cylinder have not been observed in specimens of $C_{16.6}ICl_{1.1}$ under pressure. The absence of these oscillations may admittedly be explained, because in $C_{24.8}ICl_{1.1}$ with $n = 3$ under normal conditions such oscillations are observed only in highly perfect specimens, which can hardly be expected in regions of GIC with $n = 3$ arising under pressure. It is most likely that in the $n = 3$ regions emerged the crystal structure is far from being perfect and the conditions necessary for observing high-frequency oscillations are not fulfilled, at least not in magnetic fields up to 60 kOe, in which these compounds were investigated by Brandt et al. (1983).

Let us now briefly dwell on the model for the pressure-induced staging transitions themselves in graphite intercalation compounds.

In order to account for the rearrangement of GIC structure in such transitions, Daumas and Hérold (1969) have proposed a domain model, which has been confirmed experimentally by Thomas et al. (1980). The model assumes that the enlargement of the GIC region with a new stage index involves no threshold and continues steadily with increasing pressure. Simultaneously, the volume of the region of GIC with the initial stage index steadily decreases. This transition character is quite consistent with the concept of the domain structure formed in the process of graphite intercalation itself. Being inserted from the periphery of a crystallite, the intercalant molecules may fill interlayer spaces, which are displaced relative to one another along the c-axis. In the process, domains arise, whose boundaries represent curved parts of the graphite layers. The size of the domains thus formed depends on the degree of perfection of the initial graphite material and on the conditions in which the graphite intercalation takes place. In highly

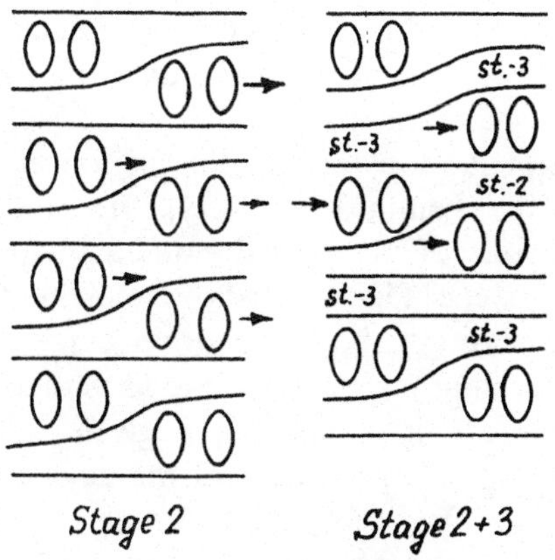

Fig. 146. Structural reconstruction of $C_{16.6}ICl_{1.1}$ under pressure resulting in the formation of stage-3 regions. (Brandt et al. 1983.)

perfect specimens the size of the domains approaches that of the crystallites. Under pressure, the domain boundaries may shift when the intercalant molecules are sufficiently mobile in the plane of the layer, and this makes it possible to explain the appearance and growth of regions of a different stage index.

This mechanism is illustrated in fig. 146 in which an example of the formation of GIC regions with $n = 3$ in a compound with $n = 2$ is shown. This mechanism can be used to explain the staging transition in $C_{16.6}ICl_{1.1}$ under pressure. As the pressure is lowered, the reverse motion, which tends to straighten the curved domain walls, becomes possible. As a result, the volume of the regions with a new stage index decreases and the initial structure of the compound is partly restored.

A theory of pressure-induced staging transitions in GIC has been developed by Miyazaki et al. (1984). It is based on finding the most stable stoichiometry of a compound for various species of intercalant atoms. Numerical results have been obtained for donor compounds with alkali metals (Li, K, Rb, Cs). In order to determine the conditions required for observation of staging transitions, they compared the ground-state energies for the various stoichiometries of a compound at $T = 0$ K.

Let us consider to a somewhat greater extent a scheme for calculations and the theoretical results obtained, taking as an example compounds with alkali metals, in the formation of which the electrons of alkali-metal atoms are transferred to the graphite layers, and occupy the conduction band of graphite. The energy gain due to intercalation is given by the sum of the conduction-electron energy and the energy of the interaction between the core electrons of the ionized intercalant atoms and the carbon atoms in the graphite. The energy associated with the conduction electrons is calculated by employing the Wigner–Seitz approximation. In order to evaluate the energy of the Coulomb interaction between the electrons and the intercalant atoms, it is necessary to solve the effective-mass equation for the conduction electrons. It employs a pairwise potential of the Lennard–Jones type to describe the interaction between the core electrons of an ionized intercalant atom and the carbon atoms of the graphite layers. The most stable stoichiometry of a sample at $T = 0\,\mathrm{K}$ is determined through the competition among these three kinds of energy. In a numerical evaluation, the total energy is taken as a function of the in-plane density of intercalant atoms for a given value of the interlayer distance d_s between the intercalate layer and the adjacent carbon layers. The effect of pressure is taken into account by varying the interlayer distance d_s. The total energy for each registered stoichiometry shows a sharp minimum when the value of d_s is varied.

If the total number of intercalant atoms in a sample is kept constant throughout the change in stoichiometry, the phase transition must be accompanied by a simultaneous change in the stage index. The transition pressure p_c is given by the ratio of the difference of the total energies per intercalant atom to that of the volumes of the Wigner–Seitz cells corresponding to the initial stage and the stage resulting from the transition. For instance, for the transition from $n = 2$ to $n = 3$ observed in $KC_{12\times2}$, the pressure p_c can be written as

$$p_c = -\frac{\varepsilon_2 - \varepsilon_3}{V_2 - V_3}, \tag{210}$$

where $\varepsilon_{2,3}$ and $V_{2,3}$ are the above-mentioned quantities for $n = 2$ and 3 (i.e., for $KC_{12\times2}$ and $KC_{8\times3}$).

A numerical calculation done for $KC_{12\times2}$ yielded a value of p_c about 4 kbar, which is in good agreement with the values of pressures (4–7 kbar) at which the staging transition in this compound was observed experimentally. The results of the calculations made by Miyazaki et al. (1984) have shown that in Li–GICs, the most stable stoichiometry, at $T = 0\,\mathrm{K}$ and ambient pressure, is $LiC_{6\times n}$, the next most stable one being $LiC_{8\times n}$. On the other hand, in GIC with heavy alkali metals $MC_{14\times n}$ and $MC_{8\times n}$ represent the most stable and the next most stable stoichiometries, respectively. At high hydrostatic pressure, $MC_{8\times n}$ is found to be more stable than $MC_{14\times n}$, however.

Thus, at present, there exists a sufficiently effective theory enabling one to estimate the stability of the stoichiometry of one or the other intercalation compound, as its interlayer distance in varied due to the uniform compression of the GIC. The theoretical calculations of stable stoichiometries carried out for donor GIC with alkali metals under pressure are in good agreement (both qualitative and quantitative) with the experimental results obtained for these

GIC under pressure. Therefore, it may be concluded that the mechanism of the pressure-induced staging transitions in graphite intercalation compounds has been elucidated and their character has been studied in sufficient detail.

10.12. Magnetic properties of GIC

Among graphite intercalation compounds, there is a group in which intercalant molecules contain magnetic ions of transition or rare-earth metals. As has been mentioned in sect. 10.10.5, at low temperatures in such compounds various magnetic transitions are observable. As the GIC whose intercalate layers contain local magnetic moments differ significantly in their magnetic properties from the remaining graphite intercalation compounds, these two groups will be considered separately.

10.12.1. GIC without magnetic ions

This class of GIC includes a large number of compounds, both donor and acceptor; for example, donor compounds with the alkali metals, acceptor compounds with halogens, inorganic acids, metal fluorides. Such compounds will henceforth be referred to as nonmagnetic, in contrast to the magnetic ones, i.e., those which contain magnetic ions in the intercalate layer.

Like graphite itself, its intercalation compounds are characterized by the anisotropy of the magnetic susceptibility, χ, which has two components, $\chi_{\parallel}$ and $\chi_{\perp}$, corresponding to two orientations of the magnetic field, viz., $\boldsymbol{H} \parallel c$ and $\boldsymbol{H} \perp c$.

The values of χ found for the nonmagnetic intercalation compounds differ significantly from those found in graphite in magnitude and sign of the components of χ, as well as in the character of their temperature dependences. The most characteristic distinctions in the behaviour of the above-mentioned components of χ in nonmagnetic compounds are:

(a) due to intercalation, the absolute value of $\chi_{\parallel}$ decreases by more than an order of magnitude;

(b) the magnetic susceptibility $\chi_{\parallel}$ is positive (paramagnetic) in donor GIC and is negative (diamagnetic) in acceptor GIC;

(c) the magnitude of the magnetic-susceptibility components (especially $\chi_{\parallel}$) in GIC depends much on the stage index. This is understandable, because one would naturally expect that, in the dilute limit, the magnetic susceptibility of GIC approaches that of graphite.

The most striking distinction of the magnetic properties of GIC from those of graphite is that, as a result of intercalation, the high diamagnetism typical of graphite ($\chi_{\parallel} = -21.1 \times 10^{-6}$ emu/g at room temperature) disappears and the magnetic susceptibility $\chi_{\parallel}$ reaches values of the order of 10^{-6} emu/g, which is typical of nonmagnetic materials.

To compare the magnetic properties of GIC with those of pure graphite, table 26 lists the values for $\chi_{\parallel}$ and $\chi_{\perp}$, as well as the average values for $\chi_{av} = \frac{1}{3}(\chi_{\parallel} + 2\chi_{\perp})$ for several donor compounds. This table also presents the values for the magnetic susceptibility χ_{pow} measured on powdered specimens of both donor and acceptor GIC, as well as the calculated values for the susceptibility components χ_0 and χ_p, which are discussed below.

The high diamagnetic susceptibility of graphite and especially synthetic quasi-two-dimensional pyrocarbons, already discussed in ch. 9, is associated with the resemblance of the electron (hole) energy spectrum in these substances to the line spectrum of a gapless semiconductor, in which the Fermi level passes near the conical band-degeneracy point and is mainly determined by the contribution of the Adams diamagnetic susceptibility of the filled bands (this contribution is large in the presence of the zero forbidden band, because it is associated with virtual interband electron transitions).

Within the rigid-band model the spectrum of GIC is equivalent to the energy spectrum of graphite with the Fermi level shifted downward (in acceptor GIC) or upward (in donor GIC) in energy, i.e., the intercalation of graphite may be regarded as its doping with donor or acceptor impurities. Accordingly, the disappearance of the high diamagnetism of pure graphite as a result of its intercalation has the same nature as the decrease of the magnetic susceptibility of graphite in doping.

In both cases, as the Fermi level departs from the band-degeneracy point in the spectrum of pure graphite the interband transition probability decreases. Sharma et al. (1974) have demonstrated that the diamagnetic contribution from interband transitions to the susceptibility may be large only in those cases in which the Fermi level ε_F lies within the π-band overlap and there are carriers of both signs in the spectrum. As soon as the Fermi level shifts beyond the π-band overlap the Adams contribution to diamagnetism sharply decreases.

The magnetic susceptibility χ of the nonmagnetic GIC is given by the sum of three main contributions:

$$\chi = \chi_{orb} + \chi_0 + \chi_p. \tag{211}$$

Here χ_0 is the magnetic susceptibility of atomic cores, χ_p is the susceptibility component corresponding to the Pauli spin paramagnetism of the Fermi electrons, and χ_{orb} is the orbital component containing contributions from both interband and intraband transitions between Landau levels (χ_{orb} also contains the contribution of the Adams filled-band diamagnetism).

The Pauli paramagnetic susceptibility χ_p is given by

$$\chi_p = \mu_B^2 \nu(\varepsilon_F), \tag{212}$$

where $\mu_B = e\hbar/2m_0c \simeq 5 \times 10^{-3}$ meV/kOe is the Bohr magneton, $\nu(\varepsilon_F)$ the density of states at the Fermi level. The values of χ_p listed in table 26 have been

Table 26
Magnetic-susceptibility components in GIC (in 10^{-6}emu/g) at room temperature.

Compound	Stage index n	$\chi_{\parallel}$	$\chi_{\perp}$	χ_{av}	χ_{pow}	χ_{0}	χ_{p}	Ref.*
graphite	∞	−21.1	−0.43	−7.32	–	−0.40	+0.016	
C_xK	1	+1.02	+0.28	+0.53	+0.62	−0.38	+0.64	[1]
	2	+1.50	+0.05	+0.53	+0.72	−0.39	+0.25	[1]
	3	+1.75	−0.02	+0.57	+0.67	−0.39	+0.21	[1]
	4	+1.87	−0.06	+0.57	+0.39	−0.40	–	[1]
C_8Rb	1	+0.66	+0.05	+0.25	+0.23	−0.33	–	[1]
C_8Cs	1	+0.48	0.00	+0.16	+0.31	−0.34	+0.34	[2]
C_6Li	1	+1.17	+0.38	+0.64	+0.64	−0.37	+0.56	[3]
$C_{16}Br_2$	2	–	–	–	−0.46	−0.42	–	[4,5]
$C_{16}Br_2$	2	–	–	–	−0.63	−0.42	–	[2]

*References
[1] Rüdorff and Schulze (1954).
[2] Delhaes (1977).
[3] Delhaes et al. (1976).
[4] McDonnel et al. (1951).
[5] Goldsmith (1950).

found by means of eq. (212) on the basis of the electronic density of states data obtained from measurements of the electronic part of heat capacity:

$$C_e(T) = \tfrac{1}{3}\pi^2 k\nu(\varepsilon_F)\,T \qquad (213)$$

(see Mizutani et al. 1978, Delhaes et al. 1976).

The values for the diamagnetic contribution from the core electrons, χ_0, were found by Delhaes (1977). Since the spin–orbit interaction in graphite and in GIC is very small and the g-factors of both electrons and holes are close to 2 (Sugihara 1984) the contributions from χ_p and χ_0 are isotropic and determine in equal measure the components $\chi_\parallel$ and $\chi_\perp$ of the magnetic susceptibility in GIC, whose anisotropy is entirely due to the anisotropy of the contribution from χ_{orb}. From table 26 one can see that the diamagnetic and paramagnetic contributions (χ_0 and χ_p) are much smaller in absolute value than χ_{orb} and that they cancel out to a significant extent. Therefore, the susceptibility $\chi_\parallel$ in GIC has an essentially orbital nature, and is close in value to the corresponding component χ_{orb}.

For $\boldsymbol{H} \perp c$, the orbital susceptibility is small and the contributions χ_0, χ_p, and χ_{orb} turn out to be close in absolute value. As a result, the sign of $\chi_\perp$ in GIC is determined by the competition between these three contributions. In donor compounds with K for $n = 3$ and 4, for example, the $\chi_\perp$ turns out to be diamagnetic, while the $\chi_\parallel$ is paramagnetic. This anomaly is not observed in compounds with Li, Rb, and Cs for $n = 1$ and with K at $n = 1$ and 2 (see table 26).

The theoretical evaluation of χ_{orb} in GIC has been undertaken by many authors. One of the early attempts is the work of Hove (1955), who calculated χ_{orb} for compounds with Br_2 on the basis of a two-dimensional model for the electron spectrum, neglecting the contribution from the Adams filled-band diamagnetism, which is rather small in GIC, in contrast to pure graphite.

In order to calculate the χ_{orb} in GIC, one can use the results of the calculations on the orbital magnetism of graphite obtained by McClure (1956, 1960) and subsequently by Sharma et al. (1974) (see ch. 9). These authors were the first to achieve quantitative agreement with the (observed) magnetic susceptibility $\chi_\parallel$ in graphite at low temperatures. Sharma et al. (1974) have also obtained the dependence of χ_{orb} on the position of the Fermi level in graphite, which enables one to explain the rapid decrease of diamagnetic susceptibility when graphite is doped with donor or acceptor impurities.

The sign reversal of $\chi_{orb\parallel}$ taking place when the Fermi level is shifted upward was predicted by Sharma et al. (1974), which proved to be of great importance for explaining the magnetic properties of GIC. According to these authors, with increasing ε_F the component $\chi_{orb\parallel}$ should decrease in absolute value, goes to zero at $\varepsilon_F \simeq 0.4\,\mathrm{eV}$ relative to the Fermi-energy value in pure graphite, and then reverses its sign. Thus the theory predicts that for $\varepsilon_F \geqslant 0.4\,\mathrm{eV}$, as is the case with donor GIC, the contribution χ_{orb} for $\boldsymbol{H} \parallel c$ should become paramagnetic. This conclusion is quite consistent with the measured magnetic susceptibility $\chi_{\parallel}$ in all GIC with alkali metals for $n \leqslant 4$, for which ε_F exceed 0.4 or 0.5 eV. In connection with this, it would be interesting to investigate the stage-index dependence in donor GIC for $n > 4$, because from the calculations made so far one may expect that in such compounds $\chi_{\parallel}$ should reverse its sign and become diamagnetic with increasing n (i.e., with decreasing ε_F). The detection of such an effect at the n corresponding to $\varepsilon_F \lesssim 0.4$ eV would conclusively prove the validity of the theory of diamagnetism of GIC developed by Sharma et al. (1974).

Needless to say, the theoretical works we have just discussed describe the magnetic behaviour of GIC with $n \geqslant 3$ reasonably well, although these theories neglect the nonuniform electron density distribution within a system of graphite layers. Taking this into account, which has been done by Safran and Di Salvo (1979), leads to a more complicated dependence of χ_{orb} on ε_F. In the paramagnetic region, this dependence, instead of steadily increasing with ε_F, is expressed by a curve with a peak on the passing of which the paramagnetic susceptibility decreases. This result allows one to explain the decrease of positive $\chi_{\parallel}$ values in C_xK with decreasing stage index (see table 26), which is accompanied by an increase in the Fermi energy ε_F.

10.12.2. GIC containing magnetic ions (magnetic GIC)

The presence of local magnetic moments in the intercalate layers leads to a temperature dependence of magnetic susceptibility in the form of the Curie law for paramagnetics:

$$\chi = Ng^2\mu_B^2\frac{J(J+1)}{3kT}, \tag{214}$$

where N is the concentration of local moments, g the effective g-factor, J the angular momentum quantum number for the local moments. This dependence of χ on T changes with decreasing temperature to a Curie–Weiss dependence: $\chi \sim (T+\theta)^{-1}$, where θ is the Curie temperature. This type of T-dependence of χ was observed by Karimov et al. (1971) and Ohhashi and Tsujikawa (1974a,b) in compounds with $FeCl_3$ and $FeCl_2$ above the magnetic ordering temperature. In these compounds, local magnetic moments are associated with the iron ions Fe^{2+} and Fe^{3+}, which have the same values for g and J as the pure initial substances $FeCl_2$ and $FeCl_3$.

Magnetic GIC are convenient objects for studying the magnetism of two-dimensional systems and, primarily, for studying magnetic phase transitions in such systems. GIC offer a unique possibility of studying the character of the phase transitions as the magnitude of the interplanar interaction between the ionic magnetic moments is altered by varying the stage index. The range of variation of the interplanar distances between the intercalate layers in GIC is

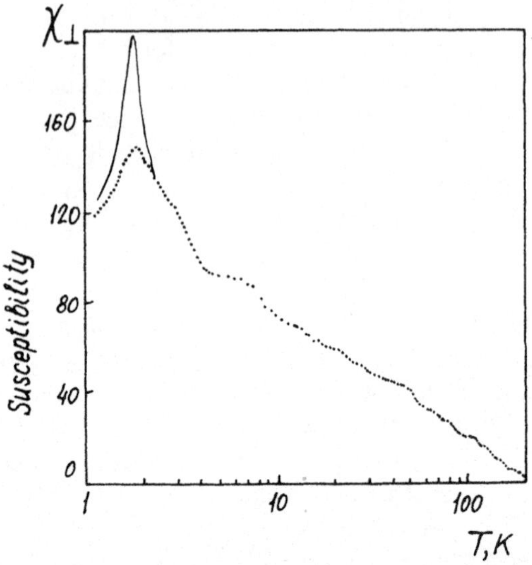

Fig. 147. In-plane susceptibility versus temperature over wide range of temperatures for stage-2 graphite–$FeCl_2$ compounds. The solid line represents the susceptibility curve in zero external DC field, while the dots represent the susceptibility curve at 80 Oe, which has been applied to quench the spin-glass peak. Notice how narrow the sharp spin-glass peak is when compared to the rest of the susceptibility curve, even when drawn on a logarithmic scale. (Millman and Zimmerman 1983.)

very wide. For instance in GIC with transition-metal chlorides these distances may vary from about 9 Å (in GIC with $n = 1$) to about 40 Å (in GIC with $n = 7$ or 8).

Although the nature of transitions in the magnetic GIC has not yet been completely elucidated and they have not yet been interpreted unambiguously, it is still possible to divide them roughly into the following groups.

10.12.2.1. Paramagnetic spin-glass transitions. The behaviour of magnetic susceptibility in GIC in transitions to a spin-glass state has been investigated theoretically by Wertheim et al. (1980). Transitions of this type are characterized by the appearance, for $H \to 0$, of a sharp peak on the curve for the temperature dependence of magnetic susceptibility $\chi_\perp$. As the constant magnetic field increases the peak on the T-dependence of $\chi_\perp$ becomes less sharp and its magnitude decreases. This has been observed in a compound with $FeCl_2$ at $n = 2$, whose T-dependence of $\chi_\perp$ is shown in fig. 147 (Millman and Zimmerman 1983). A similar behaviour of $\chi_\perp$ in the same compound at low temperatures was previously observed by Elahy et al. (1982). They have established that with increasing magnetic field the peak on the curve for $\chi_\perp(T)$ is somewhat smeared, and shifts toward a lower temperature. A detailed analysis of the experimental data, which rests on the theory developed by Wertheim et al. (1980), has made it possible to conclude that for $T = 1.7$ K in $FeCl_3$-compounds with $n = 2$ a transition from the paramagnetic state to the spin-glass state occurs. Magnetic transitions taking place at low temperatures in compounds with other transition metal chlorides appear to be of the same nature.

10.12.2.2. Transitions leading to magnetic ordering in the intercalate layer. Compounds with $CoCl_2$ are convenient objects for studying magnetic phase transitions of this type, because the structure and magnetic properties of the above-mentioned intercalant are well-known. Pure $CoCl_2$ goes over to the antiferromagnetic phase at the Neel temperature $T_N = 24.9$ K. For $T < T_N$, spins in the layers composed of Co^{2+} ions are oriented in the plane of the layers and are ordered ferromagnetically, whereas the layers themselves are ordered antiferromagnetically. The $CoCl_2$ layers in GIC represent a three-layer sandwich, consisting of a succession of planes made up by the ions of Cl^-, Co^{2+}. and again Cl^-. The intercalate-layer structure repeats the layered structure of the $CoCl_2$ lattice having very similar parameters. The constants of the intralayer and interlayer magnetic interactions between the Co^{2+} ions in $CoCl_2$, which have been determined by Hutchings (1973) from neutron scattering data, are $J = 28.5$ K and $J' = 2.2$ K, respectively. The similarity between the structure of the intercalate layer and that of the initial $CoCl_2$ enables one to expect that the magnetic interaction between Co^{2+} ions within the intercalate layer in a $CoCl_2$–GIC will closely resemble the interaction between the same ions in pure $CoCl_2$. As for the interaction between the intercalate layers separated by systems of the diamagnetic graphite layers, it is much weakened by the rather large period of the GIC structure along the *c*-axis, considerably weakening the long-range interlayer positional correlations. The results of the measurements of the differential magnetic susceptibility component $\chi_\perp$ in $CoCl_2$–GIC with $n = 2$ and 5, which have been obtained by Elahy and Dresselhaus (1982), indicate that the *T*-dependence of $\chi_\perp$ has either a sharp peak at $T = 8.2$ K (for $n = 2$) or two asymmetric peaks at $T = 7.7$ and 8.2 K (for $n = 5$), as is shown in fig. 148a,b. It has been found that, at one and the same volume of the specimens, the height of the peak observed at $n = 5$ is five times greater than that observed at $n = 2$. If the susceptibility is renormalized to one mole of the intercalant, then the difference in peak height between the GIC with $n = 2$ and $n = 5$ becomes still greater. At the same time, the value for the bulk susceptibility at the peak for compounds with $n = 5$ is close to that of a ferromagnetic, which is equal to $3/4\pi = 0.24\,\text{emu/cm}^3$ for spherical specimens. It is of interest to note that the temperatures at which the above-mentioned peaks in $CoCl_2$-compounds are observed are rather low, being not higher than one third of the Neel temperature T_N for pure $CoCl_2$.

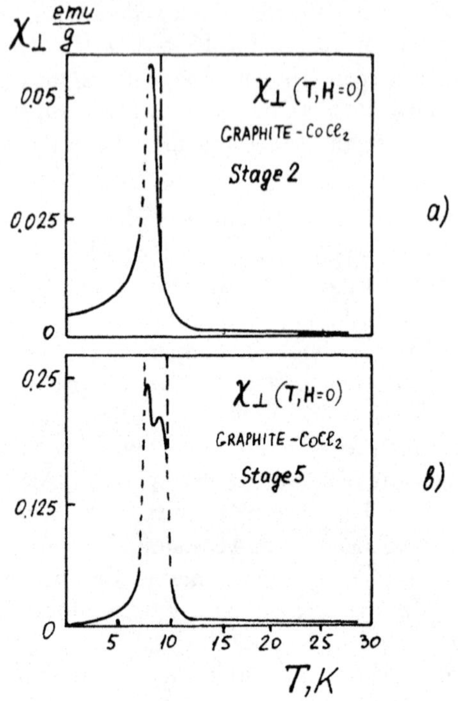

Fig. 148. The in-plane susceptibility $\chi_\perp$ of stage-2 (a) and stage-5 (b) $CoCl_2$ intercalation compounds as a function of temperature. The dashed lines are a plot of eq. (216). (Elahy and Dresselhaus 1982.)

The similarity between the structure of the $CoCl_2$ layer in a GIC and that in pure $CoCl_2$ allows one to suppose that in the temperature range within which it is possible to observe peaks on the curve for $\chi_\perp(T)$, the spins of the Co^{2+} ions in GIC lie in the plane of the layers and, therefore, the two-dimensional (planar) model can be used to describe the magnetism of the layer. The theory of such a model with continuous symmetry, which has been developed by Kosterlitz (1976), predicts the occurrence of phase magnetic transitions at a finite temperature. These transitions are characterized by the change in the form of the pair correlation function for the magnetic moments.

Within this model, the T-dependence of magnetic susceptibility near a transition has the form

$$\chi_\perp = \exp\left[\frac{(2-\eta)b'}{(t-1)\nu}\right], \quad \text{for } t > 1,$$

$$\chi_\perp \to \infty, \qquad \text{for } t < 1, \tag{215}$$

where $t = T/T_c$, $\eta = \frac{1}{4}$, $\nu = \frac{1}{2}$, $b' = \frac{2}{3}$. Pokrovskii and Vimin (1974) have shown that taking into account even a weak magnetic anisotropy in the plane of the layer leads to the appearance of a second phase transition at a temperature T_{c_2} which is somewhat lower than T_{c_1}, for the (first) phase transition discussed above. The relevant calculation of Jose et al. (1977) shows that the slopes of the susceptibility peaks facing the lower and higher temperature ranges may be described by

$$\chi_\perp = \chi_{\perp_0} + \chi_{\perp_1}\exp\left[\frac{b}{(t-1)^{1/2}}\right], \tag{216}$$

where $\chi_{\perp_0}$ coincides with χ_0 in eq. (211).

Figure 148a,b illustrates dependences of the type given by eq. (216), depicted by the dashed curves, with the parameters $\chi_{\perp_0}$, $\chi_{\perp_1}$, b T_c fitted to the experimental data. It can be seen from this figure that the above-mentioned dependence makes it possible to reproduce the experimental curve for $\chi_\perp(T)$ over a rather wide temperature range. Hence, one may conclude (Karimov 1975) that the above feature observed at T_{c_1}, probably corresponds to a transition from the paramagnetic state to the *XY*-type planar-model state, whereas the feature observed at the lower temperature T_{c_2} corresponds to a transition from the planar-model state to that with finite magnetization. Even in the absence of any interlayer interaction, the real magnitude of the magnetic susceptibility in a peak may be limited due to the finite size of the intercalant "islands". Thus the great height of the susceptibility peak in $CoCl_2$–GIC with $n = 5$, as compared to that at $n = 2$, may be attributed either to the large size of the intercalant islands or to a better realization of the two-dimensional model at $n = 5$. A similar increase in the susceptibility amplitude per one mole of magnetic ions in magnetic staging transitions in GIC was observed also in compounds with $FeCl_3$ (see Stumpp 1977).

10.12.2.3. Transitions to phases with imperfect magnetic ordering. Of great interest is the study of effects associated with spin frustration in two-dimensional (2D) antiferromagnets with triangular lattice, i.e., the study of so-called two-dimensional spin-glasses. Some magnetic GIC are very probable candidates for observing in them transitions to a spin-frustrated state in a 2D-system. It has been theoretically predicted that in such spin-frustrated 2D-systems there may exist quasi-ordered phases, e.g., a partially ordered antiferromagnetic phase in which one of three sublattices is completely frustrated (see Wada and Ishikawa 1983). One of the GIC in which the spin-frustration effect in a 2D-system appears to take place is $MnCl_2$–GIC with $n = 2$. As is known, in this compound one can observe magnetic transitions at $T_{c_1} = 1.4$ K and at $T_{cu} = 46.7$ K, which are characterized by salient points on the *T*-

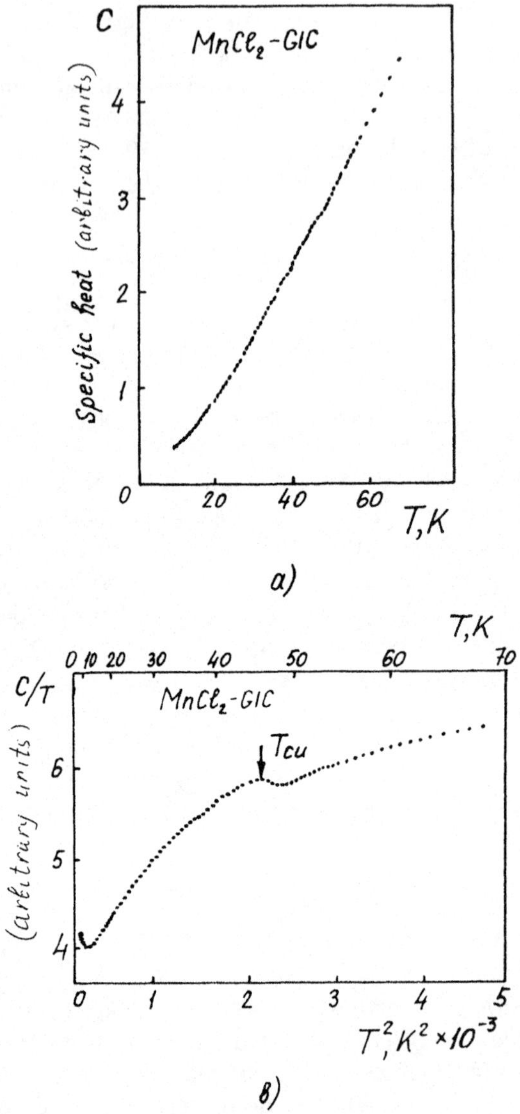

Fig. 149. (a) Temperature dependence of the specific heat C in stage-2 $MnCl_2$–GIC. (b) Plot of C/T as a function of T^2. The magnetic contribution corresponds to the deviation of the curve from a straight line; $T_{cu} = 46.7$ K. (Koga and Suzuki 1984.)

dependence of the heat capacity C (see fig. 149a,b). At the same temperatures, one can also observe anomalies on the T-dependence of differential magnetic susceptibility. It has been suggested that the partially spin-frustrated phase in $MnCl_2$–GIC with $n = 2$ may arise in the temperature range between T_{c_1}, and

T_{cu}. It should be emphasized, however, that the magnetic structures of the low-temperature or intermediate phases in these compounds have not yet been studied by direct methods (e.g., the thermal neutron scattering technique).

10.13. Superconductivity in GIC

10.13.1. General remarks

The superconductivity of GIC was first observed by Hannay et al. (1965) in compounds with alkali metals (K, Rb, Cs) at $n = 1$. Subsequently, the appearance of superconductivity in these substances has been confirmed by the measurements of their electrical conductivity (Koike et al. 1980) and magnetic susceptibility (Koike et al. 1978, 1980; Kobayashi and Tsujikawa 1979, 1981a,b). At the time of writing, interest in the superconductivity of graphite intercalation compounds has not even begun to ebb. Both the investigation of the already known superconductors and the search for new ones continue.

The great interest in this phenomenon is primarily due to its extraordinariness, because among the originally discovered superconducting GIC with alkali metals neither graphite nor the intercalants separately possess this property. Furthermore, superconducting GIC have turned out to be highly anisotropic superconductors with quasi-two-dimensional electronic structure, which offers a possibility of studying the superconductivity of practically two-dimensional systems. The study of the superconductivity in GIC also offers the hope of realizing the exciton mechanism of superconductivity proposed by Little (1964) and Ginzburg (1964). In this mechanism, the leading part is played by the effects associated with polarization of intercalant molecules by electrons displaced in the graphite layers. It is significant that the structure of superconducting GIC in principle allows such a possibility (see, e.g., the schematic structures of C_4KHg and C_8KHg shown in fig. 150).

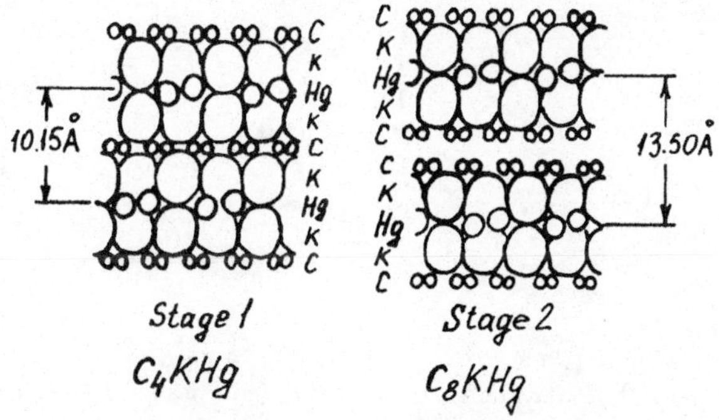

Fig. 150. Crystal structure of $C_{4n}KHg$. (Iye 1982.)

It is of interest to mention that the very capability of graphite being intercalated with a large number of different substances has been discovered in connection with the search for a realization of the exciton superconductivity mechanism in the course of which an enormous number of GIC have been synthesized.

One cannot also exclude that the combination, at the microscopic level, of metallic layers with semiconducting layers or with layers composed of complex molecules may result in the production of crystals with entirely unexpected electronic properties. In particular, there is a possibility of realizing in graphite intercalation compounds the Josephson interaction between layers, thus making it possible to explore a new type of a link of weak binding in the Josephson transition and the effects of the so-called weak superconductivity.

10.13.2. Classification of superconducting GIC

All known superconducting GIC may be divided into two groups. One of them includes compounds of $n = 1$ with alkali metals, i.e., such GIC as C_8M, where M is K, Rb, or Cs. In GIC of this group, the superconduction transition temperature T_c is about 0.1 K. An exception is the $n = 1$ compound with Li (C_6Li) which fails to become a superconductor down to several mK. No superconductivity has been found in alkali-metal-GIC with $n > 1$, either.

A second group includes ternary GIC of the $C_{4n}MHg$-type, where M is K,

Table 27
Superconduction transition temperatures T_c in the known superconducting GIC.

Compound	T_c(K)	Ref.**
C_8K*	0.55	[1]
C_8K	0.39	[1]
C_8Rb	0.020–0.151	[1]
C_8Cs	0.020–0.135	[1]
C_8K	0.128–0.198	[2]
C_8K	0.125	[3]
$KHgC_8$ ($n = 2$)	1.9	[4]
$RbHgC_8$ ($n = 2$)	1.46	[5,6]
$KHgC_4$ ($n = 1$)	$0.7 \leqslant T_c \leqslant 0.9$	[7]
$RbHgC_4$ ($n = 1$)	0.99	[7]
$KTl_{1.5}C_4$ ($n = 1$)	2.7	[8]
$KTl_{1.5}C_8$ ($n = 2$)	2.45	[8]

*Samples with excessive K
**References:
[1] Hannay et al. (1965).
[2] Koike et al. (1978, 1980).
[3] Kobayashi and Tsujikawa (1979).
[4] Alexander et al. (1980a).
[5] Pendrys et al. (1981).
[6] Alexander et al. (1981).
[7] Tanuma et al. (1981b,c).
[8] Vogel et al. (1981).

Rb, Cs (with $n = 1$ and 2) and of the $C_{4n}KTl_{1.5}$ type ($n = 1$ and 2). In this group, the temperatures of superconducting transitions vary from 1 to 3 K.

The superconduction transition temperatures in the known superconducting GIC are listed in table 27.

It is of interest to note that an excess of K in a sample of C_8K leads to an increase in the superconducting transition temperature up to 0.55 K, as compared with $T_c = 0.39$ K in a sample of the same stoichiometric composition. Therefore, it is possible that the different T_c found by different authors or by the same authors in different samples may be attributed to variations in the composition and structure of the intercalate layer.

10.13.3. Peculiarities in the superconduction properties of GIC

The main distinguishing feature of the superconduction properties of GIC, which is associated with the sharp anisotropy of their electronic spectrum, is the strong angular dependence of the superconduction parameters in the magnetic field $\boldsymbol{H}$ upon the angle θ between $\boldsymbol{H}$ and the c-axis. This is first of all true of the critical magnetic-field value. Figure 151 shows, as an illustration, the θ-dependence of a second critical field H_{c_2} in C_8KHg, which represents a type-II superconductor. One can see from this figure that in the narrow angular range $80° \leqslant \theta \leqslant 100°$ the H_{c_2} passes through a sharp peak, varying from 400 to 1500 Oe. As a result of the high anisotropy of the electron energy spectrum, in

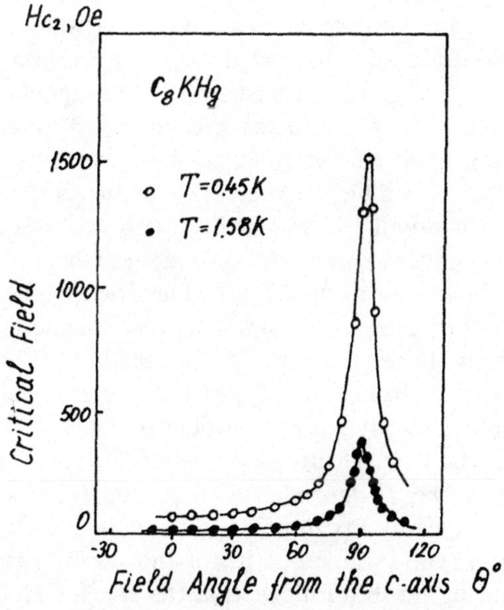

Fig. 151. Angular dependence of the critical field H_{c_2} for C_8KHg. (Iye 1982.)

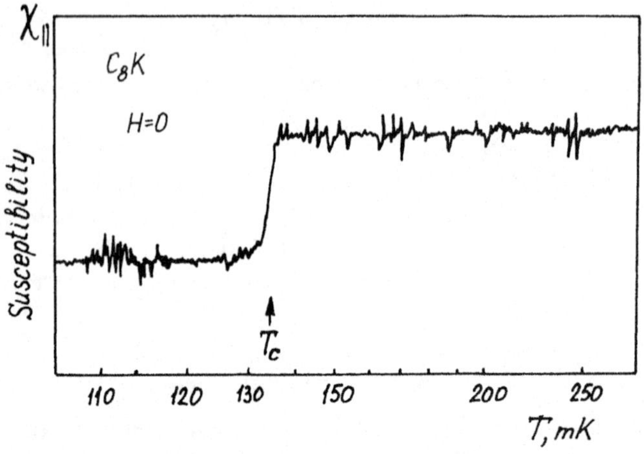

Fig. 152. The AC magnetic susceptibility of a C_8K sample as a function of temperature at a nearly zero DC field. The sudden change in susceptibility near 134 mK is identified with the superconduction transition. (Koike et al. 1980.)

certain GIC with changing orientation of the magnetic field not only the magnitude of superconduction parameters, but even the kind of superconductivity change. This unique feature has been found in C_8K, which is a type-I superconductor for $\boldsymbol{H} \| c$ and a type-II superconductor for $\boldsymbol{H} \perp c$. The existence of such superconductors was predicted by Katz (1969), C_8K being the first discovered representative of the class of substances whose kind of superconductivity depends on the magnetic field orientation. The above feature of C_8K was first observed in measuring the magnetic susceptibility of this compound for $\boldsymbol{H} \| c$ and $\boldsymbol{H} \perp c$ in the mK temperature range. Figure 152 illustrates the variation of the differential magnetic susceptibility $\chi_{\|}$ in a very weak magnetic field ($H \sim 0.5$ Oe) in the superconducting transition in C_8K, whose temperature T_c is about 134 mK. Typical dependences of the differential magnetic susceptibility of this compound on magnetic field for $\boldsymbol{H} \| c$ and $\boldsymbol{H} \perp c$ (at $T = 90$ mK) are illustrated in fig. 153a,b. The H-dependence of $\chi_{\|}$ for $\boldsymbol{H} \| c$ corresponds to the complete expulsion of the magnetic flux from the superconductor in fields lower than the critical one, and is characteristic of a type-I superconductor. It has an appreciable hysteresis and allows one to determine both the thermodynamical critical magnetic field H_c and the supercooling field H_{sc} (see fig. 153a).

On the other hand, the H-dependence of $\chi_{\perp}$ for $\boldsymbol{H} \perp c$ corresponds to a partial expulsion of magnetic flux in fields lower than the second critical field H_{c_2}, being typical of a type-II superconductor (fig. 153b). From a comparison between fig. 153a and fig. 153b it follows that the H_{c_2} for $\boldsymbol{H} \perp c$ is about twice as high as the H_c for $\boldsymbol{H} \| c$. The angular dependence of the critical fields H_c, H_{c_2}

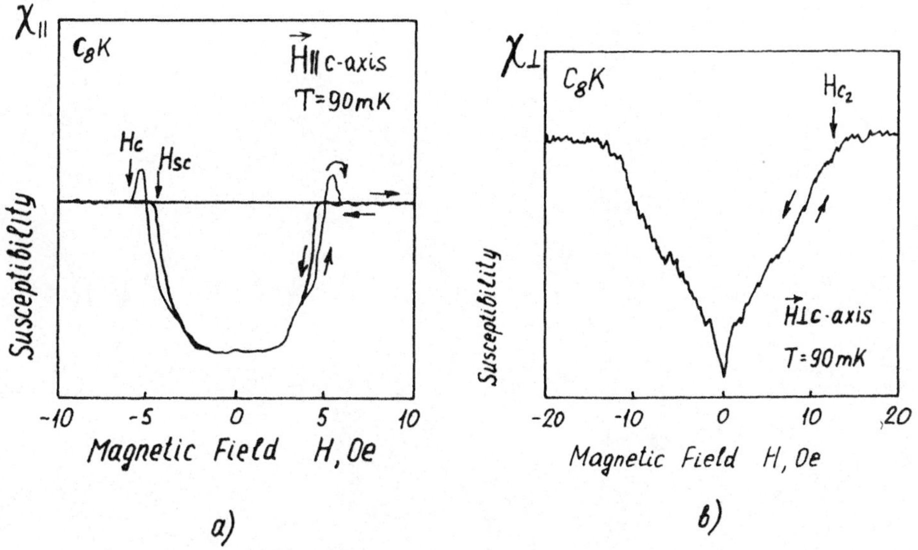

Fig. 153. The AC magnetic susceptibility measured by Koike et al. (1980) at 90 mK as a function of the DC magnetic field H for a stage-1 C_8K sample; (a) $\boldsymbol{H} \| c$-axis. From the magnetic field dependence of $\chi_{\|}$ type-I superconduction behaviour is inferred. (b) $\boldsymbol{H} \perp c$-axis. From the field dependence of $\chi_{\perp}$, type-II superconduction behaviour is inferred.

and H_{c_3} in C_8K is shown in fig. 154. It can be seen that in the range $0 \leqslant \theta \leqslant 40°$ the $H_{sc} \neq 0$ and C_8K is a type-I superconductor, while for $40° \leqslant \theta \leqslant 90°$, the $H_{sc} = 0$ and C_8K becomes a type-II superconductor.

The angular dependence of the critical field H_{c_2} can approximately be described by the relation given by Morris et al. (1972):

$$H_{c_2}(\theta) \sim (\cos^2\theta + \eta^2 \sin^2\theta)^{-1/2}, \tag{217}$$

in which the parameter η is expressed through the anisotropy of effective masses, m_c/m_a, or the anisotropy of coherence length, ξ_a/ξ_c;

$$\eta^2 = \frac{m_c}{m_a} = \left(\frac{\xi_c}{\xi_a}\right)^2, \tag{218}$$

where ξ_a, ξ_c, m_a, m_c are the coherence lengths and effective masses in the plane of the layer (subscript "a") and along the c-axis (subscript "c"). In fig. 154, the solid curve for H_{c_2} has been constructed using eq. (217) and the fitting parameter $\eta^2 = 39$. The curve agrees rather well with the experimental values of H_{c_2} in the range of existence of type-II superconductivity. The thus found anisotropy of the effective electron masses in C_8K is considerably lower than the respective anisotropy in pure graphite, where it is $m_c/m_a \simeq 170$. It is interesting to note that the effective-mass anisotropy $m_c/m_a \simeq 39$ in C_8K is close to the analogous

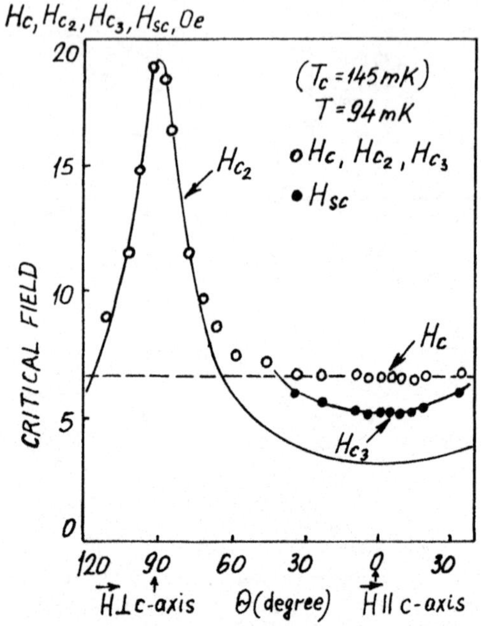

Fig. 154. Angular dependences of the critical fields H_c, H_{c_2}, H_{c_3} and the supercooling field H_{sc} for C_8K deduced from measured magnetic susceptibilities. θ is the angle between $\boldsymbol{H}$ and the c-axis. (Koike et al. 1978.)

quantity in lamellar superconductors – transition-metal dichalcogenides 2H–TaS_2, 2H–$NbSe_2$ (see Morris and Coleman 1973). Accordingly, in analogy with the lamellar dichalcogenides, superconducting GIC (at least those based on alkali metals) should probably be treated as three-dimensional highly anisotropic superconductors, in which the ratio ξ_c/ξ_a is about 6.2, rather than as substances with a purely two-dimensional superconductivity.

10.13.4. The nature of superconductivity in GIC

Recently a suggestion has been made (Iye 1982) that superconductivity in GIC results from the coexistence of two different electron systems: one in the graphite layers and the other in the intercalate layer.

Indeed, in all superconducting GIC of the C_8M type the fraction of charge transferred from the intercalant to the graphite layer is much less than unity. Thus, for example, according to the data of Inoshita et al. (1977), in C_8K, $f_{el} = 0.6$ and, therefore, in the intercalate layer, potassium has a partly filled conduction band, which exists alongside of the graphite-layer conduction band. In contrast to this, in C_6Li and in other nonsuperconducting compounds with alkali metals, the intercalant completely gives up its valence electrons to the graphite-layer conduction band ($f_{el} = 1$), so that the conduction band of the intercalant turns out to be empty (see Holzwarth et al. 1978a). For ternary superconducting GIC the transferred charge fraction f_{el} is not known exactly, but if one takes into consideration the magnitude of electron affinity in Hg and Tl, then the intercalants MHg and $MTl_{1.5}$ should be classed as weak donors, for which $f_{el} < 1$ is typical. Thus, it seems, one can assume that, as in C_8M, in ternary superconducting GIC the intercalate-layer electron system and the graphite-layer electron system coexist.

At this point, a natural question may be asked: which of the two electronic systems plays the dominant role in the formation of superconductivity in GIC – the electrons delocalized in the graphite layers or the electrons quasi-localized in the intercalate layer?

In order to find a plausible answer to this question, let us pay attention to the following peculiarities in the superconduction properties of ternary GIC, such as $C_{4n}MHg$ or $C_4KTl_{1.5}$:

(a) the superconduction transition temperature T_c in $C_{4n}MHg$ is of the same order as T_c in pure MHg;

(b) the anisotropy of the higher critical field H_{c2} is greater in C_8MHg ($n = 2$) than in C_4MHg ($n = 1$);

(c) ternary compounds $C_4KTl_{1.5}$ contain more metal than do $C_{4n}MHg$ ($n = 1$ and 2) and also have higher T_c values.

All these features taken together prompt the suggestion that in all likelihood it is the intercalate-layer electron system that plays the dominant role in the superconductivity of ternary GIC. As regards the binary superconducting compounds of the C_8M type, here the relation between the T_c and the f_{el} has not yet been established and this precludes any inferences as to the dominance of one or another electronic system in the formation of superconductivity there. A step in this direction has recently been taken by Shimizu and Kamimura (1983), whose first-principle calculation of the electron–phonon interaction in C_8K indicates that the superconductivity is due to intercalate-layer electrons and is determined by their interaction with atomic vibrations in the plane of the layer, as well as with the longitudinal acoustic and optical vibrations along the *c*-axis.

Thus the intercalate layers appear to play the fundamental role in the formation of superconductivity in GIC. It is more logical then to consider this superconductivity as a consequence of the insertion of graphite in the initial matrix of the intercalant. In such a case, the stage of intercalation is, as before, determined by the number of the graphite layers placed between the neighbouring intercalate layers; in this representation, however, pure intercalant should be assigned the stage index $n = 0$. Since the anisotropy of the superconduction properties of GIC is dependent upon their intercalation stage, it is of interest to examine the dependence of the coherence lengths ξ_c and ξ_a in

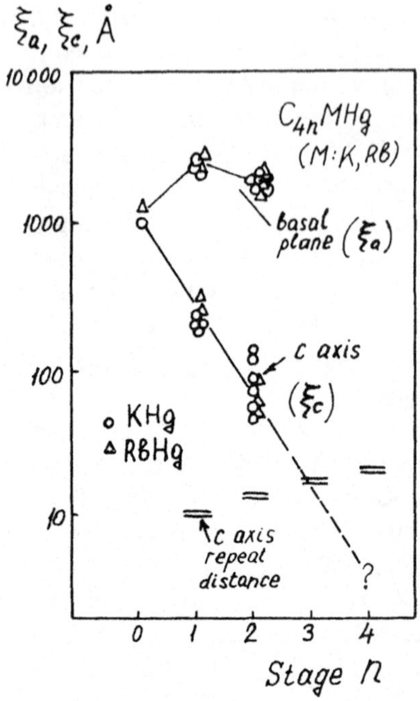

Fig. 155. Stage dependence of the coherence lengths of $C_{4n}HgM$. Stage zero denotes the pristine intercalant materials KHg and RbHg. The *c*-axis stacking period I_c is also shown in the figure for comparison. (Iye 1982.)

compounds with a superconducting intercalant. Figure 155 illustrates such dependences for compounds with KHg and RbHg, which have been obtained by Iye (1982) for 18 different samples. From the figure one can see that the coherence length in the plane of the layer ξ_a is close to 2000 Å, and depends but weakly on stage index. In contrast, ξ_c rapidly decreases as n increases from zero to two. At the same time, it should be noted that even for GIC with $n = 2$ the value of ξ_c is still considerably larger than the distance between the neighbouring intercalate layers. This fact not only serves as a substantiation of the applicability of the effective-mass model to superconducting GIC of the $C_{4n}MHg$ type but also suggests that these superconductors, though highly anisotropic, remain essentially three-dimensional, just as the binary compounds. The naive extrapolation of the dependence of ξ_c upon stage index n presented in fig. 155 allows one to expect that should compounds of the $C_{4n}MHg$-type with $n \geqslant 3$ be superconductors the behaviour of the superconductivity would be purely two-dimensional.

The lack of data still precludes an unambiguous answer to the question concerning the nature of the superconductivity in GIC. Even if the assumption about the dominant contribution from the intercalant electrons to the superconductivity of GIC appears to be to a certain extent justified for ternary compounds, it is yet unclear how much the electrons of two different systems are involved and what are the mechanisms of their respective contributions. One should note, in particular, that even if the graphite-layer electrons do not directly participate in the superconductivity, their contribution is obviously nonzero, because they play the major role in screening the intercalate layers. Moreover, the phonons corresponding to vibrations in the graphite layers may also play an important part in the superconduction electron pairing in the intercalate layers.

10.13.5. The effect of pressure on the superconduction properties of GIC

The high value of compressibility of GIC along the *c*-axis, the great anisotropy of the H_{c_2}, and the existence of pressure-induced phase transitions make it possible to suppose that the T_c and other superconducting properties of GIC should significantly depend on the magnitude of the pressure applied. The value of T_c in GIC has been measured in many experiments. Figure 156

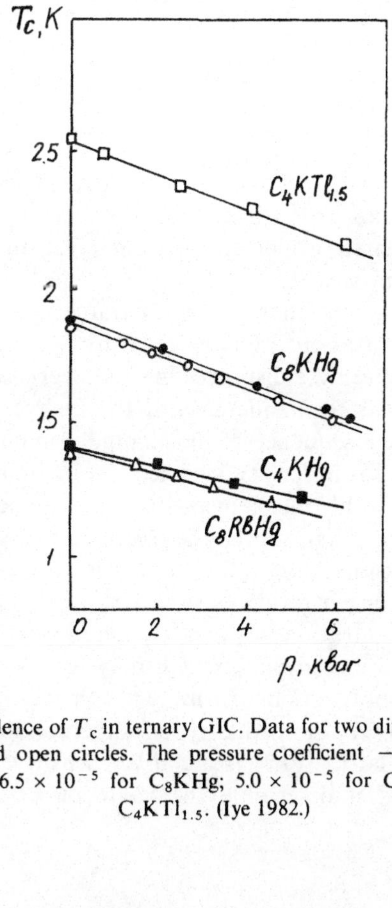

Fig. 156. Pressure dependence of T_c in ternary GIC. Data for two different samples of C_8KHg are denoted by the solid and open circles. The pressure coefficient $-dT_c/dp$ (in units of kbar) is: 3.4×10^{-5} for C_4KHg; 6.5×10^{-5} for C_8KHg; 5.0×10^{-5} for C_8RbHg; and 6.4×10^{-5} for $C_4KTl_{1.5}$. (Iye 1982.)

presents the pressure dependeces for the ternary compounds C_4KHg, C_8KHg, C_8RbHg, and $C_4KTl_{1.5}$. It can be seen that T_c linearly decreases with increasing pressure, the baric coefficients dT_c/dp (expressed in units of 10^{-5} K/bar) being −3.4, −6.5, −6.0, and −6.4, respectively. The values of these coefficients in most of the known superconducting ternary GIC are negative, are of the same order of magnitude, and are close to dT_c/dp for typical superconductors such as Hg and Sn. This behaviour of T_c in the above-mentioned GIC under pressure is in sharp contrast with that in other lamellar superconductors, such as 2H–$NbSe_2$, 2H–TaS_2, in which T_c rapidly increases with compression (Smith et al. 1972, 1975). The high positive value of dT_c/dp in these compounds is ascribed to the suppression of the charge density wave (CDW) as a result of compression (see Friend and Jerome 1979).

Let us consider in broad outline the character of the change in T_c in ternary compounds within the Bardeen–Cooper–Schrieffer (BCS) theory of superconductivity.

According to the BCS-theory, for superconductors with weak binding the relation

$$kT_c \simeq \hbar\omega_D e^{-1/\lambda} \tag{219}$$

holds, with

$$\lambda = \nu(\varepsilon_F)V = \frac{\nu(\varepsilon_F)\langle I^2\rangle}{M\langle\omega^2\rangle}, \tag{220}$$

where ω_D is the Debye frequency, $\nu(\varepsilon_F)$ is the density of states at the Fermi level, M is the ionic mass, $\langle\omega^2\rangle$ is the mean-square phonon frequency, and $\langle I^2\rangle$ is the squared matrix element of the electron–phonon interaction averaged over the Fermi surface.

What can be said about the change in the parameters contained in eqs. (219) and (220) that is brought about by compression?

Under pressure the interlayer distances in GIC decrease; the lattice becomes more "rigid", which leads to an increase in the Debye and average phonon frequencies. The other consequence of the change in interlayer distances is the redistribution of the electron density between the intercalate and the graphite layers. There are two possible mechanisms to affect the behaviour of the above parameters. The first mechanism is the increased electron wave function overlap between the atoms of graphite and those of the intercalant, which leads to an increased concentration of electrons in the graphite layers and a decreased electron concentration in the intercalate layers. The other involves a change in intercalant atom ordering, which may lead to the reverse. Nothing definite can be said about the change in $\langle I^2\rangle$. Proceeding from the existing data on the behaviour of T_c in normal type-I superconductors under pressure, one can only suppose that the $\langle I^2\rangle$ is much less sensitive to compression than the other parameters that determine the value of λ [see, e.g., Brandt and

Ginzburg 1969]. The increase in Debye frequencies in normal superconductors leads, as a rule, to a decrease in T_c. The increased carrier concentration in the layers responsible for superconductivity must lead, according to eq. (220), to an increase in λ and, therefore, in T_c. Therefore, the experimentally observed decrease of T_c in ternary GIC may be caused by two factors. The first is a decrease of λ as a result of increased $\langle\omega^2\rangle$, as is the case, for example, with Sn, In, Hg, Pb, and other metallic superconductors under pressure. The closeness of the values for dT_c/dp in GIC and in these metals testifies to the above mechanism. The second mechanism is a decrease of the electron concentration and density of states in the intercalate layer, provided that the superconducting properties of ternary compounds are due to electrons in this layer.

One should note that the character of change in the superconducting properties of the ternary compound C_4KHg ($n = 1$) under pressure is abnormal. The abnormality consists in the fact that the rather broad superconducting transitions that take place in this compound under normal pressure (the transition width δT_c being about 0.6 K) are suddenly narrowed as soon as a relatively small pressure is applied (δT_c becoming ~40 mK, see fig. 157). This is accompanied by a steady decrease of the T_c itself, as has already been mentioned above. By contrast, C_8KHg ($n = 2$) exhibits no such great narrowing under pressure. In this compound, the value of δT_c decreases but weakly, from about 0.6 K at $p = 1$ bar to ~0.4 K at $p \simeq 5$ kbar (see Yeh et al. 1982). There are several possible explanations for the abnormal change in transition width in C_4KHg for $p \leqslant 1$ kbar. We shall indicate three possibilities.

First, this may be the consequence of a phase transition of the first kind at the critical pressure $p_{cr} \leqslant 1$ kbar. The broad transitions under normal pres-

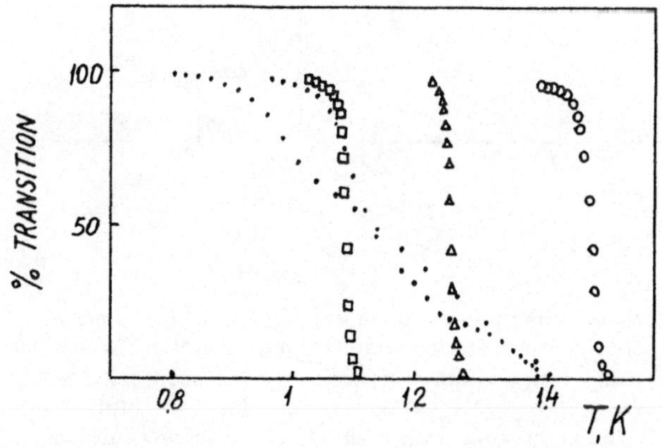

Fig. 157. Percentual completion of the inductively measured superconducting transition of $KHgC_4$ versus temperature T for various pressures p: ●, $p = 1$ bar; ○, $p = 0.8$ kbar; △, $p = 4.7$ kbar; □, $p = 8.1$ kbar. Note the sharp transition signals for $p > 0$ and the broadened transitions observed at $p = 1$ bar. (De Long and Eklund 1983.)

sure may then be explained by the fact that under these conditions C_4KHg represents a mixture of two phases, one with a high T_c, the other with a low T_c. The application of a small pressure makes the compounds single-phased which manifests itself in the narrowing of the transition.

Secondly, a high degree of disorder in the intercalate layers of C_4KHg at the normal pressure may be supposed. A relatively small compression may then result in the appearance of a long-range order in the intercalate layers or the order of layer stacking along the *c*-axis.

Thirdly, the sudden narrowing of transition width and the decrease of T_c under pressure may be due to the fact that in C_4KHg under normal pressure there is a CDW-state, which is destabilized by pressure. As has already been mentioned, a similar phenomenon can be observed in anisotropic superconducting materials, such as NbSe (see, e.g., Yeh et al. 1982).

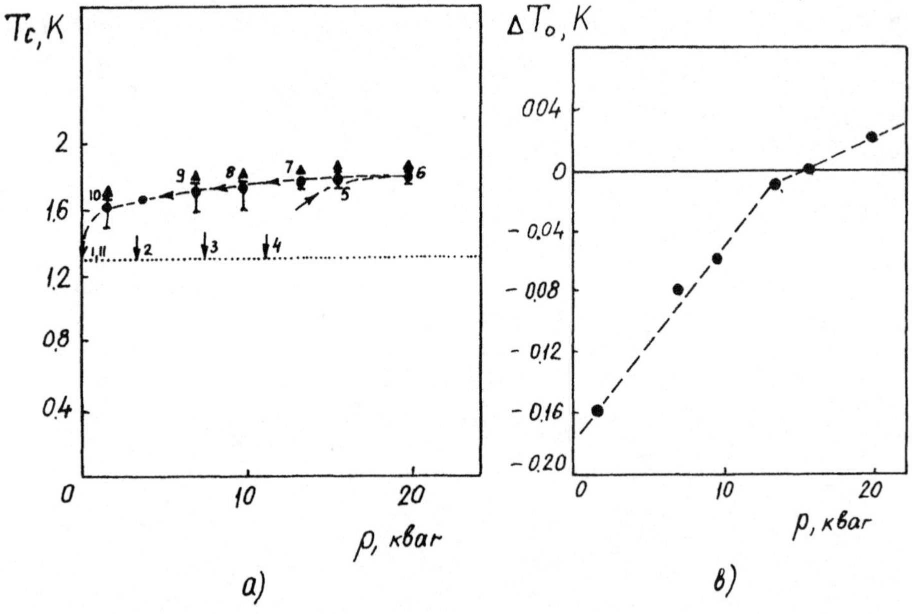

Fig. 158. (a) Superconducting transition temperature T_c versus pressure p for C_8K. The dotted horizontal line indicates the lower temperature limit for a ^4He cryostat. The data points are numbered in increasing temporal order. ▲, onset of superconductivity; ●, temperature of 50% transition completion; ---- is a guide for the eye. (b) Pressure-induced shift $\Delta T_0 \equiv T_0(P) - T_0(15.5\,\mathrm{kbar})$ of the temperature T_0 of the onset of superconductivity for C_8K as a function of the applied pressure p. --- a guide for the eye. A phase transition at room temperature has previously been identified at ~ 13 kbar. The points at $p \simeq 15.5$ and 19.5 kbar were taken as the pressure had been increased, the remaining points having been recorded upon decreasing the pressure. The sample failed to show any superconduction transition up to $T > 1.30$ K during the initial and final room-pressure measurements. (De Long and Eklund 1983.)

Unlike ternary compounds, the binary compound C_8K is characterized by a positive baric coefficient $dT_c/dp \simeq 1.3 \times 10^{-5}$ K/bar. As the measurements of its superconducting properties suggest (De Long et al. 1982), in this compound, a phase transition of the first kind occurs at a critical pressure ranging from 2 kbar to 15 kbar. The high-pressure phase in C_8K is characterized by a relatively high $T_c \geqslant 1.7$ K and a relatively low value of $dT_c/dp \simeq 0.47 \times 10^{-5}$ K/bar (see fig. 158a,b). These results are consistent with those of X-ray diffraction measurements (Wada et al. 1980b) under pressures up to 15 kbar at room temperature and also with those of conductivity measurements (Fuerst et al. 1981), which has made it possible to narrow the critical pressure range for the above transition down to 13 kbar $\leqslant p \leqslant$ 15 kbar. This range contains $p \simeq 13$ kbar, at which the curve for the p-dependence of T_c flexes, which may be ascribed to a phase transition. The character of this transition is yet obscure, although Solin (1981) has made the suggestion that it is associated with a change in layer stacking order.

The increase in T_c of C_8K under compression may be the consequence of the increasing electron concentration in the superconducting layers. If the f_{el} in C_8K under pressure increases, then the positive sign of dT_c/dp may suggest that the superconductivity of this compound is due to the electrons in the graphite layers. But how far this is true is difficult to say.

The increase in T_c in C_8K under pressure has given rise to the hopes that at high pressures it will be possible to find superconductivity in other alkali-metal compounds, in particular in C_8Li or in K-GIC of the higher stage indices. Unfortunately, as the results of the recent investigations of such materials at high pressures show, these dreams have not yet come true. Thus, according to De Long et al. (1982), at pressures to 10 kbar a K-GIC with $n = 2$ does not exhibit superconduction properties in the temperature range down to 70 mK.

CHAPTER 11

Lattice Vibrations and the Phonon Spectrum of Graphite and Graphite-Based Compounds

11.1. Models of the vibrational spectrum of graphite

All models of the vibrational spectrum of graphite are essentially based on the fact that the graphite lattice has a sharply anisotropic, layered structure. One of the earliest simple models capable of satisfactorily describing the main features of the vibrational spectrum of graphite has been developed, on the basis of a semicontinuum treatment of the spectrum, by Komatsu and Nagamiya (1951). In their model, the graphite crystal is treated as a set of elastic plates which are at an equal distance from one another and which are held together by forces preventing the crystal from being compressed, stretched, and also sheared in the plane of the layers. Except for the bending modulus, the elastic constants of each plate were determined from the force constants of the flat benzene molecule, taking account of the difference in bond length between the carbon atoms in benzene and in graphite. The compressional coupling constants were determined in the Komatsu–Nagamiya model (1951) from the known value of graphite's compressibility, while the shearing constants were estimated on the basis of the general concepts of the forces that bind together the neighbouring atoms in the graphite lattice. The undetermined parameter of the model is the bending modulus of each plate. This was determined by fitting the theoretically calculated dependence of the heat capacity of a plate on its temperature to the experimental curve obtained in the temperature range where the effect of shearing may be neglected. As has been demonstrated by Komatsu (1955), the heat capacity of graphite calculated within the semicontinuum model is in good agreement with the experimental data of De Sorbo and Tyler (1953), Bergenlid et al. (1954), and De Sorbo (1954), obtained in the temperature range from 60 K down to 1.5 K. In this range, the heat capacity is at first (from 60 down to 15–13 K) governed by the law $C \sim T^2$, typical of two-dimensional systems, but below $T = 13$ K it is already described by the relation $C \sim T^{2.4}$, which suggests that the effect of the third degree of freedom (along the c-axis) should increase.

The approximation used by Komatsu and Nagamiya (1951) in their model, in which real atomic honey-comb net planes in the graphite lattice are replaced by continuous elastic plates, ceases to be satisfactory at temperatures above 60 K when vibrations with rather large wave vectors are excited in the vibrational spectrum of graphite. In this temperature range, one should take into account the discrete atomic structure of the lattice, i.e., use the Born–von Kármán approach in one form or another.

The earliest model based on the concept of atomic interaction in the graphite lattice was proposed by Krumhansl and Brooks (1953), who introduced into their model force constants to describe the interaction between the carbon atoms. Proceeding from the same assumptions, Yoshimori and Kitano (1956) have succeeded in developing a more realistic model, applicable in a wide range of temperature. They assumed that between the carbon atoms in the graphite lattice four types of bonding forces exist which represent the restoring forces of the vibrational spectrum problem. To these forces, there correspond four force constants.

The first and the second type of these bonding forces are restoring forces of the valence-force type, the first being associated with changes in the bond lengths and the second associated with those in the bond angles of the honey-comb net planes. The third is of the central-force type, and is due to changes in the distance between those atoms which are on adjacent planes and are neighbours along the c-axis. The fourth is the bending force constant of each honey-comb net plane; this force is of the non-central force type, and may conveniently be defined as the force required for the displacement of each atom relative to the average of the normal displacement of three neighbouring atoms in the same net plane.

Apart from these forces, there are other central forces of atomic interaction, namely, interactions between pairs of atoms in the neighbouring or more distant atomic planes. Evidently, these interactions provide additional contributions to both compressional and shearing moduli. However, as has been shown by Komatsu (1955), the contribution of these forces to the compressional modulus may sufficiently accurately be described by varying the force constant associated with the force of the third type, because atomic planes in the graphite lattice are so rigid that displacement of atoms relative to one another along the c-axis in any one plane is small, as compared with the relative displacement of the planes themselves. This means that the interacting atomic planes in the graphite lattice essentially behave as rigid planes. As for the forces that oppose the relative shear of the atomic planes, they, according to Komatsu (1955), are rather weak and become significant only in the range of very low temperatures. In this range, the temperature dependence of heat capacity shows deviations from the T^2-law and the graphite lattice begins to behave as a system close to being three-dimensional, rather than a sharply anisotropic layered system.

The Yoshimori–Kitano model (1956) of atomic interactions in the graphite lattice has further been refined, as applied to a description of the vibrational spectrum, by Young and Koppel (1965), who paid attention mainly to the analysis of the high-frequency range of lattice vibrations and the temperature dependence of the heat capacity in the range from 100 to 1000 K. The Young–

Koppel model (1965), which, as the preceding model of Yoshimori and Kitano (1956), has four force constants and which is in essence fully equivalent to its predecessor, has already been discussed in sect. 3.3.2, where we have pointed out that the model is incapable of describing correctly the mechanical properties of the graphite lattice. Indeed, if the interactions between the carbon atoms are taken into account as crudely as in the Young–Koppel model, the graphite lattice turns out to be internally unstable, which manifests itself largely in the instability of the lattice to the relative displacement of the graphite layers. As a result of this, the c_{44}-component of the elastic modulus goes to zero. The completely mechanically stable graphite lattice can be described only by the rather complicated model of atomic interactions developed by Nicholson and Bacon (1977), which involves 21 force constants.

At the same time, we should note that, in spite of their crudeness and associated drawbacks apparent in describing the elastic moduli of the graphite lattice, the models of Yoshimori and Kitano (1956) and of Young and Koppel (1965) may serve as a reasonably good basis for constructing the theory of the phonon spectrum of graphite. This is due to the fact that the contribution from the (weak) interactions between the more distant carbon atoms in the lattice, which are disregarded in the above models, is small, becoming significant only in some rather narrow ranges of frequencies (and temperatures). In fact, this means that the construction of a model of atomic interactions which would ensure the complete mechanical stability of the lattice is much more difficult than the description of the most important types of vibrations and the determination of the range of the vibrational frequencies of the lattice. This fact is especially stressed here because, as comparison with experiment suggests, even the above-mentioned rather simple model of atomic interactions in the graphite lattice, containing only four force constants, makes it possible to analyze all important features of the phonon spectrum of graphite, i.e., to determine the principal modes of vibration in all branches of the spectrum, to calculate, even though approximately, the dispersion relations for each branch, as well as the partial $[D_i(\omega)]$ and total $[D(\omega)]$ phonon density of state functions. It is highly important that an analysis of the phonon spectrum using such simple models enables one to describe very clearly the principal vibrational modes.

The above considerations serve as the justification for a preliminary treatment of the phonon spectrum of graphite on the basis of this simple model involving four force constants, whereupon we shall analyze the changes and refinements provided by the more complicated models and shall make comparisons with experiment.

11.1.1. Vibrational spectrum of the graphite lattice in models with four force constants

The main feature of the vibrational spectrum of graphite in such models is that vibrations which are parallel and perpendicular to the basal plane are inde-

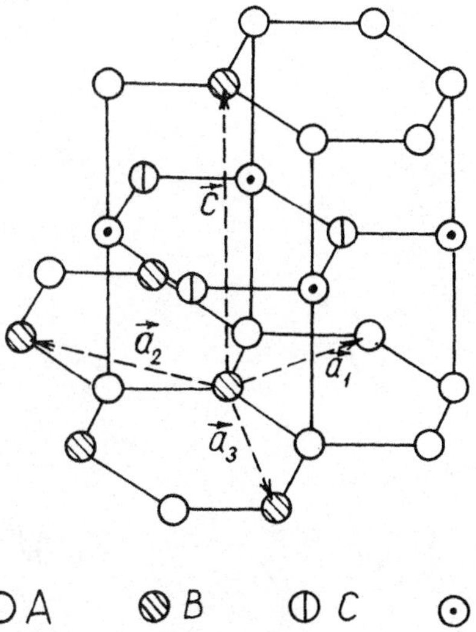

Fig. 159. Various types of atoms in a graphite lattice from the viewpoint of their interactions. $|\boldsymbol{a}| = 2.46\,Å$; $|\boldsymbol{c}| = c_0 = 6.70\,Å$.

pendent of each other, i.e., there is no coupling between them. This decoupling in a layered structure considerably simplifies the calculation of the phonon spectrum, although, as has already been mentioned above, this also leads to an inherent lattice-structure instability due to the lack of resistance to shear ($c_{44} = 0$).

Let us dwell at greater length on how the force constants μ, k, μ', k' are introduced in the simple model. To determine them, we shall consider four types of carbon atoms (A, B, C, D), which differ from one another in their position in the graphite lattice (fig. 159). The four force constants are introduced as follows:

k is the bond-stretching force constant between nearest neighbours, such as the atoms A and B; it is introduced to describe the change in the bond length Δl for each bond of the honey-comb net plane;

μ is the bond-bending force constant which describes the force needed to change the angle formed by two A–B bonds of the honey-comb net plane;

μ' is the bond-bending force constant which describes the force required to displace (by Δh) an atom along the c-axis relative to its nearest neighbours in the same plane;

k' is the bond-stretching force constant introduced to describe the change in the distance, $\Delta l'$, between two atoms, such as A and D, which are on the adjacent net planes, and are neighbours along the c-direction. The model involving the force constants k, μ, μ', k' is, in the literature, called the bond-bending and bond-stretching (BBS) model.

Such an introduction of the four force constants is illustrated in fig. 160, where $\frac{1}{2}k(\Delta l)^2$, $\frac{1}{2}\mu(l\,\Delta\psi)^2$, $\frac{1}{2}k'(\Delta l')^2$ and $\frac{1}{2}\mu'(\Delta h)^2$ represent the contributions from

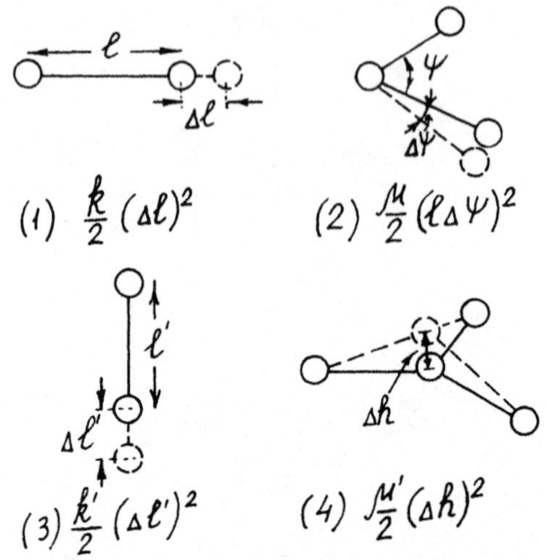

Fig. 160. Carbon-atom displacements in a graphite lattice (see fig. 159) corresponding to the force constants k (1), μ (2), k' (3), and μ' (4). (Yoshimori and Kitano 1956.)

the respective displacements of atoms to the potential energy, Φ, of the elastic deformation of a crystal, which, in the harmonic approximation, is a quadratic form of the components of the vector, $\boldsymbol{u}(m, \boldsymbol{n})$, of the displacement of the m-th carbon atom in the $\boldsymbol{n}$th unit cell from its equilibrium position (see sect. 3.3.1). The above-mentioned disturbances of the equilibrium position of carbon atoms in the graphite lattice (i.e., Δl, $\Delta\psi$, $\Delta l'$, and Δh) are expressed in terms of the components of the displacement vector $\boldsymbol{u}(m, \boldsymbol{n})$, owing to which the force constants ϕ_{ij} introduced in sect. 3.3.1. [see eq. (22)] may, in the adopted approximation, be expressed in terms of force constants k, μ, k', μ' (Young and Koppel 1965) which, in turn, are related to force constants used in other models [see, e.g., eq. (25) in sect. 3.3.2]. The equation of motion that describes the vibration of carbon atoms has the form

$$M\frac{\mathrm{d}^2\boldsymbol{u}(m, \boldsymbol{n})}{\mathrm{d}t^2} = -\mathrm{grad}_{\boldsymbol{u}(m,\boldsymbol{n})}\Phi. \tag{221}$$

The solutions of this equation are the plane waves

$$\boldsymbol{n}(m, \boldsymbol{n}) \sim \boldsymbol{e}(m, \boldsymbol{q})\mathrm{e}^{\mathrm{i}[\boldsymbol{q}\cdot\boldsymbol{n}) - \omega t]} \tag{222}$$

with the wave vector $\boldsymbol{q}$ and the frequency ω (here M denotes the mass of a carbon atom and $\boldsymbol{e}(m, \boldsymbol{q})$ denotes the polarization vector for vibrations). In accordance with the character of the force constants adopted, the secular determinant of equation (221) separates into two parts, one part giving the modes whose polarization vectors are in the honey-comb net plane (designated here as vibrations in the plane) and the other part giving the modes whose polarization vectors are perpendicular to the net plane (designated here as vibrations perpendicular to the plane). Let the unit vectors $\boldsymbol{j}_i$ and $\boldsymbol{k}$ be directed along the three primitive translation vectors $\boldsymbol{a}_i$ ($i = 1, 2, 3$) and along the $\boldsymbol{c}$ vector, as shown in fig. 159, and the unit vectors $\boldsymbol{i}_i$ ($i = 1, 2, 3$), in the directions $[\boldsymbol{j}_i \times \boldsymbol{k}]$. We shall now consider the character of the above-mentioned two types of vibrations (in, and perpendicular to, the plane) by following the argumentation of Yoshimori and Kitano (1956). For $\boldsymbol{q} \perp \boldsymbol{k}$, it follows from the secular equation that two adjacent honey-comb net planes, which are crystallographically different from each other, have the same vibrational modes in the plane. The secular equation for vibrations in the plane does not contain the vector $\boldsymbol{k}$. This means that the contours of constant frequency in $\boldsymbol{q}$-space for this mode of vibrations are cylinders whose axis is parallel to the $\boldsymbol{k}$-vector.

The dispersion relations for vibrations in the plane may be obtained in these two cases: $\boldsymbol{q} \| \boldsymbol{i}_i$ and $\boldsymbol{q} \| \boldsymbol{j}_i$. In the former case, the secular equation has four roots:

$$\omega_{1,2}^2 = \frac{1}{M}[\tfrac{3}{2}k + 9\mu \mp \sqrt{\alpha_1^2 + \beta_1^2}],$$

$$\omega_{3,4}^2 = \frac{1}{M}[\tfrac{3}{2}k + 7\mu + 2\mu \cos(\tfrac{1}{2}\sqrt{3}\, a_0 q) \mp \sqrt{\alpha_2^2 + \beta_2^2}], \tag{223}$$

where

$$\begin{aligned}
\alpha_1 &= k \cos(\tfrac{1}{3}\sqrt{3} a_0 q) + ([\tfrac{1}{2}k + 9\mu] \cos(\tfrac{1}{6}\sqrt{3} a_0 q),\\
\beta_1 &= k \sin(\tfrac{1}{3}\sqrt{3} a_0 q) - [\tfrac{1}{2}k + 9\mu] \sin(\tfrac{1}{6}\sqrt{3} a_0 q),\\
\alpha_2 &= 6\mu \cos(\tfrac{1}{3}\sqrt{3} a_0 q) + [\tfrac{3}{2}k + 3\mu] \cos(\tfrac{1}{6}\sqrt{3} a_0 q),\\
\beta_2 &= -6\mu \sin(\tfrac{1}{3}\sqrt{3} a_0 q) + [\tfrac{3}{2}k + 3\mu] \sin(\tfrac{1}{6}\sqrt{3} a_0 q),\\
a_0 &= |\boldsymbol{a}_i| = 2.46\,\text{Å}.
\end{aligned} \tag{224}$$

The four dispersion relations given by eq. (223) describe, respectively, the linearly polarized longitudinal acoustical and optical vibrations and the transverse acoustical and optical vibrations in the plane.

In the case of $\boldsymbol{q} \| \boldsymbol{j}_i$, the analogous four equations for vibrations in the plane have the form:

$$\omega_{1,2}^2 = \frac{1}{M}[2k + \tfrac{9}{2}\mu + \tfrac{1}{2}\mu \cos(a_0 q) - (\tfrac{1}{2}k - 4\mu)\cos(\tfrac{1}{2}a_0 q) \mp \sqrt{\alpha_3^2 + \beta_3^2}],$$

$$\omega_{3,4}^2 = \frac{1}{M}[k + \tfrac{21}{2}\mu + \tfrac{1}{2}\mu \cos(a_0 q) + (\tfrac{1}{2}k - 2\mu) \pm \sqrt{\alpha_4^2 + \beta_4^2}], \tag{225}$$

where

$$\begin{aligned}
\alpha_3 &= \tfrac{1}{2}(k + 6\mu) + \mu \cos(a_0 q) + (k + 5\mu)\cos(\tfrac{1}{2}a_0 q),\\
\beta_3 &= -\tfrac{1}{2}\sqrt{3}\, \mu \sin(a_0 q) + \tfrac{1}{2}\sqrt{3}\,(k - 4\mu)\sin(\tfrac{1}{2}a_0 q),\\
\alpha_4 &= \tfrac{1}{2}(k + 6\mu) - \mu \cos(a_0 q) + (k + 7\mu)\cos(\tfrac{1}{2}a_0 q),\\
\beta_4 &= \tfrac{1}{2}\sqrt{3}\, \mu \sin(a_0 q) + \tfrac{1}{2}\sqrt{3}\,(k - 8\mu)\sin(\tfrac{1}{2}a_0 q).
\end{aligned} \tag{226}$$

Here the roots ω_1^2 and ω_3^2 correspond to the longitudinal acoustical (LA) and optical (LO) vibration in the plane, while the roots ω_2^2 and ω_4^2 correspond to the transverse acoustical (TA) and optical (TO) vibrations in the plane. Equations (225) describe, in the general case, elliptically, rather than linearly, polarized vibrations, the former reducing to the latter only for small $|\boldsymbol{q}|$.

The characteristic feature of longitudinal vibrations in the plane is that both acoustical and optical branches of the spectrum degenerate in the corners of the Brillouin zone at $\boldsymbol{q} = \pm(4\pi/3a_0)\boldsymbol{j}_i$ (i.e., at the H-point of the Brillouin zone, see fig. 9). This degeneracy is the result of the symmetry of the graphite crystal and is independent of the model used to describe lattice-atom vibrations.

There are also other cases when the dispersion relation for vibrations in the plane may be written in an explicit form. Among them are the cases when the value of $\boldsymbol{q}$ is close to $\pm(4\pi/3a_0)\boldsymbol{j}_i$ and the case when the absolute value of $\boldsymbol{q}$ is small. In particular, in the latter case, the dispersion relations for arbitrarily chosen directions of $\boldsymbol{q}$ in the plane perpendicular to k have the form:

$$\omega = \left(\frac{ka_0^2}{M}\right)^{1/2}\left[\frac{k + 18\mu}{8(k + 6\mu)}\right]^{1/2}|\boldsymbol{q}| \quad \text{(for LA vibrations), and} \tag{227}$$

$$\omega = \left(\frac{ka_0^2}{M}\right)^{1/2}\left[\frac{3\mu}{2(k + 6\mu)}\right]^{1/2}|\boldsymbol{q}| \quad \text{(for TA vibrations).} \tag{228}$$

As one would expect, the expressions that ensue from eqs. (227) and (228) for the velocities of the longitudinal and transverse sonic waves travelling in the basal plane of graphite coincide with the similar expressions in the semicontinuum theory of Komatsu and Nagamiya (1951). The values of the force constants k and μ found by Yoshimori and Kitano (1956) using the force constants for the benzene molecule are

$$k = 6.71 \times 10^5 \text{ dyne/cm}, \qquad \mu = 0.48 \times 10^5 \text{ dyne/cm}. \tag{229}$$

The solutions of the secular equation for vibrations perpendicular to the plane have the form

$$\omega_{1,2}^2 = \frac{1}{M}[\tfrac{4}{3}\mu' + \tfrac{2}{9}\mu' X(\boldsymbol{q}) + \alpha_1' \mp \sqrt{(\alpha_1')^2 + |\beta'|^2}\},$$

$$\omega_{3,4}^2 = \frac{1}{M}[\tfrac{4}{3}\mu' + \tfrac{2}{9}\mu' X(\boldsymbol{q}) + \alpha_2' \mp \sqrt{(\alpha_2')^2 + |\beta'|^2}\}, \tag{230}$$

where

$$\alpha_1' = k'[1 - \cos(\tfrac{1}{2}c\boldsymbol{q})], \quad \alpha_2' = k'[1 + \cos(\tfrac{1}{2}c\boldsymbol{q})],$$

$$\beta' = \tfrac{2}{3}\mu'[X(\tfrac{1}{3}\sqrt{3}\,\boldsymbol{q}) + iY(\tfrac{1}{3}\sqrt{3}\,\boldsymbol{q})], \quad X(\boldsymbol{q}) = \sum_{i=1}^{3}\cos(\boldsymbol{a}_i\boldsymbol{q}), \quad Y(\boldsymbol{q}) = \sum_{i=1}^{3}\sin(\boldsymbol{a}_i\boldsymbol{q}) \tag{231}$$

In the case when the projection $q_\perp$ of the vector $\boldsymbol{q}$ upon the plane is small, eqs. (230) for the roots ω_1^2 and ω_3^2, to which the sign "$-$" before the radical in braces corresponds, reduce to

$$\omega_{1,3}^2 = \frac{\mu' a_0^4}{144M} q_\perp^4 + \frac{2k'}{M} \begin{Bmatrix} \sin^2(\tfrac{1}{2}\boldsymbol{cq}) \\ \cos^2(\tfrac{1}{2}\boldsymbol{cq}) \end{Bmatrix}. \tag{232}$$

When deriving this expression the terms proportional to the ratio k'/μ' raised to a higher power than one have been omitted. Equations (232) coincide with the analogous formulae obtained by Komatsu and Nagamiya (1951) within the semicontinuum model. Numerical values for k' and μ' have been obtained by Komatsu (1955) from the data on the compressibility of graphite along the c-axis and by fitting the calculated temperature dependence of the heat capacity of graphite to the experimental curve in the range from 15 to 60 K. The parameter values thus found are as follows:

$$k' = 5.57 \times 10^3 \,\text{dyne/cm}, \qquad \mu' = 2.92 \times 10^5 \,\text{dyne/cm}. \tag{233}$$

Hence one can see that the ratio $k'/\mu' \sim 2 \times 10^{-2} \ll 1$, which justifies the approximation adopted in the derivation of eq. (232). It is of interest to note that in the case when the wave vector $\boldsymbol{q}$ is perpendicular to the c-axis (i.e., $(\boldsymbol{q} \cdot \boldsymbol{c}) = 0$), the low-frequency vibration branch disobeys the standard dispersion law governing the long-wavelength phonons, $\omega = v_s|\boldsymbol{q}|$ and takes the form typical of charge carriers at the band edge: $\omega \sim |\boldsymbol{q}|^2$. Thus the BBS-model enables one to obtain the dispersion relations for all twelve (9 optical and 3 acoustical) branches of the vibrational spectrum of graphite [eqs. (223), and (230)], which corresponds to a complete set of vibrational modes of the lattice with four atoms in a unit cell described by the symmetry group D_{6h}^4.

Figure 161 illustrates the dispersion-relation curves calculated, within the BBS-model and using the force-constant values given by eqs. (229) and (233), by Nemanich and Solin (1979). The solid lines in the figure correspond to vibrations in the plane, the dots indicate the dispersion curves for modes with "out of plane" displacements. As can be seen from the figure, within the BBS-model to each wave vector there correspond only eight different frequencies out of 12 branches of the spectrum. This is due to the fact that, in the simple model we are considering, the branches of the vibrational spectrum which correspond to displacements of atoms parallel to the basal plane are pairwise degenerate. The degeneracy is removed only when the interaction between the atoms which occur in neighbouring graphite layers is taken into account more accurately.

The right-hand part of fig. 161 shows the phonon density of states function $D(\omega)$ for the BBS-model. The calculation of $D(\omega)$ on the basis of the dispersion relations for the various parts of the spectrum is, in general, a rather difficult task. In the BBS-model, this task becomes still more complicated because the $D(\omega)$ function has singularities at some frequencies. Nonetheless, even within

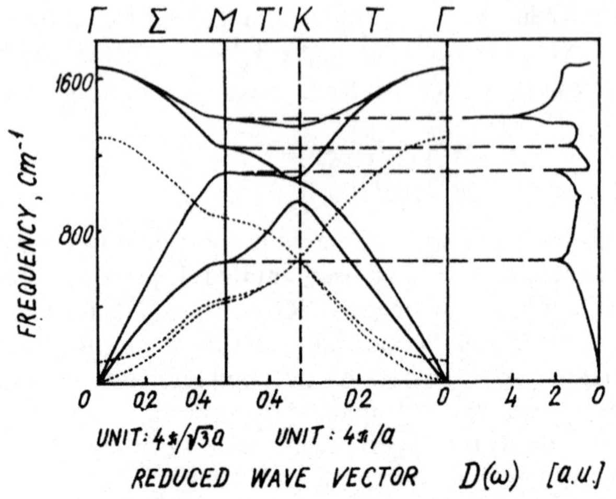

Fig. 161. Phonon dispersion curves and the density of states with "in-plane" displacements obtained with the YK-model and the YK force constants. The dotted lines indicate modes with "out-of-plane" displacements and the horizontal dashed lines show the **k**-space regions which contribute to the strong peaks in the density of states. (Nemanich and Solin 1979.)

the simple BBS-model, the complete $D(\omega)$ function and its components – the partial phonon density of states function $D_i(\omega)$ for the ith branch of the vibrational spectrum – are of considerable interest, because they allow one to obtain the overall, qualitative picture of the structure and main peculiarities of the vibrational spectrum in the graphite lattice. In connection with this, we shall dwell upon the calculation of the functions $D(\omega)$ and $D_i(\omega)$ within the BBS-model at greater length.

In order to obtain the phonon density of states function it is sufficient to consider all branches of the vibrational spectrum only within a single irreducible segment of the Brillouin zone, whose volume occupies a $\frac{1}{24}$ of the total BZ volume. The irreducible segment of the BZ projection on the basal plane is shown as the shaded portion in fig. 162.

Because of the separation of the vibrational modes parallel and perpendicular to the basal planes, one can construct two separate vibrational spectral functions corresponding to the modes: $D_{\parallel}(\omega)$ and $D_{\perp}(\omega)$. The total frequency spectrum is then given by

$$D(\omega) = \tfrac{2}{3} D_{\parallel}(\omega) + \tfrac{1}{3} D_{\perp}(\omega). \tag{234}$$

The spectra are normalized in such a way that

$$\int_0^{\omega_{\perp\max}} D_{\perp}(\omega)\,\mathrm{d}\omega = \int_0^{\omega_{\parallel\max}} D_{\parallel}(\omega)\,\mathrm{d}\omega = N, \tag{235}$$

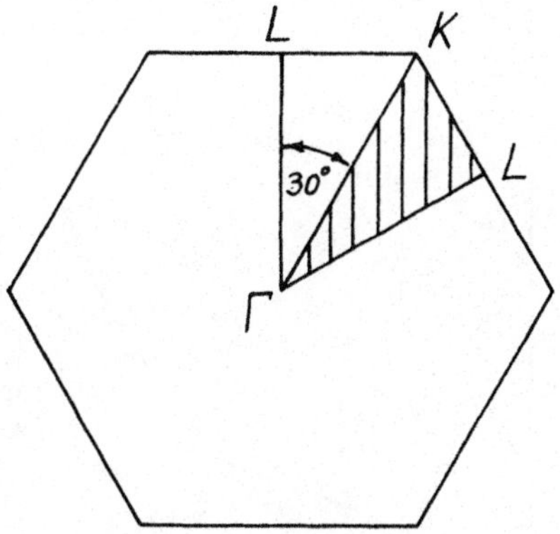

Fig. 162. The basal plane of the Brillouin zone of graphite. The shaded area is the irreducible segment, which constitutes $\frac{1}{24}$ of the area of the BZ projection upon the basal plane.

where $\omega_{\parallel \max}$ and $\omega_{\perp \max}$ are the maximum frequencies of vibrations in the plane and normal to the plane, respectively, N being the number of atoms in a unit volume.

Analytically, the function $D_{\parallel}(\omega)$ may be determined on the basis of dispersion relations (223) and (225) using the method proposed by Houston (1948). The method takes advantage of the fact that the contours of constant frequency in $\boldsymbol{q}$-space are cylinders whose axis is parallel to $\boldsymbol{k}$. One may then introduce the phonon density of states function $D_{\parallel}(\omega, \varphi)$ for a definite direction $\boldsymbol{q}$ in the basal plane, characterized by an angle φ relative, for example, to the ΓK-direction in the Brillouin zone. Since the vector $\boldsymbol{k}$ is the six-fold symmetry axis for the Brillouin zone of graphite, the function $D_{\parallel}(\omega, \varphi)$ may be expanded in a Fourier series as harmonics of $\cos(6n\varphi)$ $(n = 0, 1, \ldots)$:

$$D_{\parallel}(\omega, \varphi) = D_{\parallel}^{(0)}(\omega) + D_{\parallel}^{(1)}(\omega)\cos 6\varphi + \cdots. \tag{236}$$

For $\boldsymbol{q} \| \boldsymbol{i}_i$ and $\boldsymbol{q} \| \boldsymbol{j}_i$, dispersion relations (223) and (225) enable one to determine $D_{\parallel}(\omega, 0)$ and $D_{\parallel}(\omega, \frac{1}{6}\pi)$. Confinding ourselves to the first harmonic in series (236), we find

$$D_{\parallel}(\omega, 0) = D_{\parallel}^{(0)}(\omega) + D_{\parallel}^{(1)}(\omega), \qquad D_{\parallel}(\omega, \tfrac{1}{6}\pi) = D_{\parallel}^{(0)}(\omega) - D_{\parallel}^{(1)}(\omega). \tag{237}$$

The phonon density of states function $D_{\parallel}(\omega)$ may approximately be expressed as the average over the angle of the $D_{\parallel}(\omega, \varphi)$ function

$$\begin{aligned} D_{\parallel}(\omega) &= \frac{2\pi}{c_0}\int_0^{2\pi} D_{\parallel}(\omega, \varphi)\,\mathrm{d}\varphi = \frac{(2\pi)^2}{c} D_{\parallel}^{(0)}(\omega) \\ &= \frac{(2\pi)^2}{2c_0}\left[D_{\parallel}(\omega, 0) + D_{\parallel}(\omega, \tfrac{1}{6}\pi)\right]. \end{aligned} \tag{238}$$

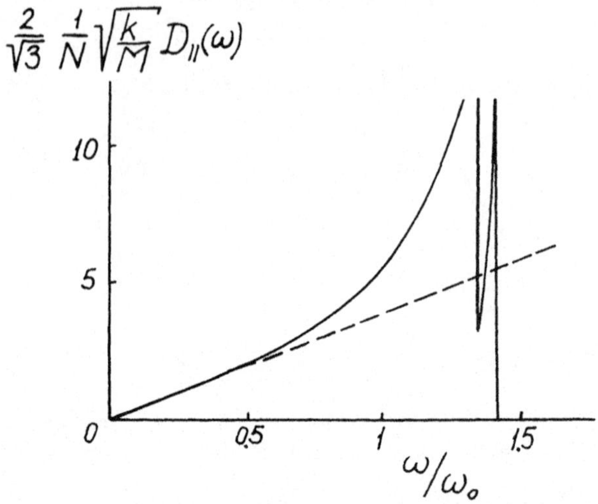

Fig. 163. The frequency distribution function for the longitudinal acoustical modes of the in-plane vibrations. The dashed line represents the values according to the semicontinuum theory. (Yoshimori and Kitano 1956.)

This approximation corresponds to the replacement of the basal hexagon in the Brillouin zone by a circle of the same area. Equation (238) makes it possible to calculate directly the $D_{\parallel}(\omega)$ on the basis of dispersion relations (223) and (225) for the directions $\varphi = 0$ and $\varphi = \frac{1}{6}\pi$.

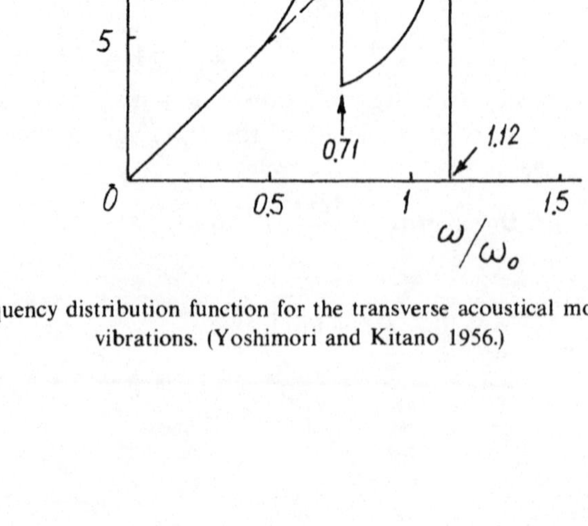

Fig. 164. The frequency distribution function for the transverse acoustical modes of the in-plane vibrations. (Yoshimori and Kitano 1956.)

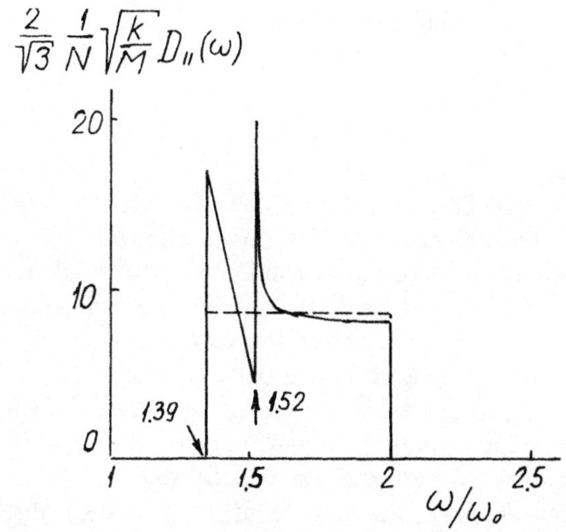

Fig. 165. The frequency distribution function for the longitudinal optical modes of the in-plane vibrations. (Yoshimori and Kitano 1956.)

The partial components of the function $D_{\|}(\omega)$ obtained in this way for four branches of the spectrum of the in-plane vibrations (longitudinal and transverse accoustical and optical) are illustrated in figs. 163–166. In these figures, the scale of abscissae is given in units of the frequency $\omega_0 = (k/M)^{1/2} = 2\pi \times 2.92 \times 10^{13}\,\mathrm{s}^{-1}$ $(\hbar\omega_0 \simeq 0.12\,\mathrm{eV})$ which corresponds to the characteristic

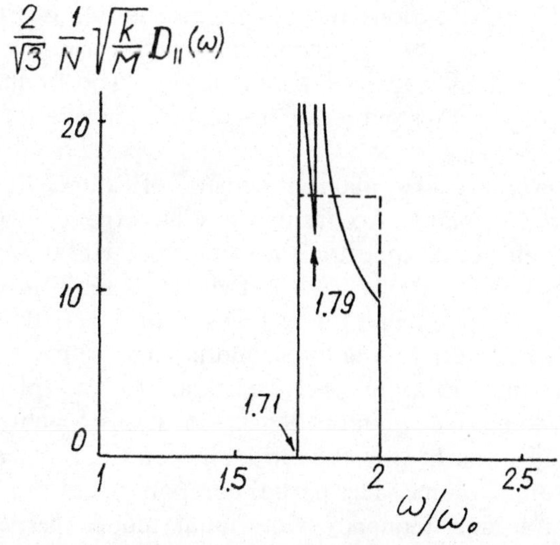

Fig. 166. The frequency distribution function for the transverse optical modes of the in-plane vibrations. (Yoshimori and Kitano 1956.)

temperature $\theta_{D_{\parallel}}$ (for in-plane vibrations, this is the Debye temperature, and is equal to $\hbar\omega_0/k \simeq 1400$ K). The partial functions $D_{\parallel_i}$ along the ordinate axis are for convenience presented as multiplied by a factor of $(2\sqrt{3}/N)\sqrt{k/M}$ with the dimension of $cm^3 s^{-1}$, which makes it possible to render the product $(2\sqrt{3}/N)\sqrt{k/M}\, D_{\parallel}(\omega)$ dimensionless.

Figures 163 and 164 show that for in-plane vibrations at frequencies $\omega \lesssim \omega_0$ in graphite the distribution functions of acoustic phonons are linear functions of frequency, and qualitatively resemble the Debye phonon distribution functions in the two-dimensional lattice $D_D^{II}(\omega) \sim \omega/v_s^2$. Thus the tangent of the slope angle of the frequency dependence of the acoustic-phonon distribution function gives the squared inverse sound velocity for the in-plane longitudinal and transverse vibrations. The dashed line in these figures shows the distribution function in the semicontinuum theory, which is essentially equivalent to Debye's two-dimensional model of the spectrum.

For frequencies $\omega \gtrsim \omega_0$, the behaviour of acoustic vibrations is greatly different from that given by the Debye model. The highest dispersion [a singularity of $D_{\parallel}(\omega)$] is observed at $\omega/\omega_0 = 1.31$ and 1.41 for the longitudinal phonons and at $\omega/\omega_0 = 0.71$ and 1.12 for the transverse phonons. The latter frequencies are also the limiting ones for the spectrum of the acoustical in-plane vibrations of the corresponding polarization.

Within the BBS-model, the spectrum of the optical in-plane vibrations of the graphite lattice extends from $1.385\omega_0$ to $2.0\omega_0$ for longitudinal and from $1.71\omega_0$ to $2.0\omega_0$ for transverse modes. Singularities in the spectrum of the optical in-plane vibrations occur at $1.518\omega_0$ for longitudinal and at $1.71\omega_0$ and $1.787\omega_0$ for transverse modes. Thus the acoustical and optical spectra of the longitudinal in-plane vibrations in the graphite lattice overlap by factor of about $0.025\omega_0$, whereas the corresponding spectra for the transverse vibrations are separated by a gap of about $0.59\omega_0$. The frequency $2\omega_0$ is the boundary frequency for the vibrational spectrum in the four-force-constant model

Let us now consider the phonon density of states, for the vibrations perpendicular to the plane. According to (230), the contours of constant frequency in the Brillouin zone for the acoustical vibrations of this polarization behave as shown in fig. 167, where the parameter X is equal to the ratio ω/ω'_0 at $\omega'_0 = (2k'/M)^{1/2} = 2\pi \times 2.76 \times 10^{12}\, s^{-1}$ ($\hbar\omega'_0 \simeq 0.0155$ eV).

The temperature corresponding to ω'_0 and having the meaning of the Debye temperature for the vibrations perpendicular to the plane is given by $\theta_{D\perp} = \hbar\omega'_0/k = 180$ K. The phonon density of states function $D_{\perp}(\omega)$ can be found analytically if, as before, the Brillouin zone is replaced by a circular cylinder of the same volume. The partial components of this function for the low-frequency and high-frequency vibrational modes perpendicular to the plane, which only conditionally may be called acoustical and optical, are illustrated in figs. 168 and 169.

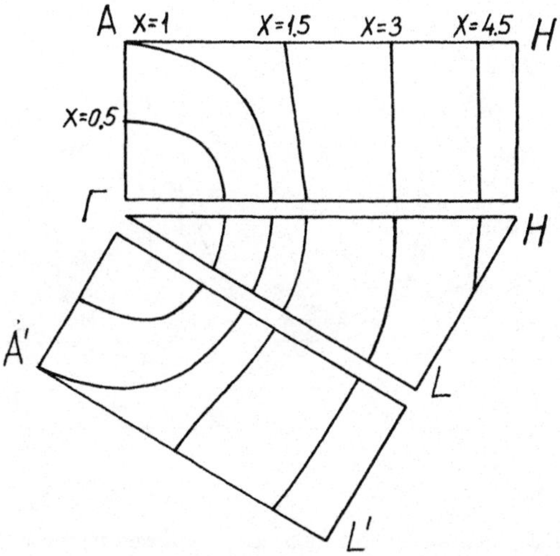

Fig. 167. Approximate contours of constant frequency for the lower frequency modes in the enlarged Brillouin zone. (Yoshimori and Kitano 1956.)

The low-frequency branch of the spectrum in the frequency range $\omega \lesssim 1.2\omega_0'$ practically coincides with the spectrum as given by the semicontinuum model, which is shown by the dashed line in fig. 168. This branch has singularities at $1.0\omega_0'$ and $3.43\omega_0'$, and is bounded from above by the frequency $5.18\omega_0'$ at which it adjoins the high-frequency branch of the spectrum extending to the frequency $10.42\omega_0'$ and having a singularity at $\omega \simeq 6.84\omega_0'$. It should be noted that integrable singularities of the functions $D_{\parallel}$ and $D_{\perp}$ arise at certain frequencies owing to the fact that they are calculated within the model in which an atom interacts only with its nearest neighbours. Taking account of interactions with the more distant neighbours eliminates these singularities, which then turn into more or less distinct peaks. Singularities in the phonon density of states can be eliminated within the four-force-constants model if in calculating the functions $D_{\parallel}(\omega)$ and $D_{\perp}(\omega)$ one uses not the approximate analytical expressions employed to obtain the spectral functions shown in figs. 163–166 and 168, 169, but numerically calculates vibrational frequencies at equally spaced points in the irreducible segment of the Brillouin zone (the so-called root sampling technique). In particular, Young and Koppel (1965) have calculated vibrational frequencies for 47 788 equally spaced points in the irreducible segment of the Brillouin zone. The spectral functions $D_{\perp}(\omega)$ and $D_{\parallel}(\omega)$ obtained in this manner are shown in fig. 170 and the total function $D(\omega)$ is presented in fig. 171. The calculation shows that the spectra of vibrations perpendicular to the plane and occurring in the plane are of the energy region from 0 to $\hbar\omega_{\max\perp} \simeq 0.164\,\text{eV}$ and $\hbar\omega_{\max\parallel} \simeq 0.209\,\text{eV}$. The spectral functions

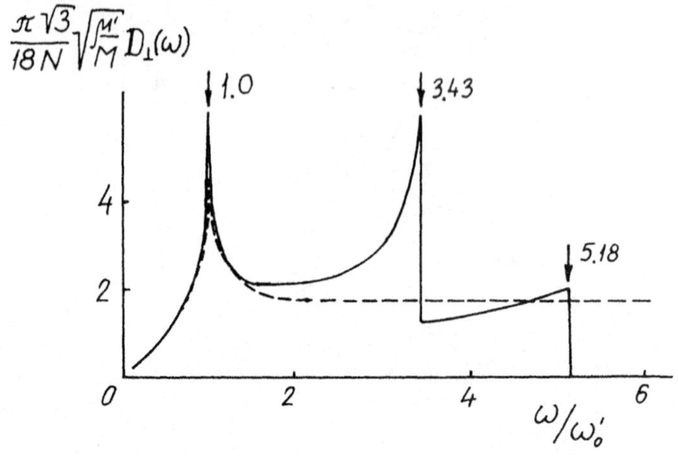

Fig. 168. The frequency distribution function for the lower frequency modes of vibrations perpendicular to the plane. The dashed line represents the values according to the semicontinuum theory. (Yoshimori and Kitano 1956.)

$D_\perp(\omega)$ and $D_\parallel(\omega)$ exhibit several sharp characteristic peaks. The most typical maxima of the function $D_\parallel(\omega)$ occur at $\hbar\omega \simeq 0.082$, 0.135, 0.155, and 0.174 eV, while those of $D_\perp(\omega)$ occur at $\hbar\omega \simeq 0.018$, 0.055, and 0.0108 eV, respectively. In addition, $D_\perp(\omega)$ has a well-pronounced deep minimum at $\hbar\omega \simeq 0.082$ eV. The total spectral function $D(\omega)$ has no minimum at this energy, because the minimum of $D_\perp(\omega)$ is cancelled by the sharp peak of $D_\parallel(\omega)$.

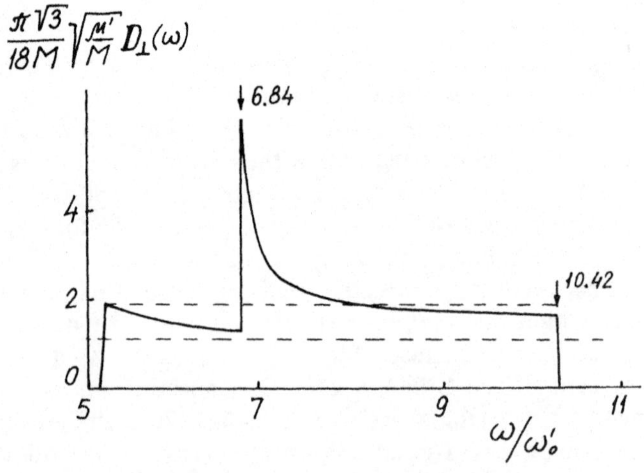

Fig. 169. The frequency distribution function for the higher frequency modes of vibrations perpendicular to the plane. (Yoshimori and Kitano 1956.)

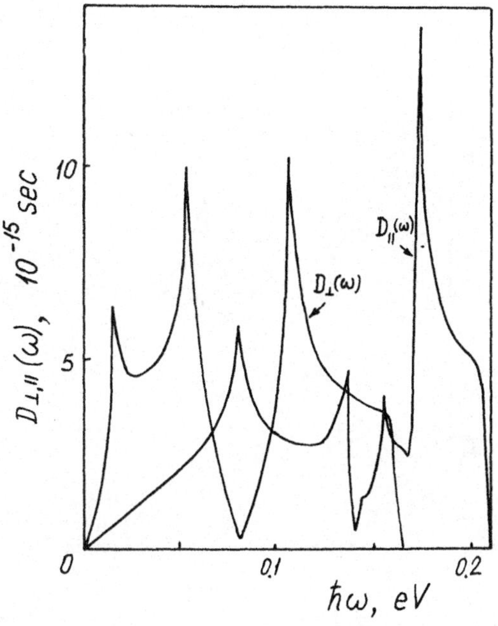

Fig. 170. Partial densities of the phonon states of graphite for lattice vibrations perpendicular [$D_{\perp}(\omega)$] and parallel [$D_{\parallel}(\omega)$] to the plane. (Young and Koppel 1965.)

Some of the features of the phonon density of states curve, which have been numerically determined by Young and Koppel (1965), can be compared with the relevant experimental data. Thus, for example, in experiments on slow-neutron scattering Haywood and Thorson (1960) observed a peak in the density of states at $\hbar\omega \simeq 0.02$ eV and a precipitous decrease of states density at $\hbar\omega \simeq 0.075$–0.08 eV, associated with the spectrum of the transverse acoustical vibrations perpendicular to the plane. Edelstaff and Harris (1963) observed a density of states peak at $\hbar\omega \simeq 0.08$ eV and a broad, poorly resolved peak at $\hbar\omega \simeq 0.175$ eV. The same workers discovered a sharp peak of $D(\omega)$ at $\hbar\omega \simeq 0.11$ eV. All these features agree with the position of the peaks in the total spectrum of $D(\omega)$ shown in fig. 171. It should be noted that the sharp, strongly pronounced peak at $\hbar\omega \simeq 0.175$ eV visible in the calculated spectrum arises due to the approximation considering the interaction of only the nearest neighbours in the plane of the graphite layer. Taking account of more distant interactions decreases the height of the peak and makes it less pronounced, which is in accordance with the experimental observation by Edelstaff and Harris (1963). Thus the partial curves for the phonon density of states $D(\omega)$ obtained within the four-force-constants BBS-model and shown in figs. 163–166 make it possible to describe qualitatively the character of the various branches of the vibrational spectrum of the graphite lattice. The partial and

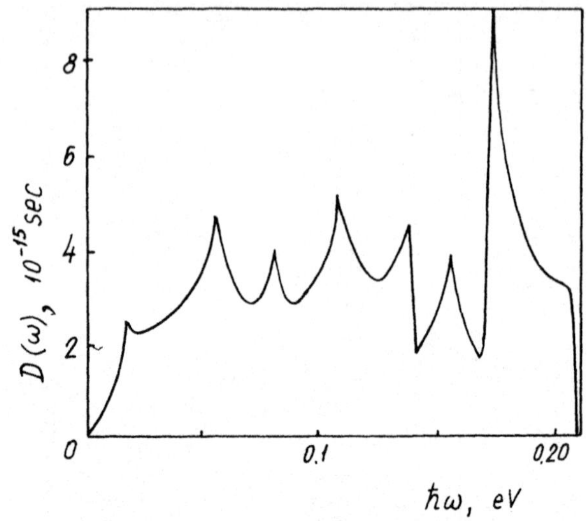

Fig. 171. Density of the phonon states of graphite according to Young and Koppel (1965).

total curves for the phonon density of states numerically calculated within the same model and shown in figs. 170 and 171 enable one to describe with sufficient quantitative accuracy some important features of the vibrational spectrum of graphite. Also, the phonon density of states curve shown in fig. 171 makes it possible to describe with high accuracy the temperature dependence of the specific heat capacity $C_V(T)$ of graphite in the temperature range between 100 and 1000 K, where practically all short-wavelength vibration modes turn out to be excited in the graphite spectrum. The theoretical dependence of C_V on T obtained in this way by Young and Koppel (1965) is in good agreement with the experimental data of Spencer (1948) and De Sorbo and Tyler (1953), whose measurements were made in the same temperature range using specimens of natural single-crystal graphite.

Notwithstanding its advantages, the four-force-constants BBS-model lacks the required accuracy to describe certain details of the phonon spectrum of graphite. In particular, this model does not allow one to make an exact quantitative calculation of the dispersion curves for the principal lattice directions or to determine quantitatively the position and relative values for the amplitudes of peaks in the phonon spectrum. The main disadvantages of the four-force-constants BBS-model are associated with the inherent inconsistency of its description of atomic interactions in the graphite lattice. Namely, the model disregards the third-neighbour interactions between the carbon atoms in the plane of honey-comb nets, but, at the same time, takes into account the interactions between the nearest-neighbour atoms in the c-direction, which are equivalent, as far as the distance between atoms is concerned, to the fourth-neighbour interactions. Thus, while taking account of the weaker interplanar interactions, the BBS-model neglects the terms which are more significant for phonon spectrum formation, i.e., those corresponding to the third-neighbour interaction in the plane.

The above-mentioned disadvantage of the BBS-model has been eliminated by Nicklow et al. (1972) who further developed the theory of the vibrational spectrum of graphite and obtained phonon dispersion relations for the principal lattice directions in graphite.

11.1.2. The axial-symmetric (AS) model of the vibrational spectrum of graphite

The interactions between the first-, second-, third-, and fourth-neighbour atoms in the graphite lattice may consistently be described by a set of 12 force constants ϕ_{ij} [see eq. (22)] to which there correspond the four upper matrices of the force-constant tensor presented in table 1 (see sect. 3.3.2).

The confinement to four (in order of increasing distance between the carbon atoms in the graphite lattice) neighbours adopted in constructing a simple, but embracing all principal types of interactions, model is logically justified. This is due to the fact that the least distance between the nearest carbon atoms in the graphite lattice along the *c*-axis exactly corresponds to the fourth-nearest-neighbour distance in the plane. Hence it follows that taking account of interactions including the fourth-nearest neighbour is an adequate approximation in describing the highly anisotropic graphite lattice.

Taking account of interactions with the fifth-nearest neighbour greatly complicates the model, because now one should consider the bonds with a large number of atoms which are both in the plane of the graphite layer and in two nearest graphite planes. Such a model, which needs as many as 21 force constants, has been discussed by Nicholson and Bacon (1977) (see table 1, sect. 3.3.2). As has already been mentioned in sect. 3.3.2, a similar model enables one to describe the mechanically equilibrated graphite lattice; yet, it can hardly be regarded as tractable in calculating the vibrational spectrum of graphite when one may, in principle, restrict oneself to considering interactions up to the fourth-nearest neighbour.

In order to pass from the general model taking into account interactions with nearest neighbours up to the fourth to the four-force-constants BBS-model, one should set the elements A_3, B_3, C_3 and A_4 in the third and fourth matrices in table 1 equal to zero. In such a case, the relation between the force constants of the BBS-model and the elements of the upper two matrices and element C_4 of the fourth (counted from top) matrix in table 1 is described by eq. (25).

From this one can clearly see the mathematical sources of incorrectness of the BBS-model. Indeed, in order to shift to this model, one has to set to zero all elements of the matrix that describes the interaction with the third-nearest neighbours and with the fourth-nearest neighbours in the plane (i.e., elements A_3, B_3, C_3, A_4). At the same time, one takes into consideration element C_4 in the fourth (from top) matrix, whose value is, generally speaking, much lower than the components of the third (from top) matrix.

The complete model that takes into account interactions between atoms up to the fourth-nearest neighbour remains too complicated for the calculation of phonon dispersion curves and the phonon spectrum of the graphite lattice. However, this model can be reasonably simplified, which makes it possible to reduce the number of force constants to be considered. At the same time, this model is free of the shortcomings inherent in the BBS-model. Such a model, qualitatively correctly (though in a simplified manner) taking into account interactions with up to the fourth-nearest neighbours, is the axial-symmetric (AS) model used by Nicklow et al. (1972). The main feature of the AS-model is that it assumes that there is a two-body potential function $V(r)$ which describes the forces between any pair of atoms and which depends solely on the distance between the atoms, a pairwise potential. This makes possible some simplification of the general model, in which interactions with atoms up to the fourth-nearest neighbour are considered.

In the AS-model, the force constants have the form

$$\phi_{\mathrm{r}}^{n} = \left(\frac{\partial^2 V}{\partial r^2}\right)_{r=r_n}, \qquad \phi_{\mathrm{t}}^{n} = \left(\frac{1}{r}\frac{\partial V}{\partial r}\right)_{r=r_n}, \tag{239}$$

where r_n is the distance between atoms which are nth nearest neighbours of each other.

The applicability of the pairwise potential and the force constants in the form given by eq. (239) has subsequently been substantiated by Ahmadieh and Rafizadeh (1973) and by Rafizadeh (1974).

In order to pass from the general model to the AS-model, one uses the following simplifying relations:

$$\begin{aligned} &\phi_{\mathrm{r}}^{1} = A_1, \quad \phi_{\mathrm{t}}^{1} = B_1 = C_1; \quad \phi_{\mathrm{r}}^{2} = B_2, \quad \phi_{\mathrm{t}}^{2} = A_2 = C_2, \quad D_2 = 0; \\ &\phi_{\mathrm{r}}^{3} = A_3, \quad \phi_{\mathrm{t}}^{3} = B_3 = C_3; \quad \phi_{\mathrm{r}}^{4} = C_4, \quad \phi_{\mathrm{t}}^{4} = A_4. \end{aligned} \tag{240}$$

Thus the AS-model contains seven force constants. The AS-model, while not as physically appealing as the BBS-model, is somewhat more general for describing vibrations perpendicular to the basal plane, because no relation between the ϕ_{t}^{n} is imposed; also, $\phi_{\mathrm{t}}^{4} \neq 0$, thereby giving a finite value of c_{44}.

The force constants of the AS-model are not entirely independent of each other; it can be demonstrated that they are related by the following relation

$$\phi_{\mathrm{t}}^{4} = \frac{a_0^2}{c_0^2}\left(\tfrac{1}{2}\phi_{\mathrm{t}}^{1} + 3\phi_{\mathrm{t}}^{2} + 2\phi_{\mathrm{t}}^{3}\right), \tag{241}$$

which reduces the number of independent force constants to six and makes it possible to separate them into groups, where these constants may be determined from independent experiments. For example, the smallest force constants ϕ_t^4 and ϕ_r^4 may be determined from data on longitudinal and transverse vibrations along the *c*-axis. The constants ϕ_t^1, ϕ_t^2, ϕ_t^3 are found from an analysis of the data on the TA- and TO-branches of the spectrum of vibrations propagating in the basal plane and polarized in the *c*-direction ($TA_\perp$, $TO_\perp$).

Finally, the constants ϕ_r^1, ϕ_r^2, ϕ_r^3 are found from the data on longitudinal vibrations in the basal plane.

It should be noted that information about some vibrational modes in graphite is still very scarce. This is true, in particular, of the longitudinal vibrations in the basal plane. Because of this, the constant ϕ_r^3 has been determined with a very large error (~100%).

The values for the force constants in the AS-model are given in table 28 (see Nicklow et al. 1972).

Table 28
Force constants for the AS-model.

Constant	Value (10^5 dyne/cm)	Constant	Value (10^5 dyne/cm)
ϕ_r^1	3.62 (±25%)	ϕ_r^3	−0.037 (±100%)
ϕ_t^1	1.99 (±3%)	ϕ_t^3	0.288 (±3%)
ϕ_r^2	1.33 (±30%)	ϕ_r^4	0.058 (±2%)
ϕ_t^2	−0.520 (±3%)	ϕ_t^4	0.0077 (±5%)

The force constants listed in table 28 have been obtained by processing the experimental data on coherent inelastic neutron scattering in graphite (Dolling and Brockhouse 1962, Nicklow et al. 1972), Raman scattering (Tuinstra and Koenig 1970, Brillson et al. 1971) and also by using the data of measurements of the elastic constants in graphite (Blackslee et al. 1970, Seldin and Nezbeda 1970).

Figure 172 shows the dispersion relations, calculated on the basis of the AS-model developed by Nicklow et al. (1972), for the principal directions in the Brillouin zone of graphite [100], [110], and [001], where [100] coincides with the vector ΓM, [001] with the vector ΓA and [010] is perpendicular to vectors [100] and [001]. In this calculation, the values for the force constants listed in table 28 have been used. In the right-hand side of fig. 172 the phonon density of states calculated within the AS-model developed by Nemanich and Solin (1979) is shown.

Attention is called to the fact that, in accordance with the lattice symmetry and the number of atoms (four) in a unit cell, the total phonon spectrum of graphite should consist of 12 branches of acoustical and optical vibrations of various polarization, i.e., for each value of the wave vector in the spectrum of

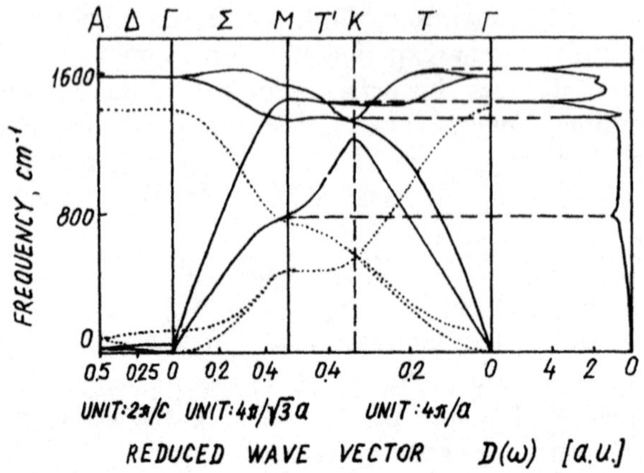

Fig. 172. Phonon dispersion curves and the density of states, $D(\omega)$, with "in-plane" displacements obtained with the NWS-model. The dotted lines in the dispersion curves represent modes with "out-of-plane" displacements. The NWS force constants have been slightly modified to account for the Raman mode at 1581 cm^{-1}. The dashed horizontal lines show $\boldsymbol{k}$-space regions which contribute to the strong peaks in the density of states. (Nemanich and Solin 1979.)

graphite there should be 12 different vibration frequencies. However, because of the very small magnitude of the force constant ϕ_t^4 in the AS-model, modes corresponding to atomic displacement parallel to the basal plane are pairwise nearly degenerate, except at low frequencies near the Γ-point of the Brillouin zone. Each line in fig. 172 which goes to any of the four highest frequencies at the M-point represents such a pair.

It is of interest to note that one consequence of the extreme anisotropy of the forces in graphite is the almost quadratic form of the dispersion relation for small $\boldsymbol{q}$ for the $TA_{\perp}$-branch. For small $\boldsymbol{q}$, this branch is described by

$$\omega^2 \sim Aq^2 + Bq^4. \tag{242}$$

Such a dependence of ω on q in this branch has already been obtained within the semicontinuum model by Komatsu (1955, 1958), who introduced a finite c_{44} when calculating the elastic properties of graphite.

For very small q, the ω varies linearly with q, as it should, because of the term $\sqrt{A}\,q$. However, the coefficient A, which is proportional to c_{44} and, hence, to the small force-constant ϕ_t^4, is by far smaller than B, which is given by a linear combination of the very large force constants ϕ_t^1, ϕ_t^2, and ϕ_t^3. Thus with increasing q the term Bq^4 soon becomes dominant in eq. (242) and gives the relation $\omega \sim q^2$.

The frequency-distribution function $D(\omega)$ was calculated by Nicklow et al. (1972) from the AS-model using the method described by Raubenheimer and Gilat (1967), appropriately modified for the graphite lattice. The results of the

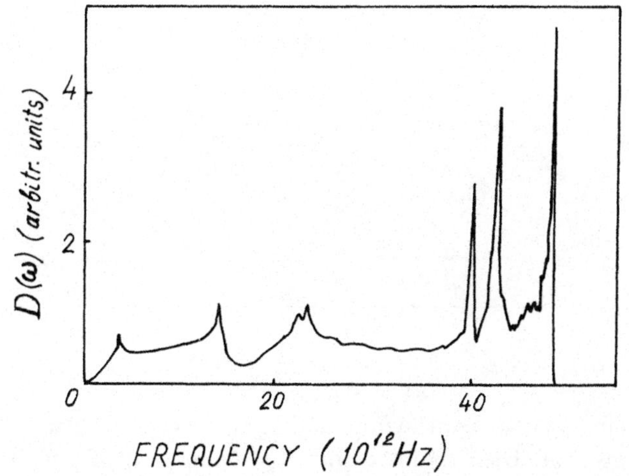

Fig. 173. Frequency distribution function for graphite, calculated with the fourth nearest-neighbour AS-model. (Nicklow et al. 1972.)

calculation are illustrated in fig. 173. It can be noticed from the figure that the $D(\omega)$ function is markedly different from the similar function obtained from the BBS-model (fig. 171) both in the position of peaks on the frequency scale and in their relative height. However, both these distributions have distinct general features at frequencies of ~ 4, ~ 14, and ~ 48 THz (i.e., at energies $\hbar\omega \sim 16$, ~ 58, and ~ 196 meV).

The difference between the phonon spectra in the BBS- and AS-models indicates that the interactions between the third- and fourth-nearest neighbours in the plane, which are disregarded in the BBS-model, significantly contribute to the graphite-lattice vibrations.

The theory of the phonon spectrum of graphite based on the concepts of interaction between the carbon atoms, similar to that assumed in the AS-model, has further been developed with success by Maeda et al. (1979). According to these authors, for the extremely anisotropic graphite lattice it is more natural to assume that the tangential force necessary to change the bond angle within a layer differs from that necessary to displace an atom along the c-axis. In order to take account of this circumstance in describing interactions between in-plane atoms, they have introduced the force constants ϕ_r^n, ϕ_{ti}^n and ϕ_{to}^n, where ϕ_r^n is the "radial" force-constant for the nth nearest neighbour, ϕ_{ti}^n is the "tangential" constant responsible for the change in bond angle within a layer, and ϕ_{to}^n is the "tangential" constant responsible for the displacement along the c-axis. On the other hand, the force constants between the nearest neighbours on adjacent planes are assumed to be axially symmetric, so that the relevant constants are represented by one "radial" and one "tangential" constant (ϕ_r^{int} and ϕ_t^{int}, respectively), as in the AS-model.

Table 29
Tangential and radial force constants for graphite in the model of broken axial symmetry.

Tangential force constant	Value (10^5 dyne/cm)	Radial force constant	Value (10^5 dyne/cm)
ϕ_{ti}^1	2.810	ϕ_r^1	3.066
ϕ_{ti}^2	−0.5269	ϕ_r^2	1.363
ϕ_{to}^1	0.8561	ϕ_r^{int}	0.0579
ϕ_{to}^2	−0.1421		
ϕ_t^{int}	0.00718		

The breaking of the axial symmetry for interactions between the in-plane carbon atoms has been pointed out by Mani and Ramani (1974). The introduction of the "broken" axial symmetry for the in-plane interactions alongside of the retention of the unbroken axial symmetry for the out-of-plane interactions has made it possible to improve significantly the agreement of the model with experiment. The set of the force constants for such a model presented in table 29 has been extracted by Maeda et al. (1979) from an analysis of the experimental data on neutron scattering (Dolling and Brockhouse 1962, Nicklow et al. 1972), Raman scattering (Tuinstra and Koenig 1970, Brillson et al. 1971, Nemanich et al. 1977c), and infra-red reflectance (Brillson et al. 1971, Friedel and Carlson 1971, Nemanich et al. 1977c).

The phonon dispersion relations for the [001], [100], and [110] principal lattice directions in graphite corresponding to the above set of force constants are illustrated in fig. 174. A comparison between fig. 174 and fig. 172 shows that the modifications of the AS-model made by Maeda et al. (1979) have left the phonon dispersion relations qualitatively unchanged. However, the introduction of the broken axial symmetry has led to a better agreement with the then latest data on phonon frequencies obtained from the study of infra-red reflection from single-crystal graphite by Nemanich et al. (1977c).

The modifications introduced into the model by Maeda et al. (1979) mainly concern the $TA_{\perp}$-branch. These modifications have reduced by almost a factor of two the values of the frequencies in this branch, as compared with those given by the models of Young and Koppel (1965) and Nicklow et al. (1972).

It should be noted that with the accumulation of the experimental data on graphite-lattice-vibration frequencies the models of the vibrational spectrum are continually refined. Thus, previous to the publication of the paper by Maeda et al. (1979), the model of the graphite's phonon spectrum had to some extent been refined by Mani and Ramani (1974) and by Nicholson and Bacon (1977). In particular, the latter authors have calculated the phonon dispersion relations within the currently most complete model taking into account

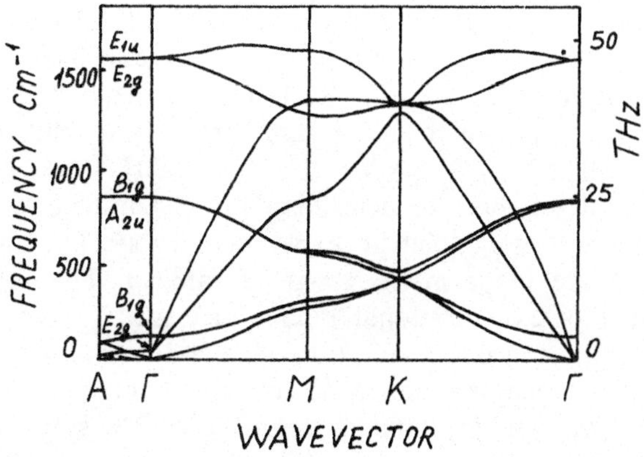

Fig. 174. Phonon dispersion curves for graphite along several high-symmetry axes. The Γ-point symmetries for the graphite structure are indicated. (Maeda et al. 1979.)

interactions between the carbon atoms up to the fifth-nearest neighbour. Quite recently, the models of the vibrational spectrum have been further elaborated by Al-Jishi and Dresselhaus (1982a) and Alldredge et al. (1984). The latter group, in particular, has discovered a previously unnoticed feature (inadvertent symmetries) in a lattice dynamical model of graphite. Taking account of this detail has made it possible to further refine the phonon-spectrum model developed by Nicklow et al. (1972) and elaborated by Maeda et al. (1979). The unexpected qualitative consequences of the above-mentioned inadvertent symmetries are the prohibition for the Rayleigh surface waves to exist on the basal plane of graphite and also a high-order degeneracy on the Brillouin-zone edge. It should be noted that the corrections and refinements made in the lattice dynamical model of graphite have not altered its principal characteristics. Therefore, one may state that by now the general idea of the phonon spectrum of graphite has been completely formed, the main features of the spectrum being described within the model taking into account atomic interactions in the lattice up to the fourth-nearest neighbour. The dispersion relations which have been derived from this model and which give the best description of the phonon spectrum were illustrated in fig. 174. The real phonon density of states in monocrystalline graphite is most accurately (both qualitatively and quantitatively) described by the curve shown in fig. 173.

11.1.3. Graphite-lattice vibrations of small wave vectors (near the Γ-point – the Brillouin-zone centre)

Thus far we have discussed the models of the phonon spectrum of graphite and the dispersion relations for the entire Brillouin zone. However, because of the absence of large single crystals of graphite, the most important investigations of the phonon spectrum have been made by optical techniques

(Raman scattering and infra-red reflectance) and cover a relatively small domain in $\boldsymbol{q}$-space surrounding the Γ-point in the Brillouin zone. The few studies of the phonon spectrum that have been carried out using the coherent inelastic scattering of slow neutrons are confined to the frequency range $\omega \lesssim 500\,\text{cm}^{-1} \simeq 0.062\,\text{eV} \simeq 15 \times 10^{12}\,\text{Hz}$ (or 15 THz) (see Nemanich et al. 1977a). As a result of this, the determination of parameters of the phonon-spectrum model by fitting them to experimental data and their comparison with experiment are made in the range of small $\boldsymbol{q}$. This makes a detailed analysis of all allowed vibrational modes near the Γ-point highly topical. Because of its importance, the question will be discussed in some detail.

As has already been mentioned above, the symmetry of the hexagonal modification of the graphite lattice is described by the space group D^4_{6h}. This group contains twenty-four elements which are the result of the direct product of the twelve elements of the C^4_{6v} space group with the group C_i, containing the identity and inversion symmetry operations. The space group C^4_{6v} contains the symmetry operations of identity E, two operations of rotation through an angle of 120° about the c-axis $[C_3(z)]$ and three operations of reflection from the planes, σ_v, one of which corresponds to that perpendicular to the vector $\boldsymbol{j}_i$ (see sect. 11.1.1) and the other two are placed 120° apart. In addition, there are three glide planes, which include the z-axis and which are located 30° away from the σ_v planes. There are also one two-fold screw axis $C_2(z)$ and two six-fold screw axes $C_6^{\pm 1}(z)$.

A feature of the graphite structure is the lack of inversion symmetry at the atomic sites, although there is a centre of inversion at the point with coordinates $(0, 0, \frac{1}{4}c_0)$.

The symmetry analysis has been made using the multiplier representations (see Maradudin et al. 1970). The reducible representation D, set up in terms of displacement of the atoms in a unit cell, can be decomposed into irreducible representations for the symmetry points in the Brillouin zone as follows [for the notation for the symmetry points see fig. 9 and the work of Slater (1965)]:

$$
\begin{aligned}
&\Gamma\colon\ D = 2(\Gamma_2^- + \Gamma_3^+ + \Gamma_6^- + \Gamma_5^+),\\
&\Delta\colon\ D = 2(\Delta_1 + \Delta_2 + \Delta_5 + \Delta_6),\\
&A\colon\ D = 2(A_1 + A_2),\\
&\Sigma\colon\ D = 4(\Sigma_1 + \Sigma_3 + \Sigma_4),\\
&T\colon\ D = 4T_1 + 2T_2 + 2T_3 + 4T_4,\\
&M\colon\ D = 2(M_1^+ + M_2^- + M_3^- + M_3^+ + M_4^- + M_4^+),\\
&K\colon\ D = K_1 + K_2 + K_3 + K_4 + 3K_5 + K_6.
\end{aligned}
\tag{243}
$$

We have already pointed out the great importance of data on the long-wavelength frequencies of the lattice-vibrations in graphite. It should be added that from these data one can determine the covalent force constants or the parameters of a potential for covalent forces.

Table 30
Character table and selection rules for crystalline graphite. Raman (R) and infra-red (IR) activities are denoted by "a" for allowed and "f" for forbidden transitions.

Representation C_{6v}	E	C_2	$2C_3$	$2C_6$	$3\sigma_v$	$3\sigma_d$	n_i	n'_i	n_τ	R	IR
A_1	1	1	1	1	1	1	3	2	1	a	a
A_2	1	1	1	1	−1	−1	0	0	0	f	f
B_1	1	−1	1	−1	1	−1	1	1	0	f	f
B_2	1	−1	1	−1	−1	1	0	0	0	f	f
E_1	2	−2	−1	1	0	0	3	2	1	a	a
E_2	2	2	−1	−1	0	0	1	1	0	a	f

The character table and the selection rules for the graphite crystal with the point group C_{6v} are given in table 30. The number of the normal modes of various types and their activity in the infra-red and Raman spectra have been obtained from the relevant relations (see, e.g., Wilson et al. 1955, Bhagavantam and Venkatarayudu 1951) and the character tables. Table 30 presents the number n_i of normal modes under a symmetry species i, the number n'_i of internal vibration modes and the number n_τ of pure translations under a symmetry species. From the table one can see that there are two active modes of species A_1 and two active modes of species E_1. The examination of the dynamical matrix at $\boldsymbol{q} = 0$ and the relative magnitudes of the force constants shows that for all practical purposes the frequencies of the two A_1-species modes are degenerate and so are the frequencies of the two E_1-species modes.

Mani and Ramani have decomposed the phonon reducible representation at the Γ-point as:

$$D = 2(A_{2u} + B_{2g} + E_{1u} + E_{2g}). \tag{244}$$

This equation is the first in the set given by eq. (243), written in the point-group notation.

The irreducible representation for the Brillouin-zone centre (Γ-point) optical modes of graphite is given by

$$\Gamma_{\text{opt}} = 2E_{2g} + E_{1u} + 2B_{2g} + A_{2u}, \tag{245}$$

A_{2u} and E_{1u} being the acoustical modes. From the properties of the point group and the requirement that all unit cells must vibrate with the same phase ($\boldsymbol{q} \to 0$ corresponds to the totally symmetric mode of the lattice translation), the symmetry properties of normal modes can be determined.

Only the E_{2g} modes are expected to be Raman (R) active as fundamentals. The symmetry of the E_{2g} modes restricts the motion of the atoms to the plane of the carbon layer. The two different E_{2g} modes in the three-dimensional analysis occur because adjacent planes can vibrate in phase or with opposite phase. The difference in energy between the two E_{2g} modes will be very small due to the relatively weak interlayer forces. In single crystals of graphite, the two E_{2g} modes will probably not show a measurable frequency separation and will appear as a single band. The Raman band observed in single crystals of graphite can be attributed to these E_{2g} bands.

The A_{2u} and E_{1u} modes are infra-red (IR) active. The B_{2g} modes are optically inactive (silent), however, one of them has been observed by neutron scattering at 127 cm^{-1} and the other occurs at a frequency near the A_{2u}-(867 cm^{-1}) mode frequency.

Thus, E-symmetry modes exhibit "in-plane" atomic displacements, while the A- and B-symmetry modes have "out-of-plane" displacements. The types of atomic displacements in the graphite lattice corresponding (for $\boldsymbol{q} \to 0$) to these modes are shown in fig. 175. The zero-frequency acoustical modes corresponding to pure translations are not shown in the figure.

Because of the weak bonds between the layers of the graphite lattice, the features of the two-dimensional spectrum should be rather prominent in the vibrational spectrum. It should be noted, however, that the IR-activity is the

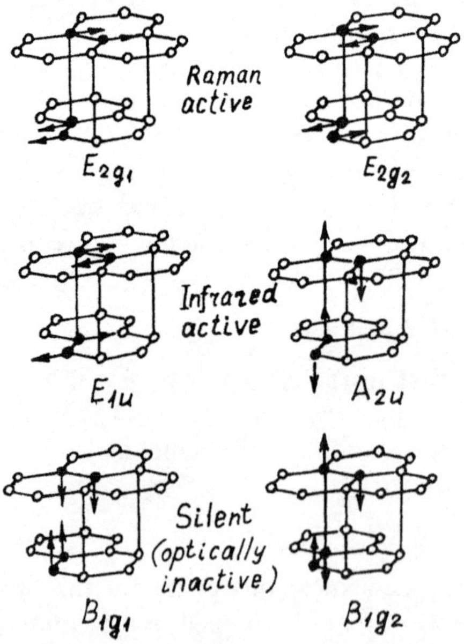

Fig. 175. Zone-centre optical lattice modes in graphite. For the in-plane modes (E_{1u}, E_{2g_1}, E_{2g_2}), only one of the degenerate pair of modes is shown. The c-axis modes (A_{2u}, B_{1g_1}, B_{1g_2}), are non-degenerate. Infrared and Raman activity are indicated. The zero-frequency acoustic modes (E_{1u}, A_{2u}), corresponding to pure translations, are not shown. (Nemanich et al. 1977a.)

direct consequence of the three-dimensional character of the graphite lattice. Let us discuss this in more detail, for which purpose we shall first consider the vibrational modes of a single graphite layer. The point symmetry group for this layer is D_{6h}. The irreducible representation for the Γ-point optical modes in graphite has the form

$$\Gamma_{opt} = E_{2g} + B_{2g}. \tag{246}$$

As the unit cell of the graphite layer contains only two atoms, the IR-active modes are completely absent, as Zallen (1968) has shown, the E_2-mode being Raman-active. The unit cell of the three-dimensional graphite lattice contains, however, four carbon atoms and only half of them are the fourth-nearest neighbours that lie either directly above or directly below one another in the adjacent layers. Therefore, the two atoms in the unit cell in each layer are not equivalent. The effect of this cell-size doubling not only increases the number of optical modes, but is also responsible for the IR-activity. Figure 176 illustrates the relationship between the vibrational modes of the single layer and those of the three-dimensional crystal. All of the single-layer optical modes become Davydov-doublets. The E_{2g}-mode generates an IR-active E_{1u}-mode and a Raman-active E_{2g}-mode. The B_{2g}-mode goes to an IR-active A_{2u}-mode and an optically inactive B_{2g}-mode. The zone-boundary acoustical modes of

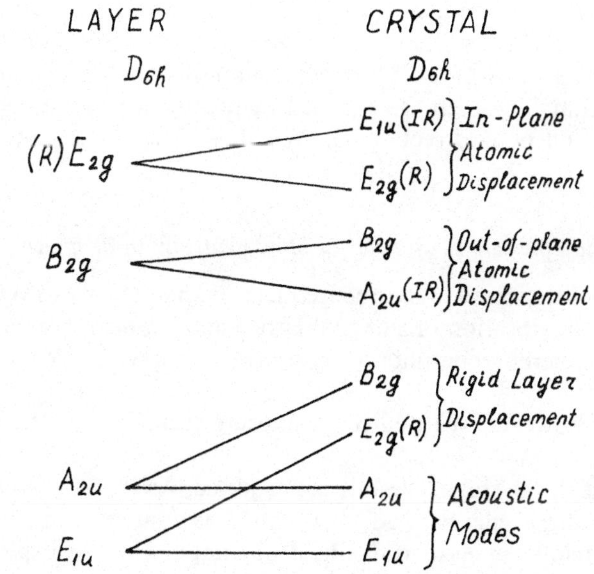

Fig. 176. Correlation of the zone-centre vibrational modes in graphite for a single layer with the three-dimensional crystal structure. Raman (R) and infrared (IR) activities are indicated in parentheses. (Nemanich et al. 1977a.)

the single layer fold back to the zone centre as rigid-layer modes, an optically inactive B_{2g}-mode and the Raman-active E_{2g}-mode. Finally, the E_{1u}-and A_{2u}-modes remain acoustical.

11.2. Experimental investigation of the phonon spectrum of graphite

The phonon spectrum of single-crystal graphite has been much investigated experimentally. However, as has already been mentioned, the experimental data cover largely the range of small wave vectors (a close vicinity of the Brillouin-zone centre), and are not sufficiently complete. Moreover, most measurements have been made using not single crystal, but specimens of pyrolytic graphite consisting of very thin (~ 1 μm) crystallites randomly oriented about a common *c*-axis.

All the experimental data directly related to the study of the graphite's phonon spectrum may be divided into two groups. One of them comprises the results of measurements of the dispersion relations in certain branches of the spectrum using the inelastic coherent scattering of slow neutrons. The other includes the results of frequency measurements at small (or zero) wave vectors using the Raman or infra-red spectra.

To these two groups of direct experimental data on the phonon spectrum one should add a third group, which contains data on the temperature dependence of the specific heat capacity of graphite – an integral characteristic of the phonon spectrum. This group also includes the data on the force constants that enter as parameters in equations of the various models of the phonon spectrum. These are determined by measuring the components of the elastic moduli tensor (see sect. 3.3.3). We shall at first examine the first two groups.

11.2.1. Inelastic coherent scattering of slow neutrons in graphite

The dispersion relation between the frequency ω and the wave vector $\boldsymbol{q}$ for the normal modes of vibration of a crystal lattice may, in principle, be obtained by studying the coherent one-phonon scattering of slow neutrons from single-crystal samples.

In practice, however, the neutron-scattering technique suffers from several limitations.

For instance, large single crystals are required. As the available flux of slow neutrons with high energy, e.g., $> 500\,\text{cm}^{-1}$ is minimal, that region of the phonon dispersion curves is difficult, if not impossible, to explore. Also, the flux limitation reduces the accuracy of the neutron measurements to $\sim \pm 5\,\text{cm}^{-1}$. Finally, an intrinsic limitation of the neutron-scattering technique is its invalidity to provide crucial symmetry information on the vibrational excitations at key points of the Brillouin zone. Thus, though neutron-scattering result are certainly necessary for a correct understanding of the lattice dynamics of a crystal, it is clear that they still may be insufficient.

In coherent one-phonon scattering, the laws of conservation of the energy and the crystal momentum hold well, i.e., we can write

$$\varepsilon_0 - \varepsilon' = \pm\hbar\omega, \qquad \boldsymbol{Q} = \boldsymbol{k}_0 - \boldsymbol{k}' = 2\pi\boldsymbol{b} - \boldsymbol{q}, \tag{247}$$

where ε_0, ε' are the energies, and $\boldsymbol{k}_0$, $\boldsymbol{k}'$ are the wave vectors, of the incident and scattered neutrons, respectively, $\boldsymbol{Q}$ is the momentum transfer vector, $\boldsymbol{b}$ is a reciprocal lattice vector.

Neutron-scattering experiments are usually conducted using the constant momentum transfer (constant $\boldsymbol{Q}$) mode of operation with fixed outgoing energy. In this mode, the analyzing spectrometer is set at a fixed energy ε' and the energy ε_0 of the incident neutrons is varied. The spectrometer is programmed so that the scattering angle and specimen orientation change with ε_0 in such a way that the wave vector transfer $\boldsymbol{Q}$ is kept constant. Energy distributions are thus measured at fixed points in reciprocal space. The centres of the neutron groups observed in the energy distribution yield the frequencies ω of lattice vibrations with wave vector $\boldsymbol{Q}$ in the extended Brillouin zone (i.e., of $\boldsymbol{q}$ in the reduced Brillouin zone).

The major difficulty presented by any such study is the lack of large single-crystal graphite samples. Accordingly, practically all known measurements of this kind have been made using pyrolytic graphite samples.

Due to the random orientation of the constituent crystallites in the basal plane, the reciprocal lattice of pyrolytic graphite consists of the reciprocal lattice of the perfect graphite structure rotated about the [001]-direction (the c-axis) to produce smeared-out "rings" of reciprocal-lattice points everywhere, except along the [001]-direction itself. The rings shown in fig. 177 represent the [110]- and [100]-type reciprocal lattice points after such a rotation. The three scattering geometries used in the measurements (labelled as I, II, III) are schematically shown in fig. 177.

In the case of geometry I, constant-$\boldsymbol{Q}$ measurements were carried out with the scattered neutron energy ε' fixed at about 25 meV. In this geometry, the frequencies of the LA and LO modes propagating along the c-axis can be found unambiguously as a function of wave vector. With geometry II, constant-$\boldsymbol{Q}$ measurements were made, with ε' also fixed at about 25 meV, to determine the frequencies of the TA and lowest-frequency TO modes propagating along the [001]-direction. As indicated by the dashed arrows in fig. 177, the wave-vector component normal to the scattering plane is not well resolved in this experiment, because any point on the [110]-ring may serve as the origin for a wave vector. In addition, the finite $\boldsymbol{q}$-resolution of the spectrometer permits observation of scattering by phonons with wave vectors in the scattering plane, but also with wave vectors along the [001]-direction.

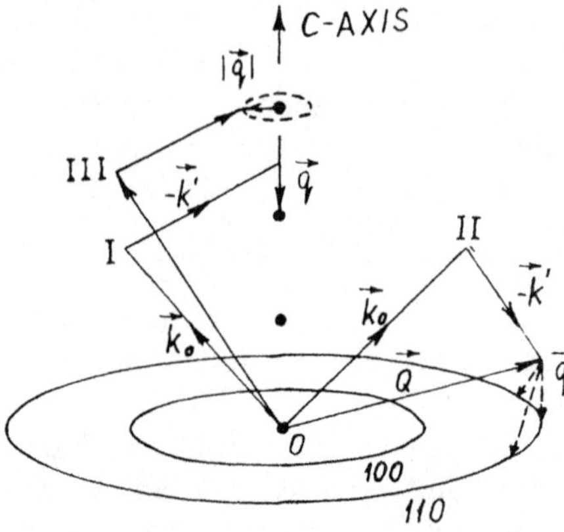

Fig. 177. Reciprocal-space diagram for pyrolytic graphite showing three neutron-scattering configurations. (Nicklow et al. 1972.)

In geometry III, the transverse modes having wave vectors in, and polarizations perpendicular to, the basal plane (the $TA_{\perp}$ and $TO_{\perp}$ modes) were investigated. Well-pronounced neutron groups are observed for small $|\boldsymbol{q}|$ and also for those $\boldsymbol{q}$ whose value is appropriate for the Brillouin-zone edge in the [100]-direction, i.e., at the M-point.

Yet another (IV) experimental geometry may be used, which is similar to geometry II except that $\boldsymbol{Q}$ (and $\boldsymbol{q}$) is now in the basal plane. In this geometry, constant-energy scans were carried out to study the LA and lowest-frequency LO modes having wave vectors in the basal plane. For reasons already

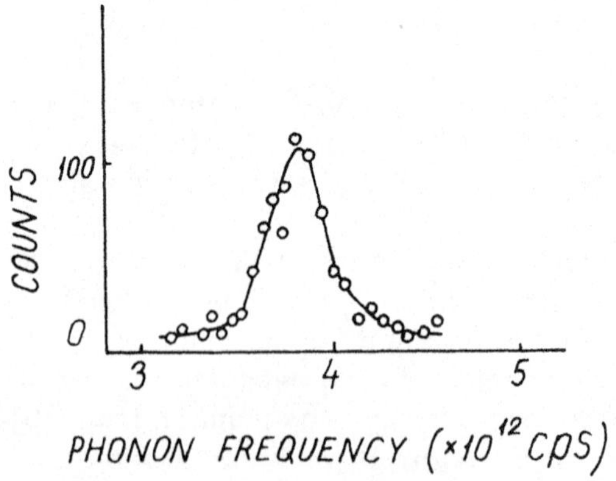

Fig. 178. Neutron group observed by means of arrangement I, at the point $c_0|\boldsymbol{Q}|/2\pi = (0, 0, 3)$. (Dolling and Brockhouse 1962.)

mentioned when discussing geometry-II experiments the phonon wave vector is not well determined in geometry IV.

To our knowledge, there have so far been only two experimental investigations of the phonon dispersion relations in graphite using neutron scattering. The LA modes propagating in the [001]-direction (*c*-axis) were studied, using geometry I, by Dolling and Brockhouse (1962), who obtained the values of LA-phonon frequencies at several regularly spaced $\boldsymbol{Q}$ values along the [001]-direction in the range $c_0|\boldsymbol{Q}|/2\pi = 2.2$–$4.4$. Figure 178 shows a typical neutron group observed in an experiment, for which $c_0|\boldsymbol{Q}|/2\pi = 3.0$. The LA-phonon dispersion relation is shown in fig. 179. The curve very closely resembles a sine curve with a maximum frequency of $(3.84 \pm 0.06) \times 10^{12}$ Hz for the zone-edge phonon of wavelength $c_0 = 6.70$ Å. The TA lattice-vibration modes have been determined less accurately by these authors, yet, the behaviour of the TA-dispersion relation is roughly similar to that for the LA-modes, the maximum frequency at the zone edge now being $(1.3 \pm 0.3) \times 10^{12}$ Hz.

Dolling and Brockhouse (1962) made their measurements using insufficiently perfect pyrolytic graphite samples having an angular spread of about

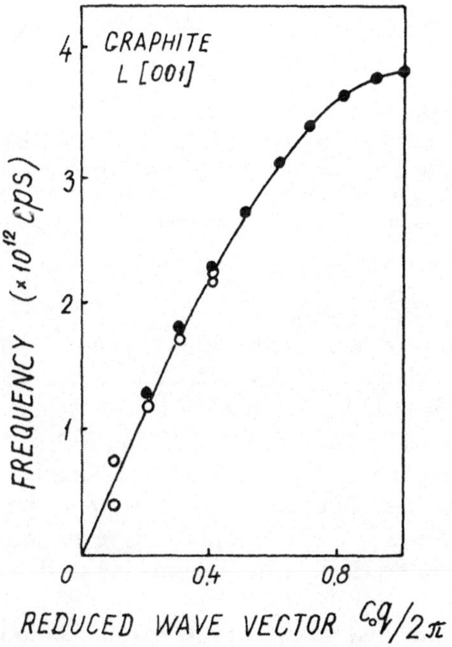

Fig. 179. Dispersion curve for the $L[001]$ modes in which the planes move as rigid units. The solid line is the best fitting (simple) sine wave to the experimental points. (Dolling and Brockhouse 1962.)

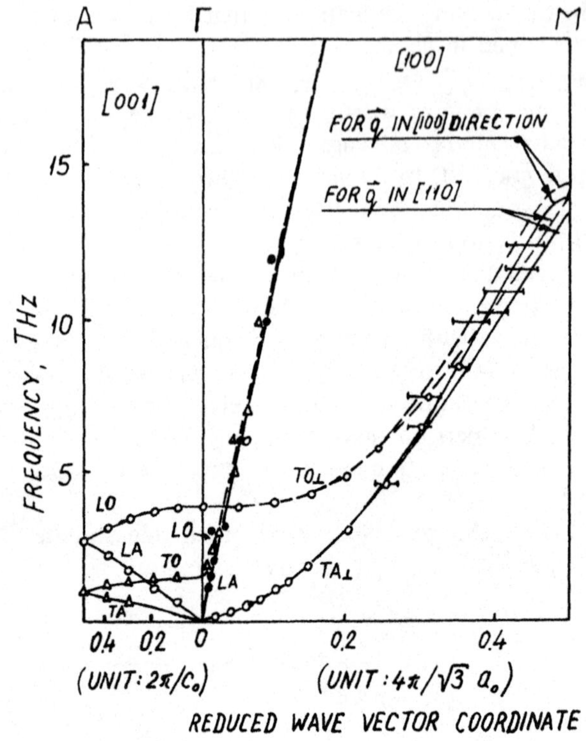

Fig. 180. Phonon frequencies measured for pyrolytic graphite. All measurements for $\boldsymbol{q}$ in the basal plane have been plotted with $\boldsymbol{q}$ expressed in units appropriate for the [100]-direction. The lines shown represent calculations with the fourth-nearest-neighbor AS-model. For the $TA_{\perp}$ and $TO_{\perp}$ branches, calculations for $\boldsymbol{q}$ in both the [100] and [110] directions are shown to illustrate the approximate isotropy of these branches. (Nicklow et al. 1972.)

5° along the c-axis. In a later experiment, Nicklow et al. (1972) used much more perfect and also larger samples with a c-axis angular spread of only 0.4°. This enabled Nicklow et al. (1972) to extract more detailed information about the dispersion relations for the LA, LO, TA, and TO phonons. The results of the measurements by Nicklow et al. (1972) are summarized in fig. 180. The best-resolved peaks of the phonon distribution are observed at low frequencies. The intensity of the peaks very rapidly decreases with frequency, so that already at $\omega \sim 12 \times 10^{12}$ Hz it is hardly possible to determine the frequency reliably.

One may thus state that complete dispersion relations have so far been determined for the LA, LO, TA, and TO phonons propagating along the c-axis, but only very small parts of the dispersion curves for the LA, LO, $TA_{\perp}$, and $TO_{\perp}$ phonons propagating in the basal plane along the [100]- and [110]-

directions have been obtained. It is of interest also to mention that the phonon frequencies obtained by Nicklow et al. (1972) are lower than those determined by Dolling and Brockhouse (1962) at the same wave-vector values. This difference is undoubtedly due to the larger mosaic spread used in the earlier work, which gave rise to contributions to the observed neutron groups for the higher frequency modes that have wave vectors slightly off the [001]-direction.

11.2.2. Raman scattering and infrared reflectance studies on the lattice dynamics of graphite

The Raman spectroscopy of graphite and pyrocarbons, which for any practical purpose are completely black materials, had been virtually impossible until powerful lasers came to be used as sources of light. Accordingly, most of the experimental Raman-scattering works on graphite appeared only in the seventies. These experiments have been performed mainly by two groups: by Dresselhaus and her co-workers at MIT and by Solin and his co-workers at the University of Chicago. These groups have succeeded in obtaining Raman (R) and infrared (IR) spectra of monocrystalline and microcrystalline graphite samples, as well as of material in the form of fibers and whiskers. In the last decade, R- and IR-spectra have been determined for many graphite intercalation compounds of both the donor and acceptor-type, and the lattice dynamics of these materials has been investigated.

One of the most important features, as well as advantages, of the optical methods of studying the lattice dynamics of graphite (such as R- and IR-spectroscopy) is that they do not require large crystals. Because of the absorption of the light in the visible and infra-red regions in graphite, the observed spectra are induced in a thin surface layer of the order of 0.05 μm. For the same reason, reflectance spectra, which carry information only about the surface parts of a crystal, can be used as IR-spectra. In most cases, the reflecting surface is in the basal plane of the graphite crystals. Therefore, it is the vibrational modes in the graphite layers that have largely been investigated by the optical methods.

A Raman spectrum represents the radiation intensity, I, scattered by a single crystal at an angle of 90° relative to the incident beam considered as a function of wavelength.

In first-order Raman scattering only one phonon participates whose wave vector is of the same order of magnitude as that of the photon, i.e., it is not greater than 10^{-3} times the Brillouin-zone size. In the first-order Raman response the scattered radiation frequency differs from the frequency of the incident light by the magnitude of the frequency of the phonon participating in the scattering.

Second-order Raman scattering is due to two-phonon states. The scattering from large single crystals must satisfy wave-vector selection rules, as does the first-order Raman scattering. For second-order Raman scattering, then, the sum of the phonon wave vectors must be equal to the wave vector transferred in the light scattering event. Because of the smallness of the photon wave vector, the two-phonon state sampled by second-order scattering has essentially equal and opposite wave vectors ($\boldsymbol{q}$ and $-\boldsymbol{q}$). If the two-phonon state is formed from phonons which belong to the same branch of the phonon dispersion curve and which have the same energy, then that state is called an overtone. If, however, the two-phonon state represents the sum or difference of two phonons with unequal energy, the state is called a combination. Therefore, due to the wave-vector conservation requirement, the spectrum of overtone states represents the density of one-phonon states convolved with the Raman matrix elément, but the frequency scale is twice that of the one-phonon density of states spectrum. In contrast, the combination spectrum shows additional peaks unobservable in the one-phonon density of states.

The first Raman-scattering studies of graphite were carried out by Tuinstra and Koenig (1970). The Raman spectrum of single crystals of graphite they obtained is shown in fig. 181. The spectrum has a single peak at a frequency of $\sim 1575\,\text{cm}^{-1}$. The position of this mode slightly depends upon the size of crystallites, shifting toward a higher frequency by $\sim 10\,\text{cm}^{-1}$ for the smallest size (Nemanich et al. 1977c). As has been established, different samples give exactly the same spectrum, different orientation of the sample with respect to the incident beam leaving the spectrum unaffected. In the first-order Raman spectrum of other graphite materials, like activated charcoal, carbon black, and vitreous carbon, a second line appears, as fig. 182 illustrates. The additional line is observed near $1355\,\text{cm}^{-1}$, its intensity increasing with an

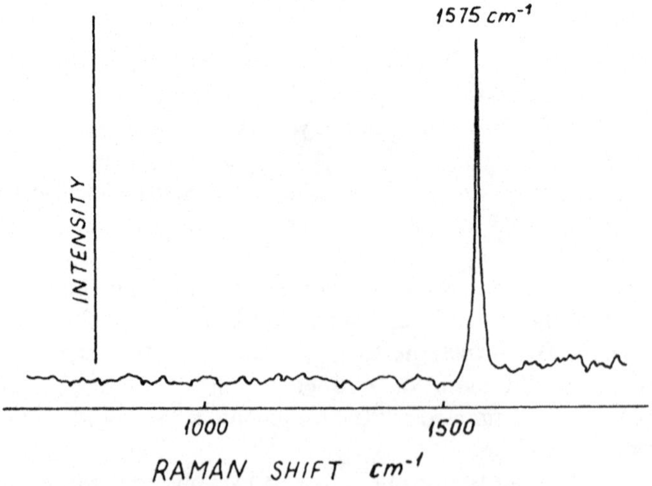

Fig. 181. Raman spectrum of a single crystal of graphite. (Tuinstra and Koenig 1970.)

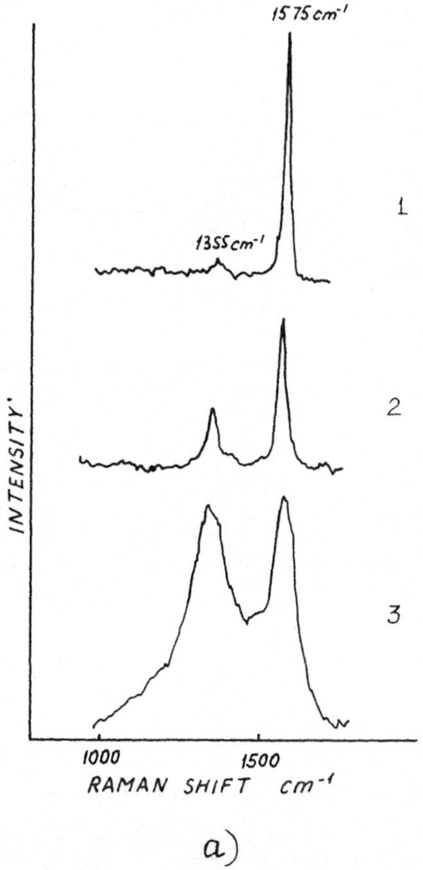

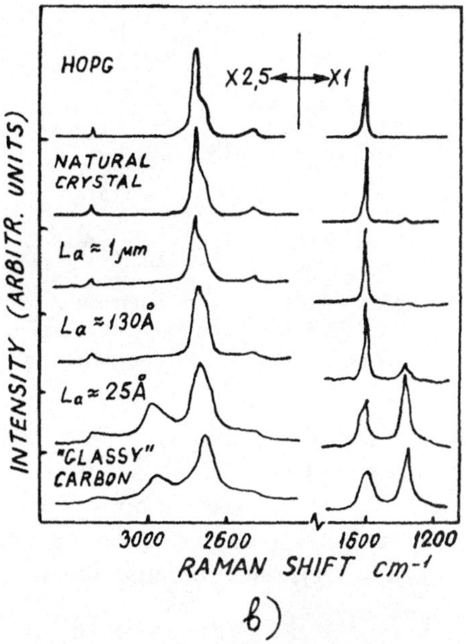

Fig. 182. (a) Comparison of the Raman spectra of: (1) stress-annealed pyrolytic graphite, (2) commercial graphite, and (3) activated charcoal. (Tuinstra and Koenig 1970.) (b) First- and second-order Raman spectra of several different forms of graphite. A spectral slit width of $\sim 10\,\mathrm{cm}^{-1}$ was used. (Nemanich and Solin 1979.)

increase in the amount of "un-organized" carbon in the sample and with a decrease of graphite crystal size. Figure 183 shows the line intensity ratio I_{1355}/I_{1575} as a function of the inverse crystallite size, L_a, in the basal plane. One can see that for $L_a \sim 30$ Å the intensity of the second line becomes comparable to that of the major peak. Nemanich et al. (1975) have subsequently refined the first-order Raman spectrum of single-crystal graphite, having found an exact value, $1581 \pm 1\,\mathrm{cm}^{-1}$, for the frequency of the sharp, intense peak and having observed an additional peak in the low-frequency range at $42\,\mathrm{cm}^{-1}$ (see fig. 184). Two years later the same authors took the IR reflectance spectra of single-crystal graphite for polarizations $\boldsymbol{E} \perp c$ and $\boldsymbol{E} \| c$,

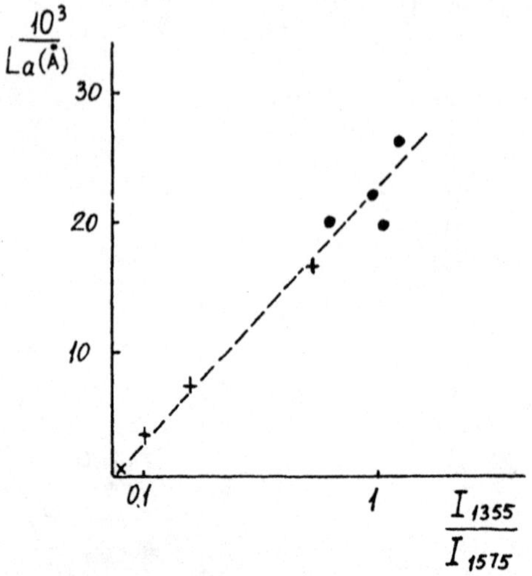

Fig. 183. Calibration of Raman intensities versus X-ray data of L_a; ●, data and samples furnished by Dr. S. L. Strong (Union Carbide); +, samples prepared following the directions in the work of Arnold (1967) ; ×, stress-annealed pyrolytic graphite. (Tuinstra and Koenig 1970.)

which exhibit features at $1588 \pm 2\,\text{cm}^{-1}$ and $868 \pm 1\,\text{cm}^{-1}$, respectively [see Nemanich et al. (1977a) and fig. 185]. Being sensitive to odd-parity phonon modes, infrared spectroscopy furnishes a complementary tool to Raman spectroscopy, which is sensitive to even-parity modes. This complementarity applies to graphite because the inversion symmetry of its crystal structure

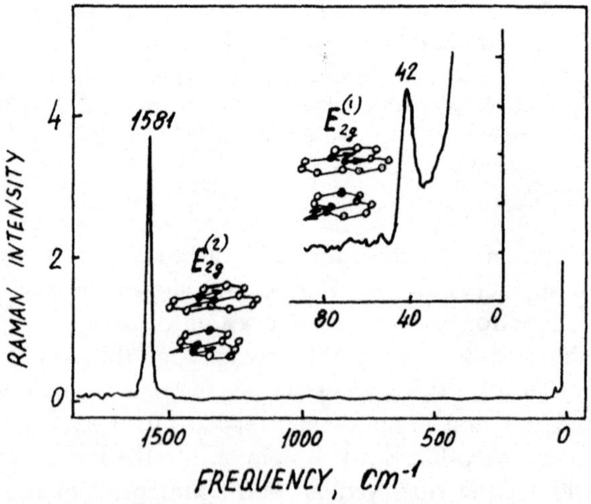

Fig. 184. The Raman spectra obtained from a cleaved surface of graphite, and the atomic displacements of the Raman-active E_{2g} modes. (Solin 1980.)

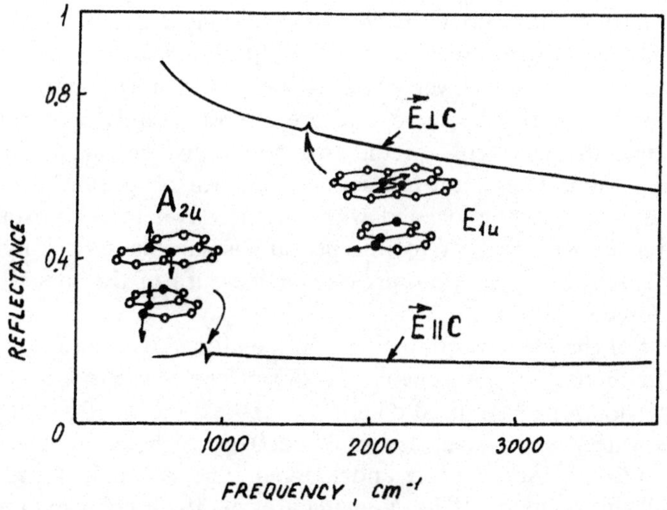

Fig. 185. The $\boldsymbol{E}\perp c$ and $\boldsymbol{E}\| c$ reflectance of graphite, and the atomic displacements of the IR modes. The $\boldsymbol{E} \perp c$ spectrum has been normalized for surface damage. (Solin 1980.)

designates each lattice mode by either an even or an odd parity. The above theoretical group analysis of lattice vibrations in graphite suggests that the sharp peak in the Raman spectrum of monocrystalline graphite should be ascribed to E_{2g}-mode vibrations. This identification of the major peak is in good agreement with the theoretical models of the vibrational spectrum of the graphite lattice. Thus, according to the BBS-model, the calculated E_{2g}-mode frequency value found with the force constants given by eqs. (229) and (233) should be $1688\,\mathrm{cm}^{-1}$, which is in reasonable agreement with the experimental value of $1581\,\mathrm{cm}^{-1}$.

Nemanich and Solin (1979) and Solin (1980) have shown that, as expected, the Raman spectrum of graphite exhibits two E_{2g} modes: one, labelled $E_{2g}^{(1)}$, is an in-plane rigid layer shear mode which occurs at low frequencies, because the layers are rigidly displaced against the weak interlayer restoring force. According to experiment, the frequency of this $E_{2g}^{(1)}$ mode is $42\,\mathrm{cm}^{-1}$. The other, labelled $E_{2g}^{(2)}$, is also an in-plane mode, but it occurs at high frequencies, because the atoms are displaced against the strong nearest-neighbour planar restoring force. The $E_{2g}^{(2)}$-mode has a vibration frequency of $1581\,\mathrm{cm}^{-1}$. The character of displacement of atoms corresponding to the $E_{2g}^{(1)}$- and $E_{2g}^{(2)}$-modes was illustrated in fig. 184.

Features of the IR reflectance spectrum of graphite were examined by Brillson et al. (1971) who demonstrated that they should correspond to the E_{1u} in-plane mode and the A_{2u} out-of-plane mode. As can be seen from a comparison between fig. 184 and fig. 185, in the E_{1u} high-frequency in-plane mode the character of atomic displacement in a layer is practically analogous to that of the $E_{2g}^{(2)}$-mode. Because of those intralayer force constants that are much larger than the interlayer force constants, the vibrational frequencies of these modes are nearly the same, and are almost entirely determined by the simple optical displacements of the two non-equivalent carbon atoms in a single layer plane. The small frequency difference of $\sim 6\,\mathrm{cm}^{-1}$ between the $E_{2g}^{(2)}$- and E_{1u}-modes is associated with the interlayer force constants arising from differences with respect to interplanar displacements (see figs. 184 and 185). This small difference thus provides a measure of the magnitude of the interlayer force constant.

The point of the greatest controversy in the lattice dynamics of graphite has been the frequency of the zone-centre out-of-plane A_{2u}-mode at $868\,\mathrm{cm}^{-1}$. Previous calculations have used different force-field models (the BBS-, the AS-, and other ones) and have assigned frequencies between $600\,\mathrm{cm}^{-1}$ and $1400\,\mathrm{cm}^{-1}$ (Zallen 1968, Young and Koppel 1965, Ahmadieh and Rafizadeh 1973, Mani and Ramani, 1974, Nemanich et al. 1977a). An interesting point about the nature of the lattice dynamics of graphite is that the A_{2u}-mode cannot be described in a simple valence force field model which includes only the usual bond-stretching and bond-bending forces. The addition of the other three-body forces does not remedy the situation. Therefore, it is necessary to consider an additional four-body force that is characterized by a "puckering" of the layer-planes.

The discrepancy between the theoretical value of the A_{2u}-mode frequency and its experimental value of $868\,\mathrm{cm}^{-1}$ has been eliminated by Maeda et al. (1979) who employed the modified Born–von Kármán model which removes the constraint of axial symmetry. The results of their calculations were presented in fig. 174, which shows the dispersion curves for the various branches of the spectrum of graphite, in particular, the dispersion curve corresponding to the A_{2u}-mode at the Γ-point ($\boldsymbol{q} \to 0$).

Let us discuss in some detail the peak at $1355\,\mathrm{cm}^{-1}$ in the first-order Raman spectrum which is observed in polycrystalline pyrocarbon materials (Tuinstra and Koenig 1970, Nemanich and Solin 1979). The increase in the intensity of this peak which is observed with decreasing crystallite size L_a (see figs. 182 and 183) indicates that this mode of vibration is connected with the presence of microcrystalline structure. A plausible explanation of the origin of this peak was offered by Tuinstra and Koenig (1970) who were the first to observe the $1355\,\mathrm{cm}^{-1}$ mode. They attribute this type of vibrations to a change in the selection rules for the Raman activity of certain phonons which are inactive in the infinite lattice, but become active in small crystallites. The new Raman mode whose activity is dependent upon the shape and size of crystallites is, in their opinion, an out-of-plane mode of the A_{1g}-type. In an infinite crystal this mode will be Raman-inactive, because the changes in polarizability cancel out over the infinite crystal. For small crystallites, this is not true and Raman activity can be observed. The A_{1g}-mode corresponds to a longitudinal acoustic mode for the infinite lattice. Therefore, the frequency of this mode should be determined solely by the bond-stretching force constant k and by the mass of the carbon atom, but it should be independent of the force constant μ involving deformation of the angle (see sect. 11.1). A calculation of the constant k based on the mode-frequency value of 1355 cm^{-1} yields $k \simeq 4.32 \times 10^5$ dyne/cm, which is rather close to the values of this constant found by Yosimori and Kitano (1956) and by Young and Koppel (1965).

As has already been mentioned above, the second-order Raman scattering makes it possible to extract information on the phonon frequency that is not limited by the $|\boldsymbol{q}| \to 0$ requirement. Second-order Raman spectroscopy can then be a useful probe into the phonon properties. It should be mentioned however that the interpretation of a second-order Raman spectrum is an arduous task, and cannot be always given unambiguously. The most complete study of the second-order Raman spectra of monocrystalline and pyrolytic graphite containing both macro- and micro-crystallites has been made by Nemanich and Solin (1977, 1979). Both prior to, and simultaneously with, their work, studies were made into the Raman spectra of various forms of graphite, including graphitizable carbon, coal and coke (Nathan et al. 1974, Nakamizo et al. 1974, Tsu et al. 1977, 1978), as well as radiation-damaged material (Wright et al. 1976, Maeta and Sato 1977). In order to relate the first- and second-order Raman features of graphite, the full Raman spectrum is shown in fig. 186.

The above examination has shown that the strong, sharp feature at

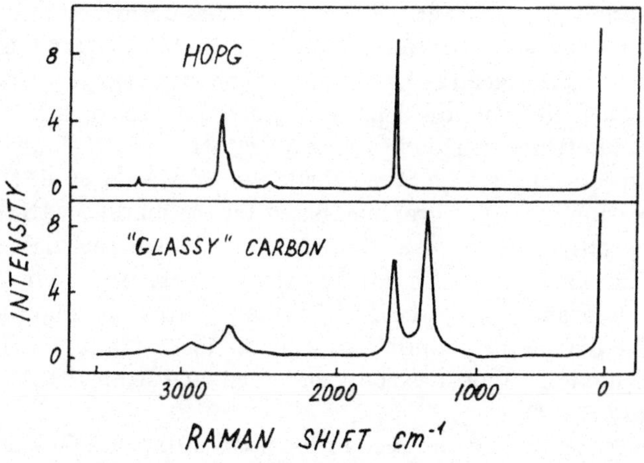

Fig. 186. First- and second-order Raman spectra of HOPG and "glassy" carbon. The spectra were recorded with a spectral slit width of ~10 cm^{-1}. For this figure the abscissa, while marked in cm^{-1}, is, however, linear in wavelength. (Nemanich and Solin 1979.)

1581 cm^{-1} is due to the E_{2g} high-frequency first-order mode. The low-frequency E_{2g}-mode, which occurs at 42 cm^{-1}, is obtainable only with special techniques (Nemanich et al. 1975, Solin 1980). The continuum scattering from 2200 to 3250 cm^{-1} represents the second-order features. The second-order spectrum exhibits three distinct groups of bands. The spectrum is dominated by the strong feature near 2735 cm^{-1}, two weaker features being observed at ~2450 and ~3250 cm^{-1}.

Let us compare the experimental second-order Raman spectra with the predictions of the existing theoretical models for the vibrational spectrum of graphite in the high-frequency range. First of all, one should mention that experiments performed on other substances with well-known phonon spectra (Si, Ge, diamond, etc.) indicate that the second-order Raman spectrum is often dominated by overtone states. Therefore, the most natural explanation of the peaks observed in the second-order Raman spectrum is their representation as features corresponding to double frequencies of the sharp peaks in the phonon-states density curve $D(\omega)$.

Since the highest-frequency out-of-plane mode A_{2u}, which is degenerate in the centre of the Brillouin zone with the B_{2g}-mode, corresponds to a frequency of 867 cm^{-1} (Nemanich et al. 1977a), it is probable that none of the features of the second-order spectrum from 2300 to 3300 cm^{-1} arise from overtones of modes with "out-of-plane" atomic displacement. Thus the above three features of the second-order Raman spectrum of graphite appear to be overtones of "in-plane" modes or the features of the combination spectrum. It should be recalled that the emergence of combination frequencies is far less probable than that of overtones.

One of the possible explanations of the features of the second-order spectrum is to describe them as overtones of frequencies corresponding to flat parts or maxima of dispersion curves which occur at $|\boldsymbol{q}| \neq 0$ (and are not seen therefore in first-order spectra). Such parts of the dispersion curves correspond to sharp peaks of the phonon density of state. The above hypothesis agrees well, both qualitatively and quantitatively, with the most realistic models of the phonon spectrum [see, e.g., the calculation of Maeda et al. (1979) on the phonon dispersion curves in graphite, based on the modified AS-model of the vibrational spectrum, figs. 174 and 172]. In particular, the highest-frequency feature at 3248 cm^{-1} may be attributed to scattering from the highest-frequency optic branch along the ΓM showing a large one-phonon density of states near the maximum phonon frequency at the Σ-point. The frequency of the observed mode is almost exactly that predicted from the AS-model (i.e., two times 1618 cm^{-1}).

The strongest feature in the second-order spectrum at 2710 cm^{-1} may be associated with the M-point of the high-frequency LA branch (2×1360 cm^{-1} $\simeq$ 2735 cm^{-1}, see fig. 172). And the third stepwise feature at ~2450 cm^{-1} can be attributed (Dresselhaus and Dresselhaus 1981, Giergiel and Eklund 1982) to the combination scattering that involves phonons from the Σ-point ($\sim 1620\,\text{cm}^{-1}$) and the Γ-point ($\sim 840\,\text{cm}^{-1}$). Thus all the features observed in the second-order Raman spectra may sufficiently correctly be associated with the peaks of the one-phonon density of states.

It is of interest to note that in graphite specimens containing microcrystallites with a size of about 130 μm or smaller, whose first-order Raman spectra exhibit a peak at $1355\,\text{cm}^{-1}$, in the second-order Raman spectra one can observe line broadening and an additional feature at $2950\,\text{cm}^{-1}$. Simultaneously, the peak that is observed in the first-order spectrum at $1582\,\text{cm}^{-1}$ splits into two peaks, the new sharp peak at $1621\,\text{cm}^{-1}$ increasing with decreasing crystallite size (see fig. 182b). Nemanich and Solin (1979) associate the broadening of the second-order peaks with the increasing uncertainty of the phonon wave vector which, by the uncertainty relation, is about $2\pi/L_a$, increasing with diminishing crystallite size. The uncertainty of the phonon wave vector in crystallites gives rise to the broadening of the features which are sharp in the spectra of HOPG and natural crystalline graphite, this being due to the overtone-type scattering of two phonons with nearly equal energy. Nemanich and Solin (1979) believe that the new feature at $2950\,\text{cm}^{-1}$ in the second-order spectra arises from a combination of the high density of states at ~ 1355 and $\sim 1620\,\text{cm}^{-1}$. Such a hypothesis seems rather plausible. Indeed, the possibility of the appearance of a feature corresponding to the combination frequency greatly increases with the increase in the uncertainty in the wave vector in small crystallites. While in an infinite crystal two phonons with approximately opposite wave vectors will scatter, the microcrystalline domain will relax the wave-vector restriction in a way similar to that described for first-order scattering. Scattering can then occur for two phonons with different wave vectors where the wave-vector difference is again limited by the uncertainty relation. However, if the wave-vector restriction is relaxed, a peak can occur at the sum of the frequencies of the two sharp features observed in the density of states. Thus an increase in wave-vector uncertainty in microcrystallites appears to lead to a qualitatively different effect, namely the emergence of combination-like features. It should be mentioned that the new feature at $2950\,\text{cm}^{-1}$ in the second-order Raman spectra has been observed also by Wright et al. (1976) and Tsu et al. (1978) in radiation-damaged carbon materials and in various forms of graphite, including graphitizable carbon, coal, and coke. In the former of the above-mentioned works, the feature at $2950\,\text{cm}^{-1}$ was also described as a combination one.

11.3. Phonon spectra of GIC

The preceding sections of this chapter have dealt in a detailed manner with the vibrational spectrum of the graphite lattice and the experimental data available on the phonon frequencies in the spectrum. Figure 174, which has been taken from the work by Maeda et al. (1979) and which summarized the most recent and reliable experimental and theoretical results, offers a modern view of all the branches of the phonon spectrum of graphite and the corresponding frequencies. Similarly, fig. 173 summarized the data on the total density of phonon states, $D(\omega)$, in graphite. Concluding, both the general features and details of the phonon spectrum of graphite are known at present.

Intercalation of graphite with various donors or acceptors leads not only to a change in the electron energy spectrum and band structure but also to a considerable change in the vibrational spectrum of the lattice. At the same time, it should be noted that the phonon spectrum of GIC retains many essential features of the spectrum of pure graphite.

Let us consider the factors which, taken together, may give rise to a change in the vibrational spectrum of the graphite lattice as a result of intercalation.

For one thing, intercalation leads to a change in free-carrier concentration within the bounding graphite layers. This, in turn, should result in a change in the bonding force between the carbon atoms in the layer and, therefore, a change in the frequencies of the various vibrational modes, both in-plane and out-of-plane.

Secondly, the emergence in the lattice of periodically repeating intercalate layers, whose constituent particles have masses and bonding forces different from those of the carbon atoms, should give rise to new frequencies, associated directly with vibrations of the intercalants, but absent in the initial graphite spectrum.

Thirdly, the bonds between the intercalate layers and the adjacent graphite layers should also contribute to the formation of new frequencies.

Fourthly, the change of the initial spectrum should be due to a change in lattice periodicity both along the c-axis (the appearance of a new identity period I_c) and in the direction perpendicular to it (the zone-folding effect). The associated change in the shape and size of the Brillouin zone, the appearance of physically equivalent points in the wave-vector space, a change in the Brillouin-zone symmetry (the disappearance of the inversion centre) – all these factors are capable of generating new branches of the original vibrational spectrum of graphite or altering the old branches.

Finally, a definite change in the vibrational spectrum of GIC may be expected when the temperature is changed, thus giving rise to a change in the structure and degree of ordering in the intercalate layer (transitions "order–disorder", "two-dimensional crystal–two-dimensional liquid", etc., in the intercalate layers).

Let us examine one by one, the factors listed above from the viewpoint of the influence they exert upon the vibrational spectrum of GIC.

11.3.1. The intercalation-induced change in the vibrational modes of pure graphite

From general considerations it is evident that the perturbation, due to intercalation, of the original graphite spectrum should weaken as the stage of intercalation, n, increases. All the principal frequencies in the spectrum should then tend to those of the graphite-lattice vibrations. On the other hand, it is the first-stage compounds, in which the intercalate and graphite layers alternate, whose phonon spectrum differs the most from that of the original graphite. Going to the $n \to \infty$ limit enables one to establish the correspondence between the frequencies in the spectrum of GIC and the equivalent frequencies in the original spectrum. From this point of view it is of interest to look at the dependence upon intercalation stage index of those individual frequencies of lattice vibrations which are typical of the graphite spectrum and also to trace how the corresponding vibrational modes are changed.

One of the most typical vibrational modes in the graphite spectrum is the Raman-active $E_{2g}^{(2)}$-mode, to which there correspond in-plane displacements of the carbon atoms in the layer at $1582\,\mathrm{cm}^{-1}$. The high-frequency $E_{2g}^{(2)}$-mode is associated with large interaction forces between carbon atoms in the graphite layer. Since the lattice constant in the graphite layers insignificantly changes in intercalation, one would expect that the $E_{2g}^{(2)}$-mode suffers but a relatively small perturbation. At the same time, when GIC with $n \geqslant 3$ are synthesized, the complete equivalence of the individual graphite layers, which takes place in pure graphite and in GIC with $n = 1$ and 2, is broken. For this reason in GIC with $n \geqslant 3$ one may expect that the $E_{2g}^{(2)}$-mode would split into a doublet. For a more detailed treatment, consider the possible nearest-layer environments of the carbon layers in a pure stage-n intercalation compound, as depicted in fig. 187 (see Solin 1980). In a stage-one compound, each carbon layer is bounded by a pair of intercalate layers (fig. 187, type A). There is only one nearest-layer environment, labelled as type A. Similarly, in a stage-two compound, each carbon layer is bounded by another carbon layer and an intercalate layer (fig. 187, type B). Again, there is only one nearest layer environment, but it differs from that in a stage-1 compound, and is labelled as type B. In a stage $n \geqslant 3$ compound, there are just two types of nearest-layer environment, one of type B, which is identical with the environment in a stage-2 compound, and the other, labelled as type C, in which a carbon layer is bounded by other carbon layers as in the case of pristine graphite (see fig. 187, type C). Evidently, for each carbon layer the intralayer $E_{2g}^{(2)}$-mode depends on the type of the environment in GIC. As in intercalation the atomic structure of carbon layers remains undistorted, it may be suggested that, in a first approximation, the atomic displacement corresponding to the $E_{2g}^{(2)}$-mode will also occur in an intercalated sample, but the force constants which determine the mode fre-

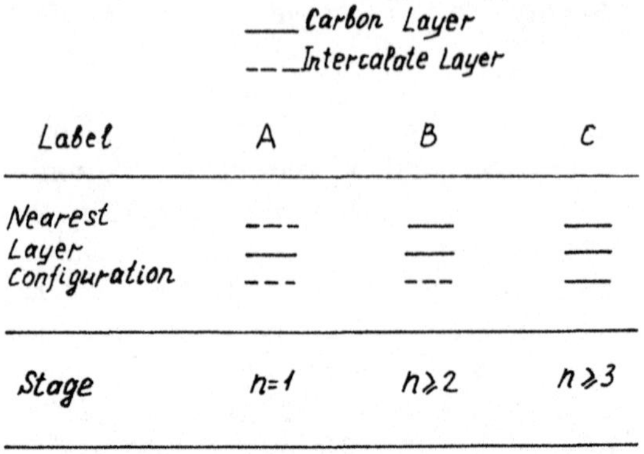

Fig. 187. Nearest-layer configurations for carbon layers in GIC. (Solin 1980.)

quency will depend on the nearest-layer environment. Proceeding from these considerations, one can predict that in a stage-1 compound there should be observed a single Raman band in the vicinity of (but displaced in frequency from) the $E_{2g}^{(2)}$-mode of pristine graphite. The same should be true of a stage-2 GIC. However for stage-3 (and higher stage) GICs one expects to observe two high-frequency Raman bands: one, in a first approximation, at the position of the $E_{2g}^{(2)}$ pristine graphite mode and the other displaced in frequency. Furthermore, the two bands which result, respectively, from the type-B and type-C environments in $n \geqslant 3$ GIC should exchange their intensities, the type-C band at the position of the $E_{2g}^{(2)}$ pristine graphite mode becoming more intense with increasing stage index n.

This latter point can be simply quantified by "counting" the number of layers having particular nearest-layer environment in a given stage-n GIC, as proposed by Solin (1977, 1980) and Dresselhaus et al. (1977) (the nearest-layer model). Using such a counting procedure, one finds that the relative integrated intensity ratio, R, of type-B and type-C derived Raman bands for stage $n \geqslant 2$ donor or acceptor compounds is

$$R = \frac{I_C}{I_B} = \frac{n-2}{2}\frac{\sigma_C}{\sigma_B}, \tag{248}$$

where I_B and I_C are, respectively, the intensities of the type-B and type-C Raman bands and σ_C, σ_B denote the Raman cross section for scattering from an $E_{2g}^{(2)}$-type intralayer mode in a type-C or type-B layer. One may suppose that the σ_C/σ_B ratio is independent of stage index. This is equivalent to the assumption that the charge density transferred from the intercalate layer to the adjacent graphite layer (or the quantity f_{el}) is independent, for one and the same intercalant, of stage index n, which seems to be quite plausible. According to (248), R is then a linear function of n which intersects the abscissa at $n = 2$, its slope-angle tangent determining the σ_C/σ_B ratio.

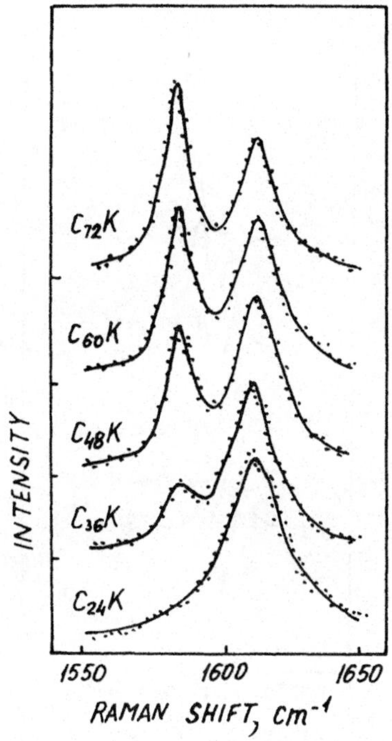

Fig. 188. Raman spectra of $C_{12n}K$ ($n = 2, \ldots, 6$). Lorentzian fit (–). Data (···). (Solin 1980.)

The above concepts of the evolution of the $E_{2g}^{(2)}$-mode during intercalation are in good agreement with the experimental data. They have first been established by Solin (1977) in donor-type graphite compounds with Rb as the intercalation stage increased from 2 to 6. Subsequently, they have been verified by Caswell and Solin (1978b) in donor-type graphite compounds with K and also by Caswell and Solin (1978a), Caswell et al. (1979), and Underhill et al. (1979) in acceptor-type graphite compounds with $FeCl_3$ for $n = 1, 2, 3, 4, 6, 11$.

As an example, figs. 188 and 189 shows the respective Raman spectra for C_xK and C_xFeCl_3 compounds with various stage indices. Notice that, as predicted, the stage-1 and stage-2 compounds exhibit a single high-frequency line in the region of the $E_{2g}^{(2)}$-mode of pristine graphite, while the higher stage compounds produce a doublet structure in this region. A similar doublet structure with a frequency separation of $\sim 20\,cm^{-1}$ is found for $n > 2$ for all the intercalants studied, including the acceptors Br, IBr and ICl (Song et al. 1976), $AlCl_3$ (Gualberto et al. 1980), $SbCl_5$ (Eklund et al. 1980a,b) and the donor Cs (Eklund et al. 1977). The lower frequency component is attributed to the $E_{2g}^{(2)}$ carbon-atom vibrations in the interior graphite layer planes and the lattice mode associated with these layers is denoted by $\mathring{E}_{2g}^{(2)}$. This identification is supported by the proximity of the $\mathring{E}_{2g}^{(2)}$-mode to the $E_{2g}^{(2)}$-mode of pristine

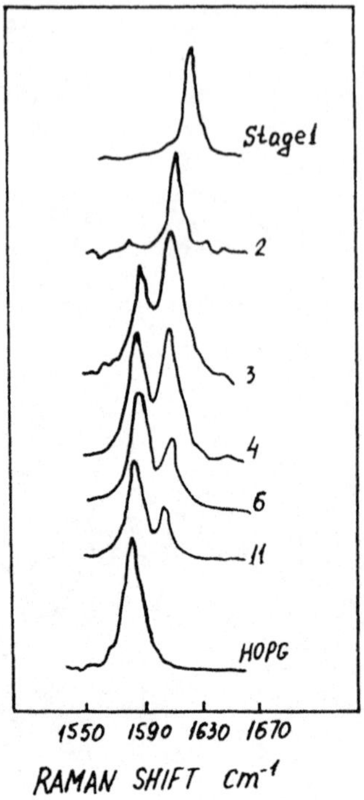

Fig. 189. The Raman spectra of stage 1, 2, 3, 4, 6, and 11 $FeCl_3$–graphite. For comparison, also HOPG is shown. (Caswell and Solin 1978a.)

graphite, by the decrease of intensity of the $\mathring{E}_{2g}^{(2)}$ with decreasing stage (increasing intercalate concentration), and by the disappearance of the $\mathring{E}_{2g}^{(2)}$ line in stage-1 and stage-2 compounds, where there are no graphite interior layers. It is also significant that for $n \geqslant 2$ the line shapes are Lorentzian and the line width of the $E_{2g}^{(2)}$ component is insensitive to intercalate concentration. This suggests that all interior graphite layers have approximately the same set of in-plane force constants. The upper frequency component of the doublet structure (see figs. 188 and 189) is identified with an $E_{2g}^{(2)}$-type graphitic mode occurring in a bounding graphite layer, and this mode is denoted by $\hat{E}_{2g}^{(2)}$. Support for this identification comes from the absence of the $\hat{E}_{2g}^{(2)}$-line in pristine graphite, the increase in its intensity with increasing intercalate concentration, and the occurrence of a single $\hat{E}_{2g}^{(2)}$-line in stage-1 and stage-2 compounds, where all graphite layers are bounding layers. The upshift in the frequency of the $\hat{E}_{2g}^{(2)}$-mode is due to the difference in force constants arising from the different environments of the carbon atoms in the graphite bounding and interior layers. The dependence on stage number of the relative intensities of the $\mathring{E}_{2g}^{(2)}$- and $\hat{E}_{2g}^{(2)}$-modes indicates that a single bounding layer on either side

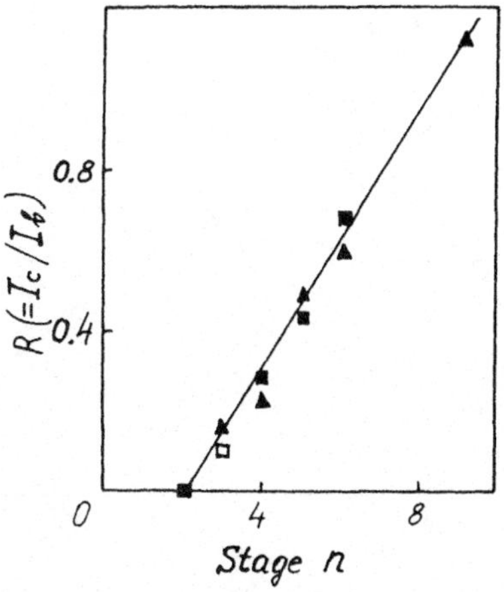

Fig. 190. The dependence of the intensity ratio R on n for $C_{8n}Rb$ (▲) and $C_{8n}K$(■). The solid line is a linear least-squares fit to the data. (Solin 1980.)

of the intercalate layer effectively screens the intercalate layer from the graphite interior layers. This behaviour is also supported by the z-dependence of the charge distribution on the basis of a Thomas–Fermi model, yielding a high charge density in the graphite bounding layers and a rapid decrease of charge density with distance from the intercalate layer.

By computer fitting each of the doublet spectra with a pair of Lorentzian spectra and correcting for instrumental broadening one can deduce accurate experimental values for I_C/I_B as a function of n. Such a procedure has been carried out by Solin (1977) and Caswell and Solin (1978b) for C_xK and C_xRb, and by Caswell and Solin (1978a) and Caswell et al. (1979) for C_xFeCl_3. The resulting dependences of I_C/I_B upon n for these substances are illustrated in figs. 190 and 191.

The data of Caswell and Solin (1978a), shown in fig. 191 by the solid circles, refer to highly inhomogeneous samples, as shown by X-ray analysis; this also accounts for their large spread. Attention is called to the fact that these dependences agree quite well with the simple linear dependence given by eq. (248), thereby confirming the validity of the nearest-layer model (at least, for such donor-type compounds as C_xK_n and C_xRb). Furthermore, fig. 190 shows that the σ_C/σ_B ratio is the same, within the accuracy of determination, for both C_xK and C_xRb, constituting 0.3 ± 0.1. The vibrations which generate the B and C members of the Raman doublet at $\simeq 1582\,\mathrm{cm}^{-1}$ have essentially identical atomic displacements. The Raman cross sections σ_C and σ_B are determined from the polarizability which, in turn, depends on the atomic displacements and the amount of polarizable charge present in the layer. Therefore, the σ_C/σ_B ratio is, in fact, a measure of charge exchange and charge

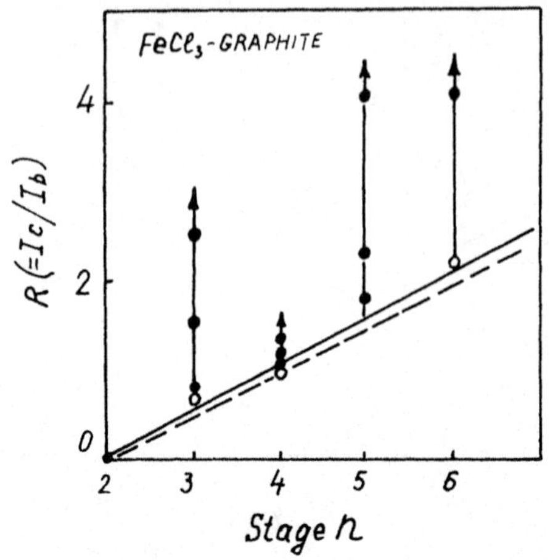

Fig. 191. The dependence of the intensity ratio R on n for $FeCl_3$–graphite; ○, Caswell et al. (1979); ●, Caswell and Solin (1978a). The solid line is a least-squares fit to the data of Caswell and Solin (1978a). The dashed line corresponds to $\sigma_C/\sigma_B \simeq 1$ in (248). (Solin 1980.)

localization, and it should be determined by the value of the fraction of the charge exchange, f_{el}. If, for donors, the B layers adjacent to the intercalate layer contain more charge than the C layers, then one would expect to find that $\sigma_B > \sigma_C$ and $\sigma_C/\sigma_B < 1$, as can be seen in fig. 190. In the case of acceptors, the B layer contains less electrons than does the C layer, i.e., $\sigma_B < \sigma_C$ and $\sigma_C/\sigma_B > 1$. This result is consistent with the data of Underhill et al. (1979) for the most homogeneous samples (see fig. 191, open circles). According to them, the σ_C/σ_B ratio is 1.16, which is in agreement with the well-known fact that in compounds with the weak acceptor $FeCl_3$ the charge exchange is much smaller than in alkali-metal donor compounds. Thus the polarizability of the B-layer is only slightly reduced relative to that of the C-layer, which loses a smaller amount of charge, and a σ_C/σ_B ratio close to 1 is obtained.

Underhill et al. (1979) also measured the IR reflectance spectra of C_xFeCl_3 for $n = 1, 2, 3, 4, 6$, and 11. They found excellent qualitative agreement with a nearest-layer model in which the E_{1u} IR-active mode of pristine graphite plays the same role as the $E_{2g}^{(2)}$-mode does in Raman scattering. From an oscillator analysis of their IR-data they concluded that the fractional charge transfer per $FeCl_3$ intercalant unit was about 0.75 for stage-1 and stage-2 and 0.95 for $n \geqq 3$ compounds. They also concluded that 70% of the exchanged charge resided in the B layers, while only 30% were transferred to the C layers independent of stage. These results are in some qualitative disagreement with those of other measurements and with the Raman analysis described above. For instance, Mele and Ritsko (1979) studied the core excitation electron loss spectrum of stage-1 $FeCl_3$ graphite and obtained a fractional charge transfer

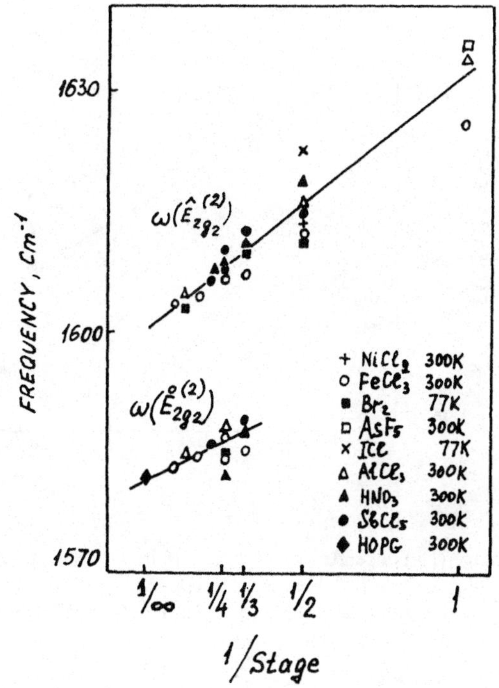

Fig. 192. Reciprocal stage ($1/n$) dependence of the Raman frequencies associated with the graphite interior layers ($\mathring{E}_{2g_2}^{(2)}$) and the graphite bounding layers ($\hat{E}_{2g_2}^{(2)}$) for various acceptor intercalants. An upshift is found for both $\omega(\mathring{E}_{2g_2}^{(2)})$ and $\omega(\hat{E}_{2g_2}^{(2)})$ versus ($1/n$) and the behaviour is similar for all acceptor intercalants. (Dresselhaus and Dresselhaus 1981.)

per intercalant unit of 0.07–0.12. Also the 70%–30% charge distribution would yield $\sigma_C/\sigma_B \simeq 2.3$, in contrast to what is observed. It is possible that these discrepancies come in part from the equation of effective charge e^*, which is determined in an oscillator analysis with the actual charge transfer.

The frequency for the $\mathring{E}_{2g}^{(2)}$ and $\hat{E}_{2g}^{(2)}$ modes also exhibits a distinctive dependence on reciprocal stage. For acceptor compounds with the intercalants $FeCl_3$, $AlCl_3$, Br_2, AsF_5, $SbCl_5$, HNO_3, both mode frequencies $\omega(\mathring{E}_{2g}^{(2)})$ and $\omega(\hat{E}_{2g}^{(2)})$ exhibit approximately the same upshift as a function of the reciprocal stage $1/n$, as is shown in fig. 192. A similar stage-dependent upshift of the two modes gives rise to a doublet separation of $22 \pm 2\,\mathrm{cm}^{-1}$ for these acceptor compounds. In contrast to this, the donor alkali-metal compounds exhibit a downshift in these mode frequencies with increasing reciprocal stage, as is shown in fig. 193. This mode softening is consistent with an in-plane lattice expansion with increasing $1/n$, as reported by Nixon and Parry (1969). The stiffening of these lattice modes in the acceptor compounds and their softening in the donor compounds have been attributed to lattice strain associated with

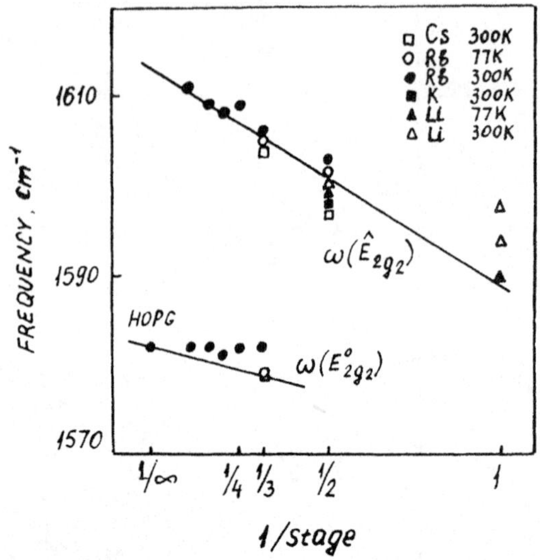

Fig. 193. Reciprocal stage ($1/n$) dependence of the Raman frequencies associated with the graphite interior layers ($\mathring{E}_{2g_2}$) and the graphite bounding layers ($\hat{E}_{2g_2}$) for various donor intercalants. A downshift is found for both $\omega(\mathring{E}_{2g_2})$ and $\omega(\hat{E}_{2g_2})$ versus $1/n$. (Dresselhaus and Dresselhaus 1981.)

stage-dependent changes in the lattice parameters (see Underhill et al. 1979). For $FeCl_3$–graphite compounds, the constancy of the line width for the $\mathring{E}^{(2)}_{2g}$ interior layer mode as a function of intercalant concentration and the identical frequency upshift for both the interior and bounding Raman modes have been interpreted by Underhill et al. (1979) in terms of an identical strain within both the bounding and interior graphite layers for a given compound, a condition which is also necessary to prevent sample fracture upon intercalation. The magnitude of the strain increases as the number of interior graphite layers decreases, which is consistent with the idea that the stress induced by intercalation is shared by fewer layers as the stage index n decreases. These Raman studies indicate that the acceptor compounds should experience an in-plane lattice contraction in the graphite layers as a function of $1/n$. Such a lattice contraction has recently been confirmed by direct X-ray measurements in $FeCl_3$-, $AlCl_3$-, AsF_5-compounds (Markiewicz et al. 1980). Thus the addition of the electron charge density expands the lattice, while its removal results in lattice contraction.

The intercalation-induced shift in the phonon frequencies corresponding to characteristic vibrational modes in pristine graphite can be observed in studying the second-order Raman spectra of GIC of various stages.

At present the second-order Raman spectra are known in detail only for donor compounds with Rb and acceptor compounds with $FeCl_3$ and $SbCl_5$ (Giergiel and Eklund 1982).

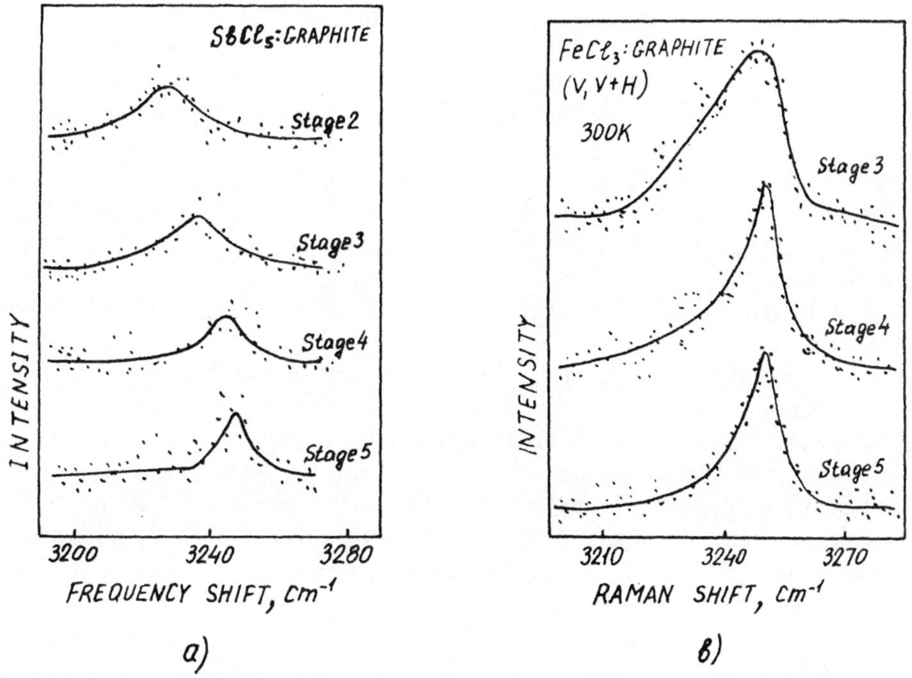

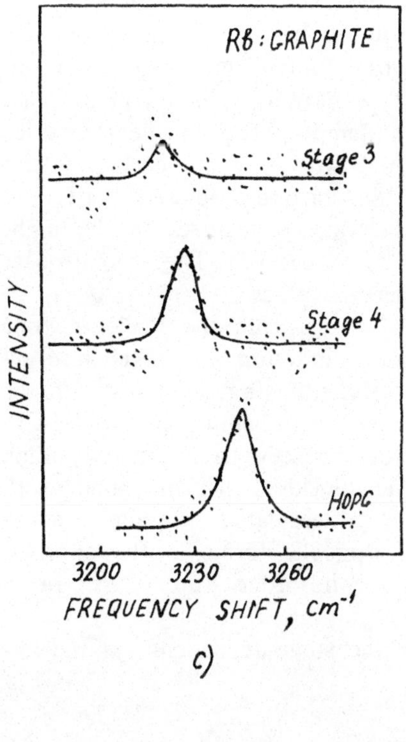

Fig. 194. Second-order Raman spectra of: (a) $SbCl_5$–, (b) $FeCl_3$– and (c) Rb–graphite in the 3250 cm^{-1} region. The solid lines in figs. 194a,c represent the result of Lorentzian-lineshape analyses. In fig. 194b ($FeCl_3$–graphite), the solid line is a guide for the eye. The spectrum of HOPG is shown for comparison. (Giergiel and Eklund 1982.)

Let us begin our analysis of the phonon frequency shift due to changing stage index with the highest-frequency feature in the second-order Raman spectra, which corresponds to the peak at $3248\,cm^{-1}$ in pristine graphite. The corresponding features in the second-order Raman spectra for compounds with $SbCl_5$, $FeCl_3$, and Rb are shown in fig. 194a,b,c, respectively. As in the case of pristine graphite, we shall proceed for the assumption that the $3248\,cm^{-1}$ features are derived from overtones of the Σ-point modes located midway along the ΓM direction in the Brillouin zone (Dresselhaus and Dresselhaus 1981).

For the reasons discussed above, one would expect that in GIC with $n \geqslant 3$, the single peak observed in the second-order Raman spectra near $3248\,cm^{-1}$ will change into a doublet. However, no distinct doublet has been observed experimentally. As can be seen from fig. 194, the features in the spectra of $SbCl_5$ GIC represent single broad symmetrical maxima with a Lorentzian lineshape. The similar features in $FeCl_3$ GIC are asymmetric, and cannot be sufficiently well described by either a Lorentzian singlet or a doublet (fig. 194b). The absence of distinctly pronounced doublets in the second-order Raman spectra for GIC with $n \geqslant 3$ may, in principle, be explained by the strong smearing of second-order peaks, as a result of which the doublets are not resolved in experiment. In spite of the difference in the shape of the curves for $SbCl_5$- and $FeCl_3$-compounds with graphite, features observed in their spectra exhibit the same frequency shifts with changing stage, namely, the softening of the Σ-point phonon modes with decreasing stage index n. This is very interesting because for the same compounds the Γ-point phonon modes ($\mathring{E}_{2g}^{(2)}$ and $E_{2g}^{(2)}$), which are observed in first-order Raman scattering, exhibit an entirely different stage dependence, i.e., the Γ-point modes stiffen with decreasing stage index (see figs. 189 and 192). The stiffening of the zone-centre modes in acceptor compounds has been associated with in-plane carbon lattice contraction resulting in hardening the carbon–carbon in-plane bonds. The stage-dependence of the frequencies of the various features in the high-frequency Raman spectra of GICs is shown in fig. 195. The second-order features are plotted at one-half of their observed values.

As is known, dispersion-curve maxima found within the Brillouin zone usually arise when second-neighbour atomic interactions are taken into account. For a one-dimensional atomic row, the stiffening of bonds with the nearest neighbours would inevitably result in strengthening the bonds with second-nearest neighbours. This statement does not seem to be generally valid. In particular, it may be violated when one considers in-plane bonds with second neighbours in the carbon layers of intercalated graphite. Such a supposition is prompted by the observed opposite forms of the phonon-frequency stage-dependence in first- and second-order Raman spectra in acceptor $SbCl_5$- and $FeCl_3$-compounds with graphite.

In contrast to the acceptor compounds, the stage-dependence of the fre-

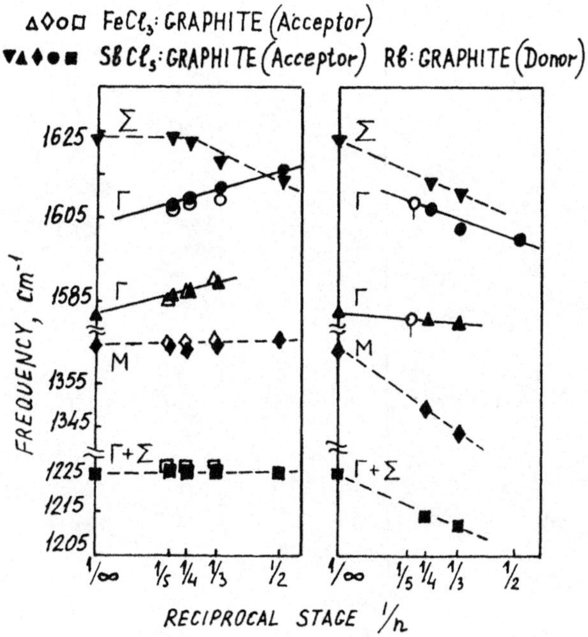

Fig. 195. Characteristic mode frequencies in $SbCl_5$–, $FeCl_3$–, and Rb–graphite plotted versus reciprocal stage ($1/n$). The open circles with tails are data from the work by Solin (1977). Second-order scattering features are plotted at one-half of their observed frequencies. The first-order features (Γ) are included for comparison. (Giergiel and Eklund 1982.)

quency of the Σ-point phonon modes in the donor compounds, as observed in second-order Raman scattering studies of Rb–graphite compounds, is similar to that of the Γ-point modes (see figs. 194c and 195). This indicates a distinctly different behaviour for the stage-dependence of the second-neighbour in-plane force constants for the acceptor and donor GIC. Attention of the reader is called to the fact that the dispersion curves for the phonon spectrum of donor GIC with Rb, K, and Cs have recently been calculated by Al-Jishi and Dresselhaus (1982c). They obtained the softening of the Σ- and Γ-point in-plane modes with increasing intercalant concentration, in agreement with the experimental dependences extracted by Giergiel and Eklund (1982), as shown in fig. 195. The results of a similar calculation by Leung et al. (1981b) for acceptor phonon spectra are at variance with the data presented in fig. 195 for the stage-dependence of the Σ-point phonon modes.

Consider now the observed stage-dependence of the $\sim 2450\,cm^{-1}$ features in second-order Raman spectra for the $SbCl_5$–, $FeCl_3$–, and Rb–graphite compounds (see fig. 196a,b,c). It should be recalled that two-phonon scattering in this frequency range is associated with the combination scattering of the Σ-point ($\sim 1620\,cm^{-1}$) and Γ-point ($\sim 850\,cm^{-1}$) phonons. As can be seen from

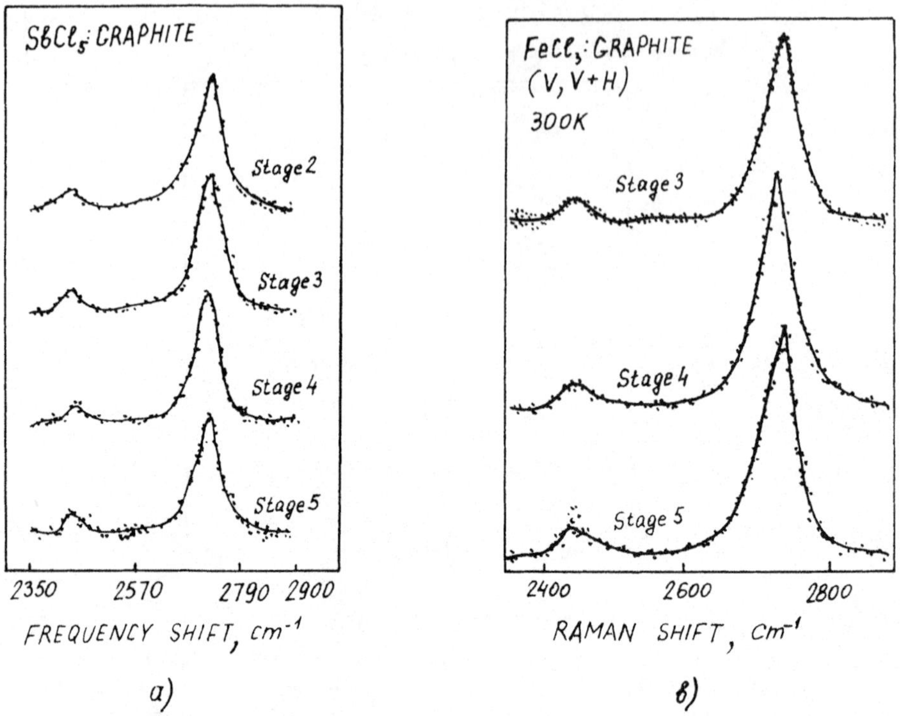

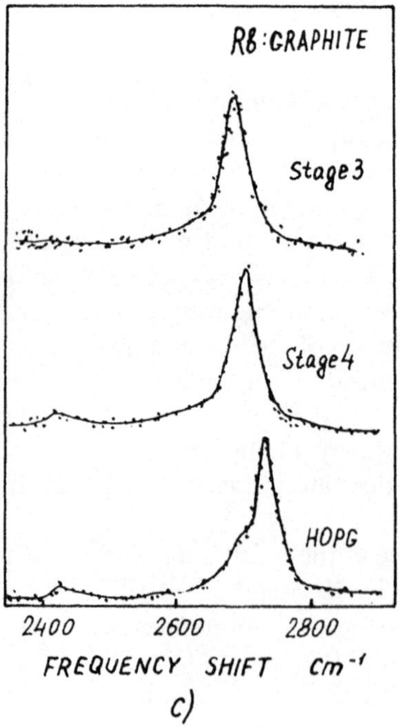

Fig. 196. Second-order Raman spectra of: (a) $SbCl_5$–, (b) $FeCl_3$– and (c) Rb–graphite in the 2400–2900 cm^{-1} region. The solid lines represent the results of Lorenztian-lineshape analyses. The spectrum of HOPG is shown for comparison. (Giergiel and Eklund 1982.)

fig. 196a,b,c in graphite intercalation compounds one observes a symmetric Lorentzian peak, whereas in pure graphite, the similar peak is distinctly asymmetric, which is indicative of dispersion in this region of frequencies.

The position of the peak for the $(\Gamma + \Sigma)$ scattering suggests that the donor (Rb) compounds should have an entirely different stage-dependence than the acceptor ($FeCl_3$, $SbCl_5$) ones (fig. 195). The figure also suggests that in the acceptor compounds the $(\Gamma + \Sigma)$ scattering modes should be relatively stage-independent, whereas in the donor compounds the frequency of the $(\Gamma + \Sigma)$ features decreases with increasing reciprocal stage index. Since the stage-dependence of the Σ-point modes is known from Raman spectra in the $\sim 3250\,cm^{-1}$ region (fig. 195), one can then determine the general trend for the stage-dependence of the Γ ($\sim 850\,cm^{-1}$) modes. One should recall that the $\sim 850\,cm^{-1}$ Γ-point modes are also IR-active and, according to the data of Nemanich et al. (1977a), in pristine graphite this mode is of the A_{2u}-type, and has a frequency of $868\,cm^{-1}$. By separating the stage-dependence of the Σ-point modes from that of the $(\Gamma + \Sigma)$ combination modes one finds that the IR-active Γ-point modes ($\sim 850\,cm^{-1}$) increase with increasing reciprocal stage index for the acceptor compounds and decrease for the donor ones. In both cases, the magnitude of the rate of change of the A_{2u} Γ-point modes with increasing reciprocal stage index $1/n$ should nearly match that of the Σ-point modes to be consistent with the stage-dependence of the $(\Gamma + \Sigma)$ combination scattering.

Finally, we shall consider the stage-dependence of the frequency shift of the $\sim 2735\,cm^{-1}$ feature in the second-order Raman spectra in GIC, which is associated with overtones $(2 \times \omega)$ of the scattering of longitudinal acoustic phonons at the M-point. As can be seen in figs. 195 and 196, the frequency of these features is stage-independent in both acceptor compounds ($SbCl_5$ and $FeCl_3$), while in donor (Rb) compounds it softens. These features are also connected with the peak on the phonon-dispersion curve and, therefore, with the in-plane second-neighbour interactions. Taking account of these interactions in phonon-spectrum calculations is rather difficult. When examining the $\sim 3250\,cm^{-1}$ feature we have already seen that a change in second-neighbour interactions in the carbon layer of a GIC is opposite to that in the nearest-neighbour interactions. In connection with this, it is not surprising that no stage-dependence of the position of the $\sim 2730\,cm^{-1}$ feature in the acceptor GIC is observed.

11.3.2. Vibrational modes in the phonon spectrum of GIC due to the intercalate layers

In addition to the high-frequency vibrational modes of GIC lattices which have been identified with carbon-layer vibrations, there have been observed numerous low-frequency modes, whose frequencies depend on the type of intercalant. Such modes usually correspond to intercalate-layer vibrations. The

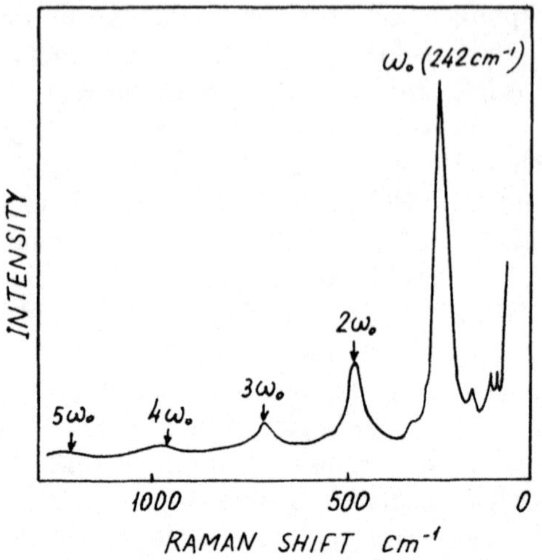

Fig. 197. Raman spectra for a graphite-Br_2 sample at 77 K in the frequency regions where the intercalate modes are prominent. The structure identified with the Br_2 stretch mode in the intercalation compound is denoted by ω_0 and its harmonics by $n\omega_0$. The harmonics of ω_0 are indicated by arrows. (Eklund et al. 1978.)

first observation of such modes was made by Song et al. (1976) in acceptor Br_2–GIC. An example of the Raman spectrum associated with intercalate-layer vibrations is that obtained by Eklund et al. (1978) for graphite intercalated with Br_2 (see fig. 197). The best-pronounced feature of the spectrum is a high and rather broad peak at $\omega_0 = 242\,cm^{-1}$ and its weaker but distinctly observable harmonics at $2\omega_0$, $3\omega_0$, $4\omega_0$ and $5\omega_0$, which are indicated by arrows in fig. 197. A correspondence has been established between the frequency of the main spectral feature at 242cm^{-1} (Cahill and Leroi 1966) and in the free Br_2 molecule at 323cm^{-1} (Herzberg 1945). This implies that, as a result of intercalation, the binding between the bromine atoms weakens. This interpretation of the 242 cm^{-1} line yields support for the molecular identity of the intercalant in $C_{8n}Br_2$ compounds, consistent with X-ray and electron diffraction results (Eeles and Turnbull 1965). The increase in the intensity of all spectral features shown in fig. 197 with increasing intercalant concentration is in accordance with such an interpretation. As suggested by this observation, the intercalate force constants are mainly determined by intralayer interactions with some perturbations introduced by the bounding graphite layers. Further support for the molecular identity of bromine in $C_{8n}Br_2$ compounds comes from the observation of the resonant Raman effect associated with a rapid increase in Raman line intensity when the laser excitation energy is close to that of an electronic transition. The observation of a strong resonant enhancement for the $\omega_0 = 242\,cm^{-1}$ intercalate line and the absence of such an enhancement for the $\hat{E}_{2g}^{(2)} = 1600\,\mathrm{cm}^{-1}$ graphite line in the bounding graphite layers indicate an electronic transition involving a bromine level. Similar resonant enhancement effects are disclosed by observations made on the $2\omega_0$ and $3\omega_0$ harmonics.

Finally, we should note that recent measurements of heat capacity in graphite–bromine residue compounds (Mizutani et al. 1982) have revealed the existence of an Einstein temperature between ~15 and ~20 K, which is associated with flat phonon modes at the corresponding frequencies.

A three-dimensional model enabling one to calculate vibration-mode frequencies for $C_{8n}Br_2$ compounds has been developed by Al-Jishi and Dresselhaus (1982d). They have shown that the zone-centre modes for the

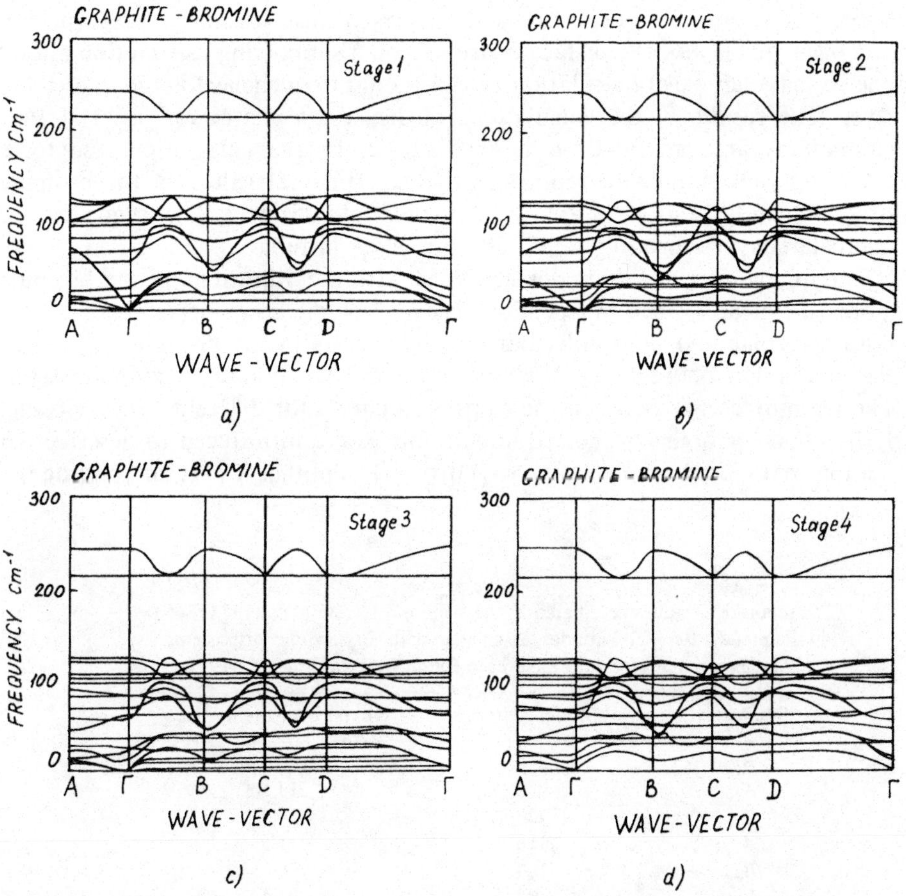

Fig. 198. The calculated low-frequency phonon dispersion curves for stage-1 (a), stage-2 (b), stage-3 (c), stage-4 (d) Br_2–GIC. The basic Br_2 modes are nearly stage independent; the n graphite layers provide three acoustic modes and $3(n - 1)$ optic modes. The total number of modes for stage n is $3(n + 4)$. (Al-Jishi and Dresselhaus 1982d.)

intercalate layer are given by the following irreducible representation

$$\Gamma = 2(A_{1g} + B_{1g} + B_{3g} + B_{1u} + B_{2u} + B_{3u}). \qquad (249)$$

The two $(B_{1u} + B_{3g})$-modes correspond to bromine-atom vibrations along the c-axis, while the two $(B_{2u} + A_{1g})$-and the two $(B_{3u} + B_{1g})$-modes correspond to those in the plane of the layer. The A_{1g}-, B_{1g}-, and B_{3g}-modes are Raman-active, whereas the B_{1u}, B_{2u} and B_{3u} are IR-active. The modes of the highest frequencies correspond to the stretching modes of the bromine molecules and have symmetries B_{1g} for the $\omega \sim 246\,\mathrm{cm}^{-1}$ mode and A_{1g} for the $\omega \sim 216\,\mathrm{cm}^{-1}$ mode. The A_{1g} and B_{1g} librational modes have been observed by Eklund et al. (1978) in Raman scattering at $\omega \simeq 95\,\mathrm{cm}^{-1}$.

In the frequency range $\omega < 300\,\mathrm{cm}^{-1}$, the detailed structure of the carbon layers is no longer important, and the layers themselves may be regarded as homogeneous layers of constant density. Such a simplifying assumption allows one to make an exact calculation of vibrational frequencies for the intercalate layer. The results of the calculations of dispersion curves for $n = 1$–4 Br_2-compounds are presented in figs 198a,b,c,d. In the calculation, the force constants were chosen to reproduce the experimental values of the bromine stretching and librational modes at $246\,\mathrm{cm}^{-1}$ and $95\,\mathrm{cm}^{-1}$. The frequencies of intercalate-layer vibrations in $C_{8n}Br_2$ are given in table 31.

The frequencies of the modes identified with the vibrations of the bromine atoms are nearly stage-independent because of the stage-dependence of the intramolecular and intermolecular force constants in the bromine layer, and the interaction between the bromine layer and the bounding graphite layer. The main difference between the dispersion curves for different stages (see fig. 198) is that for higher stages additional modes are introduced to describe the motion of the rigid graphite layers relative to each other. Curves describing the

Table 31
The theoretical values calculated by Al-Jishi and Dresselhaus (1982d) of the low-frequency Br_2–GIC modes associated with Br_2-atomic displacements. The experimental values have been found by Eklund et al. (1978).

Symmetry	Model (cm^{-1})	Experiment (cm^{-1})
B_{1g}	246	246.5
A_{1g}	216	213.5
B_{3g}	120	—
A_{1g}	116	115.5
B_{3g}	100	—
B_{1g}	95	94.5
B_{1u}	85	—
B_{2u}	65	—
B_{3u}	55	—

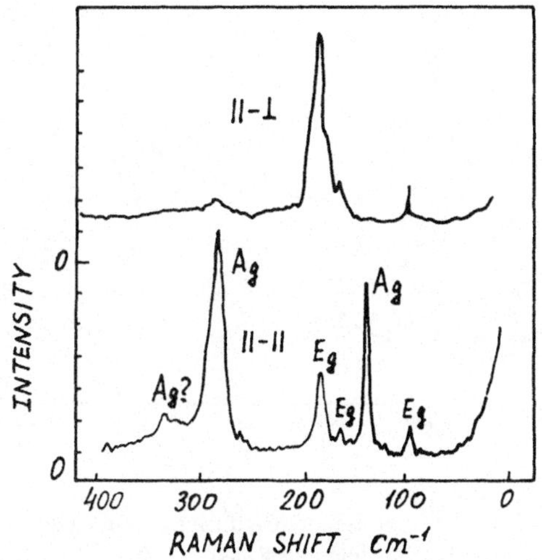

Fig. 199. Polarized Raman spectra of single-crystal $FeCl_3$. (Mele and Ritsko 1979.)

dispersion of these graphitic modes are absent for stage-1, while there are $3(n-1)$ of them for Br_2–GIC of stage $n > 1$. The calculated dispersion curves are capable of accounting for the results of the specific-heat measurements on graphite–Br_2 compounds where anomalies were found in the 15–20 K temperature range. The flat phonon modes at $\sim 10\,cm^{-1}$ and $\sim 14\,cm^{-1}$ in this calculation give rise to a high density of states at frequencies corresponding to the above temperature range.

Another typical example of an observation of vibrational modes due to atomic displacements in the intercalate layer in acceptor compounds is the study of low-frequency Raman spectra of C_xFeCl_3. Whereas bromine intercalates graphite as a molecular entity, $FeCl_3$, which crystallizes in a layer structure, enters graphite in the molecular vapour phase, but crystallizes into layer domains with a lateral extent of $\simeq 2000$ Å (Ohhashi and Tsujikawa 1974a). This is confirmed by a comparison between the polarized Raman spectra of pure cyrstalline $FeCl_3$ shown in fig. 199 and the polarized spectra of $FeCl_3$–graphite compounds, shown in fig. 200 (Mele and Ritsko 1979).

It is known that $FeCl_3$ crystallizes in the BI_3 structure and has space group symmetry $C_{3i}^2(R\bar{3})$ with two molecular units in the rhombohedral primitive cell. Its optic modes are transformed, according to Caswell and Solin (1978a), as

$$\Gamma_{opt} = 4A_g + 3A_u + 4E_g + 3E_u. \tag{250}$$

The "g"- modes are Raman-active and the "u"-modes are IR-active. Six of the eight Raman modes have been detected, and are labelled in fig. 199. The $FeCl_3$ crystal is composed of the layers whose primitive cells also contain two $FeCl_3$

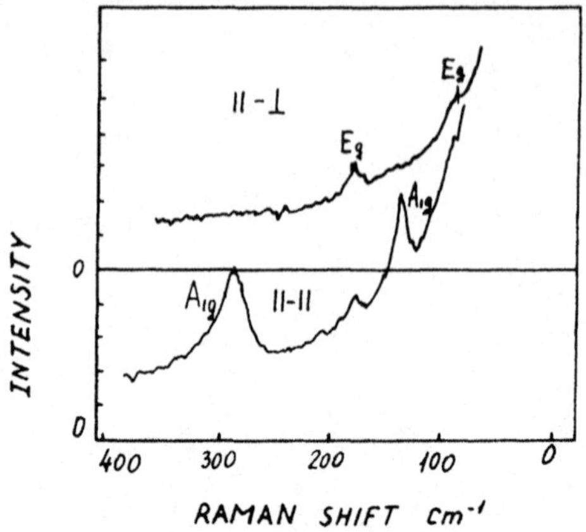

Fig. 200. Polarized Raman spectra of the intercalant intralayer modes of stage-1 $FeCl_3$-graphite. (Mele and Ritsko 1979.)

units. The layer symmetry is D_{3d} and its optic modes are transformed (Mele and Ritsko 1979) as

$$\Gamma_{opt}^{layer} = 2A_{1g} + 2A_{2g} + 4E_g + A_{1u} + 2A_{2u} + 3E_u. \qquad (251)$$

Moreover, the A_{1g}- and A_{2g}-layer modes correlate with the A_g-modes of crystal, while the E_g-layer modes correlate with the E_g-crystal modes. The two A_{1g}-modes can be detected at their expected positions in the Raman spectrum of stage-1 $FeCl_3$–graphite compound (see fig. 200), as can two of the four E_g-modes. This provides evidence for the layer structure of the $FeCl_3$ intercalant. The spectra shown in fig. 200 were obtained from a pure stage-1 region. Moreover, the relative widths of the the 139 cm^{-1} and 287 cm^{-1} modes in fig. 200 are significantly different from the corresponding features in the spectrum of crystalline $FeCl_3$ (fig. 199). Thus the spectra of fig. 200 do not arise from phase-separated $FeCl_3$ on the sample surface.

Attempts to detect molecular internal modes of other acceptor systems, e.g. AsF_5, have so far been unsuccessful.

Let us now consider vibrational modes associated with atomic displacements in the intercalate layer of the donor GIC. Such modes have been detected and studied in graphite intercalated with alkali metals (mostly K, Rb, and Cs). Most of the lattice modes of an alkali–graphite intercalation compound (AGIC) can be described as either intercalate-derived or graphite-derived. Exceptions to this rule occur for [001] longitudinal modes, in cases when graphite and intercalate layers have comparable masses per unit area, as well as for layer-bending modes at certain wave-vector values, when the alkali metal and carbon motions are strongly coupled.

In AGIC, the intercalate-derived modes include, in particular, lattice modes

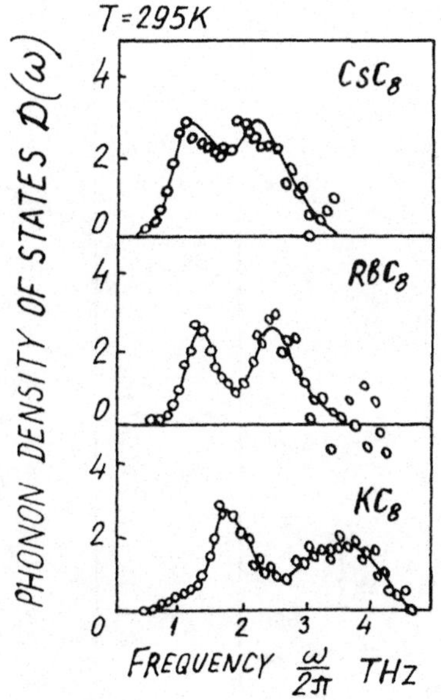

Fig. 201. Intercalate mode phonon density-of-states functions for stage-1 compounds. (Kamitakahara and Zabel 1982.)

for which wave vectors and polarization vectors are parallel to the basal plane. Such modes have been investigated by neutron scattering in AGIC samples obtained on the basis of pyrolytic graphite (Kamitakahara and Zabel 1982). The measurements were made using C_xM compounds, where $M = K$, Rb, or Cs, and $x = 8$, 24, 36. The most detailed measurements were made on samples of $C_{24}Rb$ and $C_{24}K$.

The phonon densities of states for the intercalate modes obtained for stage-1 Rb, K, and Cs compounds are shown in fig. 201. An analysis of the phonon spectra for C_8Rb, made by Kamitakahara et al. (1981), has revealed that the Rb–Rb forces in GIC are significantly greater than the analogous interactions in pure metallic Rb. Figure 201 compares the spectrum for C_8Rb with the spectra for C_8K and C_8Cs. One can see that the spectra are similar, except that higher frequencies are observed for C_8K and lower frequencies, for C_8Cs, as one might expect from the intercalate atomic mass change. Also, the spectrum for C_8Cs is less distinctly separated into two peaks, suggesting either a higher degree of anharmonicity or a structural disorder in this compound, or maybe a combined action of these two factors. The spectrum of C_8K taken at 10 K shows a resolution into two distinct peaks, perhaps indicating that anharmonicity is the more important factor of the two. A closer inspection of the spectra reveals that the C_8Cs spectrum is somewhat harder and that of C_8K slightly softer than one would expect solely from the alkali mass change:

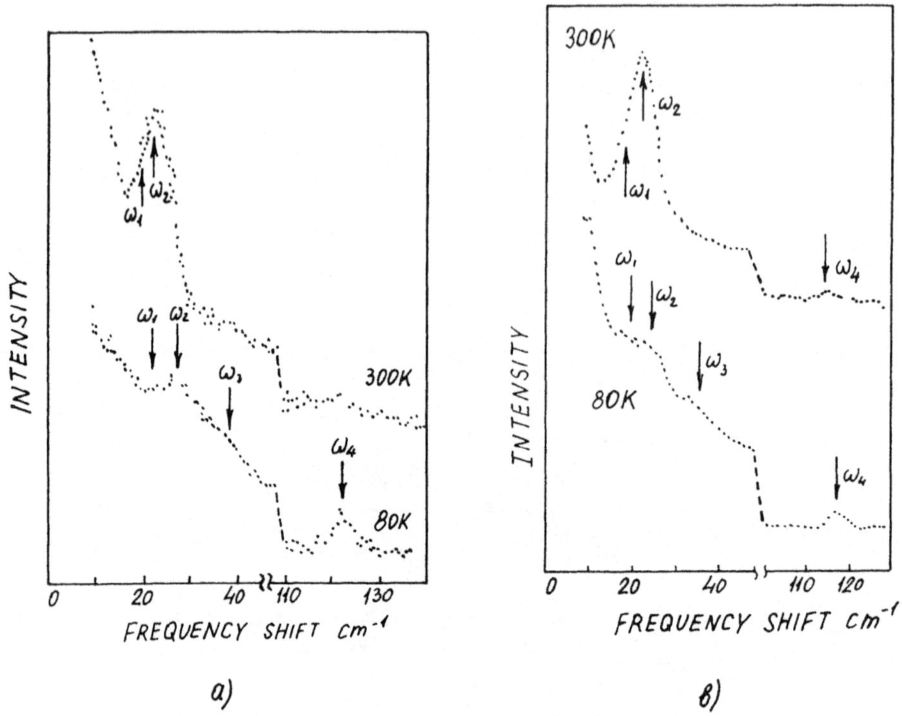

Fig. 202. (a) Low-frequency Raman scattering spectra of stage-2 graphite-K taken at 300 K and 80 K. (b) Low-frequency Raman scattering spectra of stage-2 graphite–Rb taken at 300 K and 80 K (Giergiel et al. 1982.)

i.e., $\langle\omega\rangle_K/\langle\omega\rangle_{Rb} = 1.43$ and $\langle\omega\rangle_{Cs}/\langle\omega\rangle_{Rb} = 0.87$ from experiment, whereas $(m_{Rb}/m_K)^{1/2} = 1.478$ and $(m_{Rb}/m_{Cs})^{1/2} = 0.802$, respectively ($\langle\omega\rangle$ being the average frequency in the spectrum and m is the intercalant atomic mass). One possible origin of these differences may be a changing graphite–alkali interaction. Raman scattering was used to study the vibrational spectra of stage-two graphite–Rb and graphite–K compounds in the low-frequency range $\omega < 150\,\mathrm{cm}^{-1}$. The low-temperature ($T \sim 80$ K) Raman spectra of stage-2 graphite–Rb (or –K) systems exhibit four features at 18 (19) cm^{-1}, 22 (23) cm^{-1}, 35 (35) cm^{-1}, and 118 (121) cm^{-1}, respectively, as shown in fig. 202. The low-frequency doublet (18, 22 cm^{-1}) in stage-two graphite–Rb is identified with the Γ-point phonons involving a rigid, shearing dispalcement of the carbon and rubidium layers. Thus the donor atoms (ions) generate additional Raman and IR excitations, which reflect their interactions and structural arrangements. However, for the donors studied to date, e.g., the alkali metals, such excitations would occur at very low frequencies and are experimentally inaccessible.

11.3.3. Strongly coupled phonon modes – mixed phonon modes

If a layered compound has interlayer forces that are very much weaker than the intralayer forces, some unusual phonon characteristics may result. One of these is the occurrence of modes with an almost quadratic dispersion relation $\omega \sim q^2$ at small $|\boldsymbol{q}|$, i.e., $\omega^2 = Aq^2 + Bq^4$, where A vanishes in the absence of interlayer forces [see sect. 11.1, eq. (242)]. These modes have in-plane (parallel to the layers) wave vectors $\boldsymbol{q}$, but out-of-plane displacements can be described as bending or rippling modes of the layers. It is convenient to designate these modes as $TA_{\perp}$, and similar optic modes, as $TO_{\perp}$, denoting both of them by $T_{\perp}$. The tendency towards the $\omega \sim q^2$ behaviour is best pronounced in graphite (see sect. 11.1.1) and in MoS_2 (Wakabayashi et al. 1975). Similar mixed phonon modes $T_{\perp}$ were observed by Kamitakahara et al. (1982) in alkali–graphite intercalation compounds, where both the intralayer and interlayer forces are relatively weak. Several new effects associated with the alkali intercalate have been observed, including a dramatic lowering of the layer-shearing force constant c_{44}, which can be estimated from the $|\boldsymbol{q}| \to 0$ slope of a $TA_{\perp}$ dispersion curve, and the emergence of a new alkali-like optical branch, strongly coupled to the layer bending modes of the graphite host (the optical branches are qualitatively derived from graphite-like optical branches hybridized with alkali-like branches).

The experimental data on the coupled modes have been obtained from inelastic neutron scattering studies using C_8K, C_8Rb, C_8Cs, $C_{24}K$ and $C_{24}Rb$ compounds (Kamitakahara et al. 1982). These data were interpreted on the basis of the calculations of the phonon spectra for AGIC carried out by Horie et al. (1980), Leung et al. (1981a) and Al-Jishi and Dresselhaus (1982b). The results that have thus been obtained represent the phonon dispersion relation averaged over all the directions in the basal plane. However, as Nicklow et al. (1972) have shown, the $T_{\perp}$ graphite dispersion curves are almost isotropic, even close to the Brillouin-zone boundary.

Figures 203a,b,c illustrate the dispersion curves for the C_8K, C_8Rb, and C_8Cs systems in the low-frequency range $\omega \lesssim 6\,\text{THz}$, which have been obtained from the experimental data taken at 295 K. The left-hand side of the plots shows the (001) L-branches displayed in a manner such that the zone edge in the c-direction occur at $0.25(2\pi/I_c)$ (I_c being the identity period along the c-axis). The right-hand side represents the dispersion curves for the layer bending modes. In-plane wave-vector units of $G = (4\pi/\sqrt{3})a_0^{-1} = 2.96\,\text{Å}^{-1}$ have been chosen. For each stage-1 compound, the $T_{\perp}$ branches can be described as two dispersive graphite-like branches strongly hybridized with two non-dispersive alkali-like branches. These latter are non-dispersive because the alkali layers offer very little resistance to bending or rippling deformations. In the case of C_8K, one can see a clearer separation into the alkali-like and graphite-like modes, because the K-layers are substantially lighter than the graphite layers. This leads to an alkali-like character for the

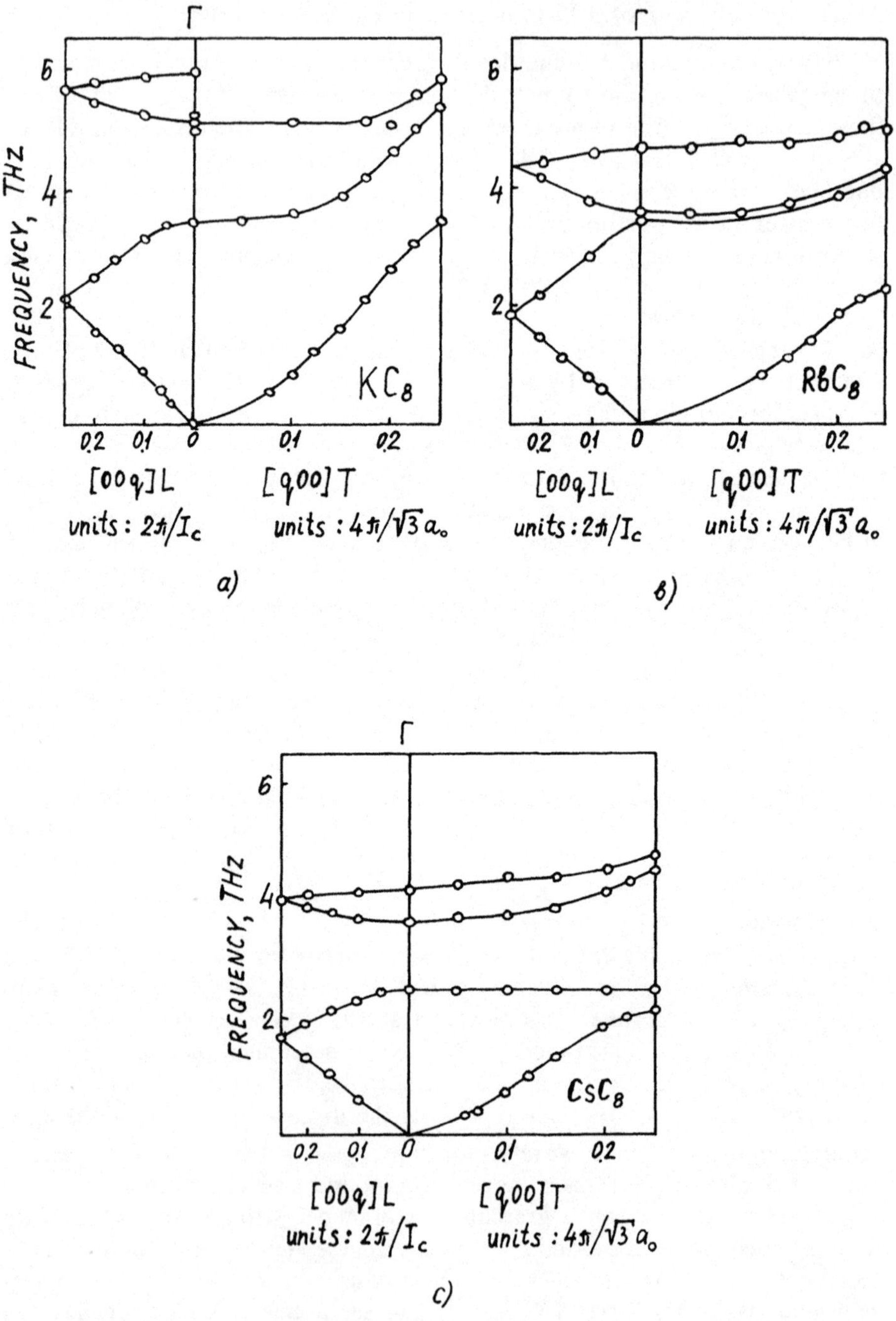

Fig. 203. Dispersion curves for stage-1 compounds at 295 K: (a) KC_8, (b) RbC_8, (c) CsC_8. (Kamitakahara et al. 1982.)

upper two branches (only one of which can actually be measured), and a graphite-like character for the lower two. In C_8Rb, the intercalate and carbon layers have comparable masses and no such distinctions are really possible. In C_8Cs, the alkali and carbon layers again have comparable masses, but at least one branch, namely, the completely flat one at 2.5 THz, can be identified as alkali-like. For the $q = 0$ mode of this branch, an elementary analysis of the modes for a diatomic chain shows that only Cs layers are capable of moving, while graphite layers are fixed, i.e., the mode is completely Cs-like. This alkali-like character apparently persists as $\boldsymbol{q}$ extends into the basal plane. A similar flat branch has been found for stage $n > 1$ AGIC by Al-Jishi and Dresselhaus (1982b). Phonon dispersion calculations for C_8K and C_8Rb have been performed by Horie et al. (1980), who found that the frequencies for $T_\perp$-type modes lie somewhat higher than the experimental frequencies, but a substantial improvement could undoubtedly be achieved by a rather simple adjustment of model parameters.

The dispersion curves measured for $C_{24}K$ and $C_{24}Rb$ at room temperature are shown in figs. 204a,b. Again, the data on the (001) L-phonon modes are

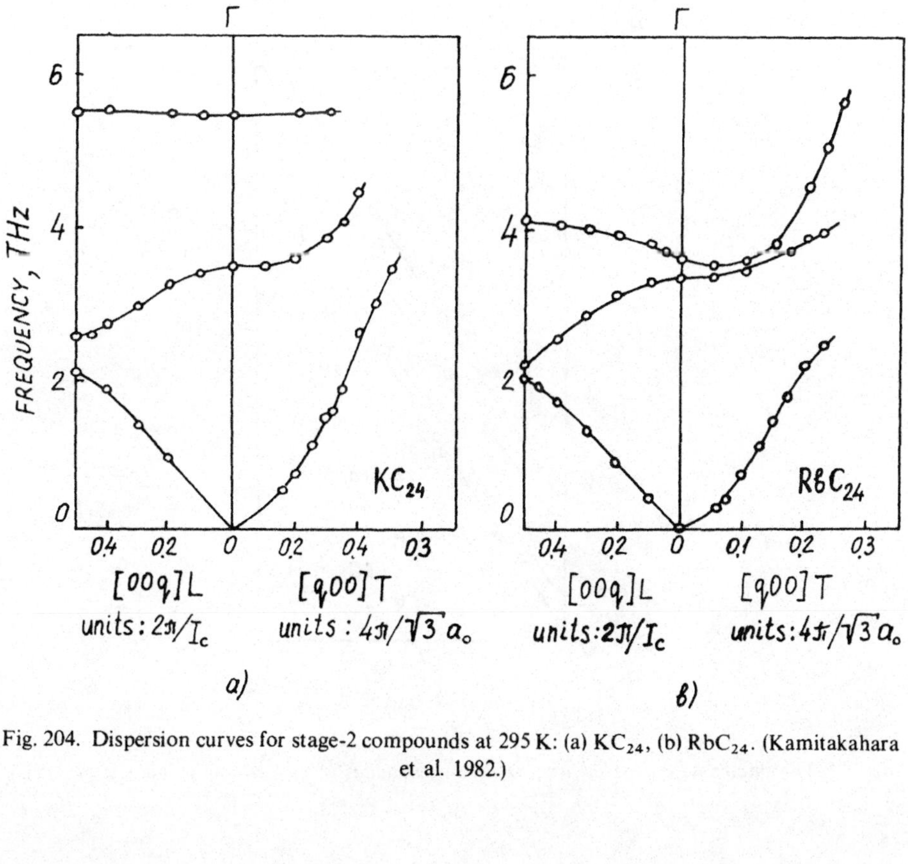

Fig. 204. Dispersion curves for stage-2 compounds at 295 K: (a) KC_{24}, (b) RbC_{24}. (Kamitakahara et al. 1982.)

plotted on the left-hand side. The wave-vector units in the c-direction are, as before, $2\pi/I_c$. The layer bending modes are plotted on the right-hand side of the figure, again using wave-vector units of $G = (4\pi/\sqrt{3})a_0^{-1}$. Qualitatively, the $T_\perp$ modes shown in fig. 204 for stage-two compounds consist of two graphite-like branches hybridized with one alkali-like non-dispersive branch. The alkali-like mode occurs at about 5.5 THz for $C_{24}K$ and at about 3.5 THz for $C_{24}Rb$.

The $TA_\perp$ branches for all the compounds studied show a pronounced upward curvature near $|\boldsymbol{q}| = 0$, indicating the tendency for the $\omega \sim q^2$ behaviour, characteristic of strongly layered compounds. At small q values, the dispersion relation has the form of $\omega^2 = Aq^2 + Bq^4$, where $A = c_{44}/\rho$, ρ being the volume density. Hence a plot of ω^2/q^2 against q^2 yields a line of slope B and intercept A (fig. 205). One can see that for all AGIC the intercepts A are smaller than for pure graphite (Nicklow et al. 1972). Especially small values are obtained for the stage-two compounds $C_{24}Rb$ and $C_{24}K$. The values of A and c_{44} derived from the plots are listed in table 32.

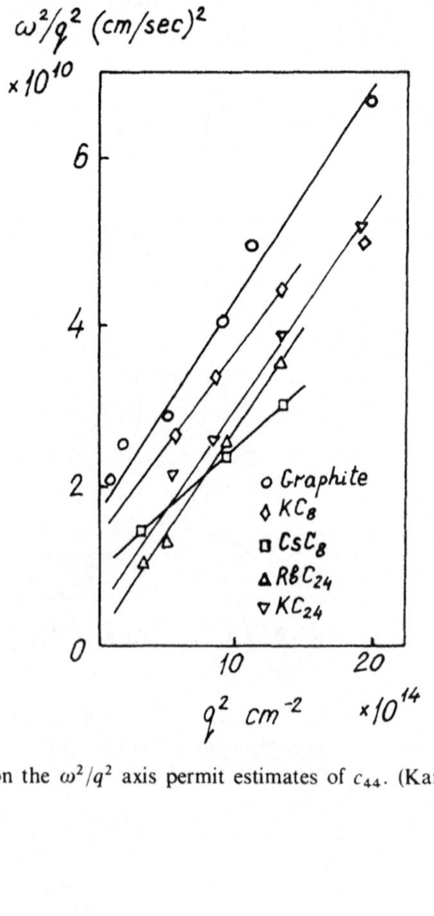

Fig. 205 The intercepts on the ω^2/q^2 axis permit estimates of c_{44}. (Kamitakahara et al. 1982.)

Table 32
Intercepts A and shear moduli c_{44} derived from the plots of fig. 205.

	graphite	C_8K	C_8Cs	$C_{24}Rb$	$C_{24}K$
$A\,(10^{10}\,cm^2/s^2)$	1.7 ± 0.2	1.4 ± 0.1	1.0 ± 0.05	0.20 ± 0.1	0.45
$c_{44}\,(10^{10}\,dyne/cm^2)$	3.84 ± 0.4	2.82 ± 0.2	3.07 ± 0.15	0.44 ± 0.2	0.89 ± 0.2

The coupling of longitudinal acoustic vibrations along the c-axis [the (001) L-branch] with the transverse Einstein-like mode represents an interesting type of mixed phonon modes observable in AGIC. An anticrossing behaviour of the acoustic L-phonon branch with a flat level predicted from the low-temperature specific heat anomaly has been observed by Suganuma et al. (1981) in $C_{24}Rb$. The origin of the flat mode and the mechanism of interaction between this mode and the collective excitation was investigated by Funahashi et al. (1982) who studied the temperature dependence of phonon excitation in the anticrossing region by measuring inelastic neutron scattering. Figure 206 shows the dispersion curves of the lowest (i.e., acoustical) L-branch measured at 40, 80, 128 and 260 K. One can see that, except for the 260 K data, the behaviour of the curves at an excitation energy of about 5 meV is a typically anticrossing one. The small gap near 5 meV can be explained by introducing a coupling of the longitudinal mode with the transverse Einstein oscillations of the rubidium atoms. The solid curves in fig. 206 have been calculated using the best fit parameters for the 80 K data. For the calculation, one uses a linear chain model which includes a bilinear coupling potential between the trans-

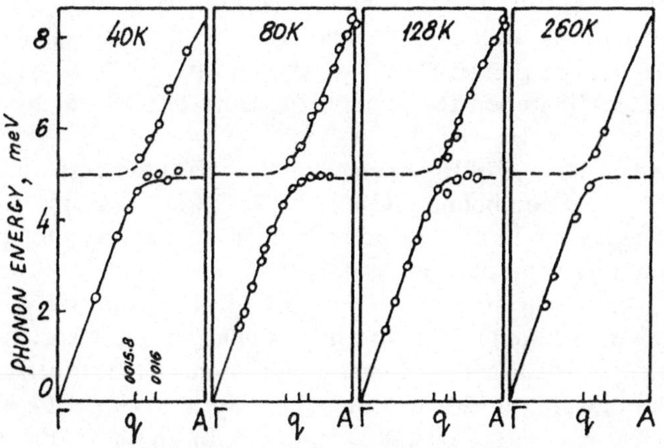

Fig. 206. Acoustic branch of the (001) L-mode phonons. The solid curves are the calculated dispersion relations. The force constants and the coupling constant are chosen to fit the curves to the data at 80 K. (Funahashi et al. 1982.)

verse displacements of the rubidium atoms and the longitudinal displacement of the neighbouring carbon atoms:

$$U = \mu v_3(u_1 - u_2), \tag{252}$$

where μ is a constant proportionality factor, v_3 an in-plane displacement of the rubidium atoms resulting from Einstein oscillations, u_1 and u_2 are displacements along the c-axis of the bounding carbon layers. This form of the potential is absent if the rubidium atoms occur at symmetric positions of the potential due to the carbon network. If the rubidium atoms are displaced to asymmetric positions, the change in interlayer distance should give rise to a transverse force acting on the rubidium atoms and the coupling potential U is expressible in the bilinear form given by eq. (252). The experimental results indicate that the rubidium atoms occur at symmetric positions in the absence of intercalate stacking order, but that they are displaced to asymmetric positions, as the in-plane order of the rubidium atoms, which accompanies the intercalate stacking order, evolves.

11.3.4. Zone-folding effect in GIC

A significant change in the phonon spectrum of pristine graphite upon intercalation is associated with the change in the period of the structure along the c-axis, as well as with the emergence of a new periodic in-plane structure as a result of increasing order in the intercalate layer. These effects have already been considered at length in the analysis of the electron spectrum of graphite intercalation compounds. Evidently, the new periodicity of a GIC lattice has the greatest influence upon the phonon spectrum of compounds of the lowest intercalation stages ($n = 1$ or 2). The new periodicity leads to a change in the shape and size of the Brillouin zone and, as a result, to the appearance of new physically equivalent points in the wave-vector space. Let us demonstrate the effect of zone-folding upon the phonon spectrum of GIC, taking several such compounds as examples.

The periodic structure of the intercalate layer has been best studied in stage-one alkali-graphite compounds (C_8K, C_8Rb, C_8Cs). In all these compounds, the intercalate layer is in an ordered crystalline state at room temperature and below. As a result of the emergence of a new periodic structure in the intercalate layer in AGIC, the edge points M and L are folded into the Γ-point of the new Brillouin zone. In order to find out the associated change in the spectrum, we shall make a direct comparison of the dispersion curves in the spectrum of C_8K (fig. 207) calculated by Horie et al. (1980), who used the Born–von Kármán force-constants model, with those in the spectrum of graphite folding into the smaller Brillouin zone of C_8K (fig. 208). Comparing the dispersion curves, we can point out that for C_8K there are six new optical modes in the low-frequency region originating from vibrations of the two

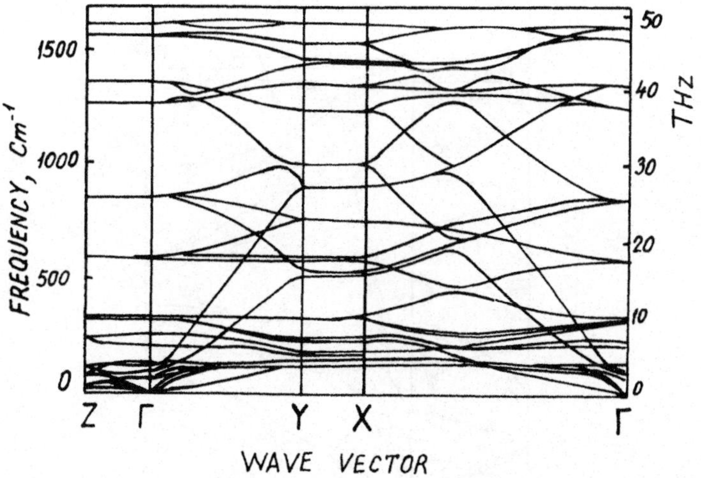

Fig. 207. Phonon dispersion relation for C_8K. (Horie et al. 1980.)

potassium atoms in a primitive cell. The remaining forty-eight modes are evidently associated with carbon-atom vibrations, and exhibit a definite correspondence with the folded dispersion relation for graphite. From fig. 209 one can see that the lowest branches, B_{2g} and B_{3g}, of the new modes associated with vibrations of the potassium atoms give rise to significant perturbations of the acoustic modes, producing a high density of states around $100\,\text{cm}^{-1}$.

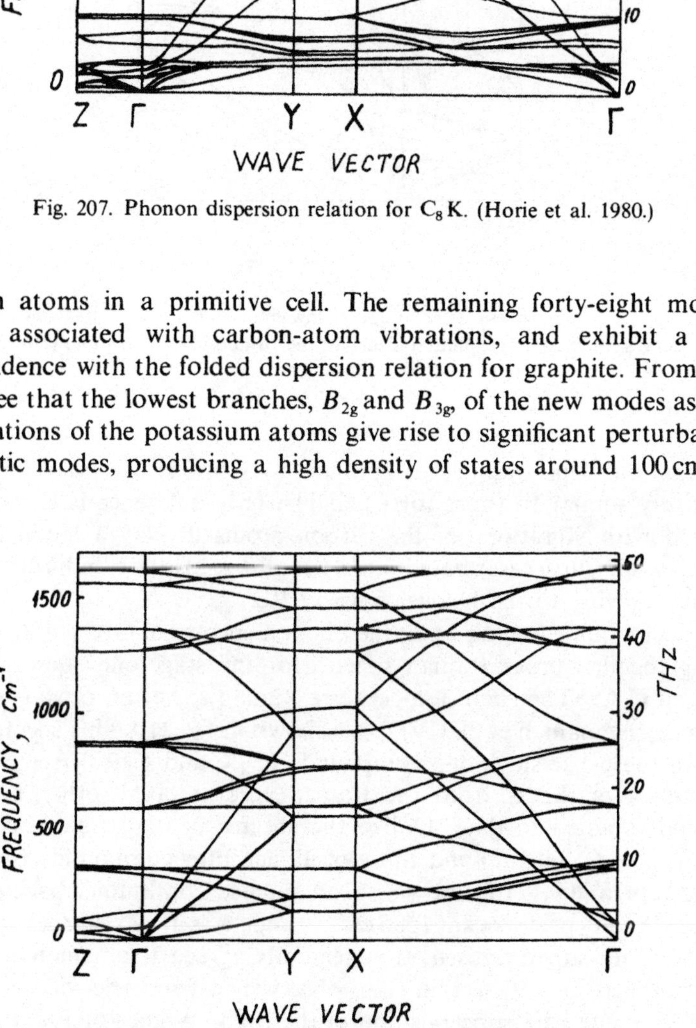

Fig. 208. Phonon dispersion relation for graphite folded into the Brillouin zone of C_8K. (Horie et al. 1980.)

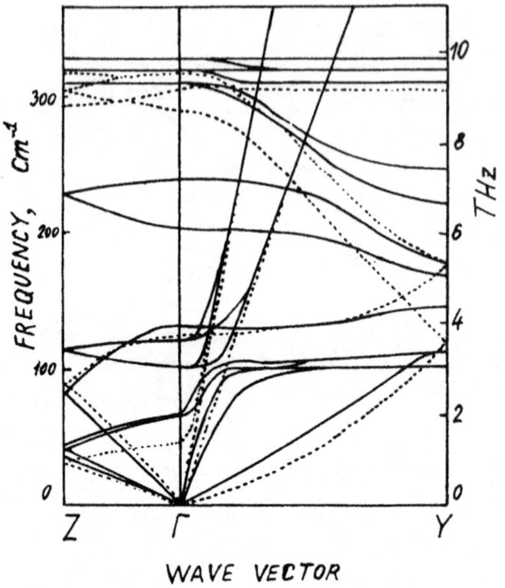

Fig. 209. Phonon dispersion curves for C_8K at low frequencies. The dotted lines represent the dispersion curves for pristine graphite folded into the Brillouin zone of C_8K. (Horie et al. 1980.)

The phonon dispersion curves obtained by Horie et al. (1980) for C_8Rb are qualitatively similar to those for C_8K. The only difference is that the modes associated with vibrations of the carbon atoms display a tendency to shift upwards, while those associated with vibrations of the rubidium atoms shift downwards owing to the heavier mass of Rb.

The effect of zone folding upon the phonon spectra of GIC can be illustrated by using the first-order Raman spectra for the stage-one compounds C_8K, C_8Rb, and C_8Cs. The room-temperature Raman spectrum of stage-one C_8Cs, obtained by Nemanich et al. (1977b), is shown in fig. 210. This spectrum is also typical of the other stage-one compounds (C_8K and C_8Rb) (see Solin 1977), but it differs markedly from the Raman spectra of all other intercalation compounds studied to date. Unlike the spectra of higher-stage donor compounds (see, e.g., fig. 188) and those of all acceptor compounds (see fig. 189), the spectrum in fig. 210 consists of an intense continuous background, on which a broad peak near $1500\,cm^{-1}$ and a set of sharper features at $\simeq 590\,cm^{-1}$ are superimposed. It is generally agreed (Nemanich et al. 1977b, Dresselhaus et al. 1977) that the observed asymmetric broad line near $1500\,cm^{-1}$ is due to a superposition of the Breit–Wigner or Fano resonances of the symmetry-allowed modes (i.e., the pristine graphite $E_{2g}^{(2)}$-like singlet) and the continuous background. The $590\,cm^{-1}$ feature, which is observed as a

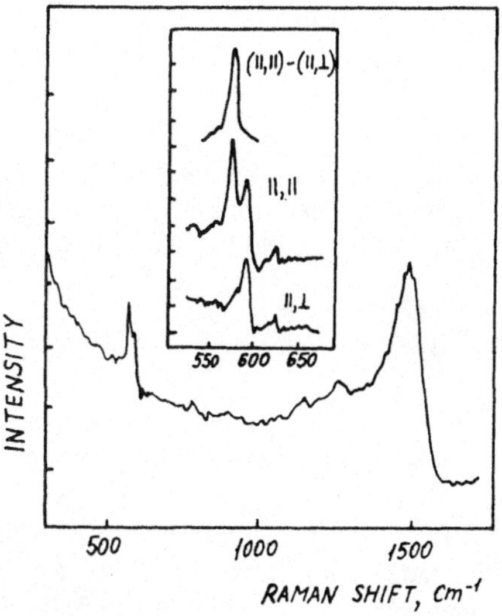

Fig. 210. Room-temperature [polarized (inset)] Raman spectrum of stage-1 C_8Cs. (Solin 1980.)

doublet is attributed (Dresselhaus et al. 1977) to first-order Raman scattering from excitations activated by zone folding or by mapping the zone-edge L- and M-modes into the modes at the zone center. Indeed, according to Horie et al. (1980), for C_8K and C_8Rb there are five vibrational modes (viz., $A_g + B_{1g}$, B_{1u}, A_u, B_{1g} and A_g) with frequencies close to $590\,cm^{-1}$. That the frequencies of these modes shift upwards in the case of C_8Rb is also in qualitative agreement with experiment.

The Raman spectrum for C_6Li differs from the spectra of C_8K, C_8Rb, and C_8Cs in two significant ways. First, the room-temperature spectrum of C_6Li exhibits (Zanini et al. 1978a,b) no sharp line in the vicinity of $590\,cm^{-1}$, while the high-frequency structure consists of a single line peaked at $1598\,cm^{-1}$. This peak lies very close to the $E_{2g}^{(2)}$ graphite line and exhibits only a slight Breit–Wigner asymmetry. The absence of any spectral feature at $590\,cm^{-1}$ for C_6Li furnishes additional evidence that the features observed at this frequency in C_8K and C_8Rb are associated with zone folding. Indeed, the symmetry of the intercalate-layer structure in C_6Li is such that, according to an in-plane zone-folding model, it is the K-point modes, rather than the L- and M-point modes, that fold into the Γ-point ($\boldsymbol{q} = 0$). As can be seen in fig. 174, which illustrates the phonon spectrum of graphite, the K-point modes occur at other frequencies than the M-point modes, and cannot produce a feature at $590\,cm^{-1}$ as a result of zone folding.

The effect this factor has upon the phonon spectrum is a general phenomenon observable for all (especially low-n) compounds whose intercalate layers exhibit a degree of ordering. This may be illustrated for acceptor GIC, taking Br_2–graphite compounds as an example. In the case of $C_{8n}Br_2$, the zone-centre frequencies can be obtained by zone folding to the Γ-point of the six points in the Brillouin zone (Al-Jishi and Dresselhaus 1982d). The in-plane zone-folding creates Raman-active modes which should be observable in Raman-scattering experiments. These modes occur at the following frequencies: $\omega = 1620$, 1500, 1450, 1390, 1350, 1250, 1050, 940 and 800 cm^{-1}, respectively.

11.3.5. Changes in intercalate-layer structure and degree of ordering with temperature

The effect of changing temperature on the phonon spectrum of GIC is distinctly observable for many compounds. Evidently, the change in the phonon spectrum resulting from intercalate-layer disordering should first of all affect the vibrational modes that have arisen as a result of in-plane zone folding. Also, the frequencies associated with intercalate-layer vibrations and the strongly coupled modes should be changed. In order to consider all the relevant aspects, we shall first discuss phase transitions of the order–disorder type in the intercalate layers of AGIC. It is known, for example, that in the temperature range from 80 to 300 K stage-two GIC with K, Rb and Cs exhibit two structural phase transitions. These phase transitions, which occur at temperatures commonly referred to as T_u and T_ℓ ($T_u > T_\ell$), have been observed in electrical resistance, specific heat, X-ray, electron diffraction, and neutron diffraction measurements (Solin 1982, Dresselhaus and Dresselhaus 1981). For stage-2 potassium–graphite, $T_u = 125$ K and $T_\ell = 96$ K. In stage-2 rubidium–graphite ($T_u = 170$ K, $T_\ell = 110$ K), two additional phase transitions have been observed in the resistivity and specific heat data of Suzuki et al. (1980a,b, 1981) at 50 K and 210 K. The temperature-induced structural phase transitions in GIC may, in principle, involve changes in the in-plane intercalate-layer order and its relation to the graphite structure. Although no clear picture of an in-plane order can be deduced because of conflicting experimental data, all stage-2 AGIC exhibit some common features in their behaviour. At low temperatures ($T < T_\ell$), a coherent intercalate-layer stacking exists, while no such order is observed above the upper transition temperature T_u. The bulk of the experimental data for stage-2 graphite–potassium suggests a two-dimensional unregistered liquid for $T > T_u$, a three-dimensional partially registered liquid in the intermediate temperature range, and some rearrangement of stacking order at T_ℓ. The low-frequency Raman spectra of stage-2 graphite–Rb and graphite–K at 300 K and 80 K has been shown in figs. 202a,b. In the 80 K-spectra, four features, labelled as ω_1, ω_2, ω_3 and ω_4, are visible. In both compounds, these features are either weak or absent in the 300 K-spectra.

Let us examine the temperature dependence of the feature at $\omega_3 \simeq 35\ cm^{-1}$.

As is evident from fig. 202a,b, it is not present in the room-temperature spectra. An inspection of the spectra taken at various temperatures reveals that in the Rb compound the ω_3-feature first appears at $T \simeq 170$ K, which is close to the upper phase-transition temperature T_u, the intensity of the ω_3-feature increasing as the temperature drops below T_u. In contrast to this, the ω_3-feature in the K-compound first appears at a much lower temperature, below T_l. As can be seen from fig. 202a,b, the main difference between the spectra for the graphite–Rb and graphite–K compounds is that in the latter the $\sim 35\,cm^{-1}$ feature is absent in the temperature range $T_l < T < T_u$. This difference is consistent with the differences observed in the spectra of the one-phonon density of states, $D(\omega)$. The $35\,cm^{-1}$ peak on the curve for the one-phonon density of states for the stage-2 graphite–rubidium compound contains a contribution from zone-edge phonons, inaccessible to the first-order light-scattering process. For graphite–Rb compounds, the observed narrow line widths and the results of lattice dynamical calculations, taken together, lead to the conclusion that the phase transition at $T_u = 170$ K folds the zone-edge modes at $\omega_3 \simeq 35\,cm^{-1}$ to the centre of the Brillouin zone. Several superlattices fold the zone-edge modes to the Γ-point. It is also generally believed that the disorder–order transition in GIC of a higher order than one (Dresselhaus and Dresselhaus 1981). This implies that the ordered phase grows as the temperature is lowered. A picture of co-existent ordered and disordered islands is consistent with the anomalous temperature dependence of the $35\,cm^{-1}$ structure in both the stage-2 potassium– and rubidium–graphite compounds. If the commensurate phase prevails below T_l (Kambe et al. 1981) and if the $35\,cm^{-1}$ mode is indeed connected with this phase, then one would anticipate an eventual decrease of the ω_3-mode intensity with decreasing temperature $T < T_l$, consistent with the temperature dependence of first-order Raman scattering. Let us now briefly discuss the ω_4-, ω_1-, ω_2-features in the spectra shown in fig. 202a,b. The ω_4-features ($114\,cm^{-1}$ for graphite–Rb and $119\,cm^{-1}$ for graphite–K) are quite consistent with the inelastic neutron scattering data, and correspond to low-frequency vibrations in the LO-phonon branch at the Γ-point. These modes are associated with the out-of-plane carbon-atom displacements, and therefore, they are not expected to be very Raman-active in the scattering geometry of Giergiel et al. (1982). The data suggest that the in-plane component of the predominantly out-of plane ω_4-mode increases with decreasing temperature which is consistent with the observed low-temperature puckering of the carbon layers in the vicinity of the registered alkali-metal atom (Caswell et al. 1980). The magnitude of this modulation is observed to increase with decreasing temperature, and can be as large as 0.2 Å at 80 K.

The two remaining features, at ω_1 and ω_2, exhibit the normal temperature dependence for first-order Raman scattering, i.e., their intensities decrease with decreasing temperature. These two features and their higher-stage equivalents

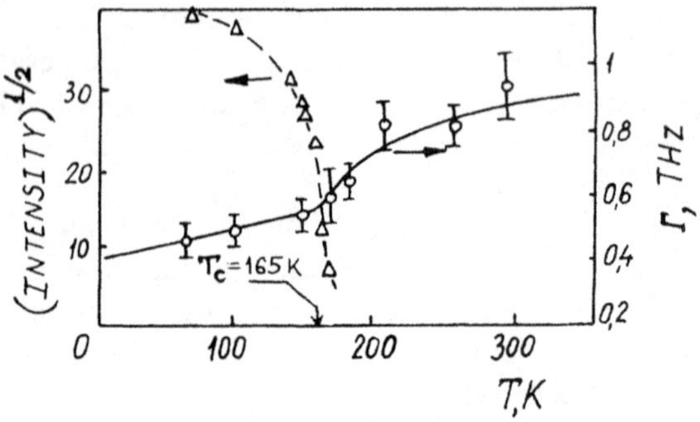

Fig. 211. T-dependence of the order parameter (left-hand scale) and inelastic width (right-hand scale) in $C_{24}Rb$. (Kamitakahara and Zabel 1982.)

have been identified with transverse modes involving rigid shearing displacements of the graphite and intercalate layers.

The influence exerted by a structural phase transition in the intercalate layer at some temperature upon the phonon spectrum of GIC may be recognized from the disappearance of the mixed phonon mode. As an example, one may indicate the change in the (001) L-phonon mode observed in stage-3 graphite–Rb compounds (Funahashi et al. 1982). As can be seen from fig. 206, the dispersion curves for the (001) L-phonons exhibit an anomaly only for $T \leqslant 128$ K. This anomaly has the shape of an energy gap, and suggests the anticrossing of the acoustic branch and the transverse Einstein oscillation level. These data are consistent with structural phase transitions in $C_{24}Rb$ at $T_u = 165$ K and $T_l = 110$ K.

The order–disorder transition in $C_{24}Rb$ does have a demonstrable influence on the vibrational spectra, as shown in fig. 211 (Kamitakahara and Zabel 1982). This figure illustrates the temperature dependence of the long-range order parameter and the width of the peaks of inelastic neutron scattering in $C_{24}Rb$ near the upper transition temperature $T_u = 165$ K. The long-range order parameter is proportional to the square root of the intensity of the corresponding Bragg peak in X-ray scattering from GIC. It can be seen that immediately above T_u the width of the peak on the curve for neutron scattering exhibits a more rapid increase with rising temperature. The observed T-dependence of the long-range order parameter in the intercalate layer suggests that above T_u the disordered phase may be regarded as a liquid-like phase.

11.4. Heat capacity of graphite and its intercalation compounds

The heat capacity, C, of a solid generally consists of three additive parts: lattice $C^{(L)}$, electronic $C^{(e)}$, and magnetic $C^{(M)}$:

$$C = C^{(\mathrm{L})} + C^{(\mathrm{e})} + C^{(\mathrm{M})}. \tag{253}$$

The $C^{(\mathrm{M})}$ contribution to the heat capacity can occur only in the case when there are localized magnetic moments in a solid. This contribution is observed in magnetic GIC, which are compounds of graphite with materials containing atoms with uncompensated magnetic moments, i.e., those of transition, rare-earth, or actinide elements. Among magnetic GIC, one may mention the compounds of graphite with $FeCl_3$, $NiCl_2$, $CoCl_2$, and others, which have already been discussed above (see sect. 10.5.5).

The electronic part $C^{(\mathrm{e})}$ in substances having a subsystem of delocalized electrons is a linear function of temperature in the electron degeneracy region. It is then given by the well-known relation

$$C^{(\mathrm{e})} = \gamma T, \tag{254}$$

where the coefficient γ depends on the density of electronic states at the Fermi level, $\nu(\varepsilon_\mathrm{F})$, and is equal to $\gamma = \frac{1}{3}\pi^2 k^2 \nu(\varepsilon_\mathrm{F})$.

Equation (254) is valid for pure graphite in the temperature range $T < 300\,\mathrm{K}$. For GIC, this expression is, in fact, valid, because of the much higher carrier density in them, over the entire range of temperatures at which a given intercalation compound may exists without noticeable decomposition.

The lattice heat capacity $C^{(\mathrm{L})}$ is wholly determined by the lattice vibrational spectrum, representing its integral characteristic. At a constant volume, the lattice component of specific heat capacity $C_V^{(\mathrm{L})}$ is given by

$$C_V^{(L)} = k \int_0^{\omega_{\max}} \left(\frac{\hbar\omega}{kT}\right)^2 \frac{e^{\hbar\omega/kT}}{(e^{\hbar\omega/kT} - 1)^2} D(\omega)\,\mathrm{d}\omega, \tag{255}$$

where $D(\omega)$ is the total density of phonon states in all the branches of the lattice vibration spectrum, $\omega_{\max}$ is the highest frequency in the spectrum.

We shall now consider the lattice contribution to the heat capacity of graphite and its intercalation compounds and also briefly discuss the magnetic contribution $C^{(\mathrm{M})}$ in some magnetic GIC in connection with the problem of separating the contributions in an analysis of measured heat-capacity values.

11.4.1. Lattice heat capacity of graphite

In pure graphite, there is no magnetic contribution $C^{(\mathrm{M})}$ to the heat capacity. The electronic contribution $C^{(\mathrm{e})}$ is rather small, as compared with that in normal metals, owing to the low concentration of carriers. It has been found (De Sorbo and Tyler 1953, Bowman and Krumhansl 1958) that for pure graphite $\gamma = 0.6 \times 10^5$ cal/(deg^2 mole). Komatsu and Nagamiya (1951) have shown that for $T \leqslant 300\,\mathrm{K}$ the electronic contribution $C^{(\mathrm{e})}$ to the heat capacity of graphite is negligibly small in comparison with the lattice contribution $C^{(\mathrm{L})}$. These data make it possible to analyze the heat capacity of graphite as if it were determined by the lattice heat capacity alone.

The earliest measurements of the graphite heat capacity in the temperature range between 100 and 1000 K were reported by Spencer (1948). Subsequently, such measurements have been extended to the lower-temperature range down to 13 K (De Sorbo and Tyler 1953, De Sorbo 1954) and even to 1.5 K (Bergenlid et al. 1954, De Sorbo and Nichols 1958). Below 2 K, the heat capacity of graphite was measured by Van der Hoeven and Keeson (1963). The data for the range of the higher temperature have been reported by Mantell (1968), who measured the heat capacity of graphite in a very wide temperature range (100–3400 K). The measurements of the heat capacity of pyrolytic graphite were made by Delhaes and Hishiyama (1970), as well as by Wostenholm and Yates (1973). The data obtained by different authors differ somewhat in value, which may be due to the different quality of the single crystals of graphite used and, for the most part, to the different size of the constituent crystallites. The most reliable data, which are well consistent with the results of heat-capacity calculations based on the most exact and up-to-date models for the graphite phonon spectrum, have been obtained by De Sorbo (1954), who used in his measurements samples of Ceylon natural graphite, whose crystallite size in the direction of the c-axis is larger than 400 Å. The data of De Sorbo and Tyler (1953), taken with samples in which the crystallite size in the direction of the c-axis was about 200 Å, yield higher heat-capacity values in the 45–300 K range, possibly due to the effect of the crystallite size.

The experimental temperature dependences of the heat capacity of graphite in the range from 1.5 to 300 K and in the range from 100 to 3400 K are illustrated in figs. 212 and 213, respectively; the data of fig. 212 having been corrected for the electronic contribution.

As is indicated by the experimental data, the heat capacity of graphite shows the T^3-behaviour, expected for any three-dimensional solid at low temperatures, only in the ultralow temperature range $T < 2$ K. The 2–10 K range is the one of transition to the typically two-dimensional behaviour $C \sim T^2$, which is observed between 13 and 54 K.

The lattice heat capacity of graphite was calculated using various vibrational spectrum models, and the model parameters were fitted by comparing the results of the calculations with the experimental data. The T-dependence of C for graphite which agrees with experiment in a wide range of temperatures (1.5–60 K) was first calculated by Komatsu (1955) on the basis of the semicontinuum model developed by Komatsu and Nagamiya (1951). In this model, the graphite crystal is treated as an assembly of equally spaced elastic plates with both compressional and shearing coupling between adjacent plates. This theory is, however, incapable of accounting satisfactorily for the observed specific heat capacity curve above 60 K, because, in this range, vibrations of rather short-wavelength are excited and the atomic discreteness of the actual graphite structure becomes an important factor.

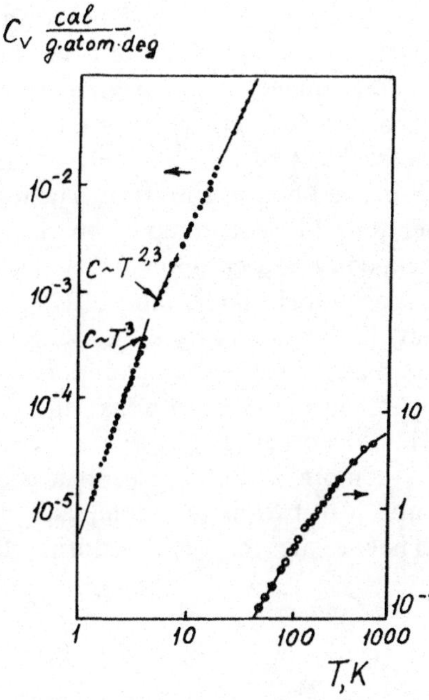

Fig. 212. Comparison of the measured specific heat and that calculated from the frequency distribution function shown in fig. 173. ●, De Sorbo and Nichols (1958); ○, De Sorbo and Tyler (1953). (Nicklow et al. 1972.)

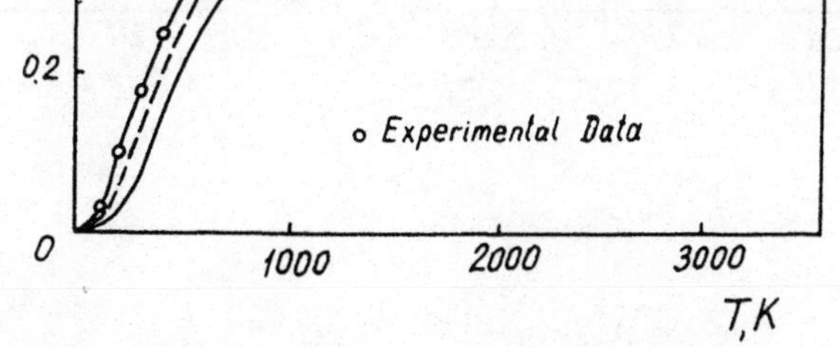

Fig. 213. Comparison of the constant-pressure heat-capacity data with the theoretical constant-volume heat capacity of graphite calculated using various frequency distribution functions. The constant-pressure specific-heat data of Mantell (1968) are represented by open circles. (Rafizadeh 1974.)

Attention of the reader is drawn to the fact that for a correct description of the high-temperature part of the dependence of C on T it is necessary to have accurate data on the high-frequency range of the lattice phonon spectrum. A rather exact calculation of the $C(T)$ curve for graphite up to a temperature of ~ 300 K was made on the basis of the relatively simple four-force-constants model (the BBS-model) of the phonon spectrum. The force constants k' and μ' [see eq. (233)] were then found from a comparison of the obtained results with the experimental heat-capacity and compressibility data for graphite.

The calculations of Yoshimori and Kitano (1956), made within the BBS-model, have shown that the heat capacity of graphite in the range from 15 to 100 K may be represented as the sum of two components, $C_1^{(L)}$ and $C_2^{(L)}$. One of them ($C_1^{(L)}$) results from in-plane vibrations and obeys a simple two-dimensional T^2-law, while the other ($C_2^{(L)}$) is due to vibrations perpendicular to the plane, and exhibits a more complex dependence on temperature. These latter vibrations are not just two-dimensional, and are characteristic of a graphite-like lattice. They relate to weak restoring forces associated with

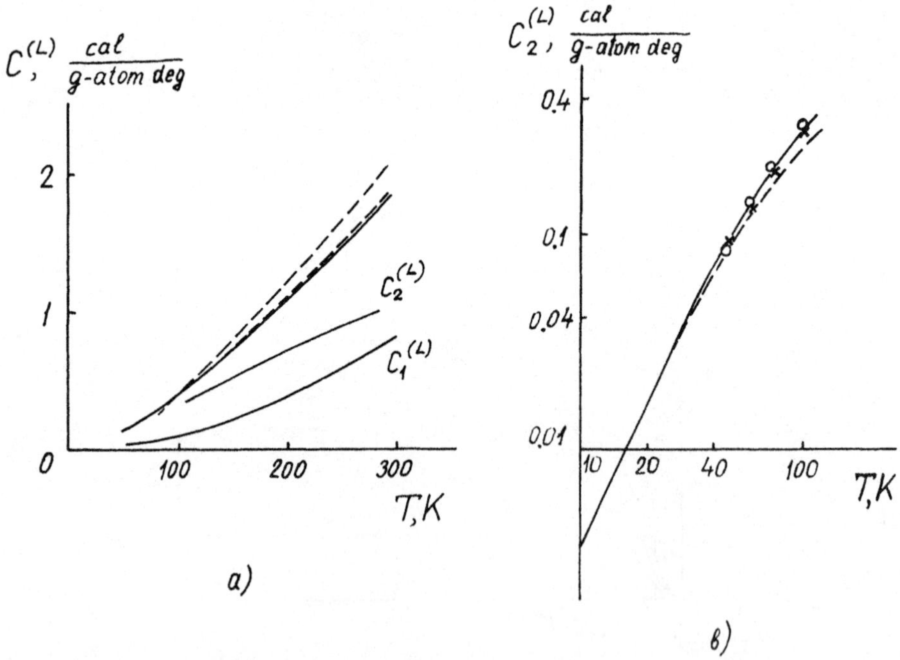

Fig. 214. (a) The components, $C_1^{(L)}$ and $C_2^{(L)}$ of the heat capacity of graphite (see text): —— calculated, ··· observed by De Sorbo (1954), - - - observed by De Sorbo and Tyler (1953). (b) The curves of $C_2^{(L)}$ (from the vibrations perpendicular to the plane): —— present calculation, - - - semicontinuum theory, ○, $C_V^{(i)} - C_1$ data by De Sorbo (1954), ×, $C_V^{(ii)} - C_1$ data by De Sorbo and Tyler (1953).

changes in distance between adjacent sets of planes and with strong bending forces for the planes. The T-dependences of the $C_1^{(L)}$ and $C_2^{(L)}$ components are illustrated in fig. 214a,b.

It should be noted that a comparison of the experimental data with the calculated results is in principle incorrect. The point is that the specific heat capacity is calculated on the basis of the phonon spectrum at a constant volume $C_V^{(L)}$, whereas experimentally it is measured for a constant pressure $C_p^{(L)}$. In spite of the fact that the difference between $C_p^{(L)}$ and $C_V^{(L)}$ is determined by the expansion coefficient α and is rather small for solids, it should be taken into account in precision measurements. It has been shown (Yoshimori and Kitano 1956) that at a constant pressure and volume in the high-temperature range the above difference depends linearly on T:

$$C_p^{(L)} - C_V^{(L)} = A(C_p^{(L)})^2 T, \tag{256}$$

where

$$A = \frac{\alpha^2 V}{(C_p^{(L)})^2 K}, \qquad \alpha = 2.2 \times 10^{-5}\,\mathrm{deg}^{-1};$$

$$C_p^{(L)} = 1.88\,\frac{\mathrm{cal}}{(\mathrm{deg\ mole})} \quad (\text{at } 300\,\mathrm{K}) \quad \text{and} \quad K = 2.98 \times 10^{-12}\,\mathrm{cm}^2/\mathrm{dyne}.$$

The calculations have shown that the correction associated with taking account of the $(C_p^{(L)} - C_V^{(L)})$-difference exceeds the experimental accuracy and for $T > 100$ K, is very important for determining the model parameters by fitting to experiment.

The data of figs. 212 and 213 indicate that the theoretical models for the phonon spectrum of graphite [BBS-model, AS-model, and other, similar in character, models described by Rafizadeh (1974)] enable one, with a suitable choice of parameters, to describe with high accuracy the temperature dependence of the lattice heat capacity of graphite from lowest temperatures up to ~1200 K. The marked difference between the calculated curves and the experimental data in the range between 1200 and 3400 K may be associated with a considerable increase, in this temperature range, of the anharmonic forces, which have not really been taken into account in any of the theoretical models. Moreover, the data of fig. 213 are not corrected for the difference $C_p^{(L)} - C_V^{(L)} \sim T$, which becomes considerable at $T > 1000$ K. Finally, at high temperatures $T \gtrsim 2000$ K an appreciable contribution to the measured heat capacity might have come from vacancies whose number increases as the temperature is raised.

From the data we have just discussed one may conclude that the behaviour of the heat capacity of graphite has been studied in detail, both experimentally and theoretically, in a wide range of temperature, from ultralow to ~1000 K. There is close agreement between theory and experiment in the whole range.

Thus the temperature dependence of the heat capacity of graphite may serve as a reference curve in studying heat capacities in various graphite intercalation compounds.

11.4.2. Heat capacity of GIC

In contrast to the case of pure graphite, when analyzing heat-capacity data for GIC the electron contribution $C^{(e)}$ [see eq. (254)] may not be neglected at practically any temperature. Moreover, for magnetic GIC it is also necessary to take into account a significant magnetic contribution $C^{(M)}$. Thus for GIC one encounters the problem of separating the contributions to the total heat capacity C. It should be mentioned that experimental measurements of the heat capacity of GIC are rather scarce. For the nonmagnetic GIC this parameter has been investigated experimentally only for GIC with alkali metals: Rb, Cs (Kondow et al. 1977, Mizutani et al. 1978, Alexander et al. 1980a,b) and Li (Delhaes et al. 1976, Ayache et al. 1980). To our knowledge, the heat capacity of the nonmagnetic acceptor GIC has not yet been measured. As for the magnetic acceptor GIC, their heat capacity has been measured in GIC with $NiCl_2$ (Karimov 1974, Karimov and Novikov 1974, Onn et al. 1982), $CoCl_2$ (Karimov 1975, Shayegan et al. 1982) and $FeCl_3$ (Onn et al. 1982).

Low-temperature specific heat capacity measurements in GIC yield valuable information on both the electronic structure and the phonon spectrum of these compounds. In general, the specific heat of the nonmagnetic GIC at low temperatures exhibits the temperature dependence

$$C = \gamma T + \beta T^3, \tag{257}$$

in which the cubic term is the lattice contribution and the linear term is the electronic contribution. The coefficient γ is determined as in (254) and, being expressed in mJ/(deg^2 mole), can be written as $\gamma = 2.357\nu(\varepsilon_F)$, where $\nu(\varepsilon_F)$ is expressed in $(eV)^{-1}$ per carbon atom. According to the general expression given by eq. (255), written for a simplified Debye model of the phonon spectrum [$D(\omega) \sim \omega^2$], the coefficient β is determined by the Debye temperature $\theta_D = \hbar\omega_{max}/k$. In units of mJ/(deg^4 mole), this coefficient is $\beta = 1944\theta_D^{-3}$ where θ_D is given in degrees Kelvin. It is commonly assumed that eq. (257) holds for temperatures $T \leqslant \frac{1}{20}\theta_D$. In order to find the coefficients γ and β, one should plot C/T versus T^2. In this case, the slope-angle tangent gives β and the intercept, γ. A typical dependence of C/T on T^2 for AGIC is depicted in fig. 215. It can be seen that the linearity of the dependence is quite good up to 25–30 K. The same figure presents an extrapolation of the results obtained by Delhaes et al. (1976) and of those obtained for HOPG under the same experimental conditions. The similar dependence obtained for HOPG has no extended linear part in coordinates of C/T and T^2. From the shape of the dependences of C/T on T^2 obtained for AGIC and graphite one may conclude

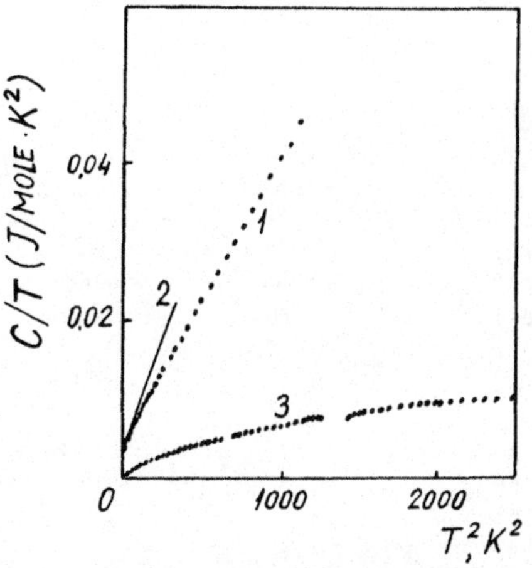

Fig. 215. Specific heat of C_6Li, C/T versus T^2 plot: (1) the results of Ayache et al. (1980) for single-crystal graphite, (2) the results of Delhaes et al. (1976), (3) the results of Ayache et al. (1980) obtained with a HOPG sample. (Ayache et al. 1980.)

that AGIC are considerably more three-dimensional than graphite, because the T^3-region in AGIC extends much higher in temperature. This is consistent with the mild conductivity anisotropy [$\sigma_a/\sigma_c \sim 10$ for C_6Li, as determined by Basu et al. (1979)], and with the results of band-structure calculations. This suggests a relatively strong coupling between the carbon and alkali-metal layers.

Table 33 presents the obtained results for the coefficients γ and β, as well as the Debye temperature values derived from the coefficient β. The coefficients listed in the table are given in terms of a molar unit defined by

$$m = \frac{xA_C + A_M}{1 + x}, \tag{258}$$

where A_C and A_M are the atomic weights of the carbon and alkali-metal atoms and x is the number of carbons per intercalant. In all cases (except for stage-one C_6Li), the Debye temperature θ_D is much lower than in graphite. For the graphite–K compounds, the θ_D values are found to exhibit an approximately linear dependence on reciprocal stage $1/n$.

Inspection of the γ values listed in table 33 shows that for the donor intercalation compounds the measured γ values are close to those obtained for the pure alkali metals and are more than one order of magnitude greater than the corresponding value in graphite. The values of the density of electronic states $\nu(\varepsilon_F)$ for AGIC derived from those of the γ coefficient [see eq. (254)] are given in table 34.

Table 33
Low-temperature specific heat coefficients in alkali metal donor compounds.

Sample	Molar weight m (g)	γ $\left(\frac{\text{mJ}}{\text{mol K}^2}\right)$	β $\left(\frac{\text{mJ}}{\text{mol K}^4}\right)$	θ_D (K)	Ref.*
Graphite	12.011	0.03	0.026	421	[1]
C_6Li	11.287	0.500	0.0054	710	[2]
C_6Li	11.287	0.430	0.0093	590	[3]
C_8K	15.0211	0.697	0.1502	234.8	[1]
$C_{24}K$	13.0946	0.241	0.0375	372.9	[1]
$C_{36}K$	12.7432	0.189	0.0339	385.2	[1]
C_8Cs	25.4436	0.63	0.049	342.1	[1]
$C_{24}Cs$	16.8468	0.25	0.085	284	[1]
$C_{36}Cs$	15.2784	0.19	0.072	300	[1]
$C_{48}Cs$	14.4782	0.16	0.060	319	[1]

*References:
[1] Mizutani et al. (1978).
[2] Ayache et al. (1980).
[3] Delhaes et al. (1976).

Table 34
Density of electronic states at the Fermi level, $\nu(\varepsilon_F)$, as found from low-temperature specific heat (γ) measurements and from band-structure calculations (theory).

Sample	$\nu(\varepsilon_F)$ in units of $10^{20}/(\text{eV cm}^3)$	
	Experiment	Theory
Graphite	14.3[a]	—
C_8K	233[a]	193[d]
$C_{24}K$	92[a]	—
$C_{36}K$	78[a]	—
C_8Cs	191[a]	—
$C_{24}Cs$	90[a]	—
$C_{36}Cs$	74[a]	—
$C_{48}Cs$	66[a]	—
C_6Li	216[b]	244[e]
	251[c]	244[e]
Cu	250	—
Ag	160	—

[a]Mizutani et al. (1978).
[b]Delhaes et al. (1976).
[c]Ayache et al. (1980).
[d]Inoshita et al. (1977).
[e]Holzwarth et al. (1978a).

An analysis of the results of electronic specific heat capacity measurements in stage-2 AGIC was made by Mizutani et al. (1978). From this analysis it has been concluded that the ionized metal sheets do not contribute significantly to the electronic specific heat, at least for the $n \geqslant 2$ compounds. From the same analysis, the effect of the electron–phonon enhancement of specific heat has been determined for C_8K, where the density of states had been calculated by Inoshita et al. (1977). Using the definition for the electron–phonon enhancement coefficient λ:

$$\frac{\nu(\varepsilon_F)_{exp}}{\nu(\varepsilon_F)_{calc}} = 1 + \lambda, \tag{259}$$

a value of $\lambda = 0.21$ was obtained.

For the case of the heavy alkali compounds (C_8K, C_8Rb and C_8Cs), a Schottky-type anomaly in the specific heat was found by Mizutani et al. (1978) and Alexander et al. (1980b), who identified it with a low-frequency optical mode associated with the motion of the intercalate layer relative to the graphite bounding layers (Kondow et al. 1977). No evidence for Schottky anomalies is found in the light alkali-compound C_6Li (Ayache et al. 1980). Einstein frequencies ω_E of 33 cm^{-1} and 24 cm^{-1} have been obtained for C_8Cs and for the higher-stage compound, respectively, using the relation $\hbar\omega_E = k\theta_E$, where θ_E corresponds to Schottky temperatures of 48 and 34 K, respectively. From the specific heat results it has been inferred that for the potassium intercalant the Einstein frequencies lie considerably higher than those in the case of Cs, which is consistent with the higher mass of the latter.

The uncommon anomaly in the T-dependence of specific heat C_p, which was found by Ayache et al. (1980) in the stage-1 C_6Li-compound near $T = 190$ K, is of interest. The anomaly consists of the presence of a small exothermal effect between 170 and 190 K followed by an endothermal effect in the 190–210 K range (figs. 216 and 217). The unusual shape of the anomaly on the $C_p(T)$ curve has so far precluded any reliable explanation. One may suggest its possible relation to a change in electron diffraction pattern and to the small anomaly in basal resistivity (Basu et al. 1979), both of them occurring approximately in the temperature range of the C_p anomaly. On the other hand, recent neutron diffraction experiments show no evidence for structural phase transitions in the C_6Li-compound in the 10–300 K range.

In conclusion, we shall briefly describe the available experimental data of specific heat measurements in the magnetic GIC based on $FeCl_3$, $NiCl_2$ and $CoCl_2$. It has already been mentioned above that the nature of magnetic ordering in all magnetic GIC has not yet been understood completely. It is of importance that in all these substances magnetic interactions are mainly two-dimensional within the intercalate layer. This is first of all indicated by the following two factors:

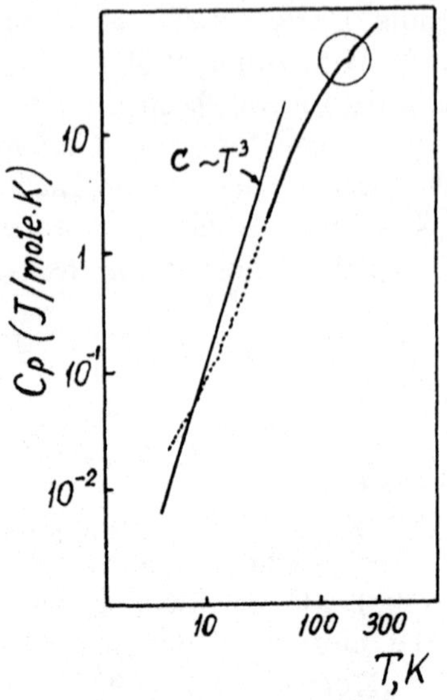

Fig. 216. Specific heat of C_6Li. The high-temperature anomaly is encircled. (Ayache et al. 1980.)

(1) the magnetic transition temperature is practically independent of intercalation stage index, and

(2) the transition temperature in GIC is lower than the transition temperature in pure intercalate material.

In studying the specific heat of magnetic GIC it is very important to separate the measured $C_p(T)$ into the magnetic $C^{(M)}$ and the nonmagnetic $(C^{(L)} + C^{(e)})$ components. This is often accomplished by employing the so-called method of corresponding states, which involves subtracting from the $C_p(T)$ dependence for a magnetic material the corresponding dependence for a nonmagnetic material with a similar crystalline and electronic structure. The method is based on the assumption that the similarly structured nonmagnetic material has the same values for $C^{(L)}$ and $C^{(e)}$ as the magnetic material. This method is, however, hardly applicable when studying the specific heat of the magnetic GIC, because it is very difficult to grow two specimens of GIC with identical structure. In the case of GIC, another method of separating the $C^{(M)}$ component is used. The method consists in measuring the $C_p(T)$ in zero magnetic field and in a strong magnetic field and subtracting the former curve from the latter. If the magnetic field (applied in the plane of the intercalate layers) is large enough to bring the intercalate layer in the spin-aligned paramagnetic state, then the magnetic contribution to specific heat will be entirely suppressed. To achieve this, one needs a saturation field $H^* \simeq kT_m/\mu_B$, where T_m is the magnetic transition temperature and $\mu_B = eh/2m_0c$ is the Bohr

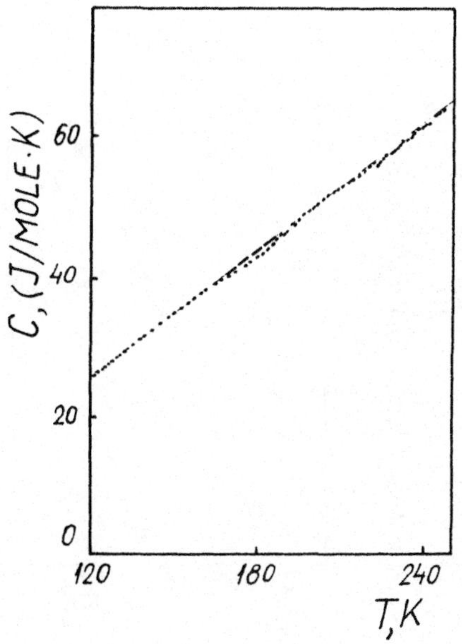

Fig. 217. Specific heat of C_6Li. A magnified linear plot of the high-temperature anomaly (see fig. 216). (Ayache et al. 1980.)

magneton. The experimental difficulties arising from the application of the method of suppression of the magnetic component are due to the fact that for high-T_m graphite intercalation compounds the magnetic saturation fields H^* become inaccessibly large. The application of the suppression method will be illustrated for $CoCl_2$–GIC. This compound is convenient for the purpose in view because, owing to its low $T_m = 9$ K, the saturation field, $H^* \simeq 13$ T, can be attained. Figure 218 illustrates the temperature dependences of the zero- and high-field heat capacity (C_p) of the stage-2 sample of graphite–$CoCl_2$ compound (Shayegan et al. 1982). The repeated measurement of C_p was made in a field $H = 14$ T. It can be seen that the zero-field heat capacity shows a broad anomaly at about 9 K, while the measured high-field heat capacity indicates a suppression of the anomaly and the downshift of zero-field values over the entire temperature range $3\,\mathrm{K} < T < 20\,\mathrm{K}$.

Attention is called to the fact that in a high magnetic field the heat capacity $C^{(L)} + C^{(e)}$ varies nearly proportionally to T^2 up to $T \sim 10$ K, while above this temperature it deviates towards a lower-power temperature dependence ($\sim T^{1.5}$). A similar quadratic dependence, typical of two-dimensional solids, is observed in some other acceptor-type magnetic GIC [e.g., in compounds with $NiCl_2$ and $FeCl_3$, as observed by Karimov (1974, 1975), Karimov and Novikov (1974), Onn et al. (1982)].

A least squares fit of the form of $\gamma T + \beta T^2$ to the high-field heat capacity

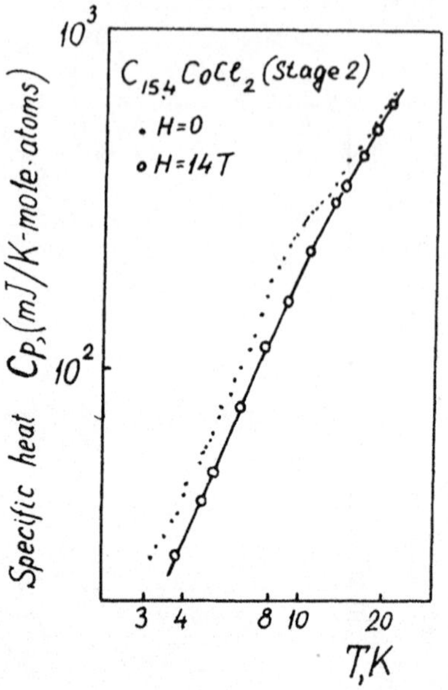

Fig. 218. Heat capacity of stage-2 $C_{15.4}CoCl_2$ at zero field and at 14 T. The solid curve is a polynomial (in temperature) fitted to the 14 T data points, and is used to determine the experimental $C_M(T)$ in fig. 219. (Shayegan et al. 1982).

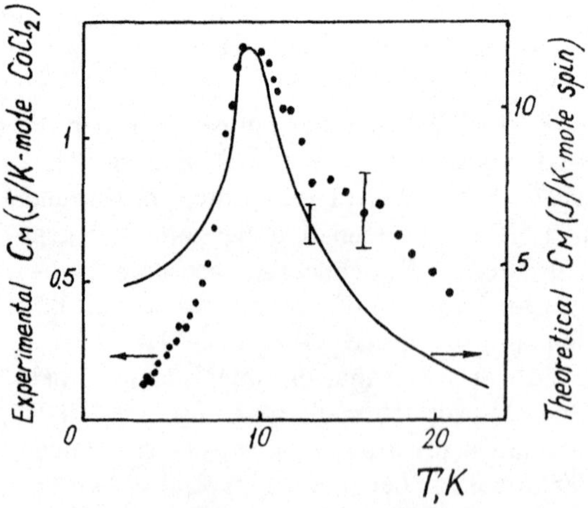

Fig. 219. Experimental (solid circles, left-hand ordinate scale) and theoretical (solid curve, right-hand ordinate scale) magnetic heat capacity C_M as a function of temperature. The theoretical C_M is the result of the calculation by Tobochnik and Chester (1979), obtained in a Monte Carlo calculation of heat capacity based on the xy-model. (Shayegan et al. 1982.)

temperature dependence of the stage-2 $CoCl_2$–graphite compound in the 3–10 K range yields $\gamma = 0.06 \pm 0.05$ mJ/(mole deg^2) and $\beta = 1.74 \pm 0.05$ mJ/(mole deg^4). Such a small value of γ is consistent with the results of other heat capacity measurements in magnetic acceptor GIC and is in good agreement with the Fermi-surface data for these compounds. For instance, almost the same value for γ may be obtained from a calculation of $\nu(\varepsilon_F)$ within the rigid-band model.

The value of β for $CoCl_2$-compounds is also in good agreement with the respective values for other magnetic acceptor GIC [e.g., in $NiCl_2$-compounds, according to Karimov (1975), $\beta = 1.54$ mJ/(mole deg^4)]. Figure 219 presents the T-dependence of the magnetic component $C^{(M)}$ for a stage-2 $CoCl_2$-compound obtained by subtracting the curves of fig. 218 from each other. The solid curve in fig. 219 represents the magnetic contribution to the heat capacity of this system obtained by a Monte Carlo calculation of Tobochnik and Chester (1979) on the basis of the XY-model for a flat magnetic layer. It can be seen that the calculated magnetic contribution has the form of a curve peaked at the magnetic transition temperature $T_m = 9$ K. Attention is drawn to the fact that the relatively broad peak of the heat capacity in fig. 219 is radically different from the narrow feature of the T-dependence of heat capacity observed in the antiferromagnetic phase transition in pure $CoCl_2$ (Chisholm and Stout 1962).

CHAPTER 12

Unsolved Problems and Applications

12.1. Research prospects for graphite, pyrocarbons and GIC

The foregoing may suggest that, on the whole, the energy spectra of electrons and phonons in monocrystalline graphite have by now been investigated in sufficient detail. How the graphite spectrum is reconstructed under the influence of external agents (strong axial or uniform compression, high magnetic fields, or doping) has also been elucidated to a significant extent. This does not mean, however, that all problems concerning the structure and properties of single-crystal graphite have been solved. Among the unsolved problems, we wish to point out the following.

12.1.1. Mechanical properties, lattice

The mechanical and elastic properties of graphite are still relatively little known. The elastic moduli data available are spread too much. This mostly concerns the small components of the c_{ij} tensor, which depend rather heavily on the degree of perfection of the graphite crystal, the presence of defects and impurities in it, as well as its heat treatment temperature.

The dependence of the c_{ij} tensor components upon these factors has not yet been studied in a systematic manner, which precludes definite conclusions about the elastic properties of single-crystal graphite by extrapolating the existing experimental data to the defectless crystal.

Models that describe the interaction between the carbon atoms in graphite lattices need considerable refining. This is true first of all of the elaboration of a model of a completely mechanically stable lattice and a detailed calculation of the phonon spectrum within such a model. The mechanical properties and elastic moduli of GIC are even less known than those of graphite, and their dependence on the degree of ordering in the intercalate layers and their variations resulting from order–disorder transitions in these layers have not been studied at all.

12.1.2. Electron and hole energy spectra in graphite

As for the electron energy spectrum of three-dimensional single-crystal graphite, an adequate description of minority groups remains the least developed problem. Experimentally, the questions concerning the number of the minority groups, the exact values of their parameters, and their location in the Brillouin zone still remain open. Theoretically, their origin is still obscure. Whereas the origination of minority holes at the *H* points of the Brillouin zone is quite consistent with the general model of the electron spectrum in three-dimensional graphite, the origin of minority electrons has not yet been settled to everyone's satisfaction, it seems. The relation of the minority electrons with the trigonal warping of the spectrum within the SWMC-model seems to be the most justified, theoretically. This view on the origin of minority electrons best agrees also with the existing (unfortunately, rather scarce) oscillational and galvanomagnetic data, which could be attributed to the minority carriers.

One of the most challenging problems in the study of monocrystalline graphite is undoubtedly that of reconstruction of its electronic spectrum in the megagauss range of magnetic fields. There are also two very interesting related classes of problems. The first is the study of the reconstruction of the spectrum of single-particle states; the observation and investigation of phase transition of the metal–dielectric and dielectric–metal type in a magnetic field brought about by a complicated motion of the bands and the character of the spin splitting of the Landau levels; the study of phenomena that accompany the magnetic freeze-out of the impurity carriers. The second is the study of the various many-body combined effects induced by a strong magnetic field, such as the charge-density waves, exciton pairing or Wigner crystallization, formation of electron–hole complexes on impurity centres in graphite. It seems to us that the study of single-crystal graphite in the extreme quantum limit of magnetic fields, especially in the megagauss range, remains the most fertile soil capable of yielding new interesting effects and phenomena. Such investigations are also of fundamental importance for solid-state physics as a whole. Here graphite, as a semimetal with a complicated lattice structure, provides a highly convenient model material, in which one may study the influence of the high anisotropy of the spectrum upon the behaviour of electrons and holes in a high magnetic field. In contrast to other anisotropic substances (for instance, bismuth and its alloys with antimony), the anisotropy of graphite is associated only with the preferred direction in its lattice (the *c*-axis), as a result of which its effect may be investigated in the most complete and pure form. Methodologically, measurements of the magnetic susceptibility and the components of the galvanomagnetic tensor seem to be the simplest in the range of ultrahigh fields.

In connection with this, we should mention that a general theory of the magnetic susceptibility of graphite, which could describe its properties in high magnetic fields at low temperatures, has not yet been developed. This primarily concerns the oscillating part of the susceptibility. A number of problems associated with the susceptibility of doped graphite have not yet been solved either.

12.1.3. Transport phenomena in graphite

In this field, many questions remain unsettled; first of all, the conductivity in graphite along the *c*-axis. Because of the quasi-two-dimensional character of the electron spectrum of graphite, small overlap integrals in the *c*-axis direction, and the short mean-free path in this direction, a description of the conductivity σ_c in terms of the kinetic equation (the classical charge transport by almost free quasi-particles) is incorrect. So far as its *c*-axis transport properties are concerned, graphite seems to belong to the intermediate domain which lies between the kinetic and the discontinuous descriptions of charge transport. This is suggested, in particular, by the unconventional temperature dependences of the σ_c, its specific relation with the size of microcrystallites, etc.

12.1.4. The energy spectrum of carriers and the magnetic properties of pyrocarbon materials

A significant number of unsolved problems is associated with the study of the properties and electron spectrum of pyrocarbon materials. Basically, these materials represent a very broad class of substances of various structure and properties, whose systematic study has just started. Among the numerous problems yet to be solved in studying these materials, here some major ones.

The most important problem is to establish the relation between the structure of pyrocarbon materials, their degree of graphitization, and the size of the constituent microcrystallites, on the one hand, and the electronic energy spectrum in these materials, on the other. Thus far, this relation has merely been outlined very broadly.

Another problem of fundamental importance is the magnetism of pyrocarbons. In connection with the discovered abnormally high diamagnetism of quasi-two-dimensional graphites (the highest among all known materials, apart from the superconductors!) this problem is also of great practical value.

Much progress has recently been made in understanding the high magnetic susceptibility of quasi-two-dimensional graphites (see, e.g., Kotosonov 1986), thus making it possible to further increase the diamagnetic susceptibility of these materials.

12.1.5. Transport properties of pyrocarbons

Finally, the major unsolved problem is the study of the transport properties of pyrocarbon materials, especially the nature of their negative magnetoresistance. The abundance of controversial models of this phenomenon, developed in the early seventies, are indicative of considerable difficulties facing the investigators. It seems to us that the theory of quantum corrections, developed to describe the electrical conductivity of disordered systems has made it possible to achieve some progress in solving the problem of charge transport in pyrocarbons. This theory may serve as a basis for elucidating electron transport in these materials. Nevertheless, there are still many unclear questions, for instance, the part played by the size effects, electron–electron interaction, the contribution from scattering by intercrystallite boundaries, the Maki–Thompson correction, etc.

12.1.6. The energy spectrum of GIC

Most of the works devoted to the study of carbon materials have very recently dealt with the properties of GIC of both the donor- and acceptor-type. A large number of GIC studies indicate the great practical value of these materials, on the one hand, and many problems to be solved in the field, on the other hand.

Some of the more important unsolved problems in the study of the structure of the energy spectrum of electron and phonons in GIC and the properties of these materials are listed below.

One of the least studied problems is that of the intercalate-layer structure and the order–disorder or order–order transitions in the layer due to varying external conditions: temperature, pressure, and (for magnetic GIC) the magnetic field. It has already been mentioned above that the structure of the intercalate layer and the character of transitions in it have so far been studied only for a very small number of GIC, mostly those with alkali donor-type metals. At the same time, for the overwhelming majority of GIC, especially those of the acceptor type, the nature of ordering in the intercalate layer still remains obscure.

Of great interest are the questions of under what conditions there arise, in the intercalate layers, structures commensurate or incommensurate with respect to those of the graphite layers and of how the f_{el} factor varies with changing intercalate-layer structure. These problems are intimately related to the elucidation of the formation mechanism of the electron–hole spectrum of GIC on the basis of the electronic systems of the original intercalant and graphite.

Analyses of the results of first-principles calculations of the electronic energy spectrum of GIC and also of the attempts made at describing the spectrum by using some or another model indicate that this fundamental problem is very far from being solved. Whereas in higher-stage GIC the electronic spectrum is, in general, close to that of initial graphite and the intercalation effect can be understood within a rigid-band model taking account of the screening of the intercalate layer by electrons in the bounding graphite layers, a description of the spectrum of lower-stage GIC presents fundamental difficulties associated with the necessity of exactly taking into account the contributions from the electronic subsystems in graphite and the intercalate. The situation is much complicated due to the violation of the equivalence of the graphite layers and the appearance of the space charge nonuniformity in the stack of the graphite layers in GIC with $n > 2$.

In fact, not only a quantitative theory of the electron energy spectrum in low-stage GIC, but even an adequate qualitative theory is lacking. This is suggested, in particular, by the disparity in the number of various oscillation frequencies observed in GIC and the number and magnitude of the calculated extremal cross sections of the Fermi surface. Detailed studies of the electron energy spectrum of various GIC and the nature of its reconstructions due to external forces (pressure, temperature, high magnetic fields) which make use of the oscillational and resonant techniques may be carried out only provided the technology of GIC synthesis is ever improved and the quality of initial graphite is raised. The current state of the problem of studying the electron energy spectrum in GIC in wholly determined by the present-day technological level of producing high-quality single crystals of these compounds.

12.1.7. Supermetallic conductivity of GIC

Of very great practical value is the problem of producing GIC with a supermetallic conductivity. Letting alone the independent, extremely complicated technological problem of utilizing this phenomenon for manufacturing electric cables, we should note that even the problem of producing a stable high-conductivity material in the form of a bulk single crystal is very far from its final solution. The few known GIC with a high conductivity along the layers are, as a rule, unstable and incapable of preserving their properties for a long time without some protective covering. The necessity of searching for stable high-conductivity GIC renders the problem of studying the transport properties in a broad class of compounds with strong donors or acceptors very topical.

We should also emphasize the scientific importance of investigating the transport properties of GIC in the direction perpendicular to the layers. The nature of conductivity along the *c*-axis, especially in perfect single crystals with an acceptor intercalant has not yet been understood. Nor is it clear whether the conductivity can be described on the basis of the kinetic equation, because the estimated mean-free-path values are less than that of the identity period in GIC. For real GIC, the effect of crystallite size and intercrystallite boundaries on the σ_c has not been completely elucidated, either.

12.1.8. Superconductivity of GIC

Of great scientific and practical value would be to solve the problem concerning the nature of superconductivity in GIC. No answer has yet been found to the most important question of which electronic subsystem (that of the graphite layers or that of the intercalate layers) is responsible for the superconducting properties and which modes in the phonon spectrum of GIC mainly cause the interelectron attraction to appear. Solution of these problems is far from easy, because, as measurements indicate, superconducting GIC seem to be anisotropic three-dimensional superconductors, and not two-dimensional ones. This means that the contributions from the two electronic subsystems to the superconductivity are closely related to each other, none of them being dominant.

Thus, along with heavy-fermion superconductors and also with magnetic and organic superconducting materials, superconducting GIC belong to the exotic superconductors, whose study has recently raised much interest.

12.2. Technological applications of graphite, pyrocarbons and GIC

Graphite and graphite-based materials are finding an extraordinarily wide application in diverse technologies, e.g. in electrical, electronic, chemical, atomic, aircraft and aerospace technology. The multi-purpose utilization of these materials is the consequence of the unique combination of their properties (heat, radiation and chemical resistance; high electrical and heat conductivity; high mechanical strength; low specific gravity; high diamagnetism) and the possibility of widely varying them by using various initial materials, different conditions of their production and heat treatment, and also by introducing various additions and impurities.

In the last decade, along with synthetic graphites, graphite intercalation compounds have been used very widely.

12.2.1. Graphite and pyrocarbon materials

At present, perfect synthetic graphites (pyrolytic graphite and kish-graphite), as well as technical quasi-two-dimensional and quasi-three-dimensional graphites are the best-known representatives of carbon materials. Merely listing the various fields of science and technology utilizing these materials would take unjustifiably much space. So we restrict ourselves to indicating only several traditional spheres of application.

These materials have long and widely been used to manufacture sliding contacts of various kinds and heat-and-erosion-resistant electrodes for arc furnaces. Another traditional sphere where graphites are utilized is the fabrication of crucibles and containers for melting metals or semiconductors, for making alloys, and for growing single crystals.

Graphites are traditionally used to lower friction between surfaces. This application is based on the layered structure of the graphite lattice and the weak bonds between the layers, owing to which graphite scales may serve as solid lubricants or may be added to liquid oils to improve their lubricating and frictional properties.

Graphites are also being used very widely as neutron moderators in nuclear reactors. In atomic technology and in atomic physics, these materials are being employed in various reflecting, polarizing and moderating systems.

Finally, graphites and carbon fibres are being used for various purposes as construction and composite materials.

The above applications of graphites and carbon materials are traditional and widely known. Of the less known spheres of their application, one can mention their utilization as magnetic suspensions in gyroscopic devices. For this purpose, quasi-two-dimensional pyrolytic graphites, which have the highest diamagnetic susceptibility among all known materials, except for superconductors, are being mainly used.

12.2.2. Graphite intercalation compounds – novel applications

An extremely broad class of carbon materials, including graphite intercalation compounds, is finding new, unconventional technological applications. These are due to the peculiar physical and chemical properties of GIC. Because of the novelty of these applications of GIC, we shall dwell on them at greater length. First at all, we shall mention some specific applications which have not yet become widespread.

Of great scientific and practical interest is the combination of metallic layers with semiconducting ones or with layers composed of complex molecules, which is realized in GIC on the microscopic scale. This combination may lead to the creation of crystals with entirely new electronic properties. In particular, GIC may find an application in instruments of the type of superconducting quantum interferometers as a weak coupling link, because these materials can readily be fashioned into very thin films (Consadori et al. 1971). The possibility of realizing the Josephson interaction between layers due to intercalation opens up highly attactive prospects for a wide application of GIC in this field.

Graphite intercalation offers unique opportunities of producing structures (superlattices) with an identity period I_c as large as 50 or even 70 Å (for high-stage GIC) at a high degree of periodicity. Such structures can be used, in particular, as low-energy neutron monochromators (Boeuf et al. 1983).

Pristine graphite itself is a material with a very high anisotropy of electrical and thermal properties. This anisotropy may be further increased by intercalating pristine graphite with appropriate intercalants. GIC with a high anisotropy of their heat conductivity may serve as heat screens. Acceptor GIC, which have a high anisotropy of the electrical conductivity, can be used as infrared polarizers, as was first demonstrated by Rupprecht et al. (1962).

Graphite intercalation may be used as a method of shifting the plasma frequency of a gas of carriers. This property is used in the fabrication of optical filters that can be tuned to a definite wavelength by changing the stage of intercalation or by varying the stress applied to a given structure (Pfluger et al. 1979). The same property allows creation of electrochemical devices in optical displays (Whittingham and Ebert 1979, Selig and Ebert 1980).

The large number of phase transitions in graphite intercalation compounds offers a versatile tool for studying critical phenomena in physics and chemistry and provides a variety of practical applications. For the latter, it is especially important that, when undergoing a phase transition, the substance is in a peculiar state, which is characterized, as a rule, by abnormal values for a whole series of physical characteristics: heat capacity, dielectric constant, electrical and heat conductivities, etc.. This may be used in various devices of modern electrical, radio, and electronic technology.

One should also mention the possibility of producing thin films of intercalated graphite with large area, which appear as a result of exfoliation when the ampoule containing the material being intercalated is rapidly cooled. Such films are highly important for producing flexible controlled-surface mirrors in adaptive optical devices, as well as materials with a highly developed sorbing surface per unit weight (up to $\sim 10^2\,m^2/g$).

An interesting application of GIC is due to the reversible processes of intercalation (so-called intercalation–deintercalation cycles). This allows GIC to be used as storages for accumulating and storing certain substances in relatively small volumes. It is of great importance that a very high concentration of an intercalant may be attained in GIC. Thus, in particular, the intercalate spatial density in stage-one alkali-metal GIC exceeds the spatial density of the atoms of the original metals!

In this application, GIC with intercalation–deintercalation cycles that may serve as storages for hydrogen are of greatest interest. Among these, one can mention, in particular, hydrogenated stage-one compounds with alkali metals in the form of ternary compounds (e.g., $C_{16}K_2H_{4/3}$, see Lagrange and Hérold 1978). Finally, it is worth mentioning that the affinity of GIC to water and oxygen makes it possible to use them for the cyclic purificiation of gases.

12.2.3. Technological and industrial uses of GIC

The above-mentioned applications of GIC are still specific enought to account for their relatively small spread. However, there are several branches of technology where even today GIC are highly promising and are being utilized on a large scale (Dresselhaus and Dresselhaus 1981).

These include the use of GIC as battery and electrode materials, as substitutes for metallic conductors, as catalysts, and as high-strength, heat-resistant fibres.

Let us discuss these applications of GIC in somewhat more detail.

12.2.3.1. GIC as battery and electrode materials. The capacity of GIC to intercalation–deintercalation cycling provides the basis for applying them as electrode materials in chemical batteries and accumulators. GIC which are used as a cathode material must allow multiple intercalation–deintercalation cycling while retaining and reproducing the structure. The conversion from chemical energy to electric and back requires that there should be ionic exchange with the electrolyte as the current flows through the external circuit. In order that a high emf of a battery be reached, it is necessary that the intercalant should have a high electron affinity. This requirement is met, in particular, by graphite fluorides, which serve as the cathode when a lithium electrode is used as the anode (Whittingham 1975). As a discharge current flows through such an accumulator, the following process occurs

$$x\mathrm{Li}^+ + xe^- + \langle \mathrm{C}_n\mathrm{AX}_y \rangle \rightarrow \langle \mathrm{Li}_x\mathrm{C}_n\mathrm{AX}_y \rangle,$$

when lithium ions pass from the anode to the electrolyte, move towards the cathode and bring about its intercalation. As the accumulator is charged, lithium ions leave the GIC and pass again onto the anode.

The operation of batteries making use of GIC as one of the electrodes has been discussed in detail by Armand and Touzain (1977). The most serious difficulties associated with this application of GIC arise when it is necessary to assure the stability of the electrolyte near the electrode surface, for this is where the GIC act as a catalyst, accelerating electrolyte decomposition.

12.2.3.2. GIC as high-conductivity materials. One of the versatile applications of GIC is due to their extremely high conductivity along the layers in conjunction with their low specific gravity. In sect. 10.7, we have already discussed the so-called supermetallic conductivity of GIC, which is manifested by the fact that some (especially acceptor) GIC have room-temperature σ_a as high as (or even higher than) the electrical conductivity of copper $\sigma_{cu} \sim 0.59 \times 10^6\,\Omega^{-1}\,\mathrm{cm}^{-1}$. At the same time, the density ratio of Cu to GIC is about four. As it follows from table 21, among the high-conductivity GIC are graphite compounds with AsF_5 ($\sigma_a = 0.63 \times 10^6\,\Omega^{-1}\,\mathrm{cm}^{-1}$ in stage-two GIC) and with SbF_5 ($\sigma_a = 0.59 \times 10^6\,\Omega^{-1}\,\mathrm{cm}^{-1}$). For practical purposes, however, it is necessary to have conductors in the form of thin wires; flexible and mechanically strong. This problem is being solved along two main lines:

(a) production of composite materials consisting of well conducting metals (e.g., copper) and GIC; and

(b) intercalation of carbon fibres with high-strength properties.

Of great interest are the efforts in the former direction made by Vogel (see, e.g., Vogel et al. 1977), who proposed and developed a technology for producing a composite wire consisting of a SbF_5–graphite compound placed in a copper envelope. The wire specimens thus obtained had a room-temperature conductivity of $0.595 \times 10^6\,\Omega^{-1}\,\mathrm{cm}^{-1}$ (i.e., exceeding that of copper), while their average density was lower than the copper density (the wire, on the whole, consisted of $\frac{2}{3}$ of copper and $\frac{1}{3}$ of GIC). The copper sheath of the composite wire not only conducts electrical currents and ensures the mechanical strength of the whole structure, but also preserves the inner-part high-conductivity GIC (usually highly reactive) by preventing their contact with ambient agents which may cause GIC's decomposition. Vogel's technology for producing composite wires is based on the traction of GIC powder-filled copper tubes until the required thickness is reached. Heat treatment of the tracted composite wire makes it possible to attain a high degree of such an orientation of the microcrystallites that their *c*-axes are for the most part perpendicular to the wire's axis.

It is true, though, that in a more recent attempt Singhal (1980) failed to reproduce Vogel's results. His composite wire composed of copper and an SbF_5–graphite compound had a conductivity of only $0.45 \times 10^6\,\Omega^{-1}\,cm^{-1}$, also with a high degree of microcrystallite *c*-axis orientation normal to the wire axis. Nevertheless, it has, in principle, been shown that composite wires can be of interest commercially and may compete, so far as their parameters are concerned, with conventional metallic conductors.

Recently, the problem of finding substitutes for metallic conductors has intensively been investigated by using graphite intercalated fibres. As the starting material, high-strength carbon fibres with a Young modulus of up to 400 GPa are being used. When well-graphitized fibres with oriented planes are intercalated, their strength does not in fact change, whereas their electrical conductivity increases. One of the first attempts at intercalating graphite fibres to make conductors was made by Herinckx et al. (1972) who have demonstrated that intercalation of the fibres with alkali metals increases their conductivity more that 20 times without deteriorating their mechanical properties. Intercalation of graphite fibres by acceptors, such as Br_2, $AlCl_3$, ICl, has been investigated by Hooley (1975) and by Hooley and Dietz (1978). The production of high-conduction fibres by intercalating graphite fibres with HNO_3 and other acids has been studied also by Vogel (1976) and by Besenhard and Fritz (1978). Significant advances have recently been achieved in producing conducting fibres by intercalation (see, e.g., Goldberg and Kalnin 1981, Oshima et al. 1983, Natarajan and Woollam 1983, Endo et al. 1983, Manini et al. 1983, Murday and Dominguez 1984). These authors studied fibres prepared by the pyrolysis of a mixture of benzene and hydrogen at 1100°C with their subsequent graphitization at 2800 to 3000°C. In the fibres thus produced, the graphite layers are ordered as concentric cylinders around the fibre axis. Intercalation with $CuCl_2$, $SbCl_5$, ICl, and HNO_3 makes it possible to obtain highly stable fibres with a low desorptive ability which seems to be assured by the tubular structure of the fibre itself. The problem of producing a sheath by sorption on the fibre surface of certain stabilizers following intercalation is also being solved.

12.2.3.3. Graphite fibres used to produce composite materials. Apart from serving as a substitution for metallic conductors, graphite fibres themselves are being used as high-strength heat resistant materials which can often replace metals quite successfully.

The high strength of the fibres is obtained by their heat treatment in tensile conditions, as a result of which the planes of the graphite layers become preferentially oriented along the fibres and the strength properties begin to be determined by the binding value for the carbon atoms in the layer.

At the present time, fibres obtained by graphitization of poly-acrylonitrile (or PAN-fibres) are the most widely used ones. They, as a rule, exhibit a radial structure of their graphite planes (so-called pitch or rayon fibres) or a ribbon structure when the graphite planes are parallel to the fibre axis. An increase in heat-treatment temperature leads, first, to an increased size of graphite microcrystallites inside the fibre and, second, to a decreased degree of graphite-plane disorientation relative to the fibre axis. At a heat-treatment temperature above 2500°C highly-oriented fibres are produced, which are called graphite fibres, but which are significantly different in their properties and structure from those obtained at the lower HTT (Bright and Singer 1979). High-strength heat-resistant carbon fibres are being used as construction materials in various fields of technology. Quite recently, carbon fibres have been increasingly used in producing composite materials in which the fibre is supplemented by a filler (e.g., rubber, epoxy, plastics).

12.2.3.4. Applications in chemical catalysis. It has been found that graphite intercalated with certain substances exhibits a high catalytic activity in chemical reactions (Whittingham and Ebert 1979). In spite of the fact that the nature of the catalytic activity of GIC has not yet been completely understood, these materials are very successfully used as catalysts, combining the advantages of both supported and metal-complex catalysts. For example, graphites intercalated with metal chlorides are capable of catalyzing many chemical reactions involving the cation mechanism, such as alkylation of aromatic compounds, hydrochlorination of olefins, cation polymerization and co-polymerization. The effect of the presence of GIC in such reactions is that of increasing the reaction selectivity. Alkali-metal GIC catalyze the ortho-hydrogen–para-hydrogen conversion and the reactions of olefin isomerization and polymerization. The excellent conduction properties of GIC and the presence of transition metals in these materials offer good prospects for utilizing them as electrodes in electrocatalytic processes (Armand and Touzain 1977). The catalytic properties of GIC have been extensively discussed in the literature. For example, the catalytic properties of compounds with alkali metals and metal chlorides for the synthesis of NH_3 out of N_2 and H_2 have been investigated by Ichikawa et al. (1972a,b) and Postnikov et al. (1975); for the formation of hydrocarbons out of H_2 and CO, by Naito et al. (1972) and Mashinskii et al. (1976); for liquid-phase oxidation reactions, by Belousov et al. (1980). Compounds with alkali metals are known to be used as catalysts for breaking the carbon–hydrogen bonds leading to substitution of deuterium for hydrogen (Ichikawa et al. 1972b). The same compounds have been used

(Beguin and Setton 1976) to catalyze organic reactions, such as the production of biphenyl in a reaction of benzene with tetrahydrofuran. Finally, alkali metal compounds have been utilized as catalysts for polymerization reactions, in particular the styren–isopren one, leading to formation of a variable co-polymer (Panayotov et al. 1975, Golé 1977, Golé et al. 1982). The catalytic properties of various GIC have been thoroughly examined by Novikov and Vol'pin (1981). Attempts to explain the nature of the catalytic activity of GIC have been made by Setton (1977) and also by Ebert and Selig (1977), who discovered chemical shifts of the NMR-lines and the appearance of asymmetry of the ESR-lines, which suggest a change in the oxidizing properties of the intercalate as a chemical reaction is catalyzed. One should also mention in this connection the results obtained by Schögl et al. (1983), who have investigated the catalytic properties of $FeCl_3$-GIC in the reactions of oxidation of SO_2 in water and have found that, in the reaction, the residual intercalate $FeCl_2$ co-exists with the surface phase of FeO(OH), which indicates that in catalysis the intercalate may take part in intermediate reactions.

The diverse properties of graphite intercalation compounds have so far been relatively little studied. Bearing in mind that these materials constitute an extremely broad class, one can justifiably expect that their potential for practical application has just begun to be used.

12.3. Graphite and carbon materials in atomic technology

12.3.1. Properties of carbons used in building nuclear reactors

One of the currently most important applications of versatile carbons is their use as construction materials in atomic technology. This is so because of the unique combination of their properties, allowing them to be successfully used in reactor cores. It is necessary to note that a material intended for use in a nuclear reactor must be capable of moderating the neutrons, but at the same time have a low neutron capture efficiency. These two characteristics, in conjuction with resistance to heat, radiation and chemicals, essentially determine whether a particular material may be used for constructing the core vessel walls. However, one should mention that a complete list of the requirements to be met by such a material is much longer. A reactor material should also have a high heat conductivity (practically as high as that of a metal) and a high heat capacity; relatively high strength retaining up to a temperature of about 1500°C; low gas permeability; increased resistance to oxidation and erosion at high temperatures; chemical resistance in reacting with metals, salts, and gases; low coefficients of linear and volume thermal expansion; capability of retaining its mechanical strength at high temperature gradients and high levels of radiation; finally, the material should be available in great enough amounts and be relatively inexpensive (Fourré et al. 1976, Fieguth et al. 1977, Beavan and Engle 1981).

Of all known materials, graphite and pyrocarbons meet the above requirements best. The problem of producing a special reactor graphite reduces to developing a technique for obtaining a carbon material with an optimal combination of the required properties regarding its prospective use.

12.3.1.1. Neutron moderating and absorbing power. Graphite has been widely used as a neutron moderator in nuclear reactors since 1942, which can be explained by its satisfactory moderating power and low neutron-capture cross section (Scott 1978, Beavan and Engle 1981). Pure graphite has a neutron capture cross section of $(0.0032 \pm 0.0002) \times 10^{-24}\,cm^2$. A suitably purified artificial graphite material is characterized by an average neutron-capture cross section of $0.0036 \times 10^{-24}\,cm^2$, the difference between these values being due to the impossibility of completely removing all impurities (such as boron, vanadium and rare-earth elements) having large cross sections for neutron capture.

The moderating power of graphite is due to its low atomic weight. Graphite lowers the energy of the fission neutrons as they are elastically scattered by the carbon atoms. In each collision, a fast neutron loses 15.8% of its initial energy, this attenuation factor is determined by the ratio of the neutron mass to that of the nucleus of the carbon atom. For carbon, the neutron scattering cross section is $4.8 \times 10^{-24}\,cm^2$. Thus, at a density of $1.65\,g/cm^3$ the moderating power of reactor graphite, δ, constitutes 0.0629 (i.e., each fast neutron loses 6.29% of its energy as it travels through 1 cm of the graphite moderator). Materials used in pile construction are customarily characterized by their relative moderating power, v, which is equal to the ratio of the moderating power δ to the total absorption coefficient λ, i.e., $v = \delta/\lambda$. Table 35, taken from the book *Nuclear Graphite* (Vjatkin et al. 1967), compares the parameters of graphite with those of some other materials used as moderators.

Graphite, as can be seen from the table, has a much lower value of relative moderating power than heavy water, but a value similar to that of beryllium

Table 35
Parameters of some moderators.

Moderator	Atomic or molecular weight	Density (g/cm^3)	Moderating power, δ (cm^{-1})	Relative moderating power, v
Graphite	12	1.65	0.0629	168
Beryllium	9	1.85	0·173	160
Beryllium oxide	25	3.00	0.130	200
water (H_2O)	18	1.00	1.330	61
heavy water (D_2O)	20	1.10	0.280	3300

and beryllium oxide, and considerably greater than that of ordinary water (Scott 1978).

Of great importance is the density of a graphite: the higher its density the higher its moderating power and the shorter the neutron migration range. The density of graphites commonly used to build reactors varies from 1.65 to 1.75 g/cm^3. An increase in their density and degree of purity allows a reactor to be made more compact.

12.3.1.2. Heat resistance and strength properties at high temperature. As has already been mentioned above, the high thermal resistance of graphite-based carbons is explained by a high binding energy of the carbon atoms in a graphite lattice (its vaporization heat being 171.4 kcal/mole). Graphite remains in the solid state even at very high temperatures, up to 4000 K. A remarkable property of this material is the increase in its strength, which is observed as graphite is heated up to 2500° or even 2600°C. Then its strength starts decreasing rather slowly, so that at a temperature of about 3000°C it only diminishes to its room-temperature value. It is worth mentioning that the plasticity of graphite increases with temperature (Awaji and Sato 1979, Prince 1979).

The increase in the strength of graphite with temperature is perhaps best illustrated by the respective change in its strength limit. According to the Industrial Graphite Engineering Handbook (1959), the limiting resistance of various graphites to stretching, which varies from 105 to 155 kg/cm^2 at room temperature, increases at a temperature of 2500°C to 182 and even 275 kg/cm^2, i.e., by 73 to 184%. For all grades of graphite, the static and dynamic elastic and shear moduli increase with temperature as the temperature is raised up to 1800–2500°C (Awaji and Sato 1979).

One of the first hypotheses advanced to explain the increase in the strength of graphite with temperature (Mrozowski 1960) is associated with the removal of the residual stresses which have arisen due to an anisotropic change in the size of the individual crystallites of graphite brought about by cooling after the process of graphitization is completed, and which have been retained in the material because of the slowness of the creep and relaxation processes in the temperature range from 20 to 2000°C. This hypothesis has been elaborated by Hove (1960), who took into account the fact that the thermal expansion coefficient of crystallites in graphite is greater along the *c*-axis than in the basal plane. As the temperature is raised, this leads to the filling of the empty space between crystallites and their interlocking, which renders the whole structure tougher, its elastic modulus, in particular, increasing with temperature. The increased strength of graphite with increasing temperature seems to be also due to the appearance of plasticity, as a result of which the effect of pores and structual inhomogeneities, i.e., the sites where stresses are concentrated, decreases.

A different hypothesis to explain the phenomenon of increased strength of graphite with temperature, put forward by Jenkins (1962), makes use of the fact that, as graphite is loaded at normal temperature, microcracks start appearing inside its grains. In the course of loading, the number of microcracks increases, thus resulting in an increased size of these cracks and, as a consequence, in the destruction of the loaded specimen. It is assumed that the formation of a lens-shaped crack in a crystallite leads to a drastic decrease of the concentration of the elastic stress energy in that region. At high temperatures, the carbon atoms begin to diffuse from the regions with a higher elastic energy concentration towards those with a lower energy concentration thereby resulting either in the healing of the crack or in the rounding off of its edges. Such a change in the shape of the crack gives rise to an increased strength, because the stress concentration has decreased. The proposed mechanism is thus determined by diffusion and, therefore, should depend on the rate and duration of the deformation process. This mechanism enables one to explain quite satisfactorily why the strength of graphite varies with temperature also as a function of the deformation rate. If, for example, the rate is about 5×10^{-3} min^{-1}, then one can observe a substantial increase in strength, reaching a maximum in the 2000–2500°C temperature range. With the greater deformation rates (about 2 min^{-1}), in the 20–2500°C range, one only observes a relatively small increase in strength, because the healing rate for cracks is then lower than that of their formation (Vjatkin et al. 1967).

Because various raw materials are used for the production of carbons their mechanical properties significantly differ, changing as functions of temperature. It is therefore hardly possible to explain by a simple mechanism the mechanical properties of various carbons whose production technology and type of initial material significantly differ.

12.3.1.3. Low penetrability. One of the highly important properties of a material used for manufacturing interfaces (which should be as nearly impenetrable as possible) between two media is a low penetrability. Experience has shown that practically all materials are to a certain extent penetrable. However, the difference in penetrability is very great for different materials. Whereas the penetrability of porous solids, characterized by the penetration coefficient, K*, may be as high as several cm^2 s^{-1}, that of metals is only on the order of 10^{-6} cm^2 s^{-1}, and the penetrability of glass is even lower, amounting to 10^{-13}–10^{-14} cm^{2}s^{-1}.

Artificial carbons are characterized by a significant open-channel porosity

* The penetration coefficient K is usually measured from the rate of increase in gas pressure, dp/dt, in a volume V, resulting from filtration through a penetrable wall shaped as a disk with area S and thickness L. The coefficient K is given by $K = (dp/dt)VL/S\Delta P$, where ΔP is the pressure difference exerted on the wall. K is given in units of cm^2 s^{-1}.

and, therefore, are highly penetrable, their penetration coefficient K ranging from 10^{-1} to $10^{-2}\,cm^2\,s^{-1}$, depending on their production method. This value can be lowered by increasing the density of a pyrocarbon and creating a uniform porosity in it. The need for carbons having a decreased penetrability is ever increasing, in connection with construction of gas-cooled nuclear reactors and development of quasi-homogeneous thermal release elements. The latter should be made of a carbon material capable of completely confining within itself all products of radioactive decay. This challenging problem may be solved in either of the following ways. First, a material may be produced with a density very close to that of perfect single-crystal graphite $\rho_{perf} = 2.267\,g/cm^3$, i.e., a material either completely free from pores or having a large but closed system of pores. Second, glassy carbon, a novel material, characterized by considerable, yet closed, porosity and extremely low penetrability, might be produced. An alternative, occupying, as it were, an intermediate position between these two courses, is to lower the penetrability of graphite by filling it with fluid organic substances which upon heat treatment change to carbon, thereby stopping the pores. Indeed, while increasing the density of the material and, therefore, decreasing its porosity, such a treatment simultaneously reduces filtration by closing pore entrances, i.e., by turning an open pore system into either a closed or "blind alley" one.

The porosity of carbons is significantly affected by increasing temperature, the effect being two-fold: the properties of both the filtrating medium (its viscosity and molecular range) and the porous material change. Although the thermal expansion of graphite is relatively small, changes in its porosity may affect its penetrability. It may be expected that heating results in submicropores being partially closed due to the thermal dilatation of graphitic crystallites. This, in particular, accounts for the low coefficient of thermal expansion of massive graphitic specimens. At the same time, the size of large pores somewhat increases with temperature as a consequence of the increased size of the specimen itself. These factors may be responsible for the overall change in the coefficient of penetration for graphite.

The penetrability of a porous body is directly related to its porosity. However, to establish a direct relation between the pore volume and the penetrability is rather difficult because of the great difference in the character of pores and their diameter, owing to which materials of the same density may have very different penetrability. Wiggs (1958) made a calculation of the coefficient of penetration by proceeding from the pore size distribution and reached a reasonably good agreement with experiment for three samples of reactor graphite differing in density and pore structure (see table 36). By measuring the penetrability one can estimate the channel porosity, pore radii, coefficient of tortuosity of pore channels, and other characterisitcs. Comparing these parameters with those established by other methods, one will be able to gain a deeper insight into the role of the pore structure of carbons.

Table 36
Coefficient of penetration of reactor graphite.

Sample	Porosity	K, coefficient of penetration (10^{-3} cm^2 s^{-1})	
		Wiggs' calculation (1958)	Experiment
1	22	220.0	211.0
2	19	13.3	11.0
3	12.5	5.0	4.6

12.3.1.4. Chemical reactivity. Carbon's place in the periodic table determines its chemical properties. Having four electrons in the outermost shell, carbon relatively readily reacts with many metals and metalloids at an elevated temperature, whereas under normal conditions it remains highly inert, even to strong oxidizers.

A chemical peculiarity of graphite, associated with its crystal structure, is its ability to form layered compounds by means of intercalation with halogens, alkali metals, acids, and salts (see sections 10.1–10.3). As has already been mentioned in ch. 10, the intercalant atoms or molecules are inserted between the planes of a graphite lattice, which is accompanied by a decrease of the free energy and an increase in the size of crystallites along the c-axis.

The application of graphite for cladding and for manufacturing the replaceable components of nuclear reactors determines the class of materials which are allowed to be in contact with graphite. These include nuclear fuels (distributed over the graphite matrix or dissolved in a heat carrier), coolants (both gaseous and liquid), protective covering (ceramic or metallic) and finally, decay products formed as a result of irradiation of the above-mentioned substances with a neutron flux (Delle et al. 1982). The temperature range in which a reactor may operate depends on its design and varies rather widely. One should always bear in mind that in graphite regions of increased temperatures may appear, both microscopic in size (due to the energy released in collisions between neutrons and atoms) and very extensive (arising from a nonuniform distribution of impurities, the field gradient, etc.). The most common coolants used in nuclear reactors are helium, nitrogen, carbon dioxide and air. Helium does not react with graphite even under neutron bombardment in a reactor. At high temperatures (about 2000°C) there may only take place the transfer of crumbling graphitic particles by the helium flow. This erosion comes to an end, however, as soon as particles whose cohesion with the main body of graphite is insufficient are removed. Various admixtures, for instance oxygen, in the helium flow radically change the situation and may give rise to corrosion of graphite (Campion 1977). At room temperature graphite practically does not react with oxygen and carbon dioxide, but at temperatures above 400°C these gases start reacting with graphite, the temperature of the onset of a reaction depending on the perfection of the graphite's crystal structure, the reaction rate increasing with temperature. With a higher order in the crystal structure, the temperature at which a reaction is initiated becomes higher. At temperatures below 500°C graphite reacts with CO_2 so slowly that, as Ubbelohde and Lewis (1960) showed, the loss of graphite constitutes about 0.1% in 20 years. Since graphite has a well-developed pore structure, its reactions with gases occur not only on the outer surface, but also inside pores (on their inner surfaces). As the HTT is raised up to 2400°C, the oxidation rate for graphite decreases. However, above that temperature it may again increase, which as Lukin and Nagorny (1964) believe, is due to the reopening of some closed pores and the resulting increase in the total area of the inner surface of the pores. The reactivity of reactor graphite is strongly affected by impurities, which may serve as catalysts for oxidation; iron, vanadium and sodium being particularly important in this respect. Of special interest is the oxidation of graphite at relatively low temperatures and a loss of weight of several percent (Delle et al. 1982), because these conditions are similar to those obtained in a nuclear reactor. Irradiation of a graphite gas system with neutrons or γ-quanta gives rise to both defects in the graphite structure (see below) and ionization of the gas molecules. These factors may significantly change the behaviour of graphite, shifting, in particular, the onset of an oxidation reaction towards lower temperatures.

Graphite practically fails to react with molecular nitrogen, but reacts rather well with atomic nitrogen, leading to formation of cyanogen, C_2N_2, the reaction equilibrium constant decreasing with increasing temperature. If hydrogen is added to the reagents at 800°C, the reaction product will be hydrocyanic acid HCN (see Nightingale 1962). Under irradiation, the oxides of nitrogen may be formed.

The main product of a reaction between graphite and hydrogen at temperatures from 300 to 1000°C is methane. The reaction can be much accelerated by a catalyst, e.g., iron (Breisacher and Marx 1963). The equilibrium of a graphite–hydrogen system is temperature dependent. As the temperature rises, the amount of methane thus formed decreases, becoming negligible at about 1000°C. Carbon may therefore be transported from the cooler region to the more heated regions where it may be precipitated. In reactions with carbon dioxide the direction of the transport is reversed, i.e. carbon is transported towards the cooler regions (Prince 1979). The most active sites in a graphite lattice are the first to be oxidized. Should these sites be occupied, e.g., due to adsorption on them of atoms of certain elements, they would not participate in the reaction any further. Various phosphorus-containing compounds have been suggested as dopants that increase the resistance of graphite to oxidation. Being decomposed on heating, such substances leave the phosphorus atoms adsorbed on the active sites of the lattice, thereby lowering the oxidation rate several times. Such doping is usually achieved by impregnating graphite articles with a phosphorus-containing substance, followed by their rapid firing at 800–1200°C in an inert atmosphere. Both organic and inorganic phosphorus-containing substances, e.g., phosphoric acid, trimethyl phosphate, tricresyl phosphate, are used as dopants. To increase the resistance of graphite to oxidation, one can use not only the individual phosphorus-containing substances, but also the mixtures of these substances which form complex compounds of the type of phosphate glasses. A phosphate glass film protects graphite rather well in the 500–950°C range. Good results have also been obtained by using solutions of ammonium phosphate, zinc chloride, and boric acid. Repeated treatment with these solutions very drastically (more than hundred-fold) decreases the oxidation rate of graphite at high (up to 950°C) temperatures.

Recently, considerable interest has been shown in nuclear reactors capable of operating at high temperatures, which are more economical. In some of these reactors, molten metals and salts, circulating directly in graphite channels, serve as heat carriers, the nuclear fuel being sometimes dissolved in them. Graphites of low penetrability are then used, which reduces the absorption of the liquid heat carrier and also secures the isolation of fission products.

Tests of graphite in a medium of molten metals, such as bismuth, tin, and others, have revealed no corrosion. It has been found, however, that if a graphite–bismuth system is cyclically heated from the temperature at which the metal is in the solid state up to about 550°C graphite will crack, which may be due to the dilatation of bismuth in graphite pores as it is congealed.

Tin, which may be used as a heat carrier, because it dissolves uranium, does not react with graphite at temperatures of 850–1000°C. Aluminium reacts with graphite to form aluminum carbide, Al_4C_3, at temperatures of 1700–1800°C. Up to 1200°C, aluminum does not dissolve graphite. Magnesium can also be in contact with graphite, without dissolving it, up to 1100°C. Both aluminum and magnesium may, however, corrode other reactor materials (see Nightingale 1962).

Halogens, acids, and alkali metals are all active intercalants in forming graphite intercalation compounds. Among the alkali metals, sodium is the slowest to react with graphite. A reaction between them is greatly accelerated in the presence of various impurities, including sodium oxide and peroxide.

With uranium, graphite forms carbides at temperatures above 1150°C. Up to 1300°C UC is formed, while at 1400°C, UC_2 is produced. Uranium is capable of diffusing through graphite, the pore-wall migration significantly exceeding the volume diffusion (Vjatkin et al. 1967). The transport of uranium through graphite by a migration mechanism is especially great for porous graphites.

Alloys rather than pure metals have more often been used as construction materials for nuclear piles. With the exception of nickel–copper alloys, nickel- and iron-based alloys are readily carbonized at 800–900°C. At the same time, the nickel–copper alloys may serve as coating for graphite up to 1000°C, because the presence of copper substantially reduces the carbonization rate.

As a heat carrier capable of dissolving nuclear fuels, it is also possible to use melts of some salts, sometimes with admixtures of metals. Lastly, it is worth noting that the presence of graphite in a nuclear reactor system may influence the corrosion of reactor components made of other materials, in particular metals, when these materials are connected by a conducting medium. When used in combination with graphite, it is titanium which shows the greatest resistance to corrosion.

12.3.2. Methods of reactor graphite production

12.3.2.1. Heat treatment and graphitization. To obtain reactor graphite, petroleum coke is mostly used as a filler, and coal-tar pitch as a cohesive. Natural graphite and soot are used as additives to the filler. Instead of tar pitch as cohesive or impregnator, it is also possible to use some synthetic resins, e.g., furan or phenolic resins. The production of reactor graphite consists of the following main technological operations which determine its structure and, ultimately, also its properties (Vjatkin et al. 1967):

(1) preliminary crushing of coke;
(2) coke firing;
(3) coke grinding and fractionating;
(4) preparation of the cohesive agent;
(5) preparation of a carbon mass;
(6) shaping the unburnt stock;
(7) firing the stock;
(8) graphitization of the burnt stock; and
(9) mechanical working of the graphitized material to give it the required size and shape.

Coke is crushed to pieces of 30–40 mm in size, and is then burnt in special furnaces at a temperature of 1300°C to give it heat resistance, reduce the amount of volatile substances present in it, raise its density as well as its electrical and thermal conductivity. After firing, coke is ground to the required size of its particles. Coke powders are batched and then mixed with sand at temperatures of 90–130°C. Upon mixing the mass is uniformly cooled to a temperature of pressing (80–100°C). The cooled material is either extruded or moulded.

To carbonize the cohesive and to cement the individual grains into a monolithic material, the stock is fired in multi-chamber furnaces for 3–5 weeks. Graphitization is the final stage of heat treatment at which the carbon material, heated to 3000°C in a furnace, changes to graphite. Depending on the particular properties required from the reactor material to be produced, some operations of the technological process outlined above may be somewhat modified. Thus, in order to obtain a denser material, the fired carbon stock is impregnated, once or several times, with tar pitch, each impregnation followed by firing, and the whole process ending in graphitization. On the other hand, in order to obtain especially pure reactor graphites, graphitization is carried out simultaneously with gas purification. It is noteworthy that the unfired coke has a disordered structure, firing bringing about a number of complicated chemical and structural changes in the material. Hydrogen, sulphur, oxygen, and nitrogen, chemically bound to carbon, are all removed from it. After firing, the coke structure becomes more ordered. With the HTT in the 2300–3000°C, range, the material undergoes changes more radical than those caused by heating to only 1300°C. At those higher temperatures, the bound hydrogen, sulphur, oxygen and nitrogen are almost completely removed and the graphitic crystal structure evolves.

The hydrogen-to-carbon abundance ratio in tar pitch suggests that it should consist of highly aromatized compounds and, therefore, can be readily graphitized. This makes pitch the most suitable binding material for manufacturing a reactor graphite. Let us dwell at greater length on the main stage in the overall technological process of producing reactor graphites – graphitization. In this process, heating a specimen to a temperature at which it is fired does not give rise to any structural changes in the material, its increased size being entirely due to thermal expansion. However, starting with the firing temperature and up to 1600–1700°C, there occurs a rearrangement in the structure of the specimen: it becomes denser and dwindles in size. Beginning from 1600°C and up to 2100°C (the precrystallization stage), the azimuthal ordering of basal planes takes place, resulting in some reduction of the interlayer distance. As a result, the size of the specimen further decreases. Above 2100°C, crystallization is accompanied by thermal expansion, characterized by the thermal expansion coefficient. When the material is heated relatively slowly, the pattern shrinkage/expansion is regular, whereas at a very rapid heating there will be neither a rearrangement of structure, nor a disruption of the transverse bond, so no shrinkage of the material is observed. In such a case, a well-developed system of closed pores is retained and the physical properties of the material change (Fialkov 1965). The cooling rate also affects, to a certain extent, the character of the change in the structure and properties of the material being graphitized. To attain a degree of graphitization approaching the limit, an exposure of several minutes at a maximum temperature is necessary (Fialkov 1965). Graphitization is terminated at temperatures between 2400 and 2800°C, depending on the required characteristics of the material. The quality of graphite thus obtained is evaluated from its density, specific electrical resistivity, impurity content, and X-ray diffraction characteristics.

12.3.2.2. Elimination of impurities. As graphite networks grow in the course of graphitization, the carbon atoms at the border usually have free bonds, and are therefore capable of linking with atoms of various elements (hydrogen, oxygen, chlorine, etc.). Removal of these elements attached to the peripheral carbon atoms leads to a further growth of the graphitic network. However, bonds formed by unsaturated carbon atoms with certain elements may prove so strong that their disruption and the removal of these elements, such as boron and nitrogen, are capable of replacing the carbon atoms in sites of a crystal lattice forming a substitutional solid solution. The incorporation of the atoms of silicon and germanium will lead to a significant deformation of a graphite lattice.

Metals may be present in graphite in the form of oxides and will reduce on heating by reacting with carbon. All these changes depend on temperature, thermal stability of intermediate products, and volatility of the metal in question and its carbide.

All impurities common to graphite may be divided, by the character of their bonding, into the following groups (Avdeenko and Bagrov 1964):

(1) O_2, H_2, Cl_2 and other similar substances form strong covalent bonds with the carbon atoms;
(2) B and N_2 may replace the carbon atoms at lattice sites;
(3) Fe, Co, Ni, Mn, Cu (and also Li, K, Na) are most likely present, at high temperatures, as substitutional solutions;
(4) Ti, Zr, Nb, Ta, U at high temperatures seem to be present in the form of their carbides;
(5) Free Mg, Sn, Zn, Pb atoms fail to form strong bonds with carbon;
(6) Chlorides of the transition and rare-earth metals of higher valency and chlorides of the elements of the third group of the periodic table appear as intercalants, forming GIC.

There are several techniques for removing impurities from graphite: thermal purification, thermal purification under vacuum, gas-thermal purification, etc. Since the evaporation rate of graphite is much lower than that of many impurities present in it, the vacuum heating of graphite above 2000°C leads, given a sufficient time of exposure, to the elimination of most of the impurities by evaporation.

The impurities listed in group (1), and also N_2, can be removed by vacuum heating to a high temperature. The impurities of group (3) are removed by evaporation in vacuum at high temperature and also by converting them into volatile compounds (chlorides) which are then swept away by a gas flow.

For the group-(4) impurities and boron, which form stable relatively non-volatile carbides, gas purification, i.e., conversion of these impurities into their volatile compounds (chlorides), is the most effective method of elimination. The impurities of group (5) are on the whole readily removed by heating. GIC formed by intercalation with impurities of group (6) can only be stable at relatively low temperatures, which suggests the method of removal.

Reactions of chlorination and fluorination offer the most effective method of removing impurities from graphite by producing volatile compounds whose boiling temperature is lower than that of the corresponding metals or of their carbides. Graphite itself does not react with either chlorine or fluorine. At the same time, chlorides and fluorides of most elements are highly stable compounds. Formation of chlorides, and especially fluorides, is accompanied by a substantial reduction of the free energy, resulting in a great shift of the system's equilibrium towards them. The compounds thus formed practically do not dissociate at the graphitization temperatures. For the metals which form stable carbides, chlorination is more advantageous than vacuum evaporation.

The kinetics of the removal of impurities from graphite is rather complicated. According to Avdeenko and Bagrov (1964), it proceeds as follows:

(1) If the impurities form a substitutional solid solution, then their atoms diffuse through a graphite lattice towards the pore surface;

(2) The impurity atoms are evaporated in the pores or react, on the pore surface, with chemically active substances (such as chlorine or fluorine) to form volatile products;

(3) The reaction products (e.g., chlorides) diffuse through the pores towards the outer surface of a graphite grain and then pass into a gas phase enveloping the grain.

As the impurities are evaporated in high vacuum, they move through the pores owing to the pressure difference. The first and third stage can, as already mentioned, be attributed to diffusion, whereas the second can be properly attributed to a chemical change. The total rate of the process is determined by the rate of its longest stage. At relatively low temperatures, the rate of the removal of impurities is limited by their diffusion in the graphite lattice and by chemical transformations. Above 2000°C, the transport of material through the pores will be the limiting factor.

Table 37

Impurity content (in 10^{-4}%) of a reactor graphite as a function of graphitization temperature.

Impurity	Graphitization temperature (°C)		
	2500	2800–3000	2800–3000 combined with gas purification
Silicon	94	46	1
Iron	310	10	2
Vanadium	30	25	0.2
Titanium	34	11	1
Aluminum	40	2.5	0.1
Calcium	320	1.47	1.4
Boron	0.5	0.4	0.06
Sulphur	175	19	10

Gas-thermal purification normally consist in supplying chlorine at a temperature of 1000°C, which is replaced by freon at 2000°C, and a subsequent blowing of nitrogen or an inert gas through the furnace as it is cooled. The last mentioned operation is necessary to prevent the reverse diffusion of the impurities and to remove the remnants of chlorine from the pores. Nitrogen has a relatively large neutron-capture cross section (0.26 mbarn). It is therefore advisable to use helium or argon as a blow gas, because nitrogen is chemisorbed by graphite in fairly large amounts and can be removed only with difficulty.

The impurity content in a typical reactor graphite [according to Nightingale (1962)] for various conditions of purification are given in table 37.

12.3.3. Radiation damage

Irradiation greatly affects the properties of the materials used in nuclear reactors. Although graphite is relatively resistant to radiation, the exacting demands placed upon the construction materials and components of the reactors, and also the necessity of securing their reliable operation over a long period of service make it necessary to consider even small variations in its characteristics when designing a reactor (Budke and Theymann 1976). Extensive investigations have elucidated the irradiation-related changes in the various properties of graphite, viz., mechanical (strength, creep resistance, hardness); thermal (thermal conductivity, thermal expansion) (see the reviews by Budke and Theymann 1976, 1977); physical (electrical resistance, size stability, crystal-lattice parameters, Hall coefficient, thermal emf, diamagnetic susceptibility, etc.); chemical (e.g., oxidation resistance), and others (Prince 1979).

The degree of these changes is primarily determined by the conditions in which the material is used, viz., the irradiation temperature, neutron energy distribution, and the magnitude of the integrated flux, which serves as a criterion for the service durability of the material under specified conditions.

12.3.3.1. Change in the crystal-lattice parameters of graphite. Under irradiation, many macroscopic characteristics of graphite (thermal conductivity, coefficient of thermal expansion, geometric size, etc.) are changed because of the deformation of the crystal lattice of a graphite. A reactor graphite irradiated with neutrons generally shows some additional disorder in its crystal-lattice structure. This manifests itself primarily in the fact that the size of a unit cell, on the average, increases along the *c*-axis, while the lattice parameter a_0 decreases in the basal plane. The dimensions of the crystallites, L_c and L_a (more exactly, the dimensions of the corresponding regions of coherent scattering), decrease (Martin and Caisley 1977).

The changes that occur in the crystal lattice of a graphite under irradiation are recorded by X-ray techniques and manifest themselves in the following:

(1) the angular position of X-ray reflections is changed;
(2) the width and shape of X-ray diffraction patterns of graphite are changed; and
(3) the intensity of small-angle scattering changes.

The greatest changes in the diffraction pattern, due to irradiation, are observed at rather low temperatures (30–100°C). Under irradiation with an integrated flux Q of thermal neutrons exceeding 5×10^{20} neutrons/cm^2 at a temperature of 30°C, absolute changes in lattice parameters along the c-axis (Δc_0) may reach 0.6 Å, the relative change in this parameter ($\Delta c_0/c_0$) in reactor graphites being 20–25% (Gain 1962). The rate at which the crystal lattice parameter c_0 changes under irradiation, dc_0/dQ, depends on the degree of disorder in the initial material (the degree of graphitization γ), radiation dose (i.e., the value of the integrated flux Q), neutron spectrum and rate of irradiation. For low-temperature irradiation the parameter c_0 changes more rapidly in graphites with a lesser degree of order (Fitzler and Vohler 1962). For highly ordered graphites, the rates of change of the lattice parameters c_0 and a_0 have opposite sign over the whole temperature range of irradiation: $dc_0/dQ > 0$ and $da_0/dQ < 0$. Absolute values for changes in the parameter a_0 are much lower than those in the parameter c_0 (see table 38).

The value of the ratio $|dc_0/dQ|/|da_0/dQ|$ is of great practical interest, because it determines the sign of the changes in the volume of a unit cell due to irradiation: at $|\Delta c_0/c_0| > 2|\Delta a_0/a_0|$ the volume increases, while at $|\Delta c_0/c_0| < 2|\Delta a_0/a_0|$ it decreases, which may result in effects of opposite sign in the change in the size of the material.

Apart from the main factors affecting the crystal lattice parameters (such as irradiation temperature, neutron, spectrum, value of the integrated flux, and intensity of irradiation), preliminary irradiation has been found to play a role. Thus, in graphite pre-irradiated at 150°C for 3 days and then irradiated at 350°C, the rates of change in lattice parameters (calculated for

Table 38
Rate of the relative change in the crystal lattice parameters of reactor graphites irradiated at 25–175°C (according to Simmons 1957).

Irradiation temperature	Rate of change in the parameters a_0 and c_0 (in %), calculated for $Q = 10^{20}$ neutrons/cm^2		
	$\frac{dc_0}{dQ}$	$\frac{da_0}{dQ}$	$\left\lvert\frac{dc_0}{dQ}\right\rvert / \left\lvert\frac{da_0}{dQ}\right\rvert$
25	1.30	−0.104	12.5
80	0.70	−0.056	12.5
175	0.14	−0.022	6.35

$Q = 10^{20}$ neutrons/cm^2) have been found to increase (Reynolds and Thrower 1964) from 0.19 to 0.28 for dc_0/dQ and from 0.044 to 0.077 for $|da_0/dQ|$.

The effect of X-ray reflection broadening in graphite under irradiation may be interpreted as a change in the size of coherent scattering regions in the lattice, associated with local deformations due to defects, leading to a decrease of the average size of the crystallites (Austin and Harrison 1957).

The sudden decrease in the rate dc_0/dQ as the irradiation temperature rises from 150°C to 230–250°C is caused either by the increased coagulation probability for defects (interstitial atoms) or by the increased annealing rate for neighbouring vacancy-displaced atom pairs in comparison with the coagulation rate for individual defects and the rate of formation of complexes by them. The line broadening induced by irradiation is reduced by half by annealing at 200°C. At higher temperatures, the rate of line narrowing decreases. This means that a significant part of the lattice distortions is already annealed at rather low temperatures, at which lattice deformations due to relatively unstable defects disappear, for a broadened line will only narrow back to its original width after annealing at 2500°C.

There are grounds to believe that the increase in the lattice constant along the *c*-axis caused by irradiation is associated with the penetration of the knocked-out carbon atoms (or their complexes) into the interlayer regions of graphite, similarly to what occurs in GIC synthesis. The lattice compression in the basal plane is probably due to its relaxation by the vacancies (Holl 1954). As the irradiation temperature is raised, the recombination rate for the pairs vacancy–interstitial atom increases. However, the coagulation of interstitials also takes place. Both these processes lead to a decrease of lattice deformation, although, because of the presence of immobile vacancies, the relative portion of lattice deformation in the basal plane (as compared with *c*-axis deformation) increases.

12.3.3.2. Defect formation. Radiation-induced defects in graphite have been investigated by both "direct" and "indirect" methods. The "direct" methods include electron microscopy, elastic scattering of cold neutrons (with a wavelength of ~7 Å and longer), small-angle X-ray scattering, etc., while the "indirect" methods use techniques based on the study of the influence exerted by defects upon the dynamical characteristics of the material, such as the temperature dependences of its thermal and electrical conductivities, elastic moduli.

The properties of graphite are greatly affected not only by the number of radiation-induced defects present in it, but also by their type, size, mobility, degree of coagulation (or aggregation), and, finally, relative position.

The early studies of the defects induced by irradiation of graphite at rather low temperatures revealed that even at 30°C and integrated fluxes of

(0.5–1) × 10^{20} neutrons/cm^2 there arise defects (interstitial atoms), representing complexes of 3–6 atoms (Goggin and Reynolds 1963). With increasing irradiation temperature, the size of the defects and the distance between them increase proportionally in the range from 30 to 600°C. In single-crystal graphite irradiated at 300°C large disks, containing up to 10^6 of interstitials, have been found (Williamson 1962). Rather large dislocation loops (with a diameter of about 1000 Å) can already be observed upon irradiation at 200°C.

At an irradiation temperature of about 650°C, a new type of defects, an aggregation of vacancies, appears in graphite. Detailed studies of defects induced in graphite by neutron irradiation and the effect that boron impurities have on defect formation at such temperatures have been carried out by Beavan (1977), Martin and Caisley (1977), Kelly (1978a, b) and Brocklehurst et al. (1979, 1980). The annealing of defects in graphite, induced by low-temperature irradiation, leads at first to an increase in their size (at temperatures of 500°C), followed by their decrease and complete disappearance. Large defects (with a diameter of about 200 Å) begin to be annealed at 1250°C, gradually decreasing in size. However, even at 1800°C, they are still as large as 100 Å across.

The character of defects is significantly different in regions of the material having different degrees of order in the lattice. As a highly perfect graphitic material is irradiated, there first appear defects in the form of dislocation loops. And as the radiation dose is increased there arise heterogeneous defects (resulting from coagulation of interstitials). The latter are also typical of relatively imperfect graphites practically for any radiation dose.

Using the techniques of neutron scattering (both elastic and inelastic), it has been possible to establish that, in graphite irradiated with an integrated flux of $\sim 10^{20}$ neutrons/cm^2, the relative number of defects constitutes 5.3%, the number of vacancies being roughly equal to that of interstitials (Martin and Caisley 1977). Vacancies which are formed in graphite as a result of irradiation will trap, almost as soon as they have emerged, two electrons, thus remaining immobile up to a temperature of 1000°C. Radiation-induced vacancies may coagulate to form aggregations on dislocations. Then a relatively stable structural distortion will arise, which is characterized by the absence of part of the atomic plane. The removal of such a distortion by means of annealing may only proceed by the self-diffusion of vacancies, requiring an activation energy of about 5 eV.

The study of the range of the carbon atoms knocked out from their lattice sites by the neutron impact and of the energy lost by the knocked-out atoms for stopping has made it possible to conclude (Martin and Henson 1964) that radiation-induced vacancies should be uniformly distributed within relatively small regions of graphite, comparable to the migration range of carbon atoms. Such regions have a diameter of about 15 Å, the maximum number of vacancies in it being six.

On the basis of the experimental data available to them, Iwata and Suzuki (1962) proposed the following model to describe the behaviour of radiation-induced defects in graphite:

(1) Individual interstitials can readily migrate at temperatures as low as that of liquid nitrogen. They are grouped together to form atomic complexes, in which atoms are separated by about 10 Å from one another (the elastically stable state);

(2) At temperatures between 80 and 110 K, atomic complexes give rise to molecules of C_2, and partially, C_3, thus leading to an increase in the electrical resistance of irradiated graphite as it is annealed in this temperature range;

(3) Molecules of C_2 can migrate, forming, at temperatures exceeding 160 K, molecular complexes, in which molecules are held together by forces somewhat weaker than the covalent forces and also by the elastic forces of the bent layers;

(4) The molecular complexes may again fall apart into separate molecules of C_2 at temperatures above 200°C. The release of their energy is associated with the emergence of the molecules to the surface or boundary of a crystallite, as a result of which C_2–C_2 complexes vanish;

(5) When molecular complexes become sufficiently large (about 10 atoms in a group), they may change to covalently linked formations characterized by a high degree of defectiveness (the so-called honeycomb structure).

The disappearance of the large molecular formations, which begins to take place at temperatures above 1000°C, is associated with the diffusion of vacancies.

The concentration of defects in graphite heated between 300 and 1400 K as a

Table 39
Energy parameters of defects in graphite.

Type and symbol of defect	Energy parameters (in eV)
Carbon atoms (C)	Formation energy 2.5 Migration energy 0.016
Atomic aggregations (C–C–)	Stable configuration with a distance of 10 Å between atoms
Molecules of C_2	Potential barrier of formation 0.6 Dissociation energy 7.0 Migration energy 0.05
Molecular formations (C_2–C_2)	Dissociation energy 1.3
Honeycomb structure $(C_2)_n$	Decomposition energy per molecule 7.0

function of radiation dose (up to 10^{22} neutrons/cm^2) has been calculated by Kelly (1976, 1977, 1978a, b), whose results are in good agreement with experiment. The data on some energy parameters of defects in graphite are given in table 39.

12.3.3.3. Change in size stability and porosity. The change of size, along with that of heat conductivity, is the most important effect that neutron bombardment has on graphite. Taking this into account in designing and constructing nuclear piles is of paramount importance.

As most other effects caused by irradiation with neutrons, the change in the size stability of graphite depends on a number of factors pertaining to both the conditions of irradiation (temperature, neutron spectrum, integrated flux) and the properties of the original material (which depend on the stock type, technology of manufacture, graphitization temperature, impregnators, etc.) (Dalessandro 1978). The change in the size of graphite is due to the deformation of its crystal lattice and an increase in its unit cell size. Apart from changes in size of this kind (they are sometimes referred to as the Wigner growth), it is also possible to observe in graphite under irradiation size changes due to loading (radiation-induced creep) (see Matheny et al. 1979, Kennedy and Eatherly 1979, Bullock 1979, Hideto and Jasuachi 1981). Where the change in lattice parameters is great, i.e., in the range of relatively low irradiation temperatures, microsizes change most intensively. Having been irradiated with a flux of $(1\text{–}3) \times 10^{21}$ neutron/cm^2 at $T \sim 30°\mathrm{C}$ (low-temperature irradiation), reactor graphites change their volume by 1 or even 2% (Bullock 1979). When graphite is irradiated at low temperatures, the anisotropy of its properties, associated with its structure, will be best pronounced (Hideto and Jasuachi 1981). For highly structured graphites, produced by extrusion, the change in size due to irradiation is different in magnitude and in sign when considered in different directions (Prince 1979). Thus, in the direction of the extrusion axis graphite will reduce in size (radiation-induced compression), while in directions normal to the extrusion axis the material will expand. The relative rate of the change in size is usually of the order of 0.02% for a flux of 10^{18} neutron/cm^2 (the neutron energy E_n being greater than 1 MeV) (see Holl 1954). The ratio of the rate of change in size along the extrusion axis to that normal to the axis usually ranges from 2.5 to 4.5.

Materials based on natural graphite and characterized by a very high degree of preferred orientation have a considerably higher anisotropy of those rates. One should mention that the change in volume is maximum for graphites with a low degree of graphitization [about 10% upon irradiation with a flux of 3×10^{21} neutrons/cm^2 for $E_n > 0.18$ MeV (see Gain 1962, Cords and Zimmermann 1978)].

An increase in irradiation temperature results in lowering the rate of change in size per unit integrated flux. This is illustrated in table 40, which presents the average rates of change in the relative length $\Delta l/l$ of reactor graphite specimens, as calculated for an integrated flux of 10^{20} neutron/cm^2 at various irradiation temperatures. At sufficiently high integrated fluxes, $Q > (3\text{–}5) \times 10^{21}$ neutron/cm^2, the dependence $\Delta l/l = f(Q)$ becomes linear. Extrapolation of the linear part of the curve to the zero value of Q yields the value of residual deformation, which is about 0.16% in the direction parallel to the extrusion axis and about 0.03% in the direction normal to it (Bullock 1979).

On the whole, the effect of the stock material upon the size stability of graphite at high temperatures of irradiation may be characterized as follows:

(1) the increase in graphitization temperature up to 3140°C leads to a decrease of compression rate in the normal direction upon irradiation of specimens at 650–850°C;

(2) additional impregnations increase the compression rate very little, whereas the introduction of a very poorly graphitizable impregnator significantly raises the compression rate;

(3) the use of raw materials with an increased graphitizability ensures the greater size stability;

(4) introduction of the more finely ground coke into the initial batch will increase the compression rate of graphite;

(5) the density of graphite insignificantly affects the compression of graphite irradiated at a high temperature.

The change in lattice parameters in the direction of the *c*-axis is a likely source of annealable effects arising from high-temperature irradiation. As in the case of low-temperature irradiation, the compression component to be annealed causes a slight increase in size in the normal direction and some compression in the parallel direction. At the same time, the presence of

Table 40
Average rate of change in relative length of reactor graphite specimens at various irradiation temperatures (Vjatkin et al. 1967).

Irradiation temperature (°C)	Rate $d(\Delta l/l)/dQ$ calculated for $Q = 10^{20}$ neutron/cm^2 and $E_n > 1$ MeV (%)
30	1.5
50	0.90
100	0.492
150	0.169
200	0.074
250	0.024
340	0.013

ungraphitized or ungraphitizable carbon predetermines unannealable compression.

The compression of graphite under irradiation may therefore be of two types:

(1) annealable compression related to the elastic deformation of crystalline regions of the material; and

(2) unannealable compression, which takes place in regions of imperfect structure and which leads to some ordering, as well as an increase in density, of these regions.

Since a reactor graphite has a degree of graphitization close to unity, it is the former type of compression that is typical of it. One may expect that the macroscopic changes in the size of a graphite will be determined by the above-mentioned types of compression at various temperatures of irradiation. In the low-temperature range, $\Delta l_c/l_c > 2\Delta l_a/l_a$ and the volume of the graphite will increase, whereas at higher temperatures of irradiation, $\Delta l_c/l_c < 2\Delta l_a/l_a$, i.e., the volume of the graphite decreases.

Accordingly, in conventional materials, which are not characterized by a high degree of graphitization, two processes should be in operation: on the one hand, the volume of a sufficiently ordered material will increase, on the other hand, in relatively disordered regions, compression will occur.

The chemical activity of graphite, as well as the degree of its permeability to gases and liquids, play an important part in utilizing the material in nuclear piles. The above characteristics are, to a significant extent, determined by the porosity of the graphitic structure (Bullock 1979). Upon irradiation the porosity changes, which can be shown by comparing the changes in density and in lattice parameters, as well as by direct methods, e.g., by determining the pore size from adsorption or electron-microscopy data (Delle et al. 1979).

It has been found (Bollman 1961) that in irradiating a reactor graphite with an integrated flux of $\sim 5 \times 10^{21}$ neutron/cm^2 at a temperature of 145°C:

(1) the X-ray density will, on average, decrease by 7 or 8%, while the density measured (by immersion in helium) will even decrease by about 15%

(2) the total volume of the pores, calculated from the difference between the density and the X-ray density of the material, will increase (on average, by 16%);

(3) the relative volume of lattice defects, which is defined as the difference between the theoretical and the X-ray density, will increase from 0.1–0.7% for the initial to 5.7–6.7% for the irradiated material;

(4) the volume of the micropores accessible to the immersion medium will substantially decrease.

The above-mentioned change of the porous structure can, to a significant extent, be annealed at $\sim$550°C. At a high temperature of irradiation (400–500°C), the specific surface of graphite may decrease, which may be due to the denser packing of the relatively disordered microscopic regions.

The decrease of the specific surface and the density measured by immersion in helium, as well as the total reduction of micropore volume, are probably a direct consequence of the expansion of the crystallites into the pores of the material, which leads to an increase in the closed pore volume as the expanding crystallites block the entrances of the pores (Bullock 1979, Delle et al. 1979).

12.3.3.4. Changes in thermal and electronic properties

Heat conductivity. As graphite is irradiated, its heat conductivity drastically changes, whereas its thermal expansion is much less affected and its heat capacity changes very little (Prince 1979).

The main factors that affect the character of these changes are the irradiation temperature, the magnitude of the integrated flux, and the neutron spectrum.

The most salient features of the change in the heat conductivity of graphite under irradiation are the following (Prince 1979):

(1) the rate of change is maximal in the first irradiation cycle;

(2) for highly perfect graphites, the relative increase in heat resistance reaches a maximum with increasing integrated flux;

(3) the change in heat conductivity is maximal at relatively low temperatures ($T \leqslant 200°C$)

As is known, the heat conductivity of graphite is mainly determined by the energy transfer by lattice vibrations (the Wiedemann–Franz relation for graphite is characterized by a value 100 times as high as the classical value for metals). As a result, heat transport in graphite very strongly depends on the degree of perfection of the graphite lattice. Because graphite has a layered structure, its thermal conductivity is also associated with the propagation of elastic waves along the layers. It may be expected that the greatest change in thermal conductivity should be due to local changes in density and elastic constants. As the results of the heat conductivity measurements at high temperatures suggest (Reynolds and Thrower 1962), the greatest decrease of the thermal conductivity of graphite is caused by vacancies. One may therefore expect that vacancies in the basal planes play the main part in decreasing the heat conductivity also for graphite under irradiation. Single vacancies should then exert the most influence, because they contribute most to the relative change of local density.

As has already been mentioned above, the greatest change in the thermal conductivity of graphite under irradiation is observed in the first cycle. As the integrated flux further increases, the change in thermal conductivity decreases, tending to some constant value. This saturation is reached, for example, at low ($\sim$30°C) temperatures after irradiating graphite with an integrated flux of $(5–6) \times 10^{20}$ neutron/cm^2. At higher integrated fluxes, the value of thermal conductivity remains practically the same for all types of reactor graphites. At irradiation temperatures ranging from 100 to 250°C, one can observe a sudden decrease of the effect of neutron radiation upon the thermal conductivity of graphite. As the temperature is raised to between 460 and 600°C, this effect becomes still less significant. Even after irradiation with a flux of about 10^{20} neutron/cm^2 ($E_n > 0.18$ MeV) the value of thermal conductivity for highly perfect graphites will be about 1 Watt/cm deg, its anisotropy being retained.

The thermal conductivity of irradiated graphites is partly restored by annealing, the effect strongly depending on the annealing temperature. Thus, annealing at 600°C leads to a reduction of the thermal resistance by several percent. The thermal conductivity is almost completely restored by annealing at temperatures of 1200–1400°C for 6 hours (Goggin and Reynolds 1963).

The results of the measurements of the thermal conductivity of graphite irradiated at low temperatures ($T < 100$°C) can be explained on the basis of the two-phase model developed by Williamson (1962). In this model, graphite is treated as a medium consisting of particles with a high degree of crystallinity, which are connected to each other by a thin isotropic layer of carbon. In such a case, the dependence of the thermal resistance $1/K$ on the temperature T has the form

$$\frac{T^3}{K} = AT + BT^2 + C, \tag{260}$$

where A is a quantity inversely proportional to the crystallite size, B is a quantity inversely proportional to the number of the internal defects of a crystallite, and C is a function of the thickness of isotropic regions. These quantities are given by

$$A = \frac{\gamma}{SL}, \qquad B = \frac{\gamma}{S} N_g^{-1}, \qquad C = \frac{C_1 - \gamma}{C_2},$$

where γ is the degree of graphitization, S an orientation parameter, N the defect density, g the Debye integral, C_1 and C_2 are constants.

A theoretical model for calculating the thermal conductivity of reactor graphites irradiated with neutrons has been advanced by Binkele (1978). The radiation-induced change in thermal conductivity may be described by introducing into the Debye heat conductivity equation a quantity for the mean free path of phonons, associated with scattering by radiation-induced defects. The mean free path of phonons l_{ph} for an irradiated graphite can be written as

$$\frac{1}{l_{ph}} = \frac{1}{l_1} + \frac{1}{l_2} + \frac{1}{l_3}, \tag{261}$$

where l_1 is the range associated with phonon–phonon scattering, l_2 is the range associated with scattering by static defects and l_3 is the range associated with scattering by radiation-induced defects. If l_3 is rather small, then the maximum heat conductivity abruptly decreases and begins to shift towards higher temperatures. This seems, in particular, to account for the negative value of dK_{irr}/dT_{meas} found in graphite irradiated at 400°C as the temperature of measurement was raised to 600°C (Thrower 1963).

For those graphites whose crystallites are small in size and whose initial heat conductivity is, therefore, low, the relative change in heat conductivity due to irradiation will be less significant – on the one hand, because of the considerable phonon scattering by grain boundaries in the initial graphitic material, and on the other hand, because defects on the surface of a crystallite will diffuse more readily. The appearance of a positive slope on the curve for $K = f(T_{meas})$ as the annealing temperature is raised to ~900°C in fine-grained graphites suggests that defects should be readily annealed even at such relatively low temperatures as a result of their diffusion towards the boundaries of the crystallites.

Electronic properties. In single-crystal graphite specimens, irradiated at room temperature and below, the electronic properties (electrical resistance, Hall coefficient, thermal emf coefficient) have been found to change rather rapidly even for relatively small integrated fluxes ($\lesssim 10^{18}$ neutron/cm^2).

The electrical resistance of graphite shows the most rapid increase in the first irradiation cycle and, as the integrated flux reaches 10^{20} neutron/cm^2, the resistance will be several times that of the initial material. In the first irradiation cycle, the change in resistance ρ is described as (Vjatkin et al. 1967)

$$\frac{1}{\rho} = \frac{A}{Q} + B, \tag{262}$$

where Q is the integrated flux, and A and B are constants.

The rate of this change is drastically reduced for $Q \gtrsim 2 \times 10^{20}$ neutron/cm^2, when, at irradiation temperatures between 100 and 200°C, the resistance is saturated.

Since the initial electrical resistance of graphite is determined by the degree of its graphitization and by its structure, the relative change in resistance due to irradiation strongly depends on the perfection of the initial material. It is of interest to note that, when graphites are irradiated at temperatures up to 200°C, the anisotropy of the electrical resistance, unlike that of the heat conductivity, will be retained, even for fluxes as high as 30×10^{20} neutron/cm^2 (Gain 1962).

The radiation-induced defects that are responsible for the increase in electrical resistance can only be annealed at a temperature of ~2400°C, although the resistance may be partially restored at lower annealing temperatures.

The Hall coefficient of an un-irradiated reactor graphite is negative, i.e., $n_e\mu_e > n_h\mu_h$. Upon irradiation there appear defects in graphite, which are capable of capturing electrons. Because of this, the rate of charge transport by electrons decreases, the value of the Hall coefficient passing through zero and becoming positive. At higher integrated fluxes ($Q \sim 10^{21}$ neutrons/cm^2), $n_e\mu_e \ll n_h\mu_h$ and the Hall coefficient decreases with increasing Q inversely proportionally to the number of holes.

The thermal emf of graphite (relative to Pb) under irradiation changes from a negative value to a positive one even for rather low integrated fluxes (Martin and Henson 1964), while with increasing Q the thermal emf decreases, tending to zero. The effect of radiation on this characteristic increases proportionally to the degree of order present in the initial material.

As an irradiated graphite is annealed, one can at first observe a sudden increase in the thermal emf and then a smooth decrease. But even at an annealing temperature as high as ~2800°C, no complete restoration of the initial thermal emf has been observed.

The diamagnetic susceptibility of graphite changes, upon irradiation, very rapidly: at $Q \sim 2 \times 10^{20}$ neutron/cm^2 it decreases by a factor of 4, while at $Q \sim 2 \times 10^{21}$ neutron/cm^2 it already decreases by a factor of 20 (Vjatkin et al. 1967).

Since the diamagnetic susceptibility of graphite depends only on the concentration of the carriers and their energy spectrum, but is independent of scattering processes, the change in this characteristic due to irradiation occurs in practically the same manner for reactor graphites of various degree of perfection (Iwata and Suzuki 1962).

The decrease in diamagnetic susceptibility under irradiation results from the electron capture by the defects. The rate of formation of such traps for Q up to 10^{18} neutron/cm^2 is about 1.4×10^{-5} traps per atom, as calculated for an integrated flux $Q = 10^{17}$ neutron/cm^2.

Energy accumulated by irradiated graphite (*Wigner energy*). As a fast neutron is slowed down in graphite, it knocks out about 2×10^4 atoms from their equilibrium position in the graphite lattice. Some of these atoms are returned to equilibrium by thermal vibrations of the lattice. However, most of the knocked-out atoms are incapable of surmounting the potential barrier to escape from the metastable state, and they are left in the lattice as interstitials, retaining some excess of energy over the equilibrium atoms, which leads to an increase in the internal energy of the irradiated graphite.

The amount of the energy thus accumulated is related to the accumulation of defects which cause a local deformation of the crystal lattice. As the temperature increases, this energy, the so-called Wigner energy, is released as heat.

The energy accumulated in graphite can be characterized by two parameters: the total accumulated energy W, which can be measured by heating specimen to a certain high temperature and the rate of release of the accumulated energy when the specimen is heated, $\mathrm{d}W/\mathrm{d}T_{\mathrm{meas}}$. The amount of accumulated energy W depends on the conditions of irradiation, i.e., the temperature and the value of the integrated flux, Q, and, to a lesser extent, on the initial graphitic structure. When graphite has been irradiated at $T \lesssim 120$ K its total accumulated energy is proportional to Q, with the proportionality coefficient about 36 cal/g, as calculated for $Q = 10^{20}$ neutron/cm^2.

For $Q \sim 3 \times 10^{21}$ neutron/cm^2, the energy accumulated in graphite may reach about 630 cal/g, its saturation attained at about 700 cal/g. The dependence of accumulated energy W on integrated flux Q may be described, in a wide range of variation of Q, by the following expression:

$$W = W_{\infty}(1 - e^{-kQ}), \tag{263}$$

where W_{∞} is the limiting accumulated energy, and k is a constant.

The values for the constants W_{∞} and k for some temperatures of irradiation are listed in table 41 (the data are from Vjatkin et al. 1967).

One may assume that the constant k is proportional to the average lifetime τ of a defect. As can be seen from table 41, the average lifetime in this case, decreases as the temperature rises to 250°C and then it again increases, which can be explained by the emergence of a new type of defects (e.g., complex defects). The rate of increase of the total accumulated energy W under irradiation at $T \sim 50$°C amounts to $\sim$42 cal/g, which is calculated for an integrated flux $Q = 10^{20}$ neutron/cm^2. As the irradiation temperature is raised to 100°C, the value of W decreases to $\sim$20 cal/g.

The energy accumulated by graphite is released in the course of annealing in a wide temperature range $T_{\text{ann}} > 2000$°C. The amount of the released energy depends on the duration and temperature of annealing, and is determined by two parameters:

(1) $(dW/dt)|_T$, the isothermal release of energy with time; and

(2) dW/dT, the rate of energy release over a constant time interval (differential enthalpy).

In some temperature intervals, the differential enthalpy dW/dT is found to increase with temperature, thereby making it possible for the graphite temperature to rise uncontrollably, endangering the reactor itself (Dalessandro 1978, Delle et al. 1982). Therefore, from a practical viewpoint, for low

Table 41

Values for W_{∞} and k at various irradiation temperatures. After Vjatkin et al. (1967).

Irradiation temperature (°C)	W_{∞} (cal/g)	k (10^{20} neutron/cm^2)
30	680	81
150	600	37
200	375	26
250	200	23
300	50	60

irradiation temperatures, the rate dW/dt is an even more important characteristic than the total accumulated energy itself.

For graphites irradiated at a low temperature ($\sim$77 K) the dependence $dW/dT = f(T)$ shows several maxima, which occur at 210–220, 300–310, 380–410, and 440–450 K (Holl 1954).

The amount of accumulated energy may, in a first approximation, be found on the basis of the elastic model (Simmons 1964) from the value of lattice deformation of the irradiated graphite $W = c_{11}\{u^2\}_{ms}$, where c_{11} is the elastic modulus of the graphite lattice (see table 2) and $\{u^2\}_{ms}$ is the mean-square displacement, determined from an analysis of the shape of the (110) line.

Agranovich and Semyonov (1964) have calculated the Wigner energy and the increase in the lattice constant along the c-axis in irradiated graphite on the basis of the elastic model, assuming that the interaction of the displaced atom with the carbon atoms in the basal planes can be described by a potential of the Lennard-Jones type:

$$\varphi(r) = \frac{A}{a_0}\left(\frac{1}{2}\frac{p_1^6}{r^{1/2}} - \frac{1}{r^6}\right). \tag{264}$$

where $A = 24.3 \times 10^{-60}\,\mathrm{erg/cm^6}$, r is the distance between the atoms in units of a_0, and $p_1 = 2.7$.

On the basis of this model, they have obtained the value of the c-axis compressibility for perfect graphite, $\kappa_{cal} = 2.59 \times 10^{-12}\,\mathrm{cm^2/dyne}$, which is in good agreement with the experimental value $\kappa_{exp} = 2.97 \times 10^{-12}\,\mathrm{cm^2/dyne}$, as well as the value for the binding energy between the graphitic planes, $E_{b.cal.} = 243\,\mathrm{erg/cm^2}$, also in reasonable agreement with experiment, $E_{b.exp.} = 260\,\mathrm{erg/cm^2}$.

Agranovich and Semyonov also succeeded in calculating the change in the lattice parameter along the c-axis, assuming that defects do not interact with each other. The parameter c_0 increases from 6.71 Å at $f = 10^{-3}$ to 7.16 Å for $f = 5 \times 10^{-2}$ (f being the number of radiation induced defects per carbon atom).

For $f = 2 \times 10^{-2}$, the calculated value for the accumulated energy W reaches 720 cal/g. A comparison between the accumulated energy and the change in the lattice parameter c_0 yields a smoothed dependence, as compared with the experimental curve. The discrepancy may be due to leaving the interaction between defects out of account and also, possibly, to the structural peculiarities and the properties of defects in real reactor graphites.

Proceedings of conferences on GIC

[1] Proceedings of the Franco–American Conference on Intercalation Compounds of Graphite, La Napoule, France, 23–27 May 1977, eds F. L. Vogel and A. Hérold (Elsevier Sequoia S.A., Lausanne, 1978), Materials Science and Engineering **31**.

[2] Proceedings of the International Conference on Layered Materials and Intercalates, Nijmegen, The Netherlands, 28–31 August 1979, eds C. F. Van Bruggen, C. Haas and H.W. Mylon (North-Holland, Amsterdam, 1980) Physica B + C **99.**

[3] Proceedings of the Second International Conference on Intercalation Compounds of Graphite, Provincetown, MA, USA, 19–23 May 1980, ed. F. L. Vogel (Elsevier Sequoia S.A., Lausanne, 1980), Synthetic Metals, **3**.

[4] Proceedings of the Yamada Conference IV on Physics and Chemistry of Layered Materials, Sendai, Japan, 8–10 September 1980, eds Y. Nisina, S. Tanuma and H.W. Myron (North-Holland, Amsterdam, 1981), Physica B + C **105** (1–3).

[5] Proceedings of the International Conference on the Physics of Intercalation Compounds, Trieste, Italy, 1981, eds L. Pietronero and E. Tosatti (Springer, Berlin, 1981).

[6] Proceedings of the Symposium on Intercalated Graphite, Boston, MA. USA, 1982, eds M.S. Dresselhaus, G. Dresselhaus, J.E. Fischer and M.J. Moran (North-Holland, Amsterdam, 1983), Materials Research Society Symposia Proceedings **20**.

[7] Proceedings of the Third International Conference on Intercalation Compounds of Graphite, Pont-a-Mousson, France, 23–27 May 1983, eds A. Hérold, D. Guérard and P. Lagrange (Elsevier Sequoia S.A., Lausanne, 1983), Synthetic Metals **7/8**.

REFERENCES

Abrikosov, A.A., 1969, Zh. Eksp. & Teor. Fiz. **56**, 1391 [Sov.Phys.-JETP **29**(4), 746].

Abrikosov, A.A., 1973, Zh. Eksp. & Teor. Fiz. **65**, 1508 [1974, Sov.Phys.-JETP **38**(4), 750].

Adams, E.N., 1953, Phys. Rev. **89**, 683.

Adkins, C.Y., S.M. Freaker and E.M. Hamilton, 1970, Philos. Mag. **2**(175), 183–188.

Avdeenko, M.A., and G.N. Bagrov, 1964, in: Konstrukzionnye uglegrafitovye materialy, Vol. 1 (Metallurgiya, Moscow) p. 34.

Agranovich, V.M., and L.P. Semyonov, 1964, in: Proc. Int. Conf. on Properties of Reactor Materials and Effects of Radiation Damage, Berkeley, CA, 1962 (Butterworths, London) p. 257.

Ahmadieh, A.A., and H.A. Rafizadeh, 1973, Phys. Rev. B **7**(10), 4527.

Al-Jishi, R., and G. Dresselhaus, 1982a, Phys. Rev. B **26**, 4514.

Al-Jishi, R., and G. Dresselhaus, 1982b, Phys. Rev. B **26**, 4523.

Al-Jishi, R., and G. Dresselhaus, 1982c, Phys. Rev. B **26**, 5250.

Al-Jishi, R., and G. Dresselhaus, 1982d, in: Intercalated Graphite, eds M.S. Dresselhaus, G. Dresselhaus, J.E. Fischer and M.J. Moran, Mater. Res. Soc. Symp. Proc. **20** (North-Holland, Amsterdam) p. 301.

Alexander, M.G., D.P. Goshorn, D.G. Onn, D. Guérard, P. Lagrange and M.E. Makrini, 1980a, Synth. Met. **2**, 203.

Alexander, M.G., D.P. Goshorn and D.G. Onn, 1980b, Phys. Rev. B **22**, 4535.

Alexander, M.G., D.P. Goshorn, D. Guérard, P. Lagrange, M.E. Makrini and D.G. Onn, 1981, Solid State Commun. **38**, 103.

Alldredge, G.P., E. de Rouffignac, B. Firey and F.W. de Wette, 1984, Phys. Rev. B **29**(6), 3712.

Altshuler, B.L., 1978, Zh. Eksp. & Teor. Fiz. **75**(10), 1330 [Sov.Phys.-JETP **48**(4), 670].

Altshuler, B.L., and A.G. Aronov, 1979, Zh. Eksp. & Teor. Fiz. **77**(11), 2028 [Sov.Phys.-JETP **50**(5), 968].

Altshuler, B.L., D.E. Khmel'nitskii, A.I. Larkin and P.A. Lee, 1980, Phys. Rev. B **22**(11), 5142.

Altshuler, B.L., A.G. Aronov, A.I. Larkin and D.E. Khmel'nitskii, 1981, Zh. Eksp. & Teor. Fiz. **81**(8), 768 [Sov.Phys.-JETP **54**(2), 411].

Altshuler, B.L., A.G. Aronov and D.E. Khmel'nitskii, 1982, Pis'ma Zh. Eksp. & Teor. Fiz. **36**(5), 157 [JETP Lett. **36**(5), 195].

Anderson, J.R., W.J. O'Sullivan, J.E. Shirber and D.E. Soule, 1967, Phys. Rev. **164**, 1038.

Arkhipov, R.G., V.V. Kechin, A.I. Likhter and Yu.A. Pospelov, 1963, Zh. Eksp. & Teor. Fiz. **44**, 1964 [Sov.Phys.-JETP **17**, 1321].

Armand, M., and P. Touzain, 1977, Mater. Sci. & Eng. **31**(14), 319.

Arnold, G.M., 1967, Carbon **5**, 33.

Arnold, G.M., and S. Mrozowski, 1968, Carbon **6**(3), 243.
Aronson, S., F.L. Salzano and D. Bellafiore, 1968, J. Chem. Phys. **49**, 434.
Aronson, S., C. Fishberg and G. Frankl, 1971, Carbon **9**, 715.
Aust, R.B., and H.G. Drickamer, 1963, Science **140**, 817.
Austin, A.E., and R.J. Harrison, 1957, in: Proc. 3rd Conf. on Carbon, Buffalo, NY, 1957 (Pergamon Press, New York, 1959) p. 585.
Awaji, H., and S. Sato, 1979, Tanso (Japanese for Carbon) **96**, 2–7.
Ayache, C., E. Bonjour, R. Laguier and J.E. Fischer, 1980, Physica B **99**, 547.
Bacon, G.E., 1951, Acta Crystallogr. **4**, 558.
Baiker, A., V. Hablegger, V.K. Sharma and W. Richarz, 1981, Carbon **19**(4) 327.
Bailey, A.C., and B. Yates, 1970, J. Appl. Phys. **41**, 5088.
Baker, C., and B.T. Kelly, 1964, Philos. Mag. **9**(102), 927.
Bartlett, N., E.M. McCarron, B.W. McQuillan and T.E. Thompson, 1980, Synth. Met. **1**(3), 221.
Baskin, J., and J. Meyer, 1955, Phys. Rev. **100**, 544.
Bassani, F., and P. Parravicini, 1967, Nuovo Cimento B **50**, 95.
Basset, J., 1941, C.R. Hebd. Séances Acad. Sci. **213**, 829.
Basu, S., C. Zeller, P. Flanders, C.D. Fuerst, W.D. Johnson and J.E. Fischer, 1979, Mater. Sci. & Eng. **38**, 275.
Batallan, F., J. Bok, I. Rosenman and J. Melin, 1978, Phys. Rev. Lett. **41**(5), 330.
Batallan, F., I. Rosenmann, C. Simon, G. Furdin and H. Fuzellier, 1980, Physica B **99**, 412.
Beavan, L.A., 1977, Trans. Nucl. Soc. **26**(1), 176–177.
Beavan, L.A., and G.B. Engle, 1981, High Temp.-High Pressure **13**(2), 139–144.
Beer, A.C., 1963, Galvanomagnetic Effects in Semiconductors, Solid State Physics, Supplement 4 (Academic Press, New York).
Beguin, F., and R. Setton, 1976, J. Chem. Soc. Chem. Commun. 611.
Bellodi, G., A. Borghesi, G. Guizetti, L. Nosenzo, E. Reguzzoni and G. Samoggia, 1975, Phys. Rev. B **12**, 5951.
Belousov, V.M., N.I. Kovtyuchova, V.A. Michailovskii, Yu.N. Novikov and M.E. Vol'pin, 1980, Dokl. Akad. Nauk SSSR Ser. Khim. **253**(2), 346.
Bender, A.S., and D.A. Young, 1971, Phys. Status Solidi b **47**, K95.
Bender, A.S., and D.A. Young, 1972, J. Phys. C **5**, 2163.
Beneslavskii, S.D., 1968, Fiz. Tverd. Tela **10**, 3162 [Sov.Phys.-Solid State **10**(10), 215].
Beneslavskii, S.D., and E. Entralgo, 1976, Fiz. Tverd. Tela **18**(10), 2915 [Sov.Phys.-Solid State **18**(10), 1701].
Beneslavskii, S.D., A.I. Buzdin and V.V. Tugushev, 1982, Fiz. Tverd. Tela **24**(4), 1202 [Sov.Phys.-Solid State **24**(4), 680].
Bergenlid, U., R.W. Hill, E.J. Webb and J. Wilks, 1954, Philos. Mag. **45**, 851.
Berker, A.N., N. Kambe, G. Dresselhaus and M.S. Dresselhaus, 1980, Phys. Rev. Lett. **45**, 1452.
Berlincourt, T.G., and M.C. Steele, 1955, Phys. Rev. **98**, 956.
Berman, R., and F.Z. Simon, 1955, Elektrochem. **59**(5), 333.
Bernal, J.D., 1924, Proc. R. Soc. London Ser. A **106**, 749.
Bernstein, H.J., 1962, Trans. Faraday Soc. **58**(480), part 12, 2285.
Besenhard, J.O., and H.P. Fritz, 1978, Z. Naturforsch. B **33**, 737.
Bhagavantam, S., and T. Venkatarayudu, 1951, Theory of Groups and its Application to the Physical Problems (Waltair Andhra University, Hyderabad, India).
Bhattacharya, S., 1959, Indian J. Phys. **33**, 407.
Billaud, D., A. Hérold and F.L. Vogel, 1980, Mater. Sci. & Eng. **45**, 55.
Binkele, L., 1978, J. Non-Equilibrium Thermodyn. **3**(4), 257–266.

Blackman, L.C.F., 1960, Research Science and its Application in Industry (London) **13**(10), 390.
Blackman, L.C.F., G. Saunder and A.R. Ubbelohde, 1961, Proc. Phys. Soc. **73**, 1048.
Blackslee, O.L., D.G. Proctor, E.J. Seldin, G.B. Spence and T. Weng, 1970, J. Appl. Phys. **41**, 3378.
Blinowski, J., and C. Rigaux, 1980, J. Phys. (France) **41**(7), 667.
Blinowski, J., Nguyen Hy Hau, C. Rigaux, J.-P. Vieren, R. Le Toullec, G. Furdin, A. Hérold and J. Melin, 1980, J. Phys. (France) **41**(1), 47.
Boardman, A.D., and D.G. Graham, 1969, J. Phys. C **2**(12), 2320.
Bobrovnikov, Yu.N., and A.S. Kotosonov, 1978, in: Konstrukzionnye Materialy na Osnove Ugleroda (Construction of Materials Based on Carbon Compounds) (Metallurgiya, Moscow) (3), 65.
Boehm, H.P., and H.Z. Hofman, 1955, Anorg. Chem. **278**, 58.
Boeuf, A., A. Freund and R. Caciuffo, 1983, Synth. Met. **8**(3/4), 307.
Bok, J., F. Batallan and I. Rosenmann, 1978, in: Proc. Int. Conf. on Applications of High Magnetic Fields to Semiconductor Physics, ed. J.P. Ryan (Clarendon Press, Oxford) p. 48.
Bollman, W., 1961, in: Proc. Int. Conf. on Properties of Reactor Materials and the Effects of Radiation Damage, Berkeley, CA, 1961 (Butterworths, London, 1962) p. 132.
Bonfiglioli, G., and A. Mojoni, 1964, J. Appl. Phys. **35**(3), 1, 683.
Bowman, J.C., and J.A. Krumhansl, 1958, J. Phys. & Chem. Solids **6**(4), 367.
Brandt, N.B., and S.M. Chudinov, 1972, Low Temp. Phys. **8**(1/2), 339.
Brandt, N.B., and N.I. Ginzburg, 1969, Usp. Fiz. Nauk **96**(1), 95 [Sov. Phys.-Usp. **12**(3), 344].
Brandt, N.B., and M.V. Semyonov, 1975, Zh. Eksp. & Teor. Fiz. **69**(9), 1072 [Sov.Phys.-JETP **42**(3), 546].
Brandt, N.B., and E.A. Svistova, 1970, Low Temp. Phys. **2**, 1.
Brandt, N.B., E.R. Ioon, S.M. Chudinov and G.A. Yakovlev, 1972, Pis'ma Zh. Eksp. & Teor. Fiz. **15**, 204 [JETP Lett. **15**(4), 141].
Brandt, N.B., G.A. Kapustin, V.G. Karavaev, A.S. Kotosonov and E.A. Svistova, 1974, Zh. Eksp. & Teor. Fiz. **67**(9), 1136 [Sov.Phys.-JETP **40**(3), 564].
Brandt, N.B., A.S. Kotosonov, S.V. Kuvshinnikov and M.V. Semyonov, 1979, Pis'ma Zh. Eksp. & Teor. Fiz. **29**(12), 784 [JETP Lett. **29**(12), 720].
Brandt, N.B., S.V. Kuvshinnikov and M.V. Semyonov, 1980a, in: Proc. XXIth Conf. on Low Temperature Physics USSR, Kharkov, Vol. 2 (Akademia Nauk SSSR) p. 225.
Brandt, N.B., A.S. Kotosonov, S.V. Kuvshinnikov and M.V. Semyonov, 1980b, Zh. Eksp. & Teor. Fiz. **79**(9), 937 [Sov.Phys.-JETP **52**(3), 476].
Brandt, N.B., S.G. Ionov, S.V. Kuvshinnikov, V.A. Mukhanov and V.V. Avdeev, 1981, Pis'ma Zh. Eksp. & Teor. Fiz. **34**(5), 256 [JETP Lett. **34**(5), 243].
Brandt, N.B., S.G. Ionov, S.V. Kuvshinnikov, V.A. Mukhanov, V.V. Avdeev and K.N. Semenenko, 1982, Pis'ma Zh. Tekh. Fiz. **8**(24), 1494 [Sov. Tech. Phys. Lett. **8**(12), 641].
Brandt, N.B., S.V. Kuvshinnikov and S.G. Ionov, 1983, Pis'ma Zh. Eksp. & Teor. Fiz. **38**(6), 275 [JETP Lett. **38**(6), 326].
Brandt, N.B., S.G. Ionov and S.V. Kuvshinnikov, 1984a, Fiz. Tverd. Tela. **26**(2), 361 [Sov.Phys.-Solid State **26**(2), 215].
Brandt, N.B., S.G. Ionov, S.V. Kuvshinnikov and V.A. Mukhanov, 1984b, Fiz. Nizk. Temp. **10**(7), 723.
Brandt, N.B., V.V. Avdeev, S.G. Ionov, V.A. Mukhanov, K.N. Semenenko, V.I. Kostikov, A.S. Kotosonov and S.V. Kuvshinnikov, 1986, Zh. Eksp. & Teor. Fiz. **91**(9), 1121 [Sov.Phys.-JETP **64**(3), 1205].
Breisacher, P., and P.C. Marx, 1963, J. Am. Chem. Soc. **85**(21), 3518.

Bridgman, P.W., 1947, J. Chem. Phys. **15**, 92.

Bridgman, P.W., 1948, Proc. Am. Acad. Arts & Sci. **76**(3), 55–87.

Bridgman, P.W., 1952, Proc. Am. Acad. Arts & Sci. **81**, 165.

Bright, A.A., 1979, Carbon **17**, 259; Phys. Rev. B **20**(12), 5142.

Bright, A.A., and L.S. Singer, 1979, Carbon **17**, 59.

Brillson, L.J., E. Burstein, A.A. Maradudin and T. Stark, 1971, in: Proc. Int. Conf. on Semimetals and Narrow Gap Semiconductors, eds D.L. Carter and R.T Bate (Pergamon Press, New York) p. 187.

Brocklehurst, J.E., B.T. Kelly and K.E. Gilchrist, 1979, in: Proc. 14th Biennial Conf. on Carbon, Extended Abstracts and Program, June 25–29, 1979 (Pennsylvania State University, University Park, PA) pp. 475, 476.

Brocklehurst, J.E., B.T. Kelly and K.E. Gilchrist, 1980, Carbon **18**(1), 68.

Brown, A.R.G., and W. Watt, 1958, in: Industrial Carbon and Graphite (Society of Chemical Industry, London) p. 86–100.

Budke, J., and W. Theymann, 1976, in: Technical meeting on radiation exposure and damage of reactor components outside the reactor core, Stuttgart FRG, ADF-Conf.-76-556-012, p. 46.

Budke, J., and W. Theymann, 1977, Reaktorstrahlenschäden an Graphit, Atomkernenergie **29**(2), 113–122.

Bullock, R.E., 1979, in: Proc. 14th Biennial Conf. on Carbon, Extended Abstracts and Program, June 25–29, 1979 (Pennsylvania State University, University Park, PA) pp. 485, 486.

Bundy, F.P., 1963, J. Chem. Phys. **38**(3), 618.

Bundy, F.P., 1980, J. Geophys. Res. **85**, 6930.

Buzdin, A.I., 1983, Fiz. Tverd. Tela **25**(11), 3504 [Sov. Phys.-Solid State **25**(11), 2019].

Buzdin, A.I., O.V. Dolgov and Yu.E. Lozovik, 1984, Fiz. Tverd. Tela **26**(3), 708 [Sov. Phys.-Solid State **26**(3), 427].

Cahill, J.E., and G.E. Leroi, 1966, J. Chem. Phys. **51**, 4514.

Campbell, L.E., L.G. Montel and G.J. Perlow, 1977, Phys. Rev. B **15**, 3318.

Campion, P., 1977, J. Inst. Nucl. Eng. **18**(2), 48–50.

Carmona, F., G. Delhaes, Y.R. Keyer and Y.P. Monceau, 1974, Solid State Commun. **14**(10), 1183.

Caswell, N., and S.A. Solin, 1978a, Solid State Commun. **27**, 961.

Caswell, N., and S.A. Solin, 1978b, Bull. Am. Phys. Soc. **23**, 218.

Caswell, N., S.A. Solin and W. Metz, 1979, Bull. Am. Phys. Soc. **24**, 375.

Caswell, N., S.A. Solin, T.M. Hayes and S.J. Hunte, 1980, Physica B **99**, 463.

Chisholm, R.C., and J.W. Stout, 1962, J. Chem. Phys. **36**, 972.

Chouteau, G., and A. Briggs, 1977, Solid State Commun. **21**, 787.

Chung, D.D.L., 1977, Ph.D. Thesis (MIT, Cambridge, MA).

Chung, D.D.L., and M.S. Dresselhaus, 1976, Solid State Commun. **19**, 227.

Chung, D.D.L., and M.S. Dresselhaus, 1977, Physica B **89**, 131.

Chung, D.D.L., G. Dresselhaus and M.S. Dresselhaus, 1977, Mater. Sci. & Eng. **31**, 107.

Clarke, R., and C. Uher, 1984, Adv. Phys. **33**(5), 469.

Clarke, R., N. Caswell and S.A. Solin, 1979a, Phys. Rev. Lett. **42**, 61.

Clarke, R., N. Caswell, S.A. Solin and P.M. Horn, 1979b, Phys. Rev. Lett. **43**(27), 2018.

Clarke, R., N. Wada and S.A. Solin, 1980, Phys. Rev. Lett. **44**(24), 1616.

Clarke, R., M. Elzinga, H. Gray, J.N. Homma, D.T. Morelli and C. Uher, 1982, Phys. Rev. B **26**, 5250.

Consadori, F., A.A. Fife, R.F. Friudt and S. Gygax, 1971, Appl. Phys. Lett. **18**, 233.

Cooper, J.D., J. Woore and D.A. Young, 1970, Nature **225**, 721.

Cooper, J.D., J.P. Smith, J. Woore and D.A. Young, 1971, J. Phys. C **4**(4), 442.
Corbato, F.Y., 1957, in: Proc. 3rd Conf. on Carbon, Buffalo, NY (Pergamon Press, New York) p. 173.
Cords, H., and R. Zimmermann, 1978, Ber. Kernforschungsanlage Jülich, Juel-1506, p. 39.
Coulson, C.A., 1961, Valence, 2nd Ed. (Oxford University Press, Oxford).
Cowley, J.M., and Y.A. Ibers, 1956, Acta Crystallogr. **9**, 421.
Dalessandro, J.A., 1978, ERA **3**(18), 4401.
Dalton, K., and S. Fujita, 1971, Acta Phys. Pol. A **40**, 136.
Daumas, N., and A. Hérold, 1969, C.R. Hebd. Séances Acad. Sci. Ser. B **268**, 373.
De Long, L.E., and P.C. Eklund, 1983, Synth. Met. **5**, 291.
De Long, L.E., V. Yeh, P.C. Eklund, V. Tonidiglia, S.E. Lambert and M.B. Maple, 1982, Phys. Rev. B **26**, 6315.
De Sorbo, W., 1954, General Electric, Tech. Rep. RL-1202.
De Sorbo, W., and G.E. Nichols, 1958, J. Phys. & Chem. Solids **6**, 352.
De Sorbo, W., and W.W. Tyler, 1953, J. Chem. Phys. **21**, 1660.
Delhaes, P., 1971, Chem. Phys. Carbon **7**, 193.
Delhaes, P., 1977, Mater. Sci. & Eng. **31**, 225.
Delhaes, P., and Y. Hishiyama, 1970, Carbon **8**(1), 31.
Delhaes, P., R. Kepper and M. Uhlrich, 1974, Philos. Mag. **29**(6), 1301.
Delhaes, P., J.C. Rouillon, J.P. Manceau, D. Guérard and A. Hérold, 1976, J. Phys. Lett. (France) **37**, L127.
Delle, W., H. Hoven and K. Frye, 1979, in: Proc. 14th Biennial Conf. on Carbon, Extended Abstracts and Program, June 25–29, 1979 (Pennsylvania State University, University Park, PA) pp. 477, 478.
Delle, W., G. Haag and H. Gortz, 1982, in: 6th Int. Conf. on Carbon and Graphite Carbon, Kernforschungsanlage Jülich, pp. 338–340.
Dillon, R.O., and I.L. Spain, 1978, Solid State Commun. **26**, 337.
Dillon, R.O., I.L. Spain and J.W. McClure, 1977, J. Phys. & Chem. Solids **38**, 635.
Dillon, R.O., I.L. Spain, J.A. Woollam and W.H. Lowrey, 1978a, J. Phys. & Chem. Solids **39**, 907.
Dillon, R.O., I.L. Spain and J.W. McClure, 1978b, J. Phys. & Chem. Solids **39**, 1071.
Doezema, R.E., W.R. Datars, H. Schaber and A. van Schyndel, 1979, Phys. Rev. B **19**, 4224.
Dolling, G., and B.N. Brockhouse, 1962, Phys. Rev. **128**, 1120.
Doni, E., and P. Parravicini, 1969, Nuovo Cimento B **64**, 117.
Dresselhaus, G., 1974, Phys. Rev. B **10**, 3602.
Dresselhaus, G., and M.S. Dresselhaus, 1965, Phys. Rev. A **140**, 401.
Dresselhaus, G., and S.Y. Leung, 1980, Solid State Commun. **35**, 819.
Dresselhaus, G., and J.G. Mavroides, 1966, Carbon **3**, 465.
Dresselhaus, M.S., and G. Dresselhaus, 1981, Adv. Phys. **30**(2), 139.
Dresselhaus, M.S., and S.Y. Leung, 1979, in: Proc. 14th Biennial Conf. on Carbon, Extended Abstracts and Program, June 25–29, 1979 (Pennsylvania State University, University Park, PA) p. 496.
Dresselhaus, M.S., and J.G. Mavroides, 1964, IBM J. Res. & Dev. **8**, 262.
Dresselhaus, M.S., G. Dresselhaus, D.D.L. Chung and P.C. Eklund, 1977, Mat. Sci. & Eng. **31**, 141.
Dresselhaus, M.S., S.Y. Leung, M. Shayegan and T.C. Chien, 1980, Synth. Met. **2**, 321.
Drude, P., 1902, The Theory of Optics (Longmans & Green, New York).
Dutta, A.K., 1953, Phys. Rev. **90**, 187.

Eberhardt, W., J. Vagas and A. Briggs, 1975, Phys. Rev. Lett. A **53**, 297.
Eberhardt, W., I.T. McGovern, E.W. Plummer and J.E. Fischer, 1980, Phys. Rev. Lett. **44**, 200.
Ebert, L.B., and H. Selig, 1977, Mater. Sci. & Eng. **31**, 177.
Edelman, V.C., 1970, Usp. Fiz. Nauk **102**(1), 55 [Sov. Phys.-Usp. **13**(5), 583].
Edelman, V.C., 1976, Adv. Phys. **25**(6), 555.
Edelman, V.C., 1977, Usp. Fiz. Nauk **123**, 257 [Sov. Phys.-Usp. **20**(10), 519].
Edelstaff, P.A., and D.H.C. Harris, 1963, Phys. Lett. **7**, 220.
Eeles, W.T., and J.A. Turnbull, 1965, Proc. R. Soc. London Ser. A **283**, 179.
Eklund, P.C., G. Dresselhaus, M.S. Dresselhaus and J.E. Fischer, 1977, Phys. Rev. B **16**, 3330.
Eklund, P.C., N. Kambe, G. Dresselhaus and M.S. Dresselhaus, 1978, Phys. Rev. B **18**, 7069.
Eklund, P.C., G. Dresselhaus, M.S. Dresselhaus and J.E. Fischer, 1980a, Phys. Rev. B **21**, 4705.
Eklund, P.C., D.S. Smith, V.R.K. Murthy and S.Y. Leung, 1980b, Synth. Met. **2**, 99.
Elahy, M., and G. Dresselhaus, 1982, in: Intercalated Graphite, eds M.S. Dresselhaus, G. Dresselhaus, J.E. Fischer and M.J. Moran, Mater. Res. Soc. Symp. Proc. **20** (North-Holland, Amsterdam) p. 207.
Elahy, M., G. Niccolini, G. Dresselhaus and G.O. Zimmerman, 1982, Solid State Commun. **41**, 289.
Endo, M., T.C. Chien, G. Timp and M.S. Dresselhaus, 1983, Synth. Met. **8**(3/4), 251.
Falardeau, E.R., L.R. Hanlon and T.E. Thompson, 1978, Inorg. Chem. **17**, 301.
Falkovskii, L.A., 1968, Usp. Fiz. Nauk **94**(1), 3.
Fenton, E.W., and R.R. Haering, 1967, Phys. Rev. **159**, 593.
Fialkov, A.C., 1965, Formation of the Structure and of the Properties of the Carbon Materials (in Russian) (Metallurgiya, Moscow).
Fieguth, P., M. Schulte-Oversohl and P. Fritz, 1977, Atomwirtsch.-Atomtech. **22**(4), 225, 226.
Finch, G.I., and H. Wilman, 1936, Proc. R. Soc. London **155**, 345.
Fischer, J.E., 1979, in: Physics and Chemistry of Materials with Layered Structures, Vol. 3, ed. F. Levy (Reidel, Dordrecht) p. 481.
Fischer, J.E., 1980, Physica B **99**, 389.
Fischer, J.E., 1981, Solid State Commun. **40**, 93.
Fischer, J.E., and T.E. Thompson, 1978, Phys. Today **31**, 36.
Fischer, J.E., M.J. Moran, J.W. Milliken and A. Briggs, 1981, Solid State Commun. **39**(3), 439.
Fischer, J.E., C.D. Fuerst and H.J. Kim, 1982, in: Intercalated Graphite, eds M.S. Dresselhaus, G. Dresselhaus, J.E. Fischer and M.J. Moran, Mater. Res. Soc. Symp. Proc. **20** (North-Holland, Amsterdam) p. 169.
Fitzler, E., and O. Vohler, 1962, in: Proc. Symp. on Radiation Damage in Reactor Materials, Venice, Italy, 1962 (IAEA, Vienna, 1963) p. 593.
Fletcher, R., 1981, J. Phys. F **11**(5), 1093.
Flood, D.J., 1969, Phys. Lett. A **30**, 168.
Foley, G.M.T., C. Zeller, E.R. Falardeau and F.L. Vogel, 1977, Solid State Commun. **24**(11), 371.
Fourré, J., A. Pernollet, W.W. Delle and G. Haag, 1976, in: Carbon, 2nd Intern. Kohlenstofftag., Baden-Baden, FRG, Prepr.S.I.s.a., pp. 353, 356.
Franklin, R.E., 1951, Acta Crystallogr. **4**, 253.
Friedberg, C.B., and R.W. Stark, 1972, in: Low Temperature Physics **4**, eds K.D. Timmerhaus, W.J. O'Sullivan and E.F. Hamlet (Proc. XIIIth Int. Conf. on Low Temperature Physics, August 21–25, 1972, Boulder, CO) (Plenum Press, New York) p. 177.
Friedel, R.A., and G.C. Carlson, 1971, J. Phys. C **4**, 1149.
Friend, H.R., and D. Jérome, 1979, J. Phys. C **12**(8), 1441.
Fröhlich, H., and B.V. Paranjable, 1956, Proc. R. Soc. London Ser. B **69**, 21.

Fuerst, C.D., D. Moses and J.E. Fischer, 1981, Phys. Rev. B **24**, 7471.

Fuerst, C.D., J.E. Fischer, J.D. Axe, J.B. Hastings and D.B. McWhan, 1983, Phys. Rev. Lett. **50**, 357.

Fujita, S., 1968, Carbon **6**, 746.

Fukuyama, H., 1971, Prog. Theor. Phys. **45**, 704.

Funahashi, S., T. Kondow and M. Iizuki, 1982, in: Intercalated Graphite, eds M.S. Dresselhaus, G. Dresselhaus, J.E. Fischer and M.J. Moran, Mater. Res. Soc. Symp. Proc. **20** (North-Holland, Amsterdam) p. 307.

Fuzellier, H., J. Melin and A. Hérold, 1977, Mater. Sci. & Eng. **31**, 91.

Gain, R., 1962, in: Proc. Symp. on Radiation Damage in Reactor Materials, Venice, Italy, 1962 (IAEA, Vienna, 1963) p. 651.

Galt, I.K., W.A. Yager and H.W. Dail, 1956, Phys. Rev. **103**, 1586.

Galt, I.K., W.A. Yager and F.R. Merrit, 1959, in: Proc. 3rd Conf. on Carbon, Buffalo, NY (Pergamon Press, New York) p. 193.

Ganguli, N., and K.S. Krishnan, 1941, Proc. R. Soc. London Ser. A **117**, 168.

Gauster, W.B., 1972, Philos. Mag. **25**, 687.

Gauster, W.B., and I.J. Fritz, 1974, J. Appl. Phys. **45**, 3309.

Giergiel, J., and P.C. Eklund, 1982, in: Intercalated Graphite, eds M.S. Dresselhaus, G. Dresselhaus, J.E. Fischer and M.J. Moran, Mater. Res. Soc. Symp. Proc. **20** (North-Holland, Amsterdam) p. 329.

Giergiel, J., P.C. Eklund, R. Al-Jishi and G. Dresselhaus, 1982, in: Intercalated Graphite, eds M.S. Dresselhaus, G. Dresselhaus, J.E. Fischer and M.J. Moran, Mater. Res. Soc. Symp. Proc. **20** (North-Holland, Amsterdam) p. 323.

Ginzburg, V.L., 1964, Phys. Lett. **13**, 101.

Giovanini, B., and F.T. Hedgecock, 1972, Solid State Commun. **11**, 367.

Goggin, P.R., and W.W. Reynolds, 1963, Philos. Mag. **8**, 265.

Goldberg, H.A., and I.L. Kalnin, 1981, Synth. Met. **3**(1/2), 169.

Goldsmith, M., 1950, J. Chem. Phys. **18**, 523.

Golé, J., 1977, Mater. Sci. & Eng. **31**, 309.

Golé, J., G. Merle and J.P. Pascault, 1982, Synth. Met. **4**(4), 269.

Gostishev, V.I., M.A. Glinskii, A.A. Drozd and S.E. Dem'yanov, 1978, Zh. Eksp. & Teor. Fiz. **74**(3), 1102 [Sov.Phys.-JETP **47**(3), 579].

Green, J.F., and I.L. Spain, 1973, J. Phys. & Chem. Solids **34**, 2177.

Green, J.F., P. Bolsaitis and I.L. Spain, 1973, J. Phys. & Chem. Solids **34**, 1927.

Greenaway, D.L., G. Harbeke, F. Bassani and E. Tosatti, 1969, Phys. Rev. **178**(3), 1340.

Gualberto, G.M., C. Underhill, S.Y. Leung and G. Dresselhaus, 1980, Phys. Rev. B **21**, 862.

Guérard, D., and A. Hérold, 1975, Carbon **13**, 337.

Guérard, D., C. Zeller and A. Hérold, 1976, C.R. Hebd. Séances Acad. Sci. Ser. C **283**, 437.

Guérard, D., G.M.T. Foley, M. Zanini and J.E. Fischer, 1977, Nuovo Cimento B **38**, 410.

Gupta, O.P., and P.R. Wallace, 1972, Phys. Status Solidi b **54**, 53.

Haering, R.R., and S. Mrozowski, 1960, Prog. Semicond. **5**, 273.

Hall, A.W., 1917, Phys. Rev. **10**, 661.

Hannay, N.B., T.H. Geballe, B.T. Matthias, K. Andres, P. Schmidt and D. McNair, 1965, Phys. Rev. Lett. **14**, 225.

Hardcastle, S.E., and H. Zabel, 1981, J. Phys. (France) C **6**, 326.

Hardcastle, S.E., M.E. Misenheimer and H. Zabel, 1982, Bull. Am. Phys. Soc. **27**, 2721.

Harker, H., 1967, Philos. Mag. **16**(144), 1193.

Harrison, W.A., 1966, Pseudopotentials in the Theory of Metals (W.A. Benjamin, New York).

Haywood, B.C., and I.M. Thorson, 1960, in: Proc. IAEA Symp. on Inelastic Scattering of Neutrons in Solids and Liquids, Vienna, Vol. 2, pp. 111–121.

Hedgecock, F.T., and T.W. Randorf, 1970, Solid State Commun. **8**, 1819.

Henning, G.R., 1965, J. Appl. Phys. **36**(4), 1482.

Henning, G.R., and L. Mayer, 1952, Phys. Rev. **87**, 439.

Herinckx, G., R. Perret and W. Ruland, 1972, Carbon **10**, 711.

Hérold, A., 1955, Bull. Soc. Chim. (France) **187**(7/8), 999.

Hérold, A., 1979, in: Physics and Chemistry of Materials with Layered Structure, Vol. 6, ed. F. Levy (Reidel, Dordrecht) p. 323.

Herring, C., 1958, J. Phys. & Chem. Solids **8**, 549.

Herring, C., 1960, J. Phys. & Chem. Solids **31**, 1939.

Herrman, F., and S. Skillman, 1963, Atomic Structure Calculations (Prentice-Hall, Englewood Cliffs, NJ).

Herzberg, G., 1945, Spectra of Diatomic Molecules (Van Nostrand, Princeton, NJ).

Hideto, M., and S. Jasuachi, 1981, J. Nucl. Mater. **101**(1/2), 232–234.

Higuchi, K., H. Suematsu and S. Tanuma, 1980, J. Phys. Soc. Jpn. **48**, 1532.

Hishiyama, Y., A. Ono, T. Tsuzuku and T. Takezawa, 1972, Jpn. J. Appl. Phys. **11**(7), 958.

Hishiyama, Y., Y. Kaburagi and A. Ono, 1978, Carbon **17**(3), 265.

Hohlwein, D., P.W. Readman, A. Chamberod and J.M.B. Coey, 1974, Phys. Status Solidi b **64**(1), 305.

Holl, G.L.J., 1954, J. Phys. & Chem. Solids **2**, 24.

Holzwarth, N.A.W., 1980, Phys. Rev. B **21**, 3665.

Holzwarth, N.A.W., and S. Rabii, 1977, Mater. Sci. & Eng. **31**, 195.

Holzwarth, N.A.W., S. Rabii and L.A. Girifalco, 1978a, Phys. Rev. B **18**, 5190.

Holzwarth, N.A.W., S. Rabii and L.A. Girifalco, 1978b, Phys. Rev. B **18**, 5206.

Honda, H., 1959, in: Proc. 3rd Carbon Conf. Buffalo, NY (Pergamon Press, New York) p. 159.

Hooley, J.G., 1972, Carbon **10**, 155.

Hooley, J.G., 1975, Carbon **11**, 225.

Hooley, J.G., and V.R. Dietz, 1978, Carbon **16**, 251.

Horie, C., M. Maeda and Y. Kuramoto, 1980, Physica B **99**, 430.

Houston, W.V., 1948, Rev. Mod. Phys. **20**, 161.

Hove, I.E., 1959, in: Progr. in Nuclear Energy, Metallurgy and Fuels, Ser. V, Vol. 2 (Pergamon Press, New York) p. 551.

Hove, I.E., 1960, in: Graphite as a High-Temperature Material, Proc. fourth conf. on Carbon, University of Buffalo, Buffalo, NY (Symposium Publication Division, Pergamon Press, New York) pp. 343, 370.

Hove, I.E., 1955, Phys. Rev. **100**, 645.

Hutchings, M.T., 1973, J. Phys. C **6**, 3143.

Hwang, D.M., 1983, Solid State Commun. **46**(2), 177.

Ichikawa, M., K. Kawase and K. Tamaru, 1972a, J. Chem. Soc. Chem. Commun. (9 February) p. 177.

Ichikawa, M., T. Kondo, K. Kawase, M. Sudo, T. Onishi and K. Tamaru, 1972b, J. Chem. Soc. Chem. Commun. (9 February) p. 176.

Ichikawa, Y.H., 1958, Phys. Rev. **109**, 653.

Industrial Graphite Engineering Handbook, 1959, (Union Carbide Corporation).

Inoshita, T., K. Nakao, H. Kamimura, 1977, J. Phys. Soc. Jpn. **43**, 1237.

Inoue, M., 1962, J. Phys. Soc. Jpn. **17**, 808.

Itskevich, E.S., 1963, Prib. Tekh. Eksp. **4**, 148 [1964, Cryogenics **4**(6), 85].

Itskevich, E.S., and L.M. Fischer, 1967, Pis'ma Zh. Eksp. & Teor. Fiz. **5**, 141 [JETP Lett. **5**, 114].

Iwata, T., and H. Suzuki, 1962, in: Proc. Symp. on Radiation Damage in Reactor Materials, Venice, Italy, 1962 (IAEA, Vienna, 1963) p. 565.

Iye, Y., 1982, in: Intercalated Graphite, eds M.S. Dresselhaus, G. Dresselhaus, J.E. Fischer and M.J. Moran, Mater. Res. Soc. Symp. Proc. **20** (North-Holland, Amsterdam) p. 185.

Iye, Y., and S. Tanuma, 1982, Tech. Rep. ISSP Tokyo Ser. A (1235) p. 44; Solid State Commun. **44**(1), 1–6; Phys. Rev. B **25**(7), 4583.

Iye, Y., and S. Tanuma, 1983, Synth. Met. **5**(3), 257.

Iye, Y., O. Takahashi and S. Tanuma, 1980, Solid State Commun. **33**(10), 1071.

Iye, Y., O. Takahashi, S. Tanuma, K. Tsuji and S. Minomura, 1981a, Synth. Met. **3**, 235.

Iye, Y., O. Takahashi, S. Tanuma, K. Tsuji and S. Minomura, 1981b, Physica B **105**(1–4), 185.

Iye, Y., O. Takahashi, S. Tanuma, K. Tsuji and S. Minomura, 1982a, J. Phys. Soc. Jpn. **51**(2), 475.

Iye, Y., T. Tedrow, G. Timp, M.S. Dresselhaus, G. Dresselhaus, A. Furukawa and S. Tanuma, 1982b, Phys. Rev. B **25**, 5478.

Jacobi, N., and J.-P. Jay-Gerin, 1977, Solid State Commun. **21**, 789.

Jenkins, G.M., 1962, J. Nucl. Mater. **5**(3), 280–286.

Johnson, D.F., 1960, Philos. Mag. **5**, 1057.

Johnson, L.C., and G. Dresselhaus, 1973, Phys. Rev. B **7**, 2275.

Jose, J.V., L. Kadanoff, S. Kirkpatrick and D.R. Nelson, 1977, Phys. Rev. B **16**, 1217.

Kabalkina, S.S., and L.F. Vershchagin, 1960, Dokl. Akad. Nauk SSSR Ser. Fiz. **131**(2), 330 [Sov. Phys.-Dokl. **5**, 273].

Kambe, N., M.S. Dresselhaus and G. Dresselhaus, 1980, Phys. Rev. B **21**, 3491.

Kambe, N., H. Mazurek, M.S. Dresselhaus and G. Dresselhaus, 1981, Physica B **105**, 272.

Kamimura, H., K. Nakao, T. Ohno and T. Inoshita, 1980, Physica B **99**, 401.

Kamitakahara, W.A., and H. Zabel, 1982, in: Intercalated Graphite, eds M.S. Dresselhaus, G. Dresselhaus, J.E. Fischer and M.J. Moran, Mater. Res. Soc. Symp. Proc. **20** (North-Holland, Amsterdam) p. 317.

Kamitakahara, W.A., N. Wada and S.A. Solin, 1981, J. Phys. (France) **42**(C6), 311.

Kamitakahara, W.A., H. Zabel and R.M. Nicklow, 1982, in: Intercalated Graphite, eds M.S. Dresselhaus, G. Dresselhaus, J.E. Fischer and M.J. Moran, Mater. Res. Soc. Symp. Proc. **20** (North-Holland, Amsterdam) p. 311.

Kapustin, G.A., and E.Z. Meilikhov, 1975, Fiz. Tverd. Tela **17**(10), 2978 [Sov. Phys.-Solid State **17**(10), 1979].

Kapustin, G.A., A.S. Lagutin and E.Z. Meilikhov, 1981, Fiz. Tverd. Tela **23**(11), 3273 [Sov. Phys.-Solid State **23**(11), 1902].

Karim, D.P., J.B. Ketterson and G.W. Crabtree, 1978, J. Low Temp. Phys. **30**(3/4), 389.

Karimov, Yu.S., 1974, Zh. Eksp. & Teor. Fiz. **66**, 1121 [1975, Sov. Phys.-JETP **39**, 547].

Karimov, Yu.S., 1975, Zh. Eksp. & Teor. Fiz. **68**, 1539 [1976, Sov. Phys.-JETP **41**, 772].

Karimov, Yu.S., and Yu.N. Novikov, 1974, Pis'ma Zh. Eksp. & Teor. Fiz. **19**(5), 268 [JETP Lett. **19**(5), 159].

Karimov, Yu.S., M.E. Vol'pin and Yu.N. Novikov, 1971, Pis'ma Zh. Eksp. & Teor. Fiz. **14**, 217 [JETP Lett. **14**(4), 142].

Kasatochkin, V.I., V.V. Kasakov and V.V. Savranskii, 1967, Dokl. Akad. Nauk SSSR Ser. Fiz. **177**(2), 358.

Katz, E.I., 1969, Zh. Eksp. & Teor. Fiz. **56**, 1675 [Sov.Phys.-JETP **29**(5), 897].

Kawamura, K., T. Saito and T. Tsuzuku, 1977, J. Phys. Soc. Jpn. **42**(2), 574.

Kechin, V.V., 1969, Fiz. Tverd. Tela **11**, 1788 [1970, Sov. Phys.-Solid State **11**, 1448].

Kechin, V.V., A.I. Likhter and G.M. Stepanov, 1968, Fiz. Tverd. Tela **10**, 1242 [Sov. Phys.-Solid State **10**, 987].

Kelly, B.T., 1964, Philos. Mag. **9**(101), 721.

Kelly, B.T., 1976, Carbon **14**(4), 239–245.

Kelly, B.T., 1977, Carbon **15**(2), 117–127.

Kelly, B.T., 1978a, Progr. Nucl. Energy **2**(4), 219–269.

Kelly, B.T., 1978b, Ber. Kernforschungsanlage Jülich, Jülich Conf. **25**, 82–94.

Kelly, B.T., and M.J. Duff, 1970, Carbon **8**, 77.

Kennedy, C.K., and W.P. Eatherly, 1979, in: Proc. 14th Biennial Conf. on Carbon, Extended Abstracts and Program, June 25–29, 1979 (Pennsylvania State University, University Park, PA) pp. 453, 454.

Keyes, R.W., and R.J. Sladek, 1956, J. Phys. & Chem. Solids **1**, 143.

Khanna, S.K., E.R. Falardeau, A.J. Heeger and J.E. Fischer, 1978, Solid State Commun. **25**, 1059.

Kittel, C., 1978, Solid State Commun. **25**, 519.

Klein, C.A., 1962, J. Appl. Phys. **33**, 3338.

Klein, C.A., 1964, J. Appl. Phys. **35**(10), 2947.

Klucker, R., M. Skibowski and W. Steinmann, 1974, Phys. Status Solidi b **65**, 703.

Kobayashi, M., and I. Tsujikawa, 1979, J. Phys. Soc. Jpn. **46**, 1945.

Kobayashi, M., and I. Tsujikawa, 1981a, J. Phys. Soc. Jpn. **50**, 3245.

Kobayashi, M., and I. Tsujikawa, 1981b, Physica B **105**, 439.

Koga, K., and M. Suzuki, 1984, J. Phys. Soc. Jpn. **53**(2), 786.

Koike, Y., N. Suematsu, K. Higuchi and S. Tanuma, 1978, Solid State Commun. **27**, 623.

Koike, Y., N. Suematsu, K. Higuchi and S. Tanuma, 1980, Physica B **99**, 503.

Komatsu, K., 1955, J. Phys. Soc. Jpn. **10**, 346.

Komatsu, K., 1958, J. Phys. & Chem. Solids **6**, 681.

Komatsu, K., 1964, J. Phys. & Chem. Solids **25**(7), 707.

Komatsu, K., and T. Nagamiya, 1951, J. Phys. Soc. Jpn. **6**, 438.

Kondow, T., U. Mizutani and T.B. Massalski, 1977, Mater. Sci. & Eng. **31**, 267.

Kosterlitz, J.M., 1976, J. Phys. C **7**, 1046.

Kotosonov, A.S., 1986, Pis'ma Zh. Eksp. & Teor. Fiz. **43**(1), 30 [JETP Lett. **43**(1), 37].

Krishnan, K.S., and N. Ganguli, 1939, Nature **144**, 667; Z. Krist. A **100**, 530.

Krumhansl, J., and H. Brooks, 1953, Chem. Phys. **21**, 1663.

Kurdyumov, A.V., 1972, Kristallografiya **17**(3), 620.

Kurdyumov, A.V., and A.N. Piljankevich, 1968, Kristallografiya **13**(2), 311.

Kuvshinnikov, S.V., 1980, Thesis (Moscow State University, Moscow).

Lagrange, P., and A. Hérold, 1978, Carbon **16**, 235.

Landau, L.D., 1930, Z. Phys. **64**, 629.

Lauginie, P., H. Estrade, J. Conard, D. Guérard, P. Lagrange and M. ElMakrini, 1980, Physica B **99**, 514.

Leung, S.Y., C. Underhill, G. Dresselhaus, T. Krapchev, R. Ogilvie and M.S. Dresselhaus, 1980, Phys. Lett. **76**, 89.

Leung, S.Y., G. Dresselhaus and M.S. Dresselhaus, 1981a, Phys. Rev. B **24**(10), 6083.

Leung, S.Y., M.S. Dresselhaus and G. Dresselhaus, 1981b, Physica B **105**, 375.

Levintovich, I.Ya., and A.S. Kotosonov, 1980, Zh. Eksp. & Teor. Fiz. **78**, 1910 [Sov.Phys.-JETP **51**(5), 959].

Levy, F., ed., 1979, Physics and Chemistry of Materials with Layered Structures, Vol. 6 (Reidel, Dordrecht).

Lifshitz, I.M., M.Ya. Azbel and M.I. Kaganov, 1971, Electronic Theory of Metals, Part I (Nauka,

Moscow) p. 98.
Likhter, A.I., and V.V. Kechin, 1964, Fiz. Tverd. Tela **5**, 3066 [Sov. Phys.-Solid State **5**, 2246].
Lipson, H., and A.P. Stokes, 1942, Proc. R. Soc. London **181**, 101.
Little, W.A., 1964, Phys. Rev. A **134**, 1416.
Lowrey, W.H., and I.L. Spain, 1977, Solid State Commun. **22**(10), 615.
Lukin, E.V., and V.G. Nagorny, 1964, in: Konstrukzionnye uglegrafitovye materialy, Vol. 1 (Metallurgiya, Moscow) p. 170.
Lutkov, A.I., T.L. Perkova and V.A. Mikhailov, 1976, in: Konstrukzionnye Materialy na Osnove Grafita (Construction of Materials Based on Graphite Compounds) (Metallurgiya, Moscow) Vol. 11, pp. 92–98.
Luttinger, J.M., and W. Kohn, 1955, Phys. Rev. **97**, 869.
Lynch, R.W., and H.G. Drickamer, 1966, J. Chem. Phys. **44**, 181.
Maeda, M., Y. Kuramoto and C. Horie, 1979, J. Phys. Soc. Jpn. **47**, 337.
Maeta, H., and Y. Sato, 1977, Solid State Commun. **23**, 23.
Mani, K.K., and R. Ramani, 1974, Phys. Status Solidi b **61**, 659.
Manini, C., J.-F. Marêché and E. McRae, 1983, Synth. Met. **8**(3/4), 261.
Mantell, C.L., 1968, in: Carbon and Graphite Handbook (Interscience, New York) p. 358.
Maradudin, A.A., E.W. Montroll, G.H. Weiss and A. Ipatova, 1970, Theory of Lattice Dynamics in the Harmonic Approximation (Academic Press, New York).
Marchand, A., and E. Dupart, 1967, Carbon **5**, 453.
Markiewicz, R.S., and C. Zahopoulos, 1983, Phys. Rev. B **27**(12), 7820.
Markiewicz, R.S., J.S. Kasper and L.V. Interrante, 1980, Synth. Met. **2**, 363.
Martin, D.G., and J. Caisley, 1977, Carbon **15**(4), 251–255.
Martin, D.G., and R.W. Henson, 1964, Philos. Mag. **9**, 659.
Mase, S., and T. Sakai, 1971, J. Phys. Soc. Jpn. **31**, 730.
Mase, S., T. Fukami, M. Mori, M. Akinaga, T. Yamaguchi and N. Shiraishi, 1980, J. Phys. Soc. Jpn. **48**, 1157 ,1166.
Mashinskii, V.I., V.A. Postnikov, Yu.N. Novikov, A.L. Lapidus, M.E. Vol'pin and Ya.I. Eidus, 1976, Izv. Akad. Nauk SSSR Ser. Khim. **253**(2), 346.
Matheny, R.A., J.C. Corelly and G.G. Trantita, 1979, J. Nucl. Mater. **83**(2), 313–448.
Matuyama, E., 1956, Nature **178**(4548), 1459.
McClure, J.W., 1956, Phys. Rev. **101**, 1642.
McClure, J.W., 1957, Phys. Rev. **108**, 612.
McClure, J.W., 1958, Phys. Rev. **112**, 715.
McClure, J.W., 1960, Phys. Rev. **119**, 606.
McClure, J.W., 1969, Carbon **7**, 425.
McClure, J.W., 1970, in: Proc. Int. Conf. on The Physics of Semimetals and Narrow-Band Semiconductors, Dallas, TX (Pergamon Press, London, 1971) p. 127.
McClure, J.W., and D. Shoenberg, 1976, J. Low Temp. Phys. **22**, 233.
McClure, J.W., and L.S. Smith, 1961, in: Proc. 5th Int. Conf. on Carbon, Vol. 2 (Pergamon Press, New York, 1962) p. 3.
McClure, J.W., and W.J. Spry, 1968, Phys. Rev. **165**, 809.
McClure, J.W., and Y. Yafet, 1961, in: Proc. 5th Int. Conf. on Carbon, Vol. 2 (Pergamon Press, New York, 1962) p. 12, 22.
McDonnell, F.R.M., R.C. Pink and A.R. Ubbelohde, 1951, J. Chem. Soc. (January) p. 191.
McLean, T.P., and E.G.S. Paige, 1961, J. Phys. & Chem. Solids **18**, 139.
McRae, E., D. Billand, J.-F. Marêché and A. Hérold, 1980, Physica B **99**, 489.
Mele, E., and J.J. Ritsko, 1979, Phys. Rev. Lett. **43**, 68.

Melin, J., and A. Hérold, 1975, Carbon **13**, 357.
Mendez, E., T.C. Chien, N. Kambe and M.S. Dresselhaus, 1980a, Solid State Commun. **33**, 837.
Mendez, E., A. Misu and M.S. Dresselhaus, 1980b, Phys. Rev. B **21**(2), 827.
Millman, S.E., and G.O. Zimmerman, 1983, J. Phys. C **16**, L89.
Misu, A., E. Mendez and M.S. Dresselhaus, 1979, J. Phys. Soc. Jpn. **47**, 199.
Miyazaki, H., Y. Kuramoto and Ch. Horie, 1984, J. Phys. Soc. Jpn. **53**(4), 1380.
Mizutani, U., T. Kondow and T.B. Massalski, 1978, Phys. Rev. B **17**, 3165.
Mizutani, U., M. Suganuma and T. Kondow, 1982, Solid State Commun. **43**(4), 303.
Moore, A.W., 1973, in: Chemistry and Physics of Carbon, Vol. 11, eds P.L. Walker Jr and P.A. Thrower (Marcel Dekker, New York) p. 69.
Moore, A.W., 1981, in: Chemistry and Physics of Carbon, Vol. 17, eds P.L. Walker Jr and P.A. Thrower (Marcel Dekker, New York) p. 233.
Morant, R.A., 1970, J. Phys. D **3**(9), 1367.
Morris, R.C., and R.V. Coleman, 1973, Phys. Rev. B **7**, 991.
Morris, R.C., R.V. Coleman and B. Bhander, 1972, Phys. Rev. B **5**, 895.
Mott, N.F., 1972, Philos. Mag. **26**, 1015.
Mrozowski, S., 1950, Phys. Rev. **77**(6), 838; **78**(5), 644; **80**(1), 123.
Mrozowski, S., 1952, Phys. Rev. **85**(4), 609.
Mrozowski, S., 1971, Carbon **9**(2), 97.
Mrozowski, S., and A. Chaberski, 1956, Phys. Rev. **104**, 74.
Mrozowski, S., 1960, in: Graphite as a High-Temperature Material, Proc. fourth Conf. on Carbon, University of Buffalo, Buffalo, NY (Symposium Publication Division, Pergamon Press, Oxford) p. 99.
Murday, J.S., and J.A. Dominguez, 1984, Synth. Met. **9**(3), 397.
Murray, J.J., and A.R. Ubbelohde, 1969, Proc. R. Soc. London Ser. A **312**, 371.
Nagayoshi, H., 1977, J. Phys. Soc. Jpn. **43**, 1760.
Nagayoshi, H., M. Tsukada, K. Nakao and Y. Uemura, 1973, J. Phys. Soc. Jpn. **35**(2), 396.
Nagayoshi, H., K. Nakao and Y. Uemura, 1976a, J. Phys. Soc. Jpn. **41**(5), 1480.
Nagayoshi, H., K. Nakao and Y. Uemura, 1976b, Solid State Commun. **18**, 225.
Naito, S., O. Igawa, M. Ichikawa and K. Tamaru, 1972, J. Chem. Soc. Chem. Commun. (6 December) p. 1226.
Nakamizo, M., R. Kammereck and P.L. Walker, 1974, Carbon **12**, 259.
Nakamura, K., T. Osada, G. Kido, N. Miura and S. Tanuma, 1983, J. Phys. Soc. Jpn. **52**, 2875.
Nakao, K., 1976, J. Phys. Soc. Jpn. **40**, 761.
Nakao, K., 1979, J. Phys. Soc. Jpn. **47**(1), 208.
Natarajan, V., and J.A. Woollam, 1983, in: Intercalated Graphite, eds M.S. Dresselhaus, G. Dresselhaus, J.E. Fischer and M.J. Moran, Mater. Res. Soc. Symp. Proc. **20** (North-Holland, Amsterdam) p. 33.
Nathan, M.I., J.E. Smith and K.N. Tu Jr, 1974, J. Appl. Phys. **45**, 2370.
Nemanich, R.J., and S.A. Solin, 1977, Solid State Commun. **23**, 417.
Nemanich, R.J., and S.A. Solin, 1979, Phys. Rev. B **20**, 392.
Nemanich, R.J., G. Lucovsky and S.A. Solin, 1975, in: Proc. Int. Conf. on Lattice Dynamics, ed. M. Balkanski (Flammarion Press, Paris) p. 619.
Nemanich, R.J., G. Lucovsky and S.A. Solin, 1977a, Solid State Commun. **23**, 117.
Nemanich, R.J., S.A. Solin and D. Guérard, 1977b, Phys. Rev. B **16**, 2965.
Nemanich, R.J., G. Lucovsky and S.A. Solin, 1977c, Mater. Sci. & Eng. **31**, 157.
Nguyen, Hy Hau, J. Blinowski, C. Rigaux and R. Letoullec, 1981, Synth. Met. **3**, 99.
Nicholson, A.P.P., and D.J. Bacon, 1977, J. Phys. C **10**, 2295.

Nicklow, R., N. Wakabayashi and H.G. Smith, 1972, Phys. Rev. B **5**, 4951.
Nightingale, R., ed., 1962, Nuclear Graphite (Academic Press, New York) pp. 427–448.
Nixon, D.E., and G.S. Parry, 1968, J. Phys. D **1**, 291.
Nixon, D.E., and G.S. Parry, 1969, J. Phys. C **2**, 1732.
Nixon, D.E., G.S. Parry and A.R. Ubellohde, 1966, Proc. R. Soc. A **291**, 324.
Noto, K., and T. Tsuzuku, 1975, Jpn. J. Appl. Phys. **14**, 46.
Novikov, Yu.N., and M.E. Vol'pin, 1981, Physica B+C **105**(1-3), 471.
Noziers, P., 1958, Phys. Rev. **109**, 1510.
Noziers, P., 1959, in: Proc. 3rd Conf. on Carbon, Buffalo, NY (Pergamon Press, New York) p. 197.
Nye, J.F., 1964, Physical Properties of Crystals and Their Representation by Tensors and Matrices, in: Lectures in Physics at the University of Bristol (Clarendon Press, Oxford).
Ohhashi, K., and I. Tsujikawa, 1974a, J. Phys. Soc. Jpn. **36**, 422.
Ohhashi, K., and I. Tsujikawa, 1974b, J. Phys. Soc. Jpn. **37**, 63.
Ohno, T., K. Nakao and H. Kamimura, 1979, J. Phys. Soc. Jpn. **47**, 1125.
Onn, D.G., G.M.T. Foley and J.E. Fischer, 1977, Mater. Sci. & Eng. **31**, 271.
Onn, D.G., G.M.T. Foley and J.E. Fischer, 1979, Phys. Rev. B **19**, 6474.
Onn, D.G., M.G. Alexander, J.J. Ritsko and S. Flandrois, 1982, J. Appl. Phys. **53**, 2751.
Ono, S., 1976, J. Phys. Soc. Jpn. **40**, 498.
Ono, S., and K. Sugihara, 1966, J. Phys. Soc. Jpn. **21**, 861.
Ono, S., and K. Sugihara, 1968, J. Phys. Soc. Jpn. **24**, 818.
Oshima, H., J.A. Woollam, A. Yavrouian and M.B. Dowell, 1983, Synth. Met. **5**(2), 113.
Painter, G.S., and D.E. Ellis, 1970, Phys. Rev. B **1**, 4747.
Panayotov, I.M., I.V. Berlinova and I.B. Rashkov, 1975, J. Polym. Sci. **43**, 2043.
Pendrys, L.A., R. Wachnik, F.L. Vogel, P. Lagrange, G. Furdin, M.E. Markini and A. Hérold, 1981, Solid State Commun. **38**, 677.
Pfluger, P., H.U. Künzi and H.J. Güntherodt, 1979, Appl. Phys. Lett. **35**, 771.
Philipp, H.R., and H. Ehrenreich, 1963, Phys. Rev. **129**, 1550.
Philipp, H.R., and E.A. Taft, 1959, Phys. Rev. **113**, 1002.
Pietronero, L., and S. Strässler, 1979, Solid State Commun. **32**(12), 1337.
Pietronero, L., and S. Strässler, 1980, J. Phys. Soc. Jpn. **49**, 895.
Pietronero, L., and S. Strässler, 1981, Synth. Met. **3**(2), 213.
Pietronero, L., S. Strässler, H.R. Zeller and M.J. Rice, 1978, Phys. Rev. Lett. **41**, 163.
Pietronero, L., S. Strässler and H.R. Zeller, 1979, Solid State Commun. **30**, 339.
Pietronero, L., S. Strässler, H.R. Zeller and M.J. Rice, 1980a, Physica B **99**, 499.
Pietronero, L., S. Strässler, H.R. Zeller and M.J. Rice, 1980b, Phys. Rev. B **22**(2), 904.
Platts, D.A., D.D.L. Chung and M.S. Dresselhaus, 1977, Phys. Rev. B **15**, 1087.
Pokrovskii, C.L., and G.V. Vimin, 1974, Zh. Eksp. & Teor. Fiz. **38**, 847 [Sov. Phys.-JETP **38**(4), 847].
Poquet, E., N. Lumbroso, J. Hoarau, A. Marchand, A. Pacault and D.E. Soule, 1960, J. Chem. Phys. **57**, 866.
Postnikov, V.A., L.M. Dmitrenko, R.F. Ivanova, N.L. Dobrolyubova, M.A. Golubeva, T.I. Gapeeva, Yu.N. Novikov, V.B. Shur and M.E. Vol'pin, 1975, Izv. Akad. Nauk SSSR Ser. Khim. **24**, 2529.
Potter, M.E., W.D. Johnson and J.E. Fischer, 1979, in: Proc. 14th Biennial Conf. on Carbon, Extended Abstracts and Program, June 25–29, 1979 (Pennsylvania State University, University Park, PA) p. 300.
Potter, M.E., W.D. Johnson and J.E. Fischer, 1981, Solid State Commun. **37**(5), 713.

Primak, W., and L.H. Fuchs, 1954, Phys. Rev. **95**, 22.
Primak, W., and L.H. Fuchs, 1956a, Phys. Rev. **103**, 541.
Primak, W., and L.H. Fuchs, 1956b, Phys. Rev. **103**, 544.
Prince, N., 1979, Nucl. Energy **18**(4), 267–275.
Rafizadeh, H.A., 1974, Physica **74**, 135.
Rashba, E.I., and V.M. Edelstein, 1969, Pis'ma Zh. Eksp. & Teor. Fiz. **9**, 475 [JETP Lett. **9**(8), 287].
Rashba, E.I., and V.M. Edelstein, 1970, Zh. Eksp. & Teor. Fiz. **58**, 1428 [Sov.Phys.-JETP **31**(4), 765].
Rasor, N.S., and J.D. McClelland, 1960, J. Phys. & Chem. Solids **15**(1/2), 17.
Raubenheimer, L.J., and G. Gilat, 1967, Phys. Rev. **157**, 586.
Reifenberger, R., F.W. Holroyd and F. Fawcett, 1980, J. Low. Temp. Phys. **38**(3/4) 421.
Reynolds, W.W., 1968, Physical Properties of Graphite (Elsevier, Amsterdam).
Reynolds, W.W., and P.A. Thrower, 1962, in: Proc. Symp. on Radiation Damage in Reactor Materials, Venice, Italy, May, 1962 (IAEA, Vienna, 1963) p. 553.
Reynolds, W.W., and P.A. Thrower, 1964, Carbon **1**, 185.
Rilay, D.P., 1945, Proc. R. Soc. London **57**(324), 486.
Rosenmann, I., F. Batallan and G. Furdin, 1979, Phys. Rev. B **20**, 2373.
Rosenmann, I., F. Batallan, C. Simon and H. Fuzellier, 1980, J. Phys. Soc. Jpn. **49**, 923.
Rüdorff, W., 1939, Z. Anorg. Chem. **42**, 174.
Rüdorff, W., 1940, Z. Phys. Chem. B **45**(1/2), 42.
Rüdorff, W., 1941, Z. Anorg. Chem. **245**, 121.
Rüdorff, W., 1959, Adv. Anorg. Chem. & Radiochem. **1**, 223.
Rüdorff, W., and U. Hoffmann, 1938, Z. Anorg. Chem. **238**, 1.
Rüdorff, W., and E. Schulze, 1954, Z. Anorg. Chem. **247**, 156.
Rüdorff, W., and H. Schulze, 1940, Z. Anorg. Chem. **245**, 303.
Rupprecht, G., D.M. Ginsberg and J.D. Leslie, 1962, J. Opt. Soc. Am. **52**, 665.
Safran, S.A., and F.J. Di Salvo, 1979, Phys. Rev. **20**, 4889.
Safran, S.A., and D.R. Hamman, 1979, Phys. Rev. **42**, 1410.
Safran, S.A., and D.R. Hamman, 1980, Physica B **99**, 469.
Safran, S.A., and D.R. Hamman, 1981, Phys. Rev. B **23**, 469.
Sagawa, T., 1966, J. Phys. Soc. Jpn. **21**, 49.
Saha, A.R., R.K. Banarjee and A.K. Das, 1970, Indian J. Phys. **44**, 438.
Samara, G.A., and H.G. Drickamer, 1962, J. Chem. Phys. **37**, 471.
Samuelson, L., I.P. Batra and C. Roetti, 1980, Solid State Commun. **33**, 817.
Sasa, T., Y. Tanahashi and T. Mukaibo, 1971, Carbon **9**, 407.
Schaffäult, P., 1841, J. Prakt. Chem. **21**, 155.
Schögl, R., W. Jones, H.-P. Boehm and W. Breimeier, 1983, Synth. Met. **8**(3/4) 323.
Schroeder, P.R., M.S. Dresselhaus and A. Javan, 1970, in: Proc. Int. Conf. on the Physics of Semimetals and Narrow-Bandgap Semiconductors, Dallas, TX (Pergamon Press, London, 1971) p. 139.
Scott, J.L., 1978, Metal Progress **114**(2), 40–49.
Seilig, H., and L.B. Ebert, 1980, Adv. Inorg. Chem. Radiochem. **23**, 281.
Seldin, E.J., 1956, in: Proc. 1st and 2nd Conf. on Carbon, Buffalo, NY, p. 117.
Seldin, E.J., and C.W. Nezbeda, 1970, J. Appl. Phys. **41**, 3389.
Setton, R., 1977, Mater. Sci. & Eng. **31**, 303.
Shapira, Y., and S. Foner, 1970, Phys. Rev. B **1**, 3088.
Sharma, M.P., L.C. Johnson and J.W. McClure, 1974, Phys. Rev. B **9**, 2467.

Shayegan, M., L. Salamanca-Riba, J. Heremans, G. Dresselhaus and J.-P. Issi, 1982, in: Intercalated Graphite, eds M.S. Dresselhaus, G. Dresselhaus, J.E. Fischer and M.J. Moran, Mater. Res. Soc. Symp. Proc. **20** (North-Holland, Amsterdam) p. 213.

Shimizu, A., and H. Kamimura, 1983, Synth. Met. **5**, 301.

Shoenberg, D., 1952, Trans. R. Soc. London Ser. A **245**, 1.

Simmons, J.H.W., 1957, in: Proc. 3rd Conf. on Carbon, Buffalo, NY, 1957 (Pergamon Press, New York, 1959) p. 559.

Simmons, J.H.W., 1964, Third United Nations Int. Conf. on the Peaceful Uses of Atomic Energy, Geneva, 1964, A (Conf.), 28/P, p. 163.

Simon, Ch., F. Batallan, I. Rosenmann and H. Fuzellier, 1981, Phys. Rev. B **23**(6), 2386.

Singhal, S.C., 1980, Physica B **99**, 536.

Slater, J.C., 1965, Quantum Theory of Molecules and Solids, Vol. **2**, (McGraw-Hill, New York).

Slater, J.C., and G.F. Koster, 1954, Phys. Rev. **94**, 1498.

Slonzewski, J.C., and P.R. Weiss, 1958, Phys. Rev. **109**(2), 272.

Slutskin, A.A., 1967, Zh. Eksp. & Teor. Fiz. **53**(8), 767 [1968, Sov.Phys.-JETP **26**(2), 474].

Slutskin, A.A., 1970, Zh. Eksp. & Teor. Fiz. **58**(3), 1098 [Sov.Phys.-JETP **31**(3), 589].

Slutskin, A.A., 1973, Zh. Eksp. & Teor. Fiz. **65**(11), 2114 [1974, Sov. Phys.-JETP **38**(5), 1057].

Smith, T.F., L.E. Delong, A.R. Moodenbaugh, T.H. Geballe and R.E. Schwall, 1972, J. Phys. C **5**, L230.

Smith, T.F., R.N. Shelton and R.E. Schwall, 1975, J. Phys. F **5**, 1713.

Smoluchowski, R., 1953, Rev. Mod. Phys. **25**, 178.

Solin, S.A., 1977, Mater. Sci. & Eng. **31**, 153.

Solin, S.A., 1980, Physica B **99**, 443.

Solin, S.A., 1981, in: Physics of Intercalation Compounds, eds L. Pietronero and E. Tosatti, Springer Series in Solid-State Sciences, Vol. 38 (Springer, Berlin) p. 214.

Solin, S.A., 1982, Adv. Chem. Phys. **49**, 45.

Song, J.J., D.D.L. Chung, P.C. Eklund and M.S. Dresselhaus, 1976, Solid State Commun. **20**, 1111.

Soule, D.E., 1958a, Phys. Rev. **112**, 698.

Soule, D.E., 1958b, Phys. Rev. **112**, 708.

Soule, D.E., 1964, IBM J. Res. Develop. **8**, 268.

Soule, D.E., J.W. McClure and L.B. Smith, 1964, Phys. Rev. A **134**, 453.

Spain, I.L., 1969, in: Proc. Conf. on Electronic Density of States (NBS Publication, Washington, DC) p. 717.

Spain, I.L., 1971, in: Proc. Int. Conf. on the Physics of Semimetals and Narrow-Bandgap Semiconductors, Dallas, TX (Pergamon Press, New York) p. 177.

Spain, I.L., 1973, in: Chemistry and Physics of Carbon, Vol. 8, eds P.L. Walker and A. Thrower (Marcel Dekker, New York) pp. 1–15.

Spain, I.L., 1980, in: Chemistry and Physics of Carbon, Vol. 16, eds P.L. Walker Jr and P.A. Thrower (Dekker, New York) p. 119.

Spain, I.L., A.R. Ubbelohde and D.A. Young, 1967, Philos. Trans. R. Soc. London Ser. A **262**, 1128.

Spence, G.B., 1961, in: Proc. 5th Conf. on Carbon, Vol. 2 (Pergamon Press, New York, 1962) p. 531.

Spencer, H.M., 1948, J. Ind. Eng. Chem. **40**, 2152.

Spry, W.J., and P.M. Scherer, 1960, Phys. Rev. **120**, 826.

Stark, R.W., and C.B. Friedberg, 1971, Phys. Rev. Lett. **26**(10), 556.

Stepanov, N.F., and V.M. Tatevskii, 1961, Zh. Strukt. Khim. **2**(2), 204.

Stumpp, E., 1977, Mater. Sci & Eng. **31**, 53.

Suematsu, H., and S. Tanuma, 1972, J. Phys. Soc. Jpn. **33**(6), 1619.

Suematsu, H., K. Higuchi and S. Tanuma, 1980, J. Phys. Soc. Jpn. **48**(5), 1541.

Suganuma, M., T. Kondow and U. Mizutani, 1981, Phys. Rev. B **24**, 6083.

Sugihara, K., 1966, J. Phys. Soc. Jpn. **21**, 324.

Sugihara, K., 1982, in: Intercalated Graphite, eds M.S. Dresselhaus, G. Dresselhaus, J.E. Fischer and M.J. Moran, Mater. Res. Soc. Symp. Proc. **20** (North-Holland, Amsterdam) p. 179.

Sugihara, K., 1984, Phys. Rev. B **29**, 5872.

Sugihara, K., and S. Ono, 1966, J. Phys. Soc. Jpn. **21**, 631.

Sugihara, K., and H. Sato, 1963, J. Phys. Soc. Jpn. **18**, 332.

Sugihara, K., T. Tsuzuku, T. Takezawa, Y. Hishiyama and A. Ono, 1972, J. Phys. & Chem. Solids **33**, 1475.

Suzuki, M., H. Ikeda, H. Suematsu, Y. Endoh, H. Shiba and M.T. Hutchings, 1980a, J. Phys. Soc. Jpn. **49**, 671.

Suzuki, M., H. Ikeda, H. Suematsu, Y. Endoh and H. Shiba, 1980b, in: Proc. IVth Yamada Conf. on Physics and Chemistry of Layered Materials, Sendai, Japan, 8–10 September 1980, eds Y. Nishina, S. Tanuma and H.W. Myron (North-Holland, Amsterdam, 1981) [see also Physica B **105**(1–3), pp. 280–284].

Suzuki, M., H. Ikeda, H. Suematsu, Y. Endoh and H. Shiba, 1981, Physica B **105**, 280.

Syme-Johnson, A.W., 1967, Acta Crystallogr. **23**, 770.

Taft, E.A., and H.R. Philipp, 1965, Phys. Rev. A **138**, 197.

Tanuma, S., 1983, in: Symp. on Recent Topics in Semiconductor Physics, eds H. Kamimura and Y. Toyozawa (World Scientific, Singapore) p. 145.

Tanuma, S., H. Suematsu, K. Higuchi, R. Inada and Y. Onuki, 1978, in: Proc. Conf. on the Applications of High Magnetic Fields in Semiconductor Physics, ed. J.E. Ryan (Clarendon Press, Oxford) p. 85.

Tanuma, S., A. Inada, A. Fuzukawa and O. Takahashi, 1981a, in: Physics in High Magnetic Field, Springer Series in Solid-State Science, Vol. 24 (Springer, Berlin) p. 316.

Tanuma, S., Y. Iye, O. Takahashi, Y. Koike and Y. Iye, 1981b, Synth. Met. **2**, 341.

Tanuma, S., Y. Iye and Y. Koike, 1981c, in: Physics of Intercalation Compounds, eds L. Pietronero and E. Tosatti, Springer Series in Solid-State Sciences, Vol. 38 (Springer, Berlin) p. 298.

Tanuma, S., H. Suematsu, K. Higuchi, R. Inada and Y. Onuki, 1983, in: Electronic Properties of Graphite Intercalation Compounds and Tantalum Disulphide at High Magnetic Fields (Institute for Solid State Physics, University of Tokyo) p. 85.

Tatewskii, V.M., 1953, Chemical Structure of Hydrocarbons and Regularities Governing Their Physical and Chemical Properties (Moscow State University, Moscow).

Thomas, J.M., G.R. Millwood, R.F. Schögl and H.-P. Boehm, 1980, Mater. Res. Bull. **15**(5), 671.

Thrower, P.A., 1963, AERE-R-4463, November 1963.

Tobochnik, J., and G.V. Chester, 1979, Phys. Rev. B **20**, 3761.

Tojoda, S., and S. Mrozowski, 1969, Carbon **7**(3), 239.

Tosatti, E., and F. Bassani, 1970, Nuovo Cimento B **65**(1), 161.

Toy, W.V., M.S. Dresselhaus and G. Dresselhaus, 1977, Phys. Rev. B **15**, 4077.

Toyozawa, Y., 1962, J. Phys. Soc. Jpn. **17**, 986.

Tsu, R., J.H. Gonzalez, J.G. Hernandez and C.A. Luengo, 1977, Solid State Commun. **24**, 809.

Tsu, R., J.H. Gonzalez, J.G. Hernandez and C.A. Luengo, 1978, Solid State Commun. **27**, 507.

Tsukada, M., K. Nakao, Y. Uemura and S. Nagai, 1972, J. Phys. Soc. Jpn. **32**, 54.

Tsuzuku, T., 1979, Carbon **17**, 293.

Tsuzuku, T., T. Takezawa, Y. Hishiyama and A. Ono, 1972, Philos. Mag. **25**, 929.

Tuinstra, F., and J.L. Koenig, 1970, J. Chem. Phys. **53**(3), 1126.
Ubbelohde, A.R., 1972, Proc. R. Soc. London Ser. A **327**, 289.
Ubbelohde, A.R., 1976, Carbon **14**(1), 1–5.
Ubbelohde, A.R., and F.A. Lewis, 1960, Graphite and Its Crystal Compounds (Clarendon Press, Oxford).
Uebersfeld, J., and E. Erb, 1959, in: Proc. 3rd Conf. on Carbon, Buffalo, NY (Pergamon Press, New York) p. 103.
Uher, C., and D.T. Morelli, 1982, in: Intercalated Graphite, eds M.S. Dresselhaus, G. Dresselhaus, J.E. Fischer and M.J. Moran, Mater. Res. Soc. Symp. Proc. **20** (North-Holland, Amsterdam) p. 163.
Underhill, J., S.Y. Leung, G. Dresselhaus and M.S. Dresselhaus, 1979, Solid State Commun. **29**, 769.
Ushio, H., T. Uda and Y. Uemura, 1972, J. Phys. Soc. Jpn. **23**(6), 1551.
Van der Hoeven, J.C., and P.H. Keeson, 1963, Phys. Rev. **130**, 1318.
Van Haeringen, W., and H.G. Junginger, 1969, Solid State Commun. **7**, 1723.
Van Vleck, J.H., 1932, Theory of Electric and Magnetic Susceptibilities (Clarendon Press, Oxford).
Vangelisti, R., and A. Hérold, 1976, Carbon **14**, 333.
Vangelisti, R., G. Furdin, B. Carton and A. Hérold, 1974, C. R. Hebd. Séances Acad. Sci. Ser. C **278**(12), 869.
Vedeneev, V.V., 1962, Bond-Breaking Energies, Ionisation Potentials, and Electron Affinity, Handbook (USSR Academy of Sciences, Moscow).
Vjatkin, S.E., A.N. Deev, V.G. Nagornyi, V.S. Ostrovskii, A.M. Sigarev and G.A. Sokker, 1967, Nuclear Graphite (Atomizdat, Moscow).
Vogel, F.L., 1976, Carbon **14**, 175.
Vogel, F.L., 1977, J. Mater. Sci. **12**(5), 982.
Vogel, F.L., 1979/1980, Synth. Met. L **2**, 79.
Vogel, F.L., 1979a, Carbon **17**, 255.
Vogel, F.L., 1979b, Intercalation compounds of graphite, in: Molecular Metals, ed. W.E. Hardfield (Plenum Press, New York) p. 261.
Vogel, F.L., 1979c, Synth. Met. **1**(3), 279.
Vogel, F.L., G.M.T. Foley, C. Zeller, E.R. Falardeau and J. Gan, 1977, Mater. Sci. & Eng. **31**(14), 261.
Vogel, F.L., H. Fuzellier, C. Zeller and E.J. McRae, 1979, Carbon **17**, 255.
Vogel, F.L., R.A. Wachnik and L.A. Pendrys, 1981, in: Physics of Intercalation Compounds, eds L. Pietronero and E. Tosatti, Springer Series in Solid-State Sciences, Vol. 38 (Springer, Berlin) p. 288.
Von Ortenberg, M.V., 1973, J. Phys. & Chem. Solids **1**, 143.
Wachnik, R.A., L.A. Pendrys and F.L. Vogel, 1982, Solid State Commun. **43**(1), 5.
Wada, N., 1981, Phys. Rev. B **24**, 1065.
Wada, N., and T. Ishikawa, 1983, J. Phys. Soc. Jpn. **52**, 1774.
Wada, N., and S.A. Solin, 1981, Physica B **105**(1–4), 268.
Wada, N., R. Clarke and S.A. Solin, 1980a, Synth. Met. **2**, 27.
Wada, N., R. Clarke and S.A. Solin, 1980b, Solid State Commun. **35**, 675.
Wagoner, G., 1960, Phys. Rev. **118**, 647.
Wakabayashi, N., H.G. Smith and R.M. Nicklow, 1975, Phys. Rev. B **12**, 659.
Wallace, P.R., 1947, Phys. Rev. **71**, 622.
Wallace, P.R., and O.P. Gupta, 1974, Solid State Commun. **15**, 1577.
Warren, B.E., 1941, Phys. Rev. **59**(9), 693.

Wertheim, G.K., P.M.Th.M. Van Attekum, H.J. Guggenheim and K.E. Clements, 1980, Solid State Commun. **33**, 809.

Whittingham, M.S., 1975, J. Electrochem. Soc. **121**, 1308.

Whittingham, M.S., and L.B. Ebert, 1979, in: Physics and Chemistry of Materials with Layered Structures, Vol. 6, ed. F. Levy (Reidel, Dordrecht) p. 533.

Wiggs, P.K.C., 1958, Carbon and Graphite (SWI, London) p. 252.

Williamson, G.K., 1962, in: Proc. Int. Conf. on Properties of Reactor Materials and the Effects of Radiation Damage, Berkeley, CA (Butterworths, London) p. 144.

Williamson, S.J., S. Foner and M.S. Dresselhaus, 1965, Phys. Rev. A **140**, 1429.

Williamson, S.J., S. Foner and M.S. Dresselhaus, 1966, Carbon **4**, 29.

Wills, R.F., B. Fenerbacher and B. Fitton, 1971, Phys. Rev. B **4**, 244.

Wills, R.F., B. Fitton and G.S. Painter, 1974, Phys. Rev. **9**, 1926.

Wilson, E.V., J.C. Decius and P.C. Cross, 1955, Molecular Vibrations (McGraw-Hill, New York).

Woollam, J.A., 1970a, Phys. Rev. Lett. **25**, 810.

Woollam, J.A., 1970b, Phys. Lett. A **32**, 115.

Woollam, J.A., 1971a, Phys. Rev. B **4**, 3393.

Woollam, J.A., 1971b, Phys. Rev. A **35**, 332.

Woollam, J.A., 1971c, Phys. Rev. B **3**, 1148.

Woollam, J.A., E.D. Haugland, M.B. Dowell, N. Kambe, E. Mendez, F. Hakimi, G. Dresselhaus and M.S. Dresselhaus, 1979, in: Proc. 14th Biennial Conf. on Carbon, Extended Abstracts and Program, June 25–29, 1979 (Pennsylvania State University, University Park, PA) p. 320.

Wostenholm, G.H., and B. Yates, 1973, Philos. Mag. **27**, 185.

Wright, R.B., R. Varma and D.M. Gruen, 1976, J. Nucl. Mater. **63**, 415.

Wu, T.C., F.L. Vogel, L.A. Pendrys and C. Zeller, 1981, Mater. Sci. & Eng. **47**(2), 161.

Wyckoff, R.W.G., 1964, Crystal Structure, Vol. 1 (Interscience, New York).

Yeh, V., L.E. De Long and P.C. Eklund, 1982, Bull. Am. Phys. Soc. **27**, 340.

Yeoman, M., and D.A. Young, 1969, J. Phys. C Ser. 2, **2**(10), 1742.

Chong, Hon Yew, 1979, Thèse (Université de Grenoble, Grenoble).

Yoshimori, A., and Y. Kitano, 1956, J. Phys. Soc. Jpn. **2**, 352.

Yoshioka, D., and H. Fukuyama, 1981, J. Phys. Soc. Jpn. **50**, 725.

Young, D.A., 1968, Carbon **6**, 135.

Young, J., and J.U. Koppel, 1965, J. Chem. Phys. **42**(1), 357.

Zabel, H., S.C. Moss, N. Caswell and S.A. Solin, 1979, Phys. Rev. Lett. **43**, 2022.

Zallen, R., 1968, Phys. Rev. **173**, 824.

Zanini, M., and J.E. Fischer, 1977, J. Mater. Sci. & Eng. **31**, 169.

Zanini, M., D. Grubisic and J.E. Fischer, 1978a, Phys. Status Solidi b **90**(1), 151.

Zanini, M., S. Basu and J.E. Fischer, 1978b, Carbon **16**, 211.

Zanini, M., L.-Y. Ching and J.E. Fischer, 1978c, Phys. Rev. B **18**, 2020.

Zeller, C., A. Denenstein and G.M.T. Foley, 1979a, Rev. Sci. Instrum. **50**, 602.

Zeller, C., L.A. Pendrys and F.L. Vogel, 1979b, J. Mater. Sci. **14**, 2241.

Zeppenfeld, K., 1968, Z. Phys. **211**, 391.

Zorin, F.I., 1979, Thesis (Moscow State University, Moscow).

Zunger, J.M., 1978, Phys. Rev. B **17**, 626.

AUTHOR INDEX

Abrikosov, A.A. 93, 103
Adams, E.N. 177
Adkins, C.Y. 70
Agranovich, V.M. 446
Ahmadieh, A.A. 340, 360
Akinaga, M., *see* Mase, S. 103
Al-Jishi, R. 345, 375, 379, 380, 385, 387, 394
Al-Jishi, R., *see* Giergiel, J. 384, 395
Alexander, M.G. 310, 402, 405
Alexander, M.G., *see* Onn, D.G. 402, 407
Alldredge, G.P. 345
Altshuler, B.L. 107
Anderson, J.R. 67, 115, 125, 161, 166, 173, 174
Andres, K., *see* Hannay, N.B. 218, 309, 310
Arkhipov, R.G. 61, 86, 160, 169, 173
Armand, M. 418, 420
Arnold, G.M. 70, 358
Aronov, A.G., *see* Altshuler, B.L. 107
Aronson, S. 206, 216
Aust, R.B. 170
Austin, A.E. 435
Avdeenko, M.A. 431, 432
Avdeev, V.V., *see* Brandt, N.B. 243, 245, 246, 251, 265, 290, 295
Awaji, H. 423
Axe, J.D., *see* Fuerst, C.D. 295
Ayache, C. 402–407
Azbel, M.Ya., *see* Lifshitz, I.M. 274

Bacon, D.J., *see* Nicholson, A.P.P. 19, 27, 28, 324, 339, 344
Bacon, G.E. 11
Bagrov, G.N., *see* Avdeenko, M.A. 431, 432
Baiker, A. 217
Bailey, A.C. 25
Baker, C. 24
Banarjee, R.K., *see* Saha, A.R. 71
Bartlett, N. 244
Baskin, J. 169
Bassani, F. 33, 41
Bassani, F., *see* Greenaway, D.L. 33, 134
Bassani, F., *see* Tosatti, E. 134
Basset, J. 21, 23, 24
Basu, S. 242, 243, 246, 403, 405
Basu, S., *see* Zanini, M. 206, 393
Batallan, F. 225, 265, 268
Batallan, F., *see* Bok, J. 201
Batallan, F., *see* Rosenmann, I. 217, 268
Batallan, F., *see* Simon, Ch. 225, 268
Batra, I.P., *see* Samuelson, L. 35
Beavan, L.A. 422, 436
Beer, A.C. 104
Beguin, F. 421
Bellafiore, D., *see* Aronson, S. 216
Bellodi, G. 135, 153, 154
Belousov, V.M. 420
Bender, A.S. 260
Beneslavskii, S.D. 94, 95, 97, 99
Bergenlid, U. 322, 398
Berker, A.N. 217
Berlincourt, T.G. 38, 124–126
Berlinova, I.V., *see* Panayotov, I.M. 421
Berman, R. 6
Bernal, J.D. 3, 9

Bernstein, H.J. 6
Besenhard, J.O. 419
Bhagavantam, S. 347
Bhander, B., *see* Morris, R.C. 313
Bhattacharya, S. 75
Billand, D., *see* McRae, E. 242, 243, 248
Billaud, D. 206
Binkele, L. 442
Blackman, L.C.F. 12, 75
Blackslee, O.L. 22, 23, 25, 341
Blinowski, J. 226, 227, 229, 231, 265, 267
Blinowski, J., *see* Nguyen, H.H. 226
Boardman, A.D. 90
Bobrovnikov, Yu.N. 71
Boehm, H.-P. 10
Boehm, H.-P., *see* Schögl, R. 421
Boehm, H.-P., *see* Thomas, J.M. 208, 296
Boeuf, A. 416
Bok, J. 201
Bok, J., *see* Batallan, F. 225, 268
Bollman, W. 440
Bolsaitis, P., *see* Green, J.F. 23, 151
Bonfiglioli, G. 12
Bonjour, E., *see* Ayache, C. 402–407
Borghesi, A., *see* Bellodi, G. 135, 153, 154
Bowman, J.C. 21, 23, 397
Brandt, N.B. 63, 78–80, 83, 91–93, 103, 113, 114, 124–126, 128, 162–166, 172, 173, 188, 190–192, 216, 221, 223, 232, 243, 245, 246, 249, 251, 261, 262, 265, 268, 270, 271, 290–293, 295–297, 318
Breimeier, W., *see* Schögl, R.F. 421
Breisacher, P. 427
Bridgman, P.W. 23, 170
Briggs, A., *see* Chouteau, G. 183, 184, 186, 187
Briggs, A., *see* Eberhardt, W. 119
Bright, A.A. 105, 420
Brillson, L.J. 135, 341, 344, 359
Brockhouse, B.N., *see* Dolling, G. 22, 23, 341, 344, 352, 353, 355
Brocklehurst, J.E. 436
Brooks, H., *see* Krumhansl, J. 323
Brown, A.R.G. 75
Budke, J. 433
Bullock, R.E. 438–441
Bundy, F.P. xi, 6
Burstein, E., *see* Brillson, L.J. 135, 341, 344, 359
Buzdin, A.I. 99, 195
Buzdin, A.I., *see* Beneslavskii, S.D. 97, 99

Caciuffo, R., *see* Boeuf, A. 416
Cahill, J.E. 378
Caisley, J., *see* Martin, D.G. 433, 436
Campion, P. 426
Carlson, G.C., *see* Friedel, R.A. 344
Carmona, F. 70
Carton, B., *see* Vangelisti, R. 214
Caswell, N. 211, 367–370, 381, 395
Caswell, N., *see* Clarke, R. 213
Chaberski, A., *see* Mrozowski, S. 104, 107
Chamberod, A., *see* Hohlwein, D. 219
Chester, G.V., *see* Tobochnik, J. 408, 409
Chien, T.C., *see* Dresselhaus, M.S. 268
Chien, T.C., *see* Endo, M. 419
Chien, T.C., *see* Mendez, E. 67, 68, 281–284
Chisholm, R.C. 409
Chouteau, G. 183, 184, 186, 187
Chudinov, S.M., *see* Brandt, N.B. 103, 221, 223
Chung, D.D.L. 217, 281
Chung, D.D.L., *see* Dresselhaus, M.S. 201, 217, 366, 392, 393
Chung, D.D.L., *see* Platts, D.A. 49, 60
Chung, D.D.L., *see* Song, J.J. 367, 378
Clarke, R. xi, 213, 253, 294
Clarke, R., *see* Wada, N. 294, 321
Clements, K.E., *see* Wertheim, G.K. 304, 305
Coey, J.M.B., *see* Hohlwein, D. 219
Coleman, R.V., *see* Morris, R.C. 313, 314
Conard, J., *see* Lauginie, P. 255
Consadori, F. 416
Cooper, J.D. 40, 124–126
Corbato, F.Y. 30
Cords, H. 438
Corelly, J.C., *see* Matheny, R.A. 438
Coulson, C.A. 4, 5
Cowley, J.M. 214
Crabtree, G.W., *see* Karim, D.P. 274
Cross, P.C., *see* Wilson, E.V. 347

Dail, H.W., *see* Galt, I.K. 137
Dalessandro, J.A. 438, 445
Dalton, K. 106

Das, A.K., *see* Saha, A.R. 71
Datars, W.R., *see* Doezema, R.E. 49, 60, 67, 68, 125, 126, 142–144, 146, 172
Daumas, N. 207, 208, 255, 296
De Long, L.E. 319–321
De Long, L.E., *see* Yeh, V. 218, 319, 320
de Rouffignac, E., *see* Alldredge, G.P. 345
De Sorbo, W. 322, 338, 397–400
de Wette, F.W., *see* Alldredge, G.P. 345
Decius, J.C., *see* Wilson, E.V. 347
Deev, A.N., *see* Vjatkin, S.E. 422, 424, 428, 429, 439, 443–445
Delhaes, G., *see* Carmona, F. 70
Delhaes, P. 104–107, 301, 302, 398, 402–404
Delle, W. 426, 427, 440, 441, 445
Delle, W.W., *see* Fourré, J. 422
Delong, L.E., *see* Smith, T.F. 318
Dem'yanov, S.E., *see* Gostishev, V.I. 274
Denenstein, A., *see* Zeller, C. 251, 253
Di Salvo, F.J., *see* Safran, S.A. 303
Dietz, V.R., *see* Hooley, J.G. 419
Dillon, R.O. 47, 67, 79, 83–87, 143, 172, 173, 185
Dmitrenko, L.M., *see* Postnikov, V.A. 420
Dobrolyubova, N.L., *see* Postnikov, V.A. 420
Doezema, R.E. 49, 60, 67, 68, 125, 126, 142–144, 146, 172
Dolgov, O.V., *see* Buzdin, A.I. 195
Dolling, G. 22, 23, 341, 344, 352, 353, 355
Dominguez, J.A., *see* Murday, J.S. 419
Doni, E. 33
Dowell, M.B., *see* Oshima, H. 419
Dowell, M.B., *see* Woollam, J.A. 265
Dresselhaus, G. 47, 48, 51, 55–59, 68, 112, 137, 146, 238
Dresselhaus, G., *see* Al-Jishi, R. 345, 375, 379, 380, 385, 387, 394
Dresselhaus, G., *see* Berker, A.N. 217
Dresselhaus, G., *see* Chung, D.D.L. 217
Dresselhaus, G., *see* Dresselhaus, M.S. xi, 201, 209, 217, 238, 239, 245, 286–288, 362, 366, 371, 372, 374, 392–395, 417
Dresselhaus, G., *see* Eklund, P.C. 367, 378, 380
Dresselhaus, G., *see* Elahy, M. 304–306
Dresselhaus, G., *see* Giergiel, J. 384, 395
Dresselhaus, G., *see* Gualberto, G.M. 367
Dresselhaus, G., *see* Iye, Y. 101, 102
Dresselhaus, G., *see* Johnson, L.C. 68, 135, 237, 238, 280
Dresselhaus, G., *see* Kambe, N. 216, 217, 395
Dresselhaus, G., *see* Leung, S.Y. 22, 23, 375, 385
Dresselhaus, G., *see* Shayegan, M. 402, 407, 408
Dresselhaus, G., *see* Toy, W.V. 39, 48, 67, 68, 137, 147, 149, 150
Dresselhaus, G., *see* Underhill, J. 201, 367, 370, 372
Dresselhaus, G., *see* Woollam, J.A. 265
Dresselhaus, M.S. xi, 53, 54, 146, 201, 209, 217, 238, 239, 245, 247, 268, 286–288, 362, 366, 371, 372, 374, 392–395, 417
Dresselhaus, M.S., *see* Berker, A.N. 217
Dresselhaus, M.S., *see* Chung, D.D.L. 217, 281
Dresselhaus, M.S., *see* Dresselhaus, G. 48, 51
Dresselhaus, M.S., *see* Eklund, P.C. 367, 378, 380
Dresselhaus, M.S., *see* Endo, M. 419
Dresselhaus, M.S., *see* Iye, Y. 101, 102
Dresselhaus, M.S., *see* Kambe, N. 216, 217, 395
Dresselhaus, M.S., *see* Leung, S.Y. 22, 23, 375, 385
Dresselhaus, M.S., *see* Mendez, E. 53, 67, 68, 148, 150, 157, 164, 168, 169, 172, 173, 281–284
Dresselhaus, M.S., *see* Misu, A. 67, 68, 152–157, 171, 173
Dresselhaus, M.S., *see* Platts, D.A. 49, 60
Dresselhaus, M.S., *see* Schroeder, P.R. 55, 57, 137, 146–148, 150, 172
Dresselhaus, M.S., *see* Song, J.J. 367, 378
Dresselhaus, M.S., *see* Toy, W.V. 39, 48, 67, 68, 137, 147, 149, 150
Dresselhaus, M.S., *see* Underhill, J. 201, 367, 370, 372
Dresselhaus, M.S., *see* Williamson, S.J. 115, 116, 124–126
Dresselhaus, M.S., *see* Woollam, J.A. 265
Drickamer, H.G., *see* Aust, R.B. 170
Drickamer, H.G., *see* Lynch, R.W. 158

Drickamer, H.G., *see* Samara, G.A. 170
Drozd, A.A., *see* Gostishev, V.I. 274
Drude, P. 132
Duff, M.J., *see* Kelly, B.T. 27
Dupart, E., *see* Marchand, A. 180, 181
Dutta, A.K. 75, 76

Eatherly, W.P., *see* Kennedy, C.K. 438
Eberhardt, W. 119, 233, 234
Ebert, L.B. 421
Ebert, L.B., *see* Seilig, H. 416
Ebert, L.B., *see* Whittingham, M.S. 416, 420
Edelman, V.C. xi
Edelstaff, P.A. 337
Edelstein, V.M., *see* Rashba, E.I. 94
Eeles, W.T. 214, 268, 378
Ehrenreich, H., *see* Philipp, H.R. 132
Eidus, Ya.I., *see* Mashinskii, V.I. 420
Eklund, P.C. 367, 378, 380
Eklund, P.C., *see* De Long, L.E. 319–321
Eklund, P.C., *see* Dresselhaus, M.S. 201, 217, 366, 392, 393
Eklund, P.C., *see* Giergiel, J. 362, 372, 373, 375, 376, 384, 395
Eklund, P.C., *see* Song, J.J. 367, 378
Eklund, P.C., *see* Yeh, V. 218, 319, 320
Elahy, M. 304–306
Ellis, D.E., *see* Painter, G.S. 34, 39, 41
ElMakrini, M., *see* Lauginie, P. 255
Elzinga, M., *see* Clarke, R. 253
Endo, M. 419
Endoh, Y., *see* Suzuki, M. 394
Engle, G.B., *see* Beavan, L.A. 422
Entralgo, E., *see* Beneslavskii, S.D. 94, 95
Erb, E., *see* Uebersfeld, J. 194
Estrade, H., *see* Lauginie, P. 255

Falardeau, E.R. 216
Falardeau, E.R., *see* Foley, G.M.T. 242, 243, 245
Falardeau, E.R., *see* Khanna, S.K. 255
Falardeau, E.R., *see* Vogel, F.L. 246
Falkovskii, L.A. xi
Fawcett, F., *see* Reifenberger, R. 274
Fenerbacher, B., *see* Wills, R.F. 39
Fenton, E.W. 95
Fialkov, A.C. 430
Fieguth, P. 422
Fife, A.A., *see* Consadori, F. 416
Finch, G.I. 10
Firey, B., *see* Alldredge, G.P. 345
Fischer, J.E. xi, 242, 243, 245, 246, 252, 253, 255
Fischer, J.E., *see* Ayache, C. 402–407
Fischer, J.E., *see* Basu, S. 242, 243, 246, 403, 405
Fischer, J.E., *see* Eberhardt, W. 233, 234
Fischer, J.E., *see* Eklund, P.C. 367
Fischer, J.E., *see* Fuerst, C.D. 295, 321
Fischer, J.E., *see* Guérard, D. 242, 243, 259
Fischer, J.E., *see* Khanna, S.K. 255
Fischer, J.E., *see* Onn, D.G. 218, 259
Fischer, J.E., *see* Potter, M.E. 246, 251
Fischer, J.E., *see* Zanini, M. 206, 393
Fischer, L.M., *see* Itskevich, E.S. 124, 125, 161, 162, 166, 173, 174
Fishberg, C., *see* Aronson, S. 206, 216
Fitton, B., *see* Wills, R.F. 39, 135
Flanders, P., *see* Basu, S. 242, 243, 246, 403, 405
Flandrois, S., *see* Onn, D.G. 402, 407
Fletcher, R. 274
Flood, D.J. 116
Foley, G.M.T. 242, 243, 245
Foley, G.M.T., *see* Guérard, D. 242, 243, 259
Foley, G.M.T., *see* Onn, D.G. 218, 259
Foley, G.M.T., *see* Vogel, F.L. 246
Foley, G.M.T., *see* Zeller, C. 251, 253
Foner, S., *see* Shapira, Y. 184
Foner, S., *see* Williamson, S.J. 115, 116, 124–126
Fourré, J. 422
Frankl, G., *see* Aronson, S. 206, 216
Franklin, R.E. 7
Freaker, S.M., *see* Adkins, C.Y. 70
Freund, A., *see* Boeuf, A. 416
Friedberg, C.B. 273, 274
Friedberg, C.B., *see* Stark, R.W. 273, 277
Friedel, R.A. 344
Friend, H.R. 318
Fritz, H.P., *see* Besenhard, J.O. 419
Fritz, I.J., *see* Gauster, W.B. 25, 151
Fritz, P., *see* Fieguth, P. 422
Friudt, R.F., *see* Consadori, F. 416

Fröhlich, H. 90
Frye, K., *see* Delle, W. 440, 441
Fuchs, L.H., *see* Primak, W. 74–76
Fuerst, C.D. 295, 321
Fuerst, C.D., *see* Basu, S. 242, 243, 246, 403, 405
Fuerst, C.D., *see* Fischer, J.E. 252, 253
Fujita, S. 106
Fujita, S., *see* Dalton, K. 106
Fukami, T., *see* Mase, S. 103
Fukuyama, H. 180
Fukuyama, H., *see* Yoshioka, D. 66, 102
Funahashi, S. 389, 396
Furdin, G., *see* Batallan, F. 265
Furdin, G., *see* Blinowski, J. 226, 227, 229, 231
Furdin, G., *see* Pendrys, L.A. 310
Furdin, G., *see* Rosenmann, I. 217
Furdin, G., *see* Vangelisti, R. 214
Furukawa, A., *see* Iye, Y. 101, 102
Fuzellier, H. 215
Fuzellier, H., *see* Batallan, F. 265
Fuzellier, H., *see* Rosenmann, I. 268
Fuzellier, H., *see* Simon, Ch. 225, 268
Fuzukawa, A., *see* Tanuma, S. 66, 101

Gain, R. 433, 438, 443
Galt, I.K. 125, 126, 137
Gan, J., *see* Vogel, F.L. 246
Ganguli, N. 179
Ganguli, N., *see* Krishnan, K.S. 74–76, 175, 178
Gapeeva, T.I., *see* Postnikov, V.A. 420
Gauster, W.B. 23, 25, 151
Geballe, T.H., *see* Hannay, N.B. 218, 309, 310
Geballe, T.H., *see* Smith, T.F. 318
Giergiel, J. 362, 372, 373, 375, 376, 384, 395
Gilat, G., *see* Raubenheimer, L.J. 342
Gilchrist, K.E., *see* Brocklehurst, J.E. 436
Ginsberg, D.M., *see* Rupprecht, G. 416
Ginzburg, N.I., *see* Brandt, N.B. 318
Ginzburg, V.L. 309
Giovanini, B. 105
Girifalco, L.A., *see* Holzwarth, N.A.W. 42, 233, 234, 315, 404
Glinskii, M.A., *see* Gostishev, V.I. 274
Goggin, P.R. 436, 442
Goldberg, H.A. 419
Goldsmith, M. 301
Golé, J. 421
Golubeva, M.A., *see* Postnikov, V.A. 420
Gonzalez, J.H., *see* Tsu, R. 361, 363
Gortz, H., *see* Delle, W. 426, 427, 445
Goshorn, D.P., *see* Alexander, M.G. 310, 402, 405
Gostishev, V.I. 274
Graham, D.G., *see* Boardman, A.D. 90
Gray, H., *see* Clarke, R. 253
Green, J.F. 23, 27, 151
Greenaway, D.L. 33, 134
Grubisic, D., *see* Zanini, M. 393
Gruen, D.M., *see* Wright, R.B. 361, 363
Gualberto, G.M. 367
Guérard, D. 206, 212, 216, 242, 243, 259
Guérard, D., *see* Alexander, M.G. 310, 402
Guérard, D., *see* Delhaes, P. 301, 302, 402–404
Guérard, D., *see* Lauginie, P. 255
Guérard, D., *see* Nemanich, R.J. 392
Guggenheim, H.J., *see* Wertheim, G.K. 304, 305
Guizetti, G., *see* Bellodi, G. 135, 153, 154
Güntherodt, H.J., *see* Pfluger, P. 416
Gupta, O.P. 55, 57, 63, 66
Gupta, O.P., *see* Wallace, P.R. 187
Gygax, S., *see* Consadori, F. 416

Haag, G., *see* Delle, W. 426, 427, 445
Haag, G., *see* Fourré, J. 422
Hablegger, V., *see* Baiker, A. 217
Haering, R.R. 30
Haering, R.R., *see* Fenton, E.W. 95
Hakimi, F., *see* Woollam, J.A. 265
Hall, A.W. 3
Hamilton, E.M., *see* Adkins, C.Y. 70
Hamman, D.R., *see* Safran, S.A. 201, 202
Hanlon, L.R., *see* Falardeau, E.R. 216
Hannay, N.B. 218, 309, 310
Harbeke, G., *see* Greenaway, D.L. 33, 134
Hardcastle, S.E. 25, 26, 255
Harker, H. 69
Harris, D.H.C., *see* Edelstaff, P.A. 337
Harrison, R.J., *see* Austin, A.E. 435
Harrison, W.A. 40
Hastings, J.B., *see* Fuerst, C.D. 295

Haugland, E.D., *see* Woollam, J.A. 265
Hayes, T.M., *see* Caswell, N. 211, 395
Haywood, B.C. 337
Hedgecock, F.T. 105
Hedgecock, F.T., *see* Giovanini, B. 105
Heeger, A.J., *see* Khanna, S.K. 255
Henning, G.R. 12, 218
Henson, R.W., *see* Martin, D.G. 436, 444
Heremans, J., *see* Shayegan, M. 402, 407, 408
Herinckx, G. 419
Hernandez, J.G., *see* Tsu, R. 361, 363
Hérold, A. 205, 215
Hérold, A., *see* Billaud, D. 206
Hérold, A., *see* Blinowski, J. 226, 227, 229, 231
Hérold, A., *see* Daumas, N. 207, 208, 255, 296
Hérold, A., *see* Delhaes, P. 301, 302, 402–404
Hérold, A., *see* Fuzellier, H. 215
Hérold, A., *see* Guérard, D. 206, 212, 216
Hérold, A., *see* Lagrange, P. 417
Hérold, A., *see* McRae, E. 242, 243, 248
Hérold, A., *see* Melin, J. 215, 268
Hérold, A., *see* Pendrys, L.A. 310
Hérold, A., *see* Vangelisti, R. 214
Herring, C. 104
Herrman, F. 41
Herzberg, G. 378
Hideto, M. 438
Higuchi, K. 264
Higuchi, K., *see* Koike, Y. 218, 309, 310, 312–314
Higuchi, K., *see* Suematsu, H. 259
Higuchi, K., *see* Tanuma, S. 238–240, 257, 258, 261, 263, 264
Hill, R.W., *see* Bergenlid, U. 322, 398
Hishiyama, Y. 70, 83, 124–126
Hishiyama, Y., *see* Delhaes, P. 398
Hishiyama, Y., *see* Sugihara, K. 77
Hishiyama, Y., *see* Tsuzuku, T. 77
Hoarau, J., *see* Poquet, E. 182
Hoffmann, U., *see* Rüdorff, W. 215
Hofman, H.Z., *see* Boehm, H.-P. 10
Hohlwein, D. 219
Holl, G.L.J. 435, 438, 446
Holroyd, F.W., *see* Reifenberger, R. 274
Holzwarth, N.A.W. 42, 233, 234, 280, 315, 404
Homma, J.N., *see* Clarke, R. 253
Honda, H. 194
Hooley, J.G. 203, 419
Horie, C. 385, 387, 390–393
Horie, C., *see* Maeda, M. 343–345, 360, 362, 364
Horie, Ch., *see* Miyazaki, H. 297, 298
Horn, P.M., *see* Clarke, R. 213
Houston, W.V. 331
Hove, I.E. 423
Hove, J.E. 177, 302
Hoven, H., *see* Delle, W. 440, 441
Hunte, S.J., *see* Caswell, N. 211, 395
Hutchings, M.T. 305
Hutchings, M.T., *see* Suzuki, M. 394
Hwang, D.M. 23–25, 27

Ibers, Y.A., *see* Cowley, J.M. 214
Ichikawa, M. 420
Ichikawa, M., *see* Naito, S. 420
Ichikawa, Y.H. 132
Igawa, O., *see* Naito, S. 420
Iizuki, M., *see* Funahashi, S. 389, 396
Ikeda, H., *see* Suzuki, M. 394
Inada, A., *see* Tanuma, S. 66, 101
Inada, R., *see* Tanuma, S. 238–240, 257, 258, 261, 263, 264
Industrial Graphite Engineering Handbook 423
Inoshita, T. 42, 213, 217, 235, 263, 264, 314, 404, 405
Inoshita, T., *see* Kamimura, H. 249
Inoue, M. 51, 55, 125, 126
Interrante, L.V., *see* Markiewicz, R.S. 372
Ionov, S.G., *see* Brandt, N.B. 216, 232, 243, 245, 246, 249, 251, 261, 262, 265, 268, 270, 271, 290–293, 295–297
Ioon, E.R., *see* Brandt, N.B. 221, 223
Ipatova, A., *see* Maradudin, A.A. 346
Ishikawa, T., *see* Wada, N. 307
Issi, J.-P., *see* Shayegan, M. 402, 407, 408
Itskevich, E.S. 124, 125, 161, 162, 166, 173, 174
Ivanova, R.F., *see* Postnikov, V.A. 420
Iwata, T. 437, 444
Iye, Y. 101, 102, 218, 236, 240, 265, 293,

295, 309, 311, 314, 316, 317
Iye, Y., *see* Tanuma, S. 310
Jacobi, N. 184–189
Jasuachi, S., *see* Hideto, M. 438
Javan, A., *see* Schroeder, P.R. 55, 57, 137, 146–148, 150, 172
Jay-Gerin, J.-P., *see* Jacobi, N. 184–189
Jenkins, G.M. 424
Jérome, D., *see* Friend, H.R. 318
Johnson, D.F. 79
Johnson, L.C. 68, 135, 237, 238, 280
Johnson, L.C., *see* Sharma, M.P. 150, 180, 182, 183, 300, 302, 303
Johnson, W.D., *see* Basu, S. 242, 243, 246, 403, 405
Johnson, W.D., *see* Potter, M.E. 246, 251
Jones, W., *see* Schögl, R.F. 421
Jose, J.V. 307
Junginger, H.G., *see* Van Haeringen, W. 41

Kabalkina, S.S. 23, 151
Kaburagi, Y., *see* Hishiyama, Y. 70
Kadanoff, L., *see* Jose, J.V. 307
Kaganov, M.I., *see* Lifshitz, I.M. 274
Kalnin, I.L., *see* Goldberg, H.A. 419
Kambe, N. 216, 217, 395
Kambe, N., *see* Berker, A.N. 217
Kambe, N., *see* Eklund, P.C. 378, 380
Kambe, N., *see* Mendez, E. 67, 68, 281–284
Kambe, N., *see* Woollam, J.A. 265
Kamimura, H. 249
Kamimura, H., *see* Inoshita, T. 42, 213, 217, 235, 263, 264, 314, 404, 405
Kamimura, H., *see* Ohno, T. 42, 233, 234
Kamimura, H., *see* Shimizu, A. 218, 315
Kamitakahara, W.A. 383, 385–388, 396
Kammereck, R., *see* Nakamizo, M. 361
Kapustin, G.A. 63–66, 100, 128, 183–186, 189, 190
Kapustin, G.A., *see* Brandt, N.B. 78–80, 83, 91–93, 114, 124–126, 128, 190
Karavaev, V.G., *see* Brandt, N.B. 78–80, 83, 91–93, 114, 124–126, 128, 190
Karim, D.P. 274
Karimov, Yu.S. 303, 307, 402, 407, 409
Kasakov, V.V., *see* Kasatochkin, V.I. 5
Kasatochkin, V.I. 5
Kasper, J.S., *see* Markiewicz, R.S. 372
Katz, E.I. 312
Kawamura, K. 74–77, 81, 82
Kawase, K., *see* Ichikawa, M. 420
Kechin, V.V. 61, 86, 160, 170, 172, 173
Kechin, V.V., *see* Arkhipov, R.G. 61, 86, 160, 169, 173
Kechin, V.V., *see* Likhter, A.I. 160, 170
Keeson, P.H., *see* Van der Hoeven, J.C. 22, 23, 398
Kelly, B.T. 24, 27, 436, 438
Kelly, B.T., *see* Baker, C. 24
Kelly, B.T., *see* Brocklehurst, J.E. 436
Kennedy, C.K. 438
Kepper, R., *see* Delhaes, P. 104–106
Ketterson, J.B., *see* Karim, D.P. 274
Keyer, Y.R., *see* Carmona, F. 70
Keyes, R.W. 93
Khanna, S.K. 255
Khmel'nitskii, D.E., *see* Altshuler, B.L. 107
Kido, G., *see* Nakamura, K. 101–103
Kim, H.J., *see* Fischer, J.E. 252, 253
Kirkpatrick, S., *see* Jose, J.V. 307
Kitano, Y., *see* Yoshimori, A. 27, 323, 324, 326–328, 332, 333, 335, 336, 361, 400, 401
Kittel, C. 202
Klein, C.A. 45, 69, 71, 72, 75
Klucker, R. 34, 134
Kobayashi, M. 309, 310
Koenig, J.L., *see* Tuinstra, F. 341, 344, 356–358, 360
Koga, K. 308
Kohn, W., *see* Luttinger, J.M. 177
Koike, Y. 218, 309, 310, 312–314
Koike, Y., *see* Tanuma, S. 310
Komatsu, K. 22–24, 322, 323, 328, 329, 342, 397, 398
Kondo, T., *see* Ichikawa, M. 420
Kondow, T. 402, 405
Kondow, T., *see* Funahashi, S. 389, 396
Kondow, T., *see* Mizutani, U. 302, 379, 402, 404, 405
Kondow, T., *see* Suganuma, M. 389
Koppel, J.U., *see* Young, J. 18, 27, 323, 324, 326, 335, 337, 338, 344, 360, 361
Koster, G.F., *see* Slater, J.C. 233
Kosterlitz, J.M. 306
Kostikov, V.I., *see* Brandt, N.B. 265, 290,

295
Kotosonov, A.S. 412
Kotosonov, A.S., *see* Bobrovnikov, Yu.N. 71
Kotosonov, A.S., *see* Brandt, N.B. 78–80, 83, 91–93, 113, 114, 124–126, 128, 162–166, 172, 173, 190–192, 265, 290, 295
Kotosonov, A.S., *see* Levintovich, I.Ya. 181, 193
Kovtyuchova, N.I., *see* Belousov, V.M. 420
Krishnan, K.S. 74–76, 175, 178
Krishnan, K.S., *see* Ganguli, N. 179
Krumhansl, J. 323
Krumhansl, J.A., *see* Bowman, J.C. 21, 23, 397
Künzi, H.U., *see* Pfluger, P. 416
Kuramoto, Y., *see* Horie, C. 385, 387, 390–393
Kuramoto, Y., *see* Maeda, M. 343–345, 360, 362, 364
Kuramoto, Y., *see* Miyazaki, H. 297, 298
Kurdyumov, A.V. 14
Kuvshinnikov, S.V. 79, 95, 96, 117, 119, 120, 124–126, 128, 163–165, 167, 190–192
Kuvshinnikov, S.V., *see* Brandt, N.B. 113, 162–166, 172, 173, 190–192, 216, 232, 243, 245, 246, 249, 251, 261, 262, 265, 268, 270, 271, 290–293, 295–297

Lagrange, P. 417
Lagrange, P., *see* Alexander, M.G. 310, 402
Lagrange, P., *see* Lauginie, P. 255
Lagrange, P., *see* Pendrys, L.A. 310
Laguier, R., *see* Ayache, C. 402–407
Lagutin, A.S., *see* Kapustin, G.A. 100, 183–186, 189, 190
Lambert, S.E., *see* De Long, L.E. 321
Landau, L.D. 176
Lapidus, A.L., *see* Mashinskii, V.I. 420
Larkin, A.I., *see* Altshuler, B.L. 107
Lauginie, P. 255
Le Toullec, R., *see* Blinowski, J. 226, 227, 229, 231
Lee, P.A., *see* Altshuler, B.L. 107
Leroi, G.E., *see* Cahill, J.E. 378
Leslie, J.D., *see* Rupprecht, G. 416
Letoullec, R., *see* Nguyen, H.H. 226
Leung, S.Y. 22, 23, 375, 385
Leung, S.Y., *see* Dresselhaus, G. 238
Leung, S.Y., *see* Dresselhaus, M.S. 245, 247, 268
Leung, S.Y., *see* Eklund, P.C. 367
Leung, S.Y., *see* Gualberto, G.M. 367
Leung, S.Y., *see* Underhill, J. 201, 367, 370, 372
Levintovich, I.Ya. 181, 193
Levy, F. xi
Lewis, F.A., *see* Ubbelohde, A.R. xi, 427
Lifshitz, I.M. 274
Likhter, A.I. 160, 170
Likhter, A.I., *see* Arkhipov, R.G. 61, 86, 160, 169, 173
Likhter, A.I., *see* Kechin, V.V. 160, 170, 172, 173
Lipson, H. 10
Little, W.A. 309
Lowrey, W.H. 91
Lowrey, W.H., *see* Dillon, R.O. 79, 83, 86, 87
Lozovik, Yu.E., *see* Buzdin, A.I. 195
Lucovsky, G., *see* Nemanich, R.J. 135, 344, 346, 348, 349, 356–358, 360, 362, 377
Luengo, C.A., *see* Tsu, R. 361, 363
Lumbroso, N., *see* Poquet, E. 182
Lutkov, A.I. 70
Luttinger, J.M. 177
Lynch, R.W. 158

Maeda, M. 343–345, 360, 362, 364
Maeda, M., *see* Horie, C. 385, 387, 390–393
Maeta, H. 361
Makrini, M.E., *see* Alexander, M.G. 310, 402
Manceau, J.P., *see* Delhaes, P. 301, 302, 402–404
Mani, K.K. 18, 22, 23, 26–28, 344, 360
Manini, C. 419
Mantell, C.L. 398, 399
Maple, M.B., *see* De Long, L.E. 321
Maradudin, A.A. 346
Maradudin, A.A., *see* Brillson, L.J. 135, 341, 344, 359
Marchand, A. 180, 181
Marchand, A., *see* Poquet, E. 182
Maréché, J.-F., *see* Manini, C. 419
Maréché, J.-F., *see* McRae, E. 242, 243, 248

Markiewicz, R.S. 277–279, 372
Markini, M.E., *see* Pendrys, L.A. 310
Martin, D.G. 433, 436, 444
Marx, P.C., *see* Breisacher, P. 427
Mase, S. 103
Mashinskii, V.I. 420
Massalski, T.B., *see* Kondow, T. 402, 405
Massalski, T.B., *see* Mizutani, U. 302, 402, 404, 405
Matheny, R.A. 438
Matthias, B.T., *see* Hannay, N.B. 218, 309, 310
Matuyama, E. 10
Mavroides, J.G., *see* Dresselhaus, G. 47, 146
Mavroides, J.G., *see* Dresselhaus, M.S. 53, 54, 146
Mayer, L., *see* Henning, G.R. 218
Mazurek, H., *see* Kambe, N. 395
McCarron, E.M., *see* Bartlett, N. 244
McClelland, J.D., *see* Rasor, N.S. 12
McClure, J.W. 35, 38, 39, 46, 48, 51, 53, 64, 67, 79, 83, 84, 86, 88, 91–93, 177–180, 188, 237, 258, 302
McClure, J.W., *see* Dillon, R.O. 67, 143, 172, 173, 185
McClure, J.W., *see* Sharma, M.P. 150, 180, 182, 183, 300, 302, 303
McClure, J.W., *see* Soule, D.E. 67, 114, 118, 124–126
McDonnell, F.R.M. 301
McGovern, I.T., *see* Eberhardt, W. 233, 234
McLean, T.P. 90
McNair, D., *see* Hannay, N.B. 218, 309, 310
McQuillan, B.W., *see* Bartlett, N. 244
McRae, E. 242, 243, 248
McRae, E., *see* Manini, C. 419
McWhan, D.B., *see* Fuerst, C.D. 295
Meilikhov, E.Z., *see* Kapustin, G.A. 63–66, 100, 128, 183–186, 189, 190
Mele, E. 370, 381, 382
Melin, J. 215, 268
Melin, J., *see* Batallan, F. 225, 268
Melin, J., *see* Blinowski, J. 226, 227, 229, 231
Melin, J., *see* Fuzellier, H. 215
Mendez, E. 53, 67, 68, 148, 150, 157, 164, 168, 169, 172, 173, 281–284
Mendez, E., *see* Misu, A. 67, 68, 152–157, 171, 173
Mendez, E., *see* Woollam, J.A. 265
Merle, G., *see* Golé, J. 421
Merrit, F.R., *see* Galt, I.K. 125, 126
Metz, W., *see* Caswell, N. 367, 369, 370
Meyer, J., *see* Baskin, J. 169
Michailovskii, V.A., *see* Belousov, V.M. 420
Mikhailov, V.A., *see* Lutkov, A.I. 70
Millman, S.E. 304
Millwood, G.R., *see* Thomas, J.M. 208, 296
Minomura, S., *see* Iye, Y. 102, 293, 295
Misenheimer, M.E., *see* Hardcastle, S.E. 255
Misu, A. 67, 68, 152–157, 171, 173
Misu, A., *see* Mendez, E. 53, 67, 68, 148, 150, 157, 164, 168, 169, 172, 173
Miura, N., *see* Nakamura, K. 101–103
Miyazaki, H. 297, 298
Mizutani, U. 302, 379, 402, 404, 405
Mizutani, U., *see* Kondow, T. 402, 405
Mizutani, U., *see* Suganuma, M. 389
Mojoni, A., *see* Bonfiglioli, G. 12
Monceau, Y.P., *see* Carmona, F. 70
Montroll, E.W., *see* Maradudin, A.A. 346
Moodenbaugh, A.R., *see* Smith, T.F. 318
Moore, A.W. xi
Morant, R.A. 14, 69, 91
Morelli, D.T., *see* Clarke, R. 253
Morelli, D.T., *see* Uher, C. 254, 255
Mori, M., *see* Mase, S. 103
Morris, R.C. 313, 314
Moses, D., *see* Fuerst, C.D. 321
Mott, N.F. 253
Mrozowski, S. 39, 70, 104, 107, 177, 193, 423
Mrozowski, S., *see* Arnold, G.M. 70
Mrozowski, S., *see* Haering, R.R. 30
Mrozowski, S., *see* Tojoda, S. 70
Mukaibo, T., *see* Sasa, T. 216
Mukhanov, V.A., *see* Brandt, N.B. 243, 245, 246, 251, 262, 265, 271, 290, 295
Murday, J.S. 419
Murray, J.J. 242, 243
Murthy, V.R.K., *see* Eklund, P.C. 367

Nagai, S., *see* Tsukada, M. 41
Nagamiya, T., *see* Komatsu, K. 322, 328, 329, 397, 398
Nagayoshi, H. 31, 33, 34, 39, 41, 42, 55, 56, 68, 168, 172, 173
Nagornyi, V.G., *see* Vjatkin, S.E. 422, 424, 428, 429, 439, 443–445
Naito, S. 420
Nakamizo, M. 361
Nakamura, K. 101–103
Nakao, K. 51, 55, 119, 135, 137, 143, 146, 153, 154, 171, 172
Nakao, K., *see* Inoshita, T. 42, 213, 217, 235, 263, 264, 314, 404, 405
Nakao, K., *see* Kamimura, H. 249
Nakao, K., *see* Nagayoshi, H. 31, 33, 34, 39, 41, 42, 55, 56, 68, 172
Nakao, K., *see* Ohno, T. 42, 233, 234
Nakao, K., *see* Tsukada, M. 41
Natarajan, V. 419
Nathan, M.I. 361
Nelson, D.R., *see* Jose, J.V. 307
Nemanich, R.J. 135, 329, 330, 341, 342, 344, 346, 348, 349, 356–363, 377, 392
Nezbeda, C.W., *see* Seldin, E.J. 22, 23, 341
Nguyen, Hy Hau 226
Nguyen, Hy Hau, *see* Blinowski, J. 226, 227, 229, 231
Niccolini, G., *see* Elahy, M. 304
Nichols, G.E., *see* De Sorbo, W. 398, 399
Nicholson, A.P.P. 19, 27, 28, 324, 339, 344
Nicklow, R.M. 18, 22–24, 26–28, 339–345, 352, 354, 355, 385, 388, 399
Nicklow, R.M., *see* Kamitakahara, W.A. 385–388
Nicklow, R.M., *see* Wakabayashi, N. 385
Nightingale, R. 427, 428, 433
Nixon, D.E. 201, 206, 209, 212, 215, 217, 371
Nosenzo, L., *see* Bellodi, G. 135, 153, 154
Noto, K. 81, 82
Novikov, Yu.N. 421
Novikov, Yu.N., *see* Belousov, V.M. 420
Novikov, Yu.N., *see* Karimov, Yu.S. 303, 402, 407
Novikov, Yu.N., *see* Mashinskii, V.I. 420
Novikov, Yu.N., *see* Postnikov, V.A. 420
Noziers, P. 48, 125, 126
Nye, J.F. 16, 17

Ohhashi, K. 219, 303, 381
Ohno, T. 42, 233, 234
Ohno, T., *see* Kamimura, H. 249
Onishi, T., *see* Ichikawa, M. 420
Onn, D.G. 218, 259, 402, 407
Onn, D.G., *see* Alexander, M.G. 310, 402, 405
Ono, A., *see* Hishiyama, Y. 70, 83, 124–126
Ono, A., *see* Sugihara, K. 77
Ono, A., *see* Tsuzuku, T. 77
Ono, S. 61, 77, 86, 89
Ono, S., *see* Sugihara, K. 62, 63, 123
Onuki, Y., *see* Tanuma, S. 238–240, 257, 258, 261, 263, 264
Osada, T., *see* Nakamura, K. 101–103
Oshima, H. 419
Ostrovskii, V.S., *see* Vjatkin, S.E. 422, 424, 428, 429, 439, 443–445
O'Sullivan, W.J., *see* Anderson, J.R. 67, 115, 125, 161, 166, 173, 174

Pacault, A., *see* Poquet, E. 182
Paige, E.G.S., *see* McLean, T.P. 90
Painter, G.S. 34, 39, 41
Painter, G.S., *see* Wills, R.F. 135
Panayotov, I.M. 421
Paranjable, B.V., *see* Fröhlich, H. 90
Parravicini, P., *see* Bassani, F. 33, 41
Parravicini, P., *see* Doni, E. 33
Parry, G.S., *see* Nixon, D.E. 201, 206, 209, 212, 215, 217, 371
Pascault, J.P., *see* Golé, J. 421
Pendrys, L.A. 310
Pendrys, L.A., *see* Vogel, F.L. 310
Pendrys, L.A., *see* Wachnik, R.A. 218, 295
Pendrys, L.A., *see* Wu, T.C. 245, 246
Pendrys, L.A., *see* Zeller, C. 251
Perkova, T.L., *see* Lutkov, A.I. 70
Pernollet, A., *see* Fourré, J. 422
Perret, R., *see* Herinckx, G. 419
Pfluger, P. 416
Philipp, H.R. 130, 132, 153
Philipp, H.R., *see* Taft, E.A. 34, 94, 130, 131, 133–135, 152
Pietronero, L. 201, 211, 212, 249, 250
Piljankevich, A.N., *see* Kurdyumov, A.V. 14

Pink, R.C., *see* McDonnell, F.R.M. 301
Platts, D.A. 49, 60
Plummer, E.W., *see* Eberhardt, W. 233, 234
Pokrovskii, C.L. 307
Poquet, E. 182
Pospelov, Yu.A., *see* Arkhipov, R.G. 61, 86, 160, 169, 173
Postnikov, V.A. 420
Postnikov, V.A., *see* Mashinskii, V.I. 420
Potter, M.E. 246, 251
Primak, W. 74–76
Prince, N. 423, 427, 433, 438, 441
Proctor, D.G., *see* Blackslee, O.L. 22, 23, 25, 341

Rabii, S., *see* Holzwarth, N.A.W. 42, 233, 234, 315, 404
Rafizadeh, H.A. 340, 399, 401
Rafizadeh, H.A., *see* Ahmadieh, A.A. 340, 360
Ramani, R., *see* Mani, K.K. 18, 22, 23, 26–28, 344, 360
Randorf, T.W., *see* Hedgecock, F.T. 105
Rashba, E.I. 94
Rashkov, I.B., *see* Panayotov, I.M. 421
Rasor, N.S. 12
Raubenheimer, L.J. 342
Readman, P.W., *see* Hohlwein, D. 219
Reguzzoni, E., *see* Bellodi, G. 135, 153, 154
Reifenberger, R. 274
Reynolds, W.W. xi, 435, 441
Reynolds, W.W., *see* Goggin, P.R. 436, 442
Rice, M.J., *see* Pietronero, L. 201, 249
Richarz, W., *see* Baiker, A. 217
Rigaux, C., *see* Blinowski, J. 226, 227, 229, 231, 265, 267
Rigaux, C., *see* Nguyen, H.H. 226
Rilay, D.P. 21, 23, 24
Ritsko, J.J., *see* Mele, E. 370, 381, 382
Ritsko, J.J., *see* Onn, D.G. 402, 407
Roetti, C., *see* Samuelson, L. 35
Rosenmann, I. 217, 268
Rosenmann, I., *see* Batallan, F. 225, 265, 268
Rosenmann, I., *see* Bok, J. 201
Rosenmann, I., *see* Simon, Ch. 225, 268
Rouillon, J.C., *see* Delhaes, P. 301, 302, 402–404
Rüdorff, W. 197–199, 207–209, 213, 215, 301
Ruland, W., *see* Herinckx, G. 419
Rupprecht, G. 416

Safran, S.A. 201, 202, 303
Sagawa, T. 34
Saha, A.R. 71
Saito, T., *see* Kawamura, K. 74–77, 81, 82
Sakai, T., *see* Mase, S. 103
Salamanca-Riba, L., *see* Shayegan, M. 402, 407, 408
Salzano, F.L., *see* Aronson, S. 216
Samara, G.A. 170
Samoggia, G., *see* Bellodi, G. 135, 153, 154
Samuelson, L. 35
Sasa, T. 216
Sato, H., *see* Sugihara, K. 88, 89
Sato, S., *see* Awaji, H. 423
Sato, Y., *see* Maeta, H. 361
Saunder, G., *see* Blackman, L.C.F. 75
Savranskii, V.V., *see* Kasatochkin, V.I. 5
Schaber, H., *see* Doezema, R.E. 49, 60, 67, 68, 125, 126, 142–144, 146, 172
Schaffäult, P. 204
Scherer, P.M., *see* Spry, W.J. 124, 125
Schmidt, P., *see* Hannay, N.B. 218, 309, 310
Schögl, R.F. 421
Schögl, R.F., *see* Thomas, J.M. 208, 296
Schroeder, P.R. 55, 57, 137, 146–148, 150, 172
Schulte-Oversohl, M., *see* Fieguth, P. 422
Schulze, E., *see* Rüdorff, W. 198, 199, 209, 301
Schulze, H., *see* Rüdorff, W. 209
Schwall, R.E., *see* Smith, T.F. 318
Scott, J.L. 422, 423
Seilig, H. 416
Seldin, E.J. 22, 23, 194, 341
Seldin, E.J., *see* Blackslee, O.L. 22, 23, 25, 341
Selig, H., *see* Ebert, L.B. 421
Semenenko, K.N., *see* Brandt, N.B. 243, 245, 246, 251, 265, 290, 295
Semyonov, L.P., *see* Agranovich, V.M. 446
Semyonov, M.V., *see* Brandt, N.B. 113, 162–166, 172, 173, 188, 190–192

Setton, R. 421
Setton, R., *see* Beguin, F. 421
Shapira, Y. 184
Sharma, M.P. 150, 180, 182, 183, 300, 302, 303
Sharma, V.K., *see* Baiker, A. 217
Shayegan, M. 402, 407, 408
Shayegan, M., *see* Dresselhaus, M.S. 268
Shelton, R.N., *see* Smith, T.F. 318
Shiba, H., *see* Suzuki, M. 394
Shimizu, A. 218, 315
Shiraishi, N., *see* Mase, S. 103
Shirber, J.E., *see* Anderson, J.R. 67, 115, 125, 161, 166, 173, 174
Shoenberg, D. 38, 124–126
Shoenberg, D., *see* McClure, J.W. 188
Shur, V.B., *see* Postnikov, V.A. 420
Sigarev, A.M., *see* Vjatkin, S.E. 422, 424, 428, 429, 439, 443–445
Simmons, J.H.W. 434, 446
Simon, C., *see* Batallan, F. 265
Simon, C., *see* Rosenmann, I. 268
Simon, Ch. 225, 268
Simon, F.Z., *see* Berman, R. 6
Singer, L.S., *see* Bright, A.A. 420
Singhal, S.C. 419
Skibowski, M., *see* Klucker, R. 34, 134
Skillman, S., *see* Herrman, F. 41
Sladek, R.J., *see* Keyes, R.W. 93
Slater, J.C. 233, 346
Slonzewski, J.C. 31, 34, 35
Slutskin, A.A. 274
Smith, D.S., *see* Eklund, P.C. 367
Smith, H.G., *see* Nicklow, R.M. 18, 22–24, 26–28, 339–345, 352, 354, 355, 385, 388, 399
Smith, H.G., *see* Wakabayashi, N. 385
Smith, J.E., *see* Nathan, M.I. 361
Smith, J.P., *see* Cooper, J.D. 124–126
Smith, L.B., *see* Soule, D.E. 67, 114, 118, 124–126
Smith, L.S., *see* McClure, J.W. 79, 86
Smith, T.F. 318
Smoluchowski, R. 177
Sokker, G.A., *see* Vjatkin, S.E. 422, 424, 428, 429, 439, 443–445
Solin, S.A. xi, 295, 321, 358, 359, 362, 365–367, 369, 370, 392–394
Solin, S.A., *see* Caswell, N. 211, 367–370, 381, 395
Solin, S.A., *see* Clarke, R. 213, 294
Solin, S.A., *see* Kamitakahara, W.A. 383
Solin, S.A., *see* Nemanich, R.J. 135, 329, 330, 341, 342, 344, 346, 348, 349, 356–363, 377, 392
Solin, S.A., *see* Wada, N. 294, 321
Song, J.J. 367, 378
Soule, D.E. 67, 74, 78, 79, 81, 83, 86, 88, 113, 114, 116, 118, 124–126
Soule, D.E., *see* Anderson, J.R. 67, 115, 125, 161, 166, 173, 174
Soule, D.E., *see* Poquet, E. 182
Spain, I.L. xi, 43, 44, 46, 75, 77, 78, 83, 86, 159–161, 170, 242, 243
Spain, I.L., *see* Dillon, R.O. 47, 67, 79, 83–87, 143, 172, 173, 185
Spain, I.L., *see* Green, J.F. 23, 27, 151
Spain, I.L., *see* Lowrey, W.H. 91
Spence, G.B. 22, 23
Spence, G.B., *see* Blackslee, O.L. 22, 23, 25, 341
Spencer, H.M. 338, 398
Spry, W.J. 124, 125
Spry, W.J., *see* McClure, J.W. 91–93, 258
Stark, R.W. 273, 277
Stark, R.W., *see* Friedberg, C.B. 273, 274
Stark, T., *see* Brillson, L.J. 135, 341, 344, 359
Steele, M.C., *see* Berlincourt, T.G. 38, 124–126
Steinmann, W., *see* Klucker, R. 34, 134
Stepanov, G.M., *see* Kechin, V.V. 160, 170, 172, 173
Stepanov, N.F. 6
Stokes, A.P., *see* Lipson, H. 10
Stout, J.W., *see* Chisholm, R.C. 409
Strässler, S., *see* Pietronero, L. 201, 211, 212, 249, 250
Stumpp, E. 307
Sudo, M., *see* Ichikawa, M. 420
Suematsu, H. 125, 126, 138–141, 259
Suematsu, H., *see* Higuchi, K. 264
Suematsu, H., *see* Suzuki, M. 394
Suematsu, H., *see* Tanuma, S. 238–240, 257, 258, 261, 263, 264
Suematsu, N., *see* Koike, Y. 218, 309, 310,

312–314
Suganuma, M. 389
Suganuma, M., *see* Mizutani, U. 379
Sugihara, K. 62, 63, 77, 88–90, 123, 259, 302
Sugihara, K., *see* Ono, S. 61, 86, 89
Suzuki, H., *see* Iwata, T. 437, 444
Suzuki, M. 394
Suzuki, M., *see* Koga, K. 308
Svistova, E.A., *see* Brandt, N.B. 63, 78–80, 83, 91–93, 114, 124–126, 128, 190
Syme-Johnson, A.W. 215

Taft, E.A. 34, 94, 130, 131, 133–135, 152
Taft, E.A., *see* Philipp, H.R. 130, 153
Takahashi, O., *see* Iye, Y. 102, 236, 240, 265, 293, 295
Takahashi, O., *see* Tanuma, S. 66, 101, 310
Takezawa, T., *see* Hishiyama, Y. 83, 124–126
Takezawa, T., *see* Sugihara, K. 77
Takezawa, T., *see* Tsuzuku, T. 77
Tamaru, K., *see* Ichikawa, M. 420
Tamaru, K., *see* Naito, S. 420
Tanahashi, Y., *see* Sasa, T. 216
Tanuma, S. xi, 66, 101, 238–240, 257, 258, 261, 263, 264, 310
Tanuma, S., *see* Higuchi, K. 264
Tanuma, S., *see* Iye, Y. 101, 102, 218, 236, 240, 265, 293, 295
Tanuma, S., *see* Koike, Y. 218, 309, 310, 312–314
Tanuma, S., *see* Nakamura, K. 101–103
Tanuma, S., *see* Suematsu, H. 125, 126, 138–141, 259
Tatevskii, V.M., *see* Stepanov, N.F. 6
Tatewskii, V.M. 5
Tedrow, T., *see* Iye, Y. 101, 102
Theymann, W., *see* Budke, J. 433
Thomas, J.M. 208, 296
Thompson, T.E., *see* Bartlett, N. 244
Thompson, T.E., *see* Falardeau, E.R. 216
Thompson, T.E., *see* Fischer, J.E. xi
Thorson, I.M., *see* Haywood, B.C. 337
Thrower, P.A. 443
Thrower, P.A., *see* Reynolds, W.W. 435, 441
Timp, G., *see* Endo, M. 419
Timp, G., *see* Iye, Y. 101, 102
Tobochnik, J. 408, 409
Tojoda, S. 70
Tonidiglia, V., *see* De Long, L.E. 321
Tosatti, E. 134
Tosatti, E., *see* Greenaway, D.L. 33, 134
Touzain, P., *see* Armand, M. 418, 420
Toy, W.V. 39, 48, 67, 68, 137, 147, 149, 150
Toyozawa, Y. 106
Trantita, G.G., *see* Matheny, R.A. 438
Tsu, R. 361, 363
Tsuji, K., *see* Iye, Y. 102, 293, 295
Tsujikawa, I., *see* Kobayashi, M. 309, 310
Tsujikawa, I., *see* Ohhashi, K. 219, 303, 381
Tsukada, M. 41
Tsukada, M., *see* Nagayoshi, H. 68
Tsuzuku, T. 77, 242, 243
Tsuzuku, T., *see* Hishiyama, Y. 83, 124–126
Tsuzuku, T., *see* Kawamura, K. 74–77, 81, 82
Tsuzuku, T., *see* Noto, K. 81, 82
Tsuzuku, T., *see* Sugihara, K. 77
Tu Jr, K.N., *see* Nathan, M.I. 361
Tugushev, V.V., *see* Beneslavskii, S.D. 97, 99
Tuinstra, F. 341, 344, 356–358, 360
Turnbull, J.A., *see* Eeles, W.T. 214, 268, 378
Tyler, W.W., *see* De Sorbo, W. 322, 338, 397–400

Ubbelohde, A.R. xi, 242, 243, 427
Ubbelohde, A.R., *see* Blackman, L.C.F. 75
Ubbelohde, A.R., *see* McDonnell, F.R.M. 301
Ubbelohde, A.R., *see* Murray, J.J. 242, 243
Ubbelohde, A.R., *see* Spain, I.L. 75, 78, 83, 86
Ubellohde, A.R., *see* Nixon, D.E. 217
Uda, T., *see* Ushio, H. 141–143
Uebersfeld, J. 194
Uemura, Y., *see* Nagayoshi, H. 31, 33, 34, 39, 41, 42, 55, 56, 68, 172
Uemura, Y., *see* Tsukada, M. 41
Uemura, Y., *see* Ushio, H. 141–143
Uher, C. 254, 255
Uher, C., *see* Clarke, R. xi, 253
Uhlrich, M., *see* Delhaes, P. 104–106

Underhill, C., *see* Gualberto, G.M. 367
Underhill, J. 201, 367, 370, 372
Ushio, H. 141–143

Vagas, J., *see* Eberhardt, W. 119
Van Attekum, P.M.Th.M., *see* Wertheim, G.K. 304, 305
Van der Hoeven, J.C. 22, 23, 398
Van Haeringen, W. 41
van Schyndel, A., *see* Doezema, R.E. 49, 60, 67, 68, 125, 126, 142–144, 146, 172
Van Vleck, J.H. 175
Vangelisti, R. 214
Varma, R., *see* Wright, R.B. 361, 363
Vedeneev, V.V. 6
Venkatarayudu, T., *see* Bhagavantam, S. 347
Vershchagin, L.F., *see* Kabalkina, S.S. 23, 151
Vieren, J.-P., *see* Blinowski, J. 226, 227, 229, 231
Vimin, G.V., *see* Pokrovskii, C.L. 307
Vjatkin, S.E. 422, 424, 428, 429, 439, 443–445
Vogel, F.L. 241–243, 245, 246, 259, 310, 418, 419
Vogel, F.L., *see* Billaud, D. 206
Vogel, F.L., *see* Foley, G.M.T. 242, 243, 245
Vogel, F.L., *see* Pendrys, L.A. 310
Vogel, F.L., *see* Wachnik, R.A. 218, 295
Vogel, F.L., *see* Wu, T.C. 245, 246
Vogel, F.L., *see* Zeller, C. 251
Vol'pin, M.E., *see* Belousov, V.M. 420
Vol'pin, M.E., *see* Karimov, Yu.S. 303
Vol'pin, M.E., *see* Mashinskii, V.I. 420
Vol'pin, M.E., *see* Novikov, Yu.N. 421
Vol'pin, M.E., *see* Postnikov, V.A. 420
Von Ortenberg, M.V. 94

Wachnik, R., *see* Pendrys, L.A. 310
Wachnik, R.A. 218, 295
Wachnik, R.A., *see* Vogel, F.L. 310
Wada, N. 294, 307, 321
Wada, N., *see* Clarke, R. 294
Wada, N., *see* Kamitakahara, W.A. 383
Wagoner, G. 64, 75
Wakabayashi, N. 385
Wakabayashi, N., *see* Nicklow, R. 18, 22–24, 26–28, 339–345, 352, 354, 355, 385, 388, 399
Walker, P.L., *see* Nakamizo, M. 361
Wallace, P.R. 30, 41, 71, 187, 226, 227
Wallace, P.R., *see* Gupta, O.P. 55, 57, 63, 66
Warren, B.E. 7
Watt, W., *see* Brown, A.R.G. 75
Webb, E.J., *see* Bergenlid, U. 322, 398
Weiss, G.H., *see* Maradudin, A.A. 346
Weiss, P.R., *see* Slonzewski, J.C. 31, 34, 35
Weng, T., *see* Blackslee, O.L. 22, 23, 25, 341
Wertheim, G.K. 304, 305
Whittingham, M.S. 416, 418, 420
Wiggs, P.K.C. 425, 426
Wilks, J., *see* Bergenlid, U. 322, 398
Williamson, G.K. 436, 442
Williamson, S.J. 115, 116, 124–126
Wills, R.F. 39, 135
Wilman, H., *see* Finch, G.I. 10
Wilson, E.V. 347
Woollam, J.A. 62, 64, 91, 114–116, 118, 123–125, 127, 265
Woollam, J.A., *see* Dillon, R.O. 79, 83, 86, 87
Woollam, J.A., *see* Natarajan, V. 419
Woollam, J.A., *see* Oshima, H. 419
Woore, J., *see* Cooper, J.D. 40, 124–126
Wostenholm, G.H. 398
Wright, R.B. 361, 363
Wu, T.C. 245, 246
Wyckoff, R.W.G. 9

Yafet, Y., *see* McClure, J.W. 64
Yager, W.A., *see* Galt, I.K. 125, 126, 137
Yakovlev, G.A., *see* Brandt, N.B. 221, 223
Yamaguchi, T., *see* Mase, S. 103
Yates, B., *see* Bailey, A.C. 25
Yates, B., *see* Wostenholm, G.H. 398
Yavrouian, A., *see* Oshima, H. 419
Yeh, V. 218, 319, 320
Yeh, V., *see* De Long, L.E. 321
Yeoman, M. 158, 159, 170
Yoshimori, A. 27, 323, 324, 326–328, 332, 333, 335, 336, 361, 400, 401
Yoshioka, D. 66, 102
Young, D.A. 77

Young, D.A., *see* Bender, A.S. 260
Young, D.A., *see* Cooper, J.D. 40, 124–126
Young, D.A., *see* Spain, I.L. 75, 78, 83, 86
Young, D.A., *see* Yeoman, M. 158, 159, 170
Young, J. 18, 27, 323, 324, 326, 335, 337, 338, 344, 360, 361

Zabel, H., *see* Hardcastle, S.E. 25, 26, 255
Zabel, H., *see* Kamitakahara, W.A. 383, 385–388, 396
Zahopoulos, C., *see* Markiewicz, R.S. 277–279
Zallen, R. 349, 360
Zanini, M. 206, 393
Zanini, M., *see* Guérard, D. 242, 243, 259
Zeller, C. 251, 253
Zeller, C., *see* Basu, S. 242, 243, 246, 403, 405
Zeller, C., *see* Foley, G.M.T. 242, 243, 245
Zeller, C., *see* Guérard, D. 212
Zeller, C., *see* Vogel, F.L. 246
Zeller, C., *see* Wu, T.C. 245, 246
Zeller, H.R., *see* Pietronero, L. 201, 249
Zeppenfeld, K. 34, 134
Zimmerman, G.O., *see* Elahy, M. 304
Zimmerman, G.O., *see* Millman, S.E. 304
Zimmermann, R., *see* Cords, H. 438
Zorin, F.I. 258
Zunger, J.M. 41

SUBJECT INDEX

activation energy of electron in magnetic field 96
anisotropic semimetals xii
average distance between graphite layers 13
average mobility of carriers
– in GIC 256, 257, 259, 260
– in graphite 81–84, 88, 89

band inversion xi
baric coefficients (in GIC) of
– concentration of carriers 293
– critical temperature of superconducting transition 317–321
– extremal cross-sections of Fermi surface 290–292
– Fermi level 294
– parameters of energy spectrum of carriers 294
baric coefficients (in graphite) of
– conductivity 170
– cyclotron masses 165, 166
– effective masses 166, 172
– extremal cross-sections of Fermi surface 161, 163, 164, 166, 167
– interlayer distance 151, 167
– SWMC-model parameters 168, 169, 171–174
battery 417, 418
binding energy per carbon atom in graphite 6
Brillouin zone for single graphite layer 29
Brillouin zone for three-dimensional graphite 35
broken axial symmetry in graphite lattice 344

carbine 5, 6
carbon fibers 419, 420
carbon layer 7
carbon materials 6
– coal 6
– coke 6
– diamond 6
– graphite x, xi
– highly oriented pyrolytic graphite (HOPG) 13
– kish-graphite 13
– natural graphite ix, xi
– soot 6
– synthetic carbon 2
carriers in graphite
– majority electrons 49, 50, 106, 107, 166
– majority holes 48, 49, 106, 107, 166
– minority electrons 49, 50, 106, 107, 110, 166
– minority holes 48–50, 106, 107, 110, 166
catalysts (GIC) 420, 421
change in porosity under irradiation 440, 441
change in the crystal lattice parameters of graphite under irradiation 433–435
change in thermal and electronic properties of graphite under irradiation 441–444
– diamagnetic susceptibility 444
– electrical resistance 443
– Hall coefficient 443, 444

– heat conductivity 441
– thermal emf 444
change in type of superconductivity in GIC 312–314
charge density wave state xi, 66, 101–103, 128
chemical reactivity of graphite 426–429
coefficient of tortuosity of pore channels 425
combined state in Raman spectra 356
composite materials 420
concentration of electrons and holes in
– GIC 228, 230, 238, 256, 257, 259, 267
– graphite 60–66, 68, 83, 86, 87, 122
conductivity of GIC
– along *c*-axis 242, 243, 250, 251
– along layers 242–250
– anisotropy 241–243
– temperature dependence 251–255
conductivity of graphite
– along *c*-axis 74–77
– along carbon layers 74–76
– anisotropy 74, 75
– temperature dependence 74–76, 87–91
coolants 426
corrosion of graphite 426
– with gaseous coolants 426, 427
– with metals 426–428
cracks 424
– lens-shaped 424
– micro 424
cut-off effect in graphite 59, 60
cut-off of Fermi level in GIC 286, 287
cyclotron masses of electrons and holes in
– GIC 228, 230, 266, 284, 285
– graphite 45, 46, 54, 68, 69, 116, 118, 125, 138, 141–143, 165, 166
cyclotron resonance in graphite 137–145

Debye temperature of graphite 88, 89
defects in graphite lattice
– foreign atoms 12
– Frenkel defects 12
– pliers 12
– Schottky defects 12
– stacking faults 11
defects induced by irradiation 435–438
deformation rate 424
degree of preferred orientation of crystallites in graphite 14
diamagnetism of
– graphite 175–183
– graphite in extreme quantum limit 183–190
– quasi-two-dimensional graphite 190–193
dielectric constant for graphite 129, 131, 132
dielectric function for graphite 129, 131, 132
differential enthalpy 445
dispersion law of carriers in GIC
– rigid-band model 235–240
– tight-binding approximation (Blinowski model) 227–232
dispersion law of carriers in graphite
– four-parameter model 42
– quadratic model 43–45
– quantizing magnetic field 51–54
– semiclassical 51, 52
– SWMC-model 38
– trigonal warping 46
dispersion law of phonons in graphite
– axial-symmetric (AS) model 341, 342
– four-force constant model (BBS model) 327–330

effective masses of carriers in graphite 46, 54, 137, 147, 148, 157, 158
elastic constants for graphite 16–18, 21–26
elastic energy for graphite 17, 18
elastic model for lattice deformation of the irradiated graphite 446
elastic moduli for GIC 389
electrode materials 417, 418
electron density oscillations in GIC 203, 204
electron-loss spectrum for GIC 370
electron-loss spectrum for graphite 130, 131
electroneutrality condition for graphite 60, 61
elimination of impurities in graphite 430–433
energy accumulated in irradiated graphite (Wigner energy) 444–446
energy-loss function for graphite 130, 131
excitonic insulator phase xi

Fermi level in
- GIC 267
- graphite 60–66, 143
Fermi surface of carriers in graphite
- cross-sections 43, 68, 118, 124
- structure 40, 43, 44, 47–49, 120, 121
-- neck 48, 111
-- outrigger pieces 47
- trigonal warping 46–49
- velocities of carriers 45
- volume 86, 122
Fermi surface of GIC
- cross-section 238–240, 266–269
-- angular dependence 270–273
- structure 227–232, 235, 238–240, 263–269
- volume 228, 230
flexible controlled surface mirrors 417
force constants 5, 15, 17–21, 27, 28, 325, 326, 328, 329, 339–341, 343, 344

g-factor of carriers in graphite 64, 66, 126, 127
galvanomagnetic effects in
- GIC 255–260
- graphite 73–88, 91–103
-- extreme quantum limit of magnetic fields 91–103, 127
-- weak magnetic fields 77–91
- pyrocarbons 104–108
gapless semiconductor 30, 31
gapless state xi, 30, 31, 65
GIC, *see* graphite intercalation compounds 197–200
glassy carbon 425
graphite x
graphite intercalation compounds x, 197
graphitization 429, 430
- degree of 434
graphitization parameter 13, 69, 74, 75

Hall effect in
- GIC 258–260
- graphite 78–81, 91–93, 95–100
Hamiltonian of SWMC-model 36, 37
harmonic approximation of elastic energy in graphite 17, 326
healing rate 424
heat capacity of GIC 402–409
- electronic contribution 402–405
- lattice contribution 402–404
- magnetic contribution 396, 405–409
heat capacity of graphite 397–402
- electronic contribution 396, 397
- lattice contribution 396–401
heat resistance of reactor graphite 423, 424
heat screens 416
heat treatment temperature (HTT) 69–72, 104, 105
hybridization (sp-types) of electron states in carbon materials 4
- digonal 4, 5
- tetragonal 4
- trigonal 4

identity period in GIC 208, 209
impurities in graphite 431–433
inelastic coherent scattering of slow neutrons in graphite 350–355
infra-red polarizers 416
infra-red spectrum of
- GIC 370
- graphite 349, 350, 359
intercalate-induced vibration modes 365–377
intercalate layer modes 377–384
isothermal release of accumulated energy 445

Landau levels in electronic energy spectrum of graphite
- central levels 57–60
- leg levels 49, 56–60, 143, 146
- ordinary levels for majority carriers 51–54, 57, 143, 146, 147
- trigonal warped levels 49, 56–60
lattice of graphite 8–11
- hexagonal (Bernal-type structure) 8, 9, 11
- rhombohedral 8, 10
layered semimetals xii
Lennard-Jones potential for interaction of atoms in graphite 26
Lifshitz–Onsager formula 50, 51
limiting accumulated energy in irradiated graphite 445
localization of electrons in high magnetic fields xi, 93, 95, 97–100

magnetic breakdown effect in
- GIC 273–280
- graphite 60
magnetic breakdown oscillations 273–280
magnetic freeze-out effect 93–100, 188–190
magnetic ordering in the intercalate layer 305–307
magnetic susceptibility of
- carbon materials 193–196
- GIC 299–304
- graphite 175–190
- quasi-two-dimensional graphite 190–193
magnetic suspensions in gyroscopic devices 416
magnetic transitions in GIC
- paramagnetic–spin-frustrated phase 307–309
- paramagnetic–spin-glass 304, 305
magneto-optical spectra of
- GIC 281–284
- graphite 135–150
magnetoresistance of
- GIC 256–260
- graphite 77, 78, 80–82, 84, 91–93, 101–103
- pyrocarbons 71, 72, 104–108
manufacturing reactor graphite 429, 430
melting heat of graphite 6
methods of reactor graphite production 429, 430
methods of synthesis of GIC 204–206
- co-intercalation technique 206
- electrochemical technique 206
- intercalation from a melt 206
- liquid-phase intercalation 205
- vapour-phase intercalation 205
mixed phonon modes 385–390
mobility of electrons and holes in graphite 80–85, 87–89, 91
models of conductivity in GIC
- layers connected in parallel 245, 247, 248
- series-connected resistance 250, 251
models of energy spectrum of carriers in GIC
- Blinowski model (tight-binding approximation) 226–232
- rigid-band model 235–240
models of energy spectrum of carriers in graphite layers
- Nagayoshi model (tight-binding appoximation) 31
- Slonzewski–Weiss model 31–35
- Wallace model 30
models of energy spectrum of carriers in pyrocarbon materials
- Klein model 71, 72
- Mrozowski model 70
- Wallace model 71
models of energy spectrum of carriers in three-dimensional graphite
- four-parameter model 42–46
- Slonzewski–Weiss–McClure (SWMC) model 35–40, 49, 50, 67–69
models of interactions in the graphite lattice
- Mani–Ramani 18, 20
- Nicholson–Bacon (many-force-constants model) 19–21
- Nicklow 18
- Young–Koppel 18, 20
models of lattice vibrations of graphite
- axial-symmetric (AS) model 339–343
- bond-bending and bond-stretching (BBS) model 324–339
- Krumhansl–Brooks 323
- Maeda (broken axial symmetry) model 343, 344
- Nicholson–Bacon 324
- semicontinuum 322
- Yoshimori–Kitano 323, 324
- Young–Koppel 323
models of structure of GIC
- classical (Rüdorff) model 207
- domain structure (Daumas and Hérold) model 207, 208
modes of Raman spectrum
- for GIC 365–369, 371–378, 380–384, 393–395
- for graphite 347–350, 355–363

negative magnetoresistance of
- GIC 257, 258
- pyrocarbon materials 71, 72, 104–108
neutron absorbing power of graphite 422
neutron capture cross section of graphite 422
neutron moderating power of graphite 422

neutron moderator in nuclear reactor 415
neutron scattering in graphite 350–355

optical spectrum of graphite 130–135
optical transitions 31–33, 135
oscillation effects in
– GIC 260–263, 268–280
– graphite 109–121, 124, 125
overtone states in Raman spectra 356, 362

pairwise valence-type potential for bonding forces in graphite 27
parameters of energy spectrum of carriers in GIC 266, 267, 280, 283
parameters of SWMC-model of energy spectrum of carriers in graphite 67, 68, 141, 143, 147–150, 153, 168, 169, 171–174
penetrability of porous body 424, 425
penetrability of reactor graphite 424
penetration coefficient 424, 425
phase transitions in GIC
– change in order of intercalate layer stacking 216
– change in stage index 216
– from commensurate to incommensurate structure 216
– order–disorder 217
– to charge-density wave state 217
– to magnetic ordering in intercalate layer 218, 219, 305–309
– to superconducting state 218, 309–314, 414, 415
phonon density of states function for graphite
– axial-symmetric model 343
– bond-bending and bond-stretching model 332, 333, 336–338
– broken axial symmetry model 344, 345
– experimental data 337–339
phonon spectra of
– GIC 363–396
– graphite 324–350, 353, 354, 357–359, 361
plasma frequency for graphite 132
plasma resonance in graphite 133
protection against corrosion 428
– with halogens 428, 429

quantum correction in conductivity of pyrocarbons 107, 108
radiation damage 433–446
Raman spectra of
– GIC 367, 368, 371–373, 375, 376, 378, 381, 382, 384, 393
– – first-order spectra 367, 368, 371, 372, 378, 381, 382, 384, 393
– – second-order spectra 373, 375, 376
– monocrystalline graphite 356–358, 361
– small crystallites of graphite 357, 358, 363
reactor core 421
reactor graphite 421
reduced metal phase xi, 103
reflection spectrum of graphite 133, 134
relative moderating power of graphite 422
relaxation time of carriers 89–91
residue compound 200

saddle points in energy spectrum of graphite 56, 110
selection rules of
– cyclotron resonance 138
– magneto-optical transitions 136, 146, 286–289
– optical transitions 32, 134, 135
size stability under irradiation 438–440
spin–orbit splitting of energy levels in graphite 48
stage index 197
staging phenomenon in GIC 197
strength limit of reactor graphite 424
strength of reactor graphite 423, 424
strongly coupled phonon modes 385–390
structural phase transitions in GIC 197, 216–218, 394–396
structure of carbon materials
– hexagonal 8, 9, 11
– rhombohedral 8, 10
– turbostratum 7
superconductivity of GIC 309–321
supermetallic conductivity of GIC 241, 243–246, 414, 418, 419

tight-binding approximation 30, 31
total absorption coefficient 422
transitions induced by magnetic field
– semimetal–semiconductor 63–67, 127

trigonal warping in
- GIC 273–280
- graphite 46–49, 84, 85
type of defects induced by irradiation 436, 437

velocities of sound in graphite 88

weak-coupling link in superconducting quantum interferometers 416

zone-centre modes of lattice vibrations in
- GIC 370, 380–382
- graphite 345–350
zone-folding effect in
- electron energy spectrum of GIC 224–226, 278
- phonon spectrum of GIC 390–394

CUMULATIVE INDEX, VOLUMES 1–21

Monographs

Agranovich, V.M., and M.D. Galanin, Electronic excitation energy transfer in condensed matter **3**

Brandt, N.B., S.M. Chudinov and Ya.G. Ponomarev, Semimetals – 1. Graphite and its compounds **20.1**

Chudinov, S.M., see Brandt, N.B. **20.1**

Galanin, M.D., see Agranovich, V.M. **3**
Gantmakher, V.F., and Y.B. Levinson, Carrier scattering in metals and semiconductors **19**
Gurevich, V.L., Transport in phonon systems **18**

Levinson, Y.B., see Gantmakher, V.F. **19**

Mashkova, E.S., and V.A. Molchanov, Medium-energy ion scattering by solid surfaces **11**
Molchanov, V.A., see Mashkova, E.S. **11**

Ponomarev, Ya.G., see Brandt, N.B. **20.1**

Wagner, M., Unitary transformations in solid state physics **15**

Contributed Volumes

Abelés, F., and T. Lopez-Rios, Surface polaritons at metal surfaces and interfaces **1**, 239
Agranovich, V.M., Effects of the transition layer and spatial dispersion in the spectra of surface polaritons **1**, 187
Agranovich, V.M., V.E. Kravtsov and T.A. Leskova, Scattering of surface polaritons by order parameter fluctuations near phase transition points **1**, 511
Agranovich, V.M., Biphonons and Fermi resonance in vibrational spectra of crystals **4**, 83
Agranovich, V.M., Surface excitons **9**, 513

Agranovich, V.M., and V.V. Kirsanov, Production of radiation defects by collision cascades in metals **13**, 117
Alexandrova, I.P., Nuclear quadrupole resonance of the incommensurate phase in Rb_2ZnCl_4 and Rb_2ZnBr_4 **14.1**, 277
Altshuler, B.L., and A.G. Aronov, Electron–electron interaction in disordered conductors **10**, 1
Al'tshuler, S.A., A.Kh. Khasanov and B.I. Kochelaev, Studies of spin systems by means of light scattering in paramagnetic crystals **21**, 607
Aoyagi, K., see Tanabe, Y. **2**, 603
Aronov, A.G., and A.S. Ioselevich, Exciton electrooptics **2**, 267
Aronov, A.G., see Altshuler, B.L. **10**, 1
Aronov, A.G., Yu.M. Gal'perin, V.L. Gurevich and V.I. Kozub, Nonequilibrium properties of superconductors (transport equation approach) **12**, 325
Aslamazov, L.G., and A.F. Volkov, Nonequilibrium phenomena in superconducting weak links **12**, 65
Axe, J.D., M. Iizumi and G. Shirane, Phase transformations in K_2SeO_4 and structurally related insulators **14.2**, 1

Bagaev, V.S., T.I. Galkina and N.N. Sibeldin, Interaction of EHD with deformation field, ultrasound and non-equilibrium phonons **6**, 267
Bak, P., see Pokrovsky, V.L. **17**, 71
Basiev, T.T., V.A. Malyshev and A.K. Przhevuskii, Spectral migration of excitations in rare-earth activated glasses **21**, 275
Basun, S.A., see Kaplyanskii, A.A. **16**, 373
Beasley, M.R., see De Lozanne, A.L. **12**, 111
Belousov, M.V., Vibrational Frenkel excitons **2**, 771
Belyakov, V.A., and E.I. Kats, Light scattering and phase transitions in liquid crystals **5**, 227
Benoit à la Guillaume, C., see Planel, R. **8**, 353
Bereznyak, P.A., see Slyozov, V.V. **13**, 575
Berkovits, V.L., see Paget, D. **8**, 381
Bernard, L., see Currat, R. **14.2**, 161
Bersuker, I.B., and V.Z. Polinger, Theoretical background of the Jahn–Teller effect **7**, 21
Beyer, J., C.J. Pethick, J. Rammer and H. Smith, Pair breaking in nonequilibrium superconductors **12**, 129
Bhatt, R.N., see Milligan, R.F. **10**, 231
Bill, H., Observation of the Jahn–Teller effect with electron paramagnetic resonance **7**, 709
Birman, J.L., Electrodynamic and non-local optical effects mediated by exciton polaritons **2**, 27
Bishop, A.R., see Steiner, M. **17**, 783
Blinc, R., P. Prelovšek, V. Rutar, J. Seliger and S. Žumer, Experimental observations of incommensurate phases **14.1**, 143
Borstel, G., see Wöhlecke, M. **8**, 423
Bron, W.E., Phonon generation, transport and detection through electronic states in solids **16**, 227
Bullough, R., and M.H. Wood, Theory of microstructural evolution **13**, 189
Burns, M.J., W.K. Liu and A.H. Zewail, Nonlinear laser spectroscopy and dephasing of molecules: An experimental and theoretical overview **4**, 301
Burstein, E., see Chen, Y.J. **1**, 587
Büttner, H., see Mertens, F.G. **17**, 723

Cailleau, H., Incommensurate phases in an aromatic molecular crystal: biphenyl **14.2**, 71
Celotta, R.J., see Pierce, D.T. **8**, 259
Challis, L.J., and A.M. de Goër, Phonon spectroscopy of Jahn–Teller ions **7**, 533
Challis, L.J., Crossing effects in phonon scattering **16**, 145
Chang, Jhy-Jiun, Kinetic equations in superconducting thin films **12**, 453
Chen, Y.J., and E. Burstein, Three wave nonlinear interactions involving surface polaritons; Raman scattering, light diffraction and parametric mixing **1**, 587
Clarke, J., Experiments on charge imbalance in superconductors **12**, 1
Clinard Jr, F.W., and L.W. Hobbs, Radiation effects in non-metals **13**, 387
Cone, R.L., and R.S. Meltzer, Ion–ion interactions and exciton effects in rare earth insulators **21**, 481
Cottam, M.G., and A.A. Maradudin, Surface linear response functions **9**, 1
Cummins, H.Z., Brillouin scattering studies of phase transitions in crystals **5**, 359
Currat, R., see Durand, D. **14.2**, 101
Currat, R., see Dénoyer, F. **14.2**, 129
Currat, R., L. Bernard and P. Delamoye, Incommensurate phase in β-$ThBr_4$ **14.2**, 161

Davydov, A.S., Solitons in biology **17**, 1
de Goër, A.M., see Challis, L.J. **7**, 533
De Lozanne, A.L., and M.R. Beasley, Time-dependent superconductivity in SNS bridges: an example of TDGL theory **12**, 111
de Wolff, P.M., and F. Tuinstra, The incommensurate phase of Na_2CO_3 **14.2**, 253
Delamoye, P., see Currat, R. **14.2**, 161
deMartini, F., see Shen, Y.R. **1**, 629
Dénoyer, F., see Durand, D. **14.2**, 101
Dénoyer, F., and R. Currat, Modulated phases in thiourea **14.2**, 129
Dmitriev, V.M., V.N. Gubankov and F.Ya. Nad', Experimental study of enhanced superconductivity **12**, 163
Dolgopolov, V.T., and I.L. Landau, Resistive states in type-I superconductors **12**, 641
Dolino, G., The incommensurate phase of quartz **14.2**, 205
Durand, D., F. Dénoyer, R. Currat and M. Lambert, Incommensurate phase in $NaNO_2$ **14.2**, 101
Dyakonov, M.I., and V.I. Perel', Theory of optical spin orientation of electrons and nuclei in semiconductors **8**, 11

Echegut, P., see Gervais, F. **14.1**, 337
Efros, A.L., and B.I. Shklovskii, Coulomb interaction in disordered systems with localized electronic states **10**, 483
Ehrhart, P., K.H. Robrock and H.R. Schober, Basic defects in metals **13**, 3
Elesin, V.F., see Galitskii, V.M. **12**, 377
Eliashberg, G.M., and B.I. Ivlev, Theory of superconductivity stimulation **12**, 211
Errandonéa, G., see Tolédano, J.C. **14.2**, 233

Fayer, M.D., Exciton coherence **4**, 185
Fischer, B., see Lagois, J. **1**, 69
Fleisher, V.G., and I.A. Merkulov, Optical orientation of the coupled electron–nuclear spin system of a semiconductor **8**, 173
Fleury, P.A., and K.B. Lyons, Central peaks near structural phase transitions **5**, 449
Flores, F., and F. Garcia-Moliner, Electronic surface excitations **9**, 441
Fukuyama, H., Interaction effects in the weakly localized regime of two- and three-dimensional disordered systems **10**, 155

Furrer, R., Electron paramagnetic resonance in the excited states of rare earth ions in crystals **21**, 641

Galaǐko, V.P., and N.B. Kopnin, Theory of the resistive state in narrow superconducting channels **12**, 543
Galitskii, V.M., V.F. Elesin and Yu.V. Kopaev, The kinetic theory of superconductors with excess quasiparticles **12**, 377
Galkina, T.I., see Bagaev, V.S. **6**, 267
Gal'perin, Yu.M., see Aronov, A.G. **12**, 325
Garcia-Moliner, F., see Flores, F. **9**, 441
Gershenzon, E.M., A.P. Mel'nikov and R.I. Rabinovich, H^--like impurity centers, molecular complexes and electron delocalization in semiconductors **10**, 483
Gervais, F., and P. Echegut, Infrared studies of incommensurate systems **14.1**, 337
Ginzburg, V.L., A.A. Sobyanin and A.P. Levanyuk, General theory of light scattering near phase transitions in ideal crystals **5**, 3
Gladkii, V.V., Macroscopic electric quadrupole moment in the incommensurate phase in ferroelectrics **14.1**, 309
Glinchuck, M.D., Paraelectric resonance of off-center ions **7**, 819
Goldburg, W.I., Light scattering investigations of the critical region in fluids **5**, 531
Goldman, A.M., see Kadin, A.M. **12**, 253
Golovko, V.A., and A.P. Levanyuk, Light scattering from incommensurate phases **5**, 169
Gor'kov, L.P., Disorder and interactions in the system of quasi one-dimensional electrons **10**, 619
Grun, J.B., B. Hönerlage and R. Lévy, Biexcitons in CuCl and related systems **2**, 459
Gubankov, V.N., see Dmitriev, V.M. **12**, 163
Gurevich, V.L., see Aronov, A.G. **12**, 325
Guseva, M.I., and Yu.V. Martynenko, Blistering **13**, 621

Haarer, D., and M.R. Philpott, Excitons and polarons in organic weak charge transfer crystals **4**, 27
Hamano, K., Thermal hysteresis phenomena in incommensurate systems **14.1**, 365
Hanson, D.M., J.S. Patel, I.C. Winkler and A. Morrobel-Sosa, Effects of electric fields on the spectroscopic properties of molecular solids **4**, 621
Harley, R.T., Spectroscopic studies of Jahn–Teller phase transitions in rare-earth crystals **21**, 557
Hermann, C., and C. Weisbuch, Optical detection of conduction electron spin resonance in semiconductors and its application to $\boldsymbol{k}$, $\boldsymbol{p}$ perturbation theory **8**, 463
Hesselink, W.H., and D.A. Wiersma, Theory and experimental aspects of photon echoes in molecular solids **4**, 249
Hizhnyakov, V.V., and N.N. Kristoffel, Jahn–Teller mercury-like impurities in ionic crystals **7**, 383
Hobbs, L.W., see Clinard Jr, F.W. **13**, 387
Hönerlage, B., see Grun, J.B. **2**, 459
Horovitz, B., Solitons in charge and spin density wave systems **17**, 691
Huber, D.L., Energy transfer in crystals **21**, 251

Iizumi, M., see Axe, J.D. **14.2**, 1
Ioselevich, A.S., see Aronov, A.G. **2**, 267
Ishibashi, Y., Phenomenology of incommensurate phases in the A_2BX_4 family **14.2**, 49
Ivchenko, E.L., Spatial dispersion effects in the exciton resonance region **2**, 141
Ivchenko, E.L., see Pikus, G.E. **2**, 205
Ivlev, B.I., see Eliashberg, G.M. **12**, 211

Janssen, T., Microscopic theories of incommensurate crystal phases **14.1**, 67
Jeffries, C.D., see Wolfe, J.P. **6**, 431
Judd, B.R., Group theoretical approaches **7**, 87

Kadin, A.M., and A.M. Goldman, Dynamical effects in nonequilibrium superconductors: some experimental perspectives **12**, 253
Kalia, R.K., see Vashishta, P. **6**, 1
Kamimura, H., Electron–electron interactions in the Anderson-localised regime near the metal–insulator transition **10**, 555
Kaplyanskii, A.A., and S.A. Basun, Multiple resonant scattering of the 29 cm^{-1} acoustic phonons in optically excited ruby **16**, 373
Kaplyanskii, A.A., and A.I. Ryskin, P.P. Feofilov and the spectroscopy of activated crystals **21**, 1
Kats, E.I., see Belyakov, V.A. **5**, 227
Keldysh, L.V., and N.N. Sibeldin, Phonon wind in highly excited semiconductors **16**, 455
Khasanov, A.Kh., see Al'tshuler, S.A. **21**, 607
Kholodar, G.A., see Vinetskii, V.L. **13**, 283
Kind, R., and P. Muralt, Unique incommensurate–commensurate phase transitions in a layer-structure perovskite **14.2**, 301
Kirsanov, V.V., see Agranovich, V.M. **13**, 117
Kivelson, S., Soliton model of polyacetylene **17**, 301
Klein, M.V., Light scattering studies of incommensurate transitions **5**, 503
Kochelaev, B.I., see Al'tshuler, S.A. **21**, 607
Kolb, D.M., The study of solid–liquid interfaces by surface plasmon polariton excitation **1**, 299
Kopaev, Yu.V., see Galitskii, V.M. **12**, 377
Kopelman, R., Energy transport in mixed molecular crystals **4**, 139
Kopnin, N.B., see Galaĭko, V.P. **12**, 543
Kosevich, A.M., Dynamical and topological solitons in ferromagnets and antiferromagnets **17**, 555
Koteles, E.S., Investigation of exciton-polariton dispersion using laser techniques **2**, 83
Kozub, V.I., see Aronov, A.G. **12**, 325
Kravtsov, V.E., see Agranovich, V.M. **1**, 511
Kristoffel, N.N., see Hizhnyakov, V.V. **7**, 383
Kulakovskii, V.D., and V.B. Timofeev, Thermodynamics of electron–hole liquid in semiconductors **6**, 95
Kuznetsov, E.A., see Zakharov, V.E. **17**, 503

Lagois, J., and B. Fischer, Surface exciton polaritons from an experimental viewpoint **1**, 69
Lambert, M., see Durand, D. **14.2**, 101
Landau, I.L., see Dolgopolov, V.T. **12**, 641
Larkin, A.I., and Yu.N. Ovchinnikov, Vortex motion in superconductors **12**, 493
Lemaistre, J., see Zewail, A.H. **2**, 665
Leskova, T.A., see Agranovich, V.M. **1**, 511
Levanyuk, A.P., see Ginzburg, V.L. **5**, 3
Levanyuk, A.P., A.S. Sigov and A.A. Sobyanin, Light scattering anomalies due to defects **5**, 129
Levanyuk, A.P., see Golovko, V.A. **5**, 169
Levanyuk, A.P., General ideas about incommensurate phases **14.1**, 1
Levinson, Y.B., Phonon propagation with frequency down-conversion **16**, 91

Lévy, R., see Grun, J.B. **2**, 459
Litster, J.D., Scattering spectroscopy of liquid crystals **5**, 583
Liu, W.K., see Burns, M.J. **4**, 301
Lopez-Rios, T., see Abelès, F. **1**, 239
Loudon, R., see Ushioda, S. **1**, 535
Loudon, R., Ripples on liquid interfaces **9**, 589
Luhman, T., see Snead Jr, C.L. **13**, 345
Lushchik, Ch.B., Free and self-trapped excitons in alkali halides: spectra and dynamics **2**, 505
Lushchik, Ch.B., Creation of Frenkel defect pairs by excitons in alkali halides **13**, 473
L'vov, V.S., Solitons and nonlinear phenomena in parametrically excited spin waves **17**, 241
Lyons, K.B., see Fleury, P.A. **5**, 449
Lyuksyutov, I.F., A.G. Naumovets and Yu.S. Vedula, Solitons and surface diffusion **17**, 605

Macfarlane, R.M., and R.M. Shelby, Coherent transient and holeburning spectroscopy of rare earth ions in solids **21**, 51
Maki, K., Solitons in superfluid ^{3}He **17**, 435
Malkin, B.Z., Crystal field and electron–phonon interaction in rare-earth ionic paramagnets **21**, 13
Malyshev, V.A., see Basiev, T.T. **21**, 275
Maradudin, A.A., Interaction of surface polaritons and plasmons with surface roughness **1**, 405
Maradudin, A.A., see Cottam, M.G. **9**, 1
Maris, H.J., Phonon focusing **16**, 51
Markiewicz, R.S., and T. Timusk, Interaction of electromagnetic radiation with electron–hole droplets **6**, 543
Martynenko, Yu.V., see Guseva, M.I. **13**, 621
Meier, F., and D. Pescia, Spin-polarized photoemission by optical orientation **8**, 295
Mel'nikov, A.P., see Gershenzon, E.M. **10**, 483
Meltzer, R.S., see Cone, R.L. **21**, 481
Merkulov, I.A., see Fleisher, V.G. **8**, 173
Mertens, F.G., and H. Büttner, Solitons on the Toda lattice: thermodynamical and quantum-mechanical aspects **17**, 723
Mikhailov, A.V., Integrable magnetic models **17**, 623
Milligan, R.F., T.F. Rosenbaum, R.N. Bhatt and G.A. Thomas, A review of the metal–insulator transition in doped semiconductors **10**, 231
Mills, D.L., Surface spin waves on magnetic crystals **9**, 379
Mirlin, D.N., Surface phonon polaritons in dielectrics and semiconductors **1**, 3
Mirlin, D.N., Optical alignment of electron momenta in GaAs-type semiconductors **8**, 133
Morrobel-Sosa, A., see Hanson, D.M. **4**, 621
Moskalova, M.A., see Zhizhin, G.N. **1**, 93
Muralt, P., see Kind, R. **14.2**, 301

Nad', F.Ya., see Dmitriev, V.M. **12**, 163
Natadze, A.L., A.I. Ryskin and B.G. Vekhter, Jahn–Teller effects in optical spectra of II–VI and III–V impurity crystals **7**, 347
Naumovets, A.G., see Lyuksyutov, I.F. **17**, 605
Nelson, R.J., Excitons in semiconductor alloys **2**, 319
Nizzoli, F., see Stegeman, G.I. **9**, 195

Ortuño, M., see Pollak, M. **10**, 287
Osad'ko, I.S., Theory of light absorption and emission by organic impurity centers **4**, 437
Ovchinnikov, Yu.N., see Larkin, A.I. **12**, 493
Ovsyankin, V.V., Spectroscopy of collective states and cooperative transitions in disordered rare-earth activated solids **21**, 343

Paget, D., and V.L. Berkovits, Optical investigation of hyperfine coupling between electronic and nuclear spins **8**, 381
Patel, J.S., see Hanson, D.M. **4**, 621
Pearlstein, R.M., Excitons in photosynthetic and other biological systems **2**, 735
Pedersen, N.F., Solitons in Josephson transmission lines **17**, 469
Perel', V.I., and B.P. Zakharchenya, Major physical phenomena in the optical orientation and alignment in semiconductors **8**, 1
Perel', V.I., see Dyakonov, M.I. **8**, 11
Perlin, Yu.E., and M. Wagner, Introduction **7**, 1
Perlin, Yu.E., and B.S. Tsukerblat, Optical bands and polarization dichroism of Jahn–Teller centers **7**, 251
Permogorov, S., Optical emission due to exciton scattering by LO phonons in semiconductors **2**, 177
Personov, R.I., Site selection spectroscopy of complex molecules in solutions and its applications **4**, 555
Pescia, D., see Meier, F. **8**, 295
Pethick, C.J., see Beyer, J. **12**, 129
Philpott, M.R., see Haarer, D. **4**, 27
Pick, R.M., see Poulet, H. **14.1**, 315
Pierce, D.T., and R.J. Celotta, Applications of polarized electron sources utilizing optical orientation in solids **8**, 259
Pikin, S.A., Incommensurate structures in liquid crystals **14.2**, 319
Pikus, G.E., and E.L. Ivchenko, Optical orientation and polarized luminescence of excitons in semiconductors **2**, 205
Pikus, G.E., and A.N. Titkov, Spin relaxation under optical orientation in semiconductors **8**, 73
Planel, R., and C. Benoit à la Guillaume, Optical orientation of excitons **8**, 353
Pokrovskii, Ya.E., Transport phenomena in electron–hole liquid **6**, 509
Pokrovsky, V.L., A.L. Talapov and P. Bak, Thermodynamics of two-dimensional soliton systems **17**, 71
Polinger, V.Z., see Bersuker, I.B. **7**, 21
Pollak, M., and M. Ortuño, The effect of Coulomb interactions on electronic states and transport in disordered insulators **10**, 287
Pooler, D.R., Numerical diagonalization techniques in the Jahn–Teller effect **7**, 199
Poulet, H., and R.M. Pick, Light scattering in incommensurate systems **14.1**, 315
Prelovšek, P., see Blinc, R. **14.1**, 143
Przhevuskii, A.K., see Basiev, T.T. **21**, 275

Rabinovich, R.I., see Gershenzon, E.M. **10**, 483
Raether, H., Surface plasmons and roughness **1**, 331
Rammer, J., see Beyer, J. **12**, 129
Rashba, E.I., Self-trapping of excitons **2**, 543
Reik, H.G., Non-adiabatic systems: analytical approach and exact results **7**, 117
Renk, K.F., Detection of high-frequency phonons by phonon-induced fluorescence **16**, 277
Renk, K.F., Optical generation and detection of 29 cm^{-1} phonons in ruby **16**, 317

Reznichenko, E.A., see Zelensky, V.F. **13**, 527
Robrock, K.H., see Ehrhart, P. **13**, 3
Rosenbaum, T.F., see Milligan, R.F. **10**, 231
Rubenchik, A.M., see Zakharov, V.E. **17**, 503
Rutar, V., see Blinc, R. **14.1**, 143
Ryskin, A.I., see Natadze, A.L. **7**, 347
Ryskin, A.I., see Kaplyanskii, A.A. **21**, 1

Sannikov, D.G., Phenomenological theory of the incommensurate–commensurate phase transition **14.1**, 43
Schneck, J., see Tolédano, J.C. **14.2**, 233
Schneider, T., and E. Stoll, Spin dynamics of Heisenberg chains **17**, 129
Schneider, T., Classical statistical mechanics of lattice dynamic model systems **17**, 389
Schober, H.R., see Ehrhart, P. **13**, 3
Schön, G., Collective modes in superconductors **12**, 589
Scott, A.C., Experimental observation of a Davydov-like soliton **17**, 53
Scott, J.F., Raman spectroscopy of structural phase transitions **5**, 291
Scott, J.F., Statics and dynamics of incommensurate $BaMnF_4$ **14.2**, 283
Seliger, J., see Blinc, R. **14.1**, 143
Shelby, R.M., see Macfarlane, R.M. **21**, 51
Sheleg, A.U., and V.V. Zaretskii, List of incommensurate crystals **14.2**, 367
Shen, Y.R., and F. deMartini, Nonlinear wave interaction involving surface polaritons **1**, 629
Shirane, G., see Axe, J.D. **14.2**, 1
Shklovskii, B.I., see Efros, A.L. **10**, 483
Shomina, E.V., see Zhizhin, G.N. **1**, 93
Shustin, O.A., see Yakovlev, I.A. **5**, 605
Sibeldin, N.N., see Bagaev, V.S. **6**, 267
Sibeldin, N.N., see Keldysh, L.V. **16**, 455
Sigmund, E., Phonon scattering at Jahn–Teller defects in semiconductors **7**, 495
Sigov, A.S., see Levanyuk, A.P. **5**, 129
Silbey, R., Theories of energy transport **4**, 1
Silin, A.P., Electron–hole liquid in a magnetic field **6**, 619
Singwi, K.S., see Vashishta, P. **6**, 1
Sipe, J.E., and G.I. Stegeman, Nonlinear optical response of metal surfaces **1**, 661
Slyozov, V.V., and P.A. Bereznyak, Irradiation creep in metals **13**, 575
Small, G.J., Persistent nonphotochemical hole burning and the dephasing of impurity electronic transitions in organic glasses **4**, 515
Smith, D.D., see Zewail, A.H. **2**, 665
Smith, H., see Beyer, J. **12**, 129
Snead Jr, C.L., and T. Luhman, Radiation damage and stress effects in superconductors: Materials for high-field applications **13**, 345
Sobyanin, A.A., see Ginzburg, V.L. **5**, 3
Sobyanin, A.A., see Levanyuk, A.P. **5**, 129
Stegeman, G.I., see Sipe, J.E. **1**, 661
Stegeman, G.I., and F. Nizzoli, Surface vibrations **9**, 195
Steiner, M., and A.R. Bishop, Nonlinear effects in low-dimensional magnets **17**, 783
Stoll, E., see Schneider, T. **17**, 129
Sturge, M.D., Introduction **2**, 1
Sugakov, V.I., Excitons in strained molecular crystals **2**, 709

Talapov, A.L., see Pokrovsky, V.L. **17**, 71
Tanabe, Y., and K. Aoyagi, Excitons in magnetic insulators **2**, 603
Thewalt, M.L.W., Bound multiexciton complexes **2**, 393
Thomas, G.A., see Milligan, R.F. **10**, 231
Timofeev, V.B., Free many particle electron–hole complexes in an indirect gap semiconductor **2**, 349
Timofeev, V.B., see Kulakovskii, V.D. **6**, 95
Timusk, T., see Markiewicz, R.S. **6**, 543
Titkov, A.N., see Pikus, G.E. **8**, 73
Tolédano, J.C., J. Schneck and G. Errandonéa, Incommensurate phase of barium sodium niobate **14.2**, 233
Tsukerblat, B.S., see Perlin, Yu.E. **7**, 251
Tuinstra, F., see de Wolff, P.M. **14.2**, 253

Ulrici, W., Manifestations of the Jahn–Teller effect in the optical spectra of transition metal impurities in crystals **7**, 439
Ushioda, S., and R. Loudon, Raman scattering by surface polaritons **1**, 535

Vashishta, P., R.K. Kalia and K.S. Singwi, Electron–hole liquid: theory **6**, 1
Vedula, Yu.S., see Lyuksyutov, I.F. **17**, 605
Vekhter, B.G., see Natadze, A.L. **7**, 347
Vinetskii, V.L., and G.A. Kholodar, Quasichemical reactions involving point defects in irradiated semiconductors **13**, 283
Vinogradov, E.A., G.N. Zhizhin and V.I. Yudson, Thermally stimulated emission of surface polaritons **1**, 145
Volkov, A.F., see Aslamazov, L.G. **12**, 65

Wagner, M., see Perlin, Yu.E. **7**, 1
Wagner, M., Unitary transformation methods in vibronic problems **7**, 155
Weis, O., Phonon radiation across solid/solid interfaces within the acoustic mismatch model **16**, 1
Weisbuch, C., see Hermann, C. **8**, 463
Westervelt, R.M., Kinetics of electron–hole drop formation and decay **6**, 187
Wiedersich, H., Phase stability and solute segregation during irradiation **13**, 225
Wiersma, D.A., see Hesselink, W.H. **4**, 249
Winkler, I.C., see Hanson, D.M. **4**, 621
Wöhlecke, M., and G. Borstel, Spin-polarized photoelectrons and crystal symmetry **8**, 423
Wolfe, J.P., and C.D. Jeffries, Strain-confined excitons and electron–hole liquid **6**, 431
Wood, M.H., see Bullough, R. **13**, 189

Yakovlev, I.A., and O.A. Shustin, Light scattering in quartz and ammonium chloride and its peculiarities in the vicinity of phase transition of crystals. A retrospective view and recent results **5**, 605
Yakovlev, V.A., see Zhizhin, G.N. **1**, 93
Yakovlev, V.A., see Zhizhin, G.N. **1**, 275
Yen, W.M., Experimental studies of energy transfer in rare earth ions in crystals **21**, 185
Yudson, V.I., see Vinogradov, E.A. **1**, 145

Zakharchenya, B.P., see Perel', V.I. **8**, 1
Zakharov, V.E., E.A. Kuznetsov and A.M. Rubenchik, Soliton stability **17**, 503
Zapasskii, V.S., Optical detection of spin-system magnetization in rare-earth-activated crystals and glasses **21**, 673

Zaretskii, V.V., see Sheleg, A.U. **14.2**, 367
Zelensky, V.F., and E.A. Reznichenko, Irradiation growth of metals and alloys **13**, 527
Zewail, A.H., D.D. Smith and J. Lemaistre, Dynamics of molecular excitons: disorder, coherence and dephasing **2**, 665
Zewail, A.H., see Burns, M.J. **4**, 301
Zhizhin, G.N., M.A. Moskalova, E.V. Shomina and V.A. Yakovlev, Surface electromagnetic wave propagation on metal surfaces **1**, 93
Zhizhin, G.N., see Vinogradov, E.A. **1**, 145
Zhizhin, G.N., and V.A. Yakovlev, Resonance of transition layer excitations with surface polaritons **1**, 275
Žumer, S., see Blinc, R. **14.1**, 143

Printed in Dunstable, United Kingdom

8434R00293